Advances in Controlled/
Living Radical Polymerization

ACS SYMPOSIUM SERIES **854**

Advances in Controlled/ Living Radical Polymerization

Krzysztof Matyjaszewski, EDITOR

Carnegie Mellon University

Sponsored by the ACS Division of Polymer Chemistry, Inc.

American Chemical Society, Washington, DC

Library of Congress Cataloging-in-Publication Data

Advances in controlled/living radical polymerization / Krzysztof Matyjaszewski, editor.

    p. cm.—(ACS symposium series ; 854)

    "Developed from a symposium sponsored by the Division of Polymer Chemistry, Inc. at the 224th National Meeting of the American Chemical Society, Boston, Massachusetts, August 17–22, 2002."

    Includes bibliographical references and index.

    ISBN 0–8412–3854–5

    1. Polymerization—Congresses. 2. Free radical reactions—Congresses.

    I. Matyjaszewski, K. (Krzysztof), 1950- II. American Chemical Society. Division of Polymer Chemistry, Inc. III. American Chemical Society. Meeting (224th : 2002 : Boston, Mass.) IV. Series.

QD281.P6A3   2003
547′.28—dc21
                                2003045316

The paper used in this publication meets the minimum requirements of American National Standard for Information Sciences—Permanence of Paper for Printed Library Materials, ANSI Z39.48–1984.

PRINTED IN THE UNITED STATES OF AMERICA

Foreword

The ACS Symposium Series was first published in 1974 to provide a mechanism for publishing symposia quickly in book form. The purpose of the series is to publish timely, comprehensive books developed from ACS sponsored symposia based on current scientific research. Occasionally, books are developed from symposia sponsored by other organizations when the topic is of keen interest to the chemistry audience.

Before agreeing to publish a book, the proposed table of contents is reviewed for appropriate and comprehensive coverage and for interest to the audience. Some papers may be excluded to better focus the book; others may be added to provide comprehensiveness. When appropriate, overview or introductory chapters are added. Drafts of chapters are peer-reviewed prior to final acceptance or rejection, and manuscripts are prepared in camera-ready format.

As a rule, only original research papers and original review papers are included in the volumes. Verbatim reproductions of previously published papers are not accepted.

ACS Books Department

Contents

Atom Transfer Radical Polymerization: Mechanisms

Atom Transfer Radical Polymerization: Materials and Applications

Nitroxide-Mediated Polymerization and Stable Free Radical Polymerization

RAFT and Other Degenerative Transfer Processes

Indexes

Preface

Controlled/living radical polymerization (CLRP) is among the most rapidly developing areas of chemistry. CLRP employs many concepts from conventional free radical polymerization, which is used for the commercial production of approximately 50% of all polymers and is responsible for >3% of GNP in the United States. However, by adopting concepts of controlled/living polymerization CLRP overcomes the main deficiency of conventional radical polymerization, a lack of macromolecular control and difficulties in forming well-defined copolymers.

This book comprises the topical reviews and specialists' contributions presented at the American Chemical Society Symposium entitled *Advances in Controlled/Living Radical Polymerization* that was held in Boston, Massachusetts August 15–18, 1997. The Boston Meeting was a sequel to the previous ACS Symposia held in San Francisco, California in 1997 and in New Orleans, Louisiana in 1999, which were summarized in the ACS Symposium Series Volume 685, *Controlled Radical Polymerization* and Volume 768, *Controlled/Living Radical Polymerization: Progress in ATRP, NMP and RAFT*. The Boston Meeting was very successful with 80 lectures and 79 posters presented, which represents a significant growth in comparison with New Orleans (50 lectures and 50 posters) and with the San Francisco Meetings (32 lectures and 35 posters).

The first three chapters provide a general introduction to controlled/living radical polymerization, including kinetics and mechanisms of various CLRP. Topics related to kinetics, stereochemistry, and electron spin resonance (ESR) studies of conventional radical polymerization are summarized in four chapters. Twenty-one chapters cover atom transfer radical polymerization (ATRP) and are separated into mechanistic features (11 chapters) and materials made by ATRP (10 chapters). Chemistry and materials made by nitroxide-mediated polymerization (NMP) are presented in 7 chapters and systems based on degenerative transfer such as reversible-addition fragmentation transfer (RAFT) are summarized in 8 chapters.

Chapters published in this book give a strong evidence that controlled/living radical polymerization has made a significant progress within the past several years. New systems have been discovered; substantial progress has been achieved in understanding the reactions involved in ATRP, NMP and RAFT; and in quantitative measurements of the rate and equilibrium constants as well as concentrations of the involved species. The first commercial applications of CRLP were announced at the Boston Meeting and it is expected that many others will quickly follow.

This book is addressed to chemists who are interested in radical processes and especially in controlled/living radical polymerization. It provides an introduction to and summarizes the most recent accomplishments in the field.

The financial support for the symposium from the following organizations is gratefully acknowledged: ACS Division of Polymer Chemistry, Inc., ACS Petroleum Research Foundation, Bayer, BYK, Ciba, Atofina, Henkel, Kaneka, Mitsubishi, Mitsui Chemicals, Motorola, Noveon, PPG, Rohmax, Teijin, and Xerox.

Krzysztof Matyjaszewski
Department of Chemistry
Carnegie Mellon University
4400 Fifth Avenue
Pittsburgh, PA 15213

Advances in Controlled/ Living Radical Polymerization

Fundamentals

Chapter 1

Controlled/Living Radical Polymerization: State of the Art in 2002

Krzysztof Matyjaszewski

Center for Macromolecular Engineering, Department of Chemistry, Carnegie Mellon University, 4400 Fifth Avenue, Pittsburgh, PA 15213

Controlled/Living Radical Polymerization is one of the most rapidly developing areas of polymer science. Atom transfer radical polymerization (ATRP), nitroxide mediated polymerization (NMP) and various degenerate transfer processes, including reversible addition fragmentation transfer (RAFT), enable the preparation of new materials from readily available monomers under undemanding conditions. Some of the relative advantages and limitations of each of these systems and some future challenges are discussed.

2

Controlled/Living Radical Polymerization (CLRP) is amongst the most rapidly developing areas of polymer science. Figure 1 illustrates how the number of publications on this topic have increased dramatically over the last decade. The search (using SciFinder Scholar 2002) for terms: controlled radical polymerization, living radical polymerization and both concepts combined, indicates that a nearly equal number of papers have been published using each term and that there is a limited overlap in the use of the terms. The issues related to terminology have been discussed in details in a special issue of "J. Polym. Sci., Polym. Chem. Ed." May 15, 2000, and will not be discussed here. The purpose of graph 1 is just to illustrate the very dynamic (or nearly explosive) development of this field. More than 10 papers per week are presently being published in CLRP!

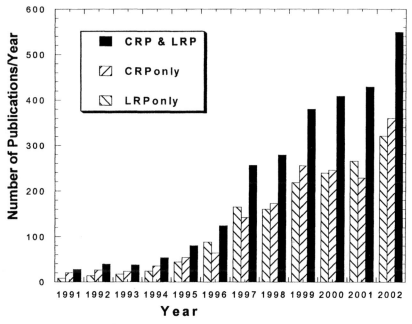

Figure 1. Annual number of publication using term controlled radical polymerization (CRP only), living radical polymerization (LRP only) and both term combined, according to SciFinder Scholar 2002, February 1, 2003

Figure 2 illustrates the annual changes in the number of publications in more specific areas of CLRP, namely atom transfer radical polymerization (ATRP), nitroxide mediated polymerization (NMP and stable free radical polymerization, SFRP) and various degenerate transfer processes, including reversible addition fragmentation transfer (RAFT) and catalytic chain transfer. The number of publications also continuously increases each year. The number of papers in the specific area of ATRP nearly matches those using a the more general CLRP term. In fact there is an increasing number of papers that do not use the terms CRLP, ATRP, NMP or RAFT in either the abstracts, titles or keywords and therefore can not be identified by computer based searches. This would indicate that for some polymer chemists CLRP is moving from a "specialty" research topic into the realm of commonly applied practice for the synthesis of materials.

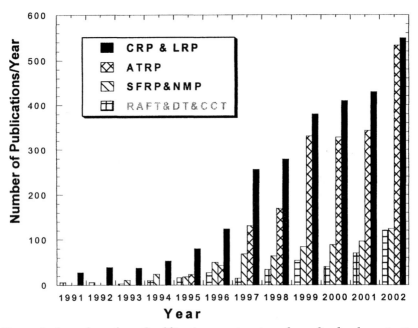

Figure 2. Annual number of publication on atom transfer radical polymerization (ATRP), nitroxide mediated polymerization (SFRP&NMP), degenerative transfer systems (RAFT&DT&CCT) and CRP&LRP, according to SciFinder Scholar 2002, February 1, 2003

What are the origins for such a dramatic increase of popularity of CLRP methodology? Conventional free radical polymerization is perhaps the most important commercial process for preparing high molecular weight polymers because it can be applied to the polymerization of many vinyl monomers under mild reaction conditions, and, although requiring the absence of oxygen, is tolerant of water and can be conducted over a wide temperature range (-80 to 250 °C). In addition, many monomers can be easily copolymerized via a radical route, leading to the preparation of an infinite number of copolymers with properties dependent on the proportion of the incorporated comonomers. The main limitation of conventional radical systems is the poor control over some of the key elements of engineered macromolecular structures such as molecular weight, polydispersity, end functionality, chain architecture and composition. Well-defined polymers with precisely controlled structural parameters are accessible by ionic living polymerization processes, however, ionic living polymerizations require stringent conditions and are limited to a relatively small number of monomers. It has always been desirable to prepare, by a free radical mechanism, well-defined block and graft copolymers, gradient and periodic copolymers, stars, combs, networks, end-functional polymers and many other materials under mild conditions from a larger range of monomers than available for ionic living polymerizations. This emergent ability to prepare long desired materials is perhaps the main reason why we have witnessed a real explosion of academic and industrial research on CLRP during the last five years, as evidenced by over three thousand papers and hundreds of patents devoted to this area. The industrial interest in CLRP may be explained by a recent estimate from Bob Matheson (DuPont) who projected that CLRP may affect a market of $20 billion/year in such areas as coatings, adhesives, surfactants, dispersants, lubricants, gels, additives, thermoplastic elastomers as well as many electronic and biomedical applications.

Development of Controlled/Living RP

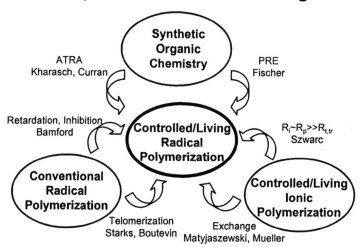

Figure 3. Areas contributing to development of CLRP

Research in CLRP really took off in the mid 90's after utilization of TEMPO as a mediator to styrene polymerization, introduction of ATRP and a subsequent discovery of RAFT/MADIX systems. Nevertheless, the roots of CLRP extend back over 50 years and are found in synthetic and physical organic chemistry (Kharsch, Minisci, Curran, Belus, Fischer), in controlled or living ionic reactions (Szwarc, Kennedy, Penczek, Mueller) in addition to the many attempts to control rates and functionalities in conventional radical polymerization (Bamford, Starks, Boutevin), as illustrated in Figure 3.

As already mentioned, ATRP, NMP and RAFT are currently the three most commonly used methods for CLRP but there are several others, such as iniferters, other degenerative transfer systems, including alkyl iodides, oligomers with methacrylate functionality, various types of non-nitroxide stable free radicals, and other transition metal mediators, which are also very efficient under specific conditions. It seems that each of the three major systems (and perhaps others) have some relative advantages and limitations, depending on the monomers used, the particular synthetic targets, and additional requirements concerning functionality, purity, process such as bulk, solution or biphasic, and perhaps the cost of the final product. Figure 4 attempts to illustrate some areas in which ATRP, NMP or RAFT may be easier / simpler / more precise / less expensive or more versatile. It must be recognized that this qualitiative picture should change with developments of new control agents and improvement in polymerization catalysts, conditions and perhaps development of entirely new systems.

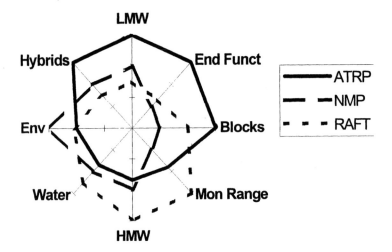

Figure 4. Relative advantages and limitations of ATRP, NMP and RAFT as applied to the synthesis of low (LMW) and high molecular weight polymers (HMW), range of polymerizable monomers (Mon Range), block copolymers (Blocks), end-functional polymers (End Funct), hybrids (Hybrids), aqueous systems (Water) and some environmental issues (Env).

Current research in CLRP encompasses a range of mechanistic, synthetic and materials aspects. Perhaps most interesting may be preparation of new materials with enhanced control of molecular weight, structural uniformity, and microstructure as well as topology, composition and functionality. In addition, the ease of mechanistic transformation between CLRP and various other systems based on ionic, coordination and even step-growth systems allows incorporation of any synthetic polymeric material into a CLRP. Another very exciting topic is the preparation of new hybrid materials involving inorganics, ceramics and natural products and well-defined vinyl polymers synthesized by CLRP methods.

Figure 5, below, illustrates some of the many possible structures which can be prepared by copolymerization (also including macromonomers) and leads to the question "what is the effect the chain architecture on properties?" The correlation between molecular structure and macroscopic properties of well-defined polymers prepared by CRLP has just started and may become the most challenging and rewarding area for future research since it may lead to rational design of new commercially important materials. We can perhaps also speculate about which other reseach areas will be developing in a future. This may be related to current trends, needs and expectations.

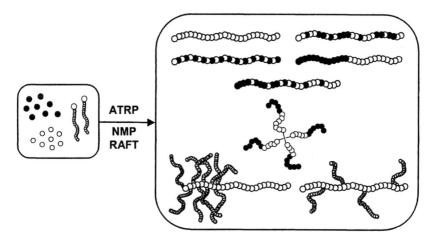

Figure 5. Examples of controlled topology and composition in CLRP copolymerization.

We now realize that none of the CLRP processes are true living polymerization procedures and that termination reactions, which limit efficiency of blocking reactions and chain end functionalities, can never be completely avoided, although they can be minimized. However, the effect of differing levels of termination and other side reactions on the properties of the (co)polymers obtained in a specific reaction remains to be established.

There are a few other important issues for CLRP that have not yet been adequately addressed and they may be easier to resolve if first approached in conventional radical systems. CLRP will always benefit from developments in conventional macromolecular and small molar mass radical systems. For example:

- An increase of k_p/k_t ratios will increase the selectivity of chain propagation and allow one to carry out CLRP at a much faster rate. The precise effect of chain length and viscosity on k_p and k_t should also be established. They might be selectively affected in compartmentalized systems, e.g. (micro/mini)emulsions, zeolites or other inclusion complexes.

- An examination of radical systems has shown that complexation with solvents, Lewis acids or other additives can affect not only the above rate constants but can also influence their chemo- and stereoselectivities. CLRP methodology can exploit these concepts and perhaps enhance their effects because of the slower controlled chain growth operating under the various exchange reactions. The slower rate of polymerization could also increase the precision for some template systems.

- Homo- and copolymerization of non-polar olefins by radical means is very challenging and enhanced chemoselectivity and perhaps stereoselectivity in radical based olefin polymerization could enable production materials that would compete with metallocene and Ziegler-Natta systems. On the other hand, synthesis of block/grafts copolymers of polyolefins with well defined segments of polyacrylates and other polar monomers prepared by CLRP can generate a new class of additives for commodity polyolefins.

Nevertheless, complete mechanistic understanding and optimization of the current processes and the aforementioned structure-property correlation remain perhaps most important items for immediate CLRP research:

- Complete mechanistic understanding, including structure-reactivity correlation for all CLRP systems. This will help define, and plausibly expand, the scope of each method with respect to range of polymerizable monomers, preservation of end-functionalities and more specifically to a reduction in the amount of ATRP catalyst required for a reaction. It may also, perhaps, enable control of microstructure in terms of tacticity and sequence distribution.

- Expanding control to encompass polymerization in various media (emulsion, suspension, gas-phase), surface polymerization, designing more robust and perhaps continuous processes, and general process optimization, catalyst recycling, etc.

- Construction of a comprehensive structure-property correlation which will allow preparation materials with new properties for targeted applications. This will require synthesis of many new materials with controlled, and systematically varied molecular weight, polydispersity, and a broad spectrum of chain architectures, perhaps even including microstructure.

Continuation and expansion of fundamental kinetic, mechanistic and characterization studies is needed to solve these challenging problems. This will also require efficient collaborations of synthetic polymer chemists with theoreticians, organic chemists, inorganic/coordination chemists, kineticists, physical organic chemists, polymer physical chemists, physicists and engineers. We hope some of these goals will be reached in a near future and results will be presented at the next ACS Symposium on Controlled/Living Radical Polymerization.

Chapter 2

Criteria for Livingness and Control in Nitroxide-Mediated and Related Radical Polymerizations

Hanns Fischer

Physikalisch-Chemisches Institut, University of Zuerich, Winterthurerstrasse 190, CH 8059 Zuerich, Switzerland

In nitroxide mediated radical polymerizations large living and controlled polymer fractions at high conversions require equilibrium constants of the polymeric alkoxyamine dissociation–coupling cycle $K = k_d/k_c$ that are smaller than $K^{max} = k_p[I]_0/2\ln(10)k_t$. Perfect livingness is reached if $K \approx K^{max}/100$. These conditions imply minimum monomer conversion times. Small polydispersity indices at moderate to large conversions result if $k_d k_c$ is close to or above a limit $k_d k_c^{perf} = 9(k_p/2\ln(10))^3[M]_0^2/[I]_0 k_t$. The limits depend on the propagation and termination constants (k_p, k_t), that is, on the monomer, and on the initiator concentration $[I]_0$. Five cases with different time evolutions of conversion, livingness and molecular weight distribution are distinguished. They explain the course of reported polymerizations. Effects of initially present persistent radicals, self- and external initiation and reaction temperature are discussed. Further, the criteria are translated to apply also to atom transfer radical polymerizations.

Introduction

Living and controlled radical polymerizations are presently of high academic and industrial interest. Several reviews are available.[1-5] One successful process is based on the reactions shown in Scheme 1. Dormant polymeric alkoxyamines dissociate with the rate constant k_d into persistent nitroxide and propagating radicals. The latter recouple with the nitroxide (k_c), add to monomer (k_p) and undergo termination (k_t) to unreactive polymer. In several cases polymers have been obtained with high nitroxide endgroup functionalities, molecular weights that are inversely proportional to the initial alkoxyamine concentration and that increase with monomer conversion, and polydispersity indices that decrease with increasing conversion to values close to unity.

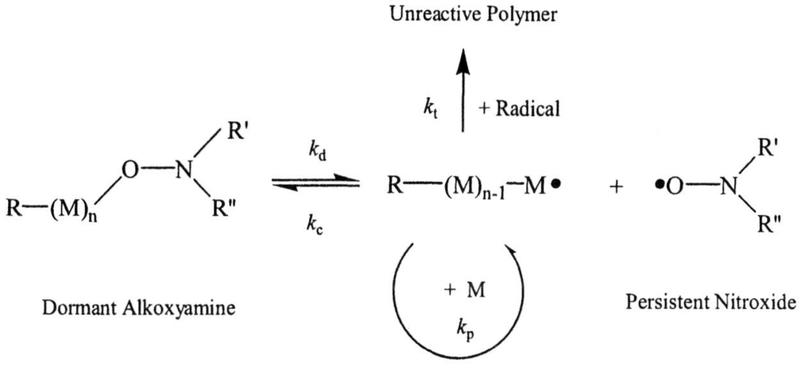

Scheme 1

Theoretical formulations have been developed by Fukuda et. al.[6] and by this author and coworkers[5,7,8] both for the reactions of Scheme 1 and for systems involving the initial presence of nitroxide, self- or conventional initiation, the concurrent formation of hydroxylamines and polymeric alkenes and for a limited stability of the nitroxide. They agree with experimental data[9] and have recently been tested rigorously by absolute comparisons.[10]

Here, we use the theoretical results to derive criteria for livingness and control in the form of easy to apply equations. The reactions follow Scheme 1 and start from a dormant alkoxyamine that contains one or more monomer units in the initial absence of nitroxide. Usually, after a short time a quasi-equilibrium of the reversible alkoxyamine decay is established. It is characterized by slowly growing nitroxide and slowly decreasing propagating radical concentrations, and it ends when termination has converted most of the alkoxyamine to unreactive products and nitroxide.[5-8]

The criteria extend previous ones,[5,8a,11] and they appear quite natural. They explain many features of reported nitroxide mediated polymerizations. Also, we present numerical predictions and mention strategies for improvement. Finally, the criteria are formulated for atom transfer radical polymerizations (ATRP). Throughout, chain length independent rate constants are implied.

Results and Discussion

Livingness. The polymers are called living when a large fraction of chains carries nitroxide endgroups. Hence, the decay-addition-recoupling cycle of Scheme 1 must be dominant during monomer conversion, and this implies (a) the existence of the quasi-equilibrium, (b) negligible monomer conversion before its establishment and (c) large conversion before its end. Condition (a) holds if the rate constants obey[7,8a]

$$K = k_d / k_c \ll k_c [I]_0 / k_t \tag{1}$$

where $[I]_0$ is the initial alkoxyamine concentration, and condition (b) requires

$$k_p \ll 3k_c \tag{2}$$

for $k_c < k_t$.[8a] Insertion of known rate constants (Tables 3, 4) shows that these conditions are fulfilled in most cases. Condition (c) means that the time for large monomer conversion is smaller than the approximate time at which the quasi-equilibrium ends, that is, smaller than [7,8a]

$$t^{end} = \frac{[I]_0}{3K^2 k_t} . \tag{3}$$

In the quasi-equilibrium regime the monomer concentration [M] obeys[5-8]

$$\ln[M]_0 /[M] = \frac{3}{2} k_p \left(\frac{K[I]_0}{3k_t} \right)^{1/3} t^{2/3} . \tag{4}$$

Taking 90% as a large monomer conversion, equations (3) and (4) yield

$$K = \frac{k_d}{k_c} < K^{\max} = \frac{k_p[I]_0}{2\ln(10)k_t} . \tag{5}$$

If the equilibrium constant is larger than $K^{\max}$, termination becomes dominant. Then, the alkoxyamine acts like a conventional initiator. It produces mainly unreactive and uncontrolled products, and the monomer conversion may remain limited.

As perfectly living we consider a polymer formed by $\geq$ 90% monomer conversion before 10% of the alkoxyamine I is converted to nitroxide Y• and unreactive products. The nitroxide concentration obeys[5-8]

$$[Y\bullet] = (3k_t K^2 [I]_0^2)^{1/3} t^{1/3} . \tag{6}$$

Combination with equation (4) shows that perfect livingness requires[8a]

$$K = \frac{k_d}{k_c} \leq K^{\text{perf}} = \frac{k_p[I]_0}{200\ln(10)k_t} . \tag{7}$$

K^{perf} is smaller than $K^{\max}$ by a factor of 100.

The time for a given monomer conversion depends inversely on the square root of the equilibrium constant (eq. (4)). Therefore, $K^{\max}$ and K^{perf} lead to the times required for 90% monomer conversion

$$t^{\min} > \frac{1}{3} \left(\frac{2\ln(10)}{k_p} \right)^2 \frac{k_t}{[I]_0} \qquad \text{and} \tag{8}$$

$$t^{\text{perf}} = \frac{10}{3} \left(\frac{2\ln(10)}{k_p} \right)^2 \frac{k_t}{[\text{I}]_0}. \tag{9}$$

The limiting values of the equilibrium constant and of the conversion time are determined by the monomer specific propagation and termination constants and the applied initial initiator concentration. Table 1 shows t^{min} and t^{perf} as calculated for different propagation constants, a common termination constant $k_t = 5 \cdot 10^8$ $M^{-1}s^{-1}$ and $[\text{I}]_0 = 5 \cdot 10^{-2}$ M which corresponds to a final molecular weight of about 20'000 g/mol. For small propagation constants the times are long. They will be shorter in the presence of self- or additional conventional initiation but at the expence of livingness. The fast conversions calculated for large k_p will lead to overheating. This can be avoided by a retarding initial addition of nitroxide.[2,5,8c,9]

Table 1. Times for 90% Monomer Conversion for a Monomer/Initiator Ratio of 200 and $k_t = 5 \cdot 10^8$ M^{-1} s^{-1}

$k_p/M^{-1}s^{-1}$	200	500	1'000	2'000	5'000	10'000	20'000	50'000
t^{min}/h	500	78	20	5	48 min	12 min	3 min	30 sec
t^{perf}/h	5'000	780	200	50	8	2	30 min	5 min

Control. Polymerizations are usually called controlled when (a) the number average degree of polymerization DP (and the molecular weight) increases linearly with increasing monomer conversion

$$DP = ([\text{M}]_0 - [\text{M}])/[\text{I}]_0, \tag{10}$$

and (b) the polydispersity index decreases in time and approaches $PDI = 1$ at high monomer conversion.

Since $DP = m_1/m_0$, where $m_1 = [\text{M}]_0 - [\text{M}]$ and m_0 denote the first and zeroths moments of the polymer molecular weight distribution, equation (10) implies that m_0, the concentration of alkoxyamine chains with one or more monomer units, is constant during the polymerization and equals the initial initiator concentration. This is not the case if most of the conversion occurs at the end of the quasi-equilibrium regime because then the alkoxyamine concentration decreases sharply.[7,8a] Consequently, control also requires that condition (5) holds. If this is fulfilled, equation (10) will always be obeyed if one uses a (macro)-initiator which contains monomer units. For a monomer deficient initiator it will hold only if this is transformed to a polymeric alkoxyamine in a time which is very short compared to the time for large, say 90%, conversion, that is, if $k_d \ll (t_{90})^{-1}$.

The desired properties of the polydispersity index *PDI* require many dissociation-propagation-recoupling cycles before a large conversion is reached. Supported by simulations we suggest $[M]_0/[I]_0$ cycles, the ideal DP_∞ for 100% conversion. This gives

$$k_d \geq \frac{[M]_0}{[I]_0} \frac{1}{t_{90}} .$$
(11)

Calculation of t_{90} with equation (4) provides condition (12) which is similar to a relation given earlier.[8a]

$$k_d k_c \geq k_d k_c^{\text{perf}} = 9 \left(\frac{k_p}{2\ln(10)} \right)^3 \frac{[I]_0}{k_t} DP_\infty^2$$
(12)

Livingness versus Control. In logarithmic forms the conditions (5), (7) and (12) read

$$\log(k_d) < -\log(2\ln(10)) + \log(k_p) + \log([I]_0/k_t) + \log(k_c)$$
(13)

$$\log(k_d) \leq -\log(200\ln(10)) + \log(k_p) + \log([I]_0/k_t) + \log(k_c)$$
(14)

$$\log(k_d) \geq -3\log(2\ln(10)/3) + 3\log(k_p) + \log([I]_0/k_t) + 2\log(DP_\infty) - \log(k_c).$$
(15)

In a $\log(k_d)$ *vs.* $\log(k_c)$ plot they are represented by straight lines with locations depending on k_p, k_t and the concentrations. Figure 1 shows these lines, denoted by K^{max}, K^{perf} and $k_d k_c^{\text{perf}}$, in experimentally established[5-11] ranges of k_d and k_c for $k_p = 2'000$ $M^{-1}s^{-1}$, $k_t = 5 \cdot 10^8$ $M^{-1}s^{-1}$, $[I]_0 = 0.05$ M and $DP_\infty = 200$. For pairs of rate constants (k_d, k_c) above K^{max} in region X condition (5) does not hold so that high livingness and control at large conversions are prohibited.

Pairs (k_d, k_c) below K^{max} provide different degrees of livingness and control. The space of useful parameters is divided by the orthogonal lines representing equations (14) and (15) into four sections: A for both perfect livingness and control at $\geq 90\%$ conversion in times $\geq t^{\text{perf}}$, B for perfect livingness but less

16

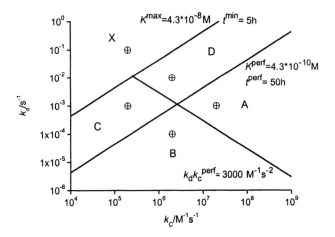

Figure 1. Graphical representation of the criteria for livingness and control for $k_p = 2'000\ M^{-1}s^{-1}$, $k_t = 5 \cdot 10^8\ M^{-1}s^{-1}$, $[I]_0 = 0.05\ M$ and $DP_\infty = 200$.

Table 2. Times, Polydispersity Indices and Alkoxyamine End Group Fractions at 90% Monomer Conversion for $k_p= 2000\ M^{-1}s^{-1}$, $k_t= 5 \cdot 10^8\ M^{-1}s^{-1}$, Oligomeric (MC, DP = 3) and Low Molecular Weight Model (MD) Initiators (Initial Ratio $[M]_0/[I]_0$ = 200) and Different k_d and k_c.

Case	Rate Constants	Initiator	t_{90}/h	PDI_{90}	% NO-groups
A	$k_d= 10^{-3}\ s^{-1}$	MC	155	1.03	98.2
	$k_c = 2 \cdot 10^7\ M^{-1}s^{-1}$	MD	154	1.03	98.2
B	$k_d= 10^{-4}\ s^{-1}$	MC	155	1.14	97.8
	$k_c = 2 \cdot 10^6\ M^{-1}s^{-1}$	MD	154	1.14	97.9
C	$k_d= 10^{-3}\ s^{-1}$	MC	18.2	1.22	80.6
	$k_c = 2 \cdot 10^5\ M^{-1}s^{-1}$	MD	18.0	1.22	80.9
D	$k_d= 10^{-2}\ s^{-1}$	MC	18.4	1.16	82.7
	$k_c = 2 \cdot 10^6\ M^{-1}s^{-1}$	MD	18.3	1.16	83.0
X	$k_d= 10^{-1}\ s^{-1}$	MC	145	17.3	6.0
	$k_c = 2 \cdot 10^5\ M^{-1}s^{-1}$	MD	135	16.5	6.3

control in similar times, D for high control but less livingness in times $\leq t^{perf}$, and C for both less control and livingness.

To check the validity of the conditions we calculated[12] the concentrations of all species and the first and second moments of the molecular weight distribution[7] for the parameters denoted by circles in Figure 1, the other parameters as given before, termination by disproportionation only and (a) a monomer containing oligomeric alkoxyamine initiator (MC) with 3 monomer units and (b) a monomer deficient low molecular weight model alkoxyamine (MD) that exhibit the same rate constants.

Table 2 shows that the different initiators provide the same results for 90% conversion. As expected, the conversion times and the living (mass) fractions are equal in cases A and B because K is the same, and B yields the larger *PDI*. C and D with equally smaller K reveal the same shorter conversion times and less livingness than A and B, and control is better for D. In case X, both livingness and control are reduced. Interestingly, the time for 90% conversion longer than for case D although K is larger. This occurs because after a fast conversion to about 20% in 6 minutes only a very slow reaction takes place while the quasi-equilibrium breaks down. The conversion is limited to a maximum of 95%.

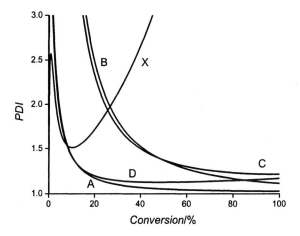

Figure 2. Polydispersity index as function of conversion for (k_d, k_c) given in Table 2, $k_p = 2'000 \ M^{-1}s^{-1}$, $k_t = 5 \cdot 10^8 \ M^{-1}s^{-1}$, $[I]_0 = 0.05 \ M$, $DP_\infty = 203$ and a trimeric alkoxyamine initiator.

Figures 2 and 3 display the *PDI* versus conversion. For the oligomeric initiator (Figure 2) *PDI* decreases for A-D initially nearly monotonically and follows the relation[5,6,7] $PDI = 1 + 1/DP + 8/3k_d t$ which holds up to moderate conversions. For B and C the *PDI* is high up to moderate conversions, and for D and C it increases slightly at large conversions. In case X, the *PDI* first behaves as in case A. Then it increases dramatically. For the model initiator one finds essentially the same behavior at large conversions(Figure 3). However, at small to intermediate conversions the ongoing transformation of the initiator to a polymeric alkoxyamine lowers the *PDI*. During this transformation the initially large *DP* decreases (Figure 4) before it follows equation (10). For the oligomeric

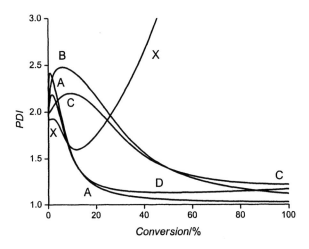

Figure 3. Polydispersity index versus conversion for (k_d,k_c) of Table 2, k_p, k_t, [I]$_0$, DP_∞ of Figure 2 and a low molecular weight model alkoxyamine initiator.

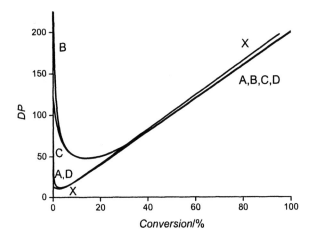

Figure 4. Number average degree of polymerization versus conversion for a low molecular weight model alkoxyamine initiator. Parameters as for Figure 3.

initiator equation (10) is obeyed exactly in cases A - D. A linear increase is also calculated for case X but the slope of the line is about 5% larger than given by equation (10).

Improvements of Conversion Rate, Livingness and Control. Long times for large conversions are predicted for many systems represented by cases A and B (Table 1). They are shortened by monomer self-initiation or by the addition of a conventional initiator. As long as the rate of these initiations is much smaller than the internal radical formation rate $k_d[I]_0$, detrimental effects on livingness and control remain tolerable.[8c] Similar rate enhancements result from a decay of the nitroxide into a transient radical.[8c] In cases C and D, livingness and control can be improved by initially added nitroxide, under retardation of the conversion, of course.[7,8c] The limits of such improvements have been discussed.[8c] In case X only very large amounts of added nitroxide lead to moderate improvements of livingness and *PDI*, and they can reduce the conversion time. In addition, the formation of hydroxylamine and a macroalkene by either a direct decomposition of the alkoxyamine or radical disproportionation can be detrimental.[7,8b] It stops the conversion at $t \approx 2/f_D k_d$ where f_D is the fraction of hydroxylamine formation.

Classification of Polymerizations. Table 3 shows k_d, k_c, K and $k_d k_c$ for several polymeric alkoxyamines based on the nitroxides TEMPO and DEPN (Scheme 2) at 120 °C. Because of chain length and reaction medium effects,[5,11] the data may describe those of actual experimental systems only within a factor of 5. Nevertheless, their large spread makes polymerizations involving the different polyalkoxyamines belong to the different cases introduced above. This classification is shown in Table 4 for a macroinitiator in a 0.05 M concentration aiming at $DP_\infty = 200$ and $k_t = 5 \cdot 10^8$ M^{-1}s^{-1}.

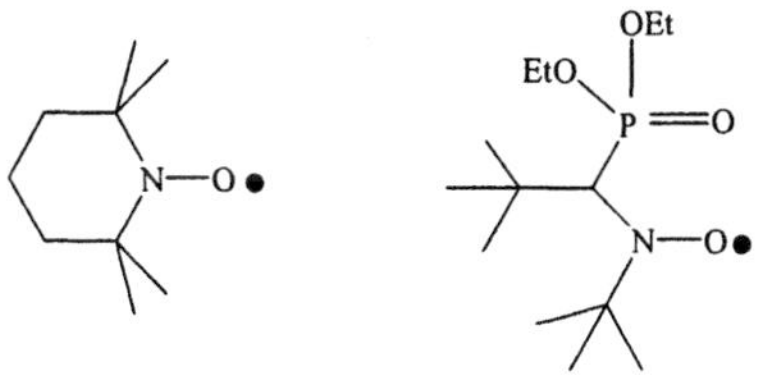

TEMPO   DEPN

Scheme 2

The perfect case A is realized by TEMPO/styrene, but the experimental time for 90% conversion is much shorter than calculated because of the styrene self-initiation.[2,5] DEPN/styrene belongs to case D in which livingness can be markedly improved by the addition of small amounts of free nitroxide.[2] Theoretically, TEMPO/MMA also belongs to case D but in this system the large fraction of disproportionation allows only very little monomer conversion.[2-6,11a,16] DEPN/*n*-butyl acrylate belongs to case B. Very fast conversion and little control are predicted. Addition of DEPN[2-6,9] retards the process and eliminates thermal overheating, and it is also required for small polydispersities. In principle, TEMPO/*n*-butyl acrylate should also conform according to case B but little success has been reported.[20] Presumably, there is little control, as expected for B, but also only low conversion. The latter points to disproportionation.[11a] Otherwise, the addition of TEMPO could be helpful on the expence of increased conversion times. Finally, at 120 °C DEPN/MMA violates condition (5) and represents the unfavourable case X.

Table 3. Kinetic Parameters of Some Polymeric Alkoxyamines

System	$k_{d,120}/s^{-1}$	$k_c/M^{-1}s^{-1}$	K/M	$k_d k_c/M^{-1}s^{-2}$
DEPN-p-*n*BuA[11,13]	$3\,10^{-3}$	$3\,10^7$	$1.0\,10^{-10}$	$9\,10^4$
DEPN-p-Sty[10,11,14]	$8.6\,10^{-3}$	$7.2\,10^5$	$1.2\,10^{-8}$	$6.2.10^3$
DEPN-p-MMA[11,15]	2.4	$5\,10^5$	$4.8\,10^{-6}$	$1.2\,10^6$
TEMPO-p-*n*BuA[11]	$7\,10^{-4}$	$5\,10^8$	$1.4\,10^{-12}$	$3.5\,10^5$
TEMPO-p-St[5,9,11]	$1\,10^{-3}$	$7.6\,10^7$	$1.3\,10^{-11}$	$7.6\,10^4$
TEMPO-p-MMA[11]	$3\,10^{-2}$	$2\,10^7$	$1.5\,10^{-10}$	$6\,10^5$

Table 4. Classification of Nitroxide Mediated Bulk Polymerizations of *n*-Butyl Acrylate, Styrene and Methyl Methacrylate at 120 °C.

System	$k_p/M^{-1}s^{-1}$	K vs. K^{max}/M K vs. K^{perf}/M	$k_d k_c$ vs. $k_d k_c^{perf}/M^{-1}s^{-2}$	Case t_{90}/h
DEPN/*n*BuA	88'000[17]	$1.0\,10^{-10} < 1.9\,10^{-6}$ $< 1.9\,10^{-8}$	$9.0\,10^4 < 2.5\,10^8$	B 0.35
DEPN/Sty	2'050[18]	$1.2\,10^{-8} < 4.5\,10^{-8}$ $> 4.5\,10^{-10}$	$6.3\,10^3 > 3.2\,10^3$	D 9.0
DEPN/MMA	2'800[19]	$4.8\,10^{-6} > 6.2\,10^{-8}$ $> 6.2\,10^{-10}$	$1.2\,10^6 > 8.1\,10^3$	X 0.28
TEMPO/*n*BuA	88'000	$1.4\,10^{-12} < 1.9\,10^{-6}$ $< 1.9\,10^{-8}$	$3.5\,10^5 < 2.5\,10^8$	B 3.0
TEMPO/Sty	2'050	$1.3\,10^{-11} < 4.5\,10^{-8}$ $< 4.5\,10^{-10}$	$7.6\,10^4 > 3.2\,10^3$	A 270
TEMPO/MMA	2'800	$1.5\,10^{-9} < 6.2\,10^{-8}$ $> 6.2\,10^{-10}$	$6.0\,10^5 > 8.1\,10^3$	D 16

Effects of k_p and Initiator Concentration. According to equations (5), (7) and (12) K^{max}, K^{perf} and $k_d k_c^{perf}$ increase with increasing propagation constant k_p. All lines in Figure 1 move up, the line for $k_d k_c^{perf}$ three times more than the others. Hence, regarding livingness, the ranges of useful parameters (k_d,k_c) are wider for monomers with large k_p than for monomers with small k_p. On the other hand, low polydispersities are more difficult to achieve for fast propagating monomers.

With increasing attempted final degrees of polymerization $[M]_0/[I]_0$, that is, decreasing initiator concentrations, K^{max} and K^{perf} move down whereas $k_d k_c^{perf}$ moves up. Consequently, it is generally more difficult to achieve livingness and control for high than for low to moderate final molecular weights.

Temperature Effects. The activation energies of the dissociation constants k_d are close to the alkoxyamine bond dissociation energies, and are typically larger than about 110 kJ/mol.[21] This causes a large temperature dependence. On the other hand, the coupling constants k_c are only weakly temperature dependent.[22] Therefore, the temperature dependence of K and $k_d k_c$ is determined by that of k_d. The propagation and termination rate constants on the right hand sides of equations (5), (7) and (12) have smaller activation energies below about 40 kJ/mol.[17-19,23] Hence, k_d increases more with increasing temperature than K^{max}, K^{perf} and $k_d k_c^{perf}$. This means that the point (k_d, k_c) for a system representing the cases A to D may shift up in Figure 1 into the unfavourable region X. On the other hand, if (k_d, k_c) is in region X at high temperatures it may fall into a more favourable region at lower temperature. In fact, DEPN/MMA seems to be such a case.[24]

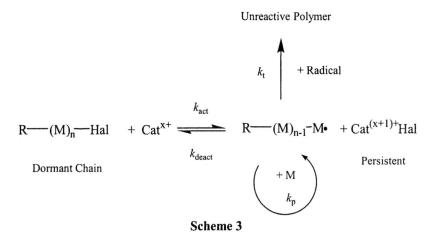

Scheme 3

Relations for ATRP. Atom transfer radical polymerizations differ from those treated above because the activation reaction is bimolecular instead of monomolecular. The basic mechanism is shown in Scheme 3. Cat denotes a transition metal complex catalyst which reacts reversibly with a dormant halogen terminated polymer with rate constant k_{act}. The reaction kinetics is very similar to that of nitroxide mediated polymerizations and has also been tested quite rigorously.[25] In the equations presented above, the rate constant of alkoxyamine cleavage k_d has thus to be replaced by $k_{act}[Cat]_0$, k_c by the rate constant k_{deact} for the reverse atom transfer reaction, and K by $k_{act}[Cat]_0/k_{deact}$ whereas the notation I can be kept for the dormant chains.

Concluding Remark

Even if detrimental side reactions are absent, living and controlled polymerizations of a given monomer to high conversion are impossible if the equilibrium constant for decay and reformation of the dormant polymeric chains exceeds an upper limit. This limit is imposed by the monomer specific propagation and termination constants and the applied initiator concentration. If the above condition is fulfilled imperfections may still occur because high livingness and high control are not correlated. A polymerization can be called living and controlled only if both features are experimentally proven.[26] In general, optimal living and controlled polymerizations of a specific monomer demand the use of a tailor made nitroxide or another persistent radical which leads to the appropriate rate constants of the dormant chains. In addition, it must show little propensity for disproportionation.

Acknowledgement. The author thanks P. Tordo and D. Bertin, University of Marseille, for their hospitality in the period when most of the presented ideas were developed and C. Jablon, TotalFinaElf, Paris, for financial support.

References and Notes

1. (a) Matyjaszewski, K. Ed. *ACS Symp. Ser.* **1998**, *685*; **2000**, *768*. (b) Matyjaszewski, K.; Davis, T. P., Eds. *Handbook of Radical Polymerization*; Wiley, New York, **2002**.
2. Hawker, C. J.; Bosman, A. W.; Harth, E. *Chem. Rev.* **2001**, *101*, 3661.
3. Kamigato, M.; Ando, T.; Sawamoto, M. *Chem. Rev.* **2001**, *101*, 3689.
4. Matyjaszewski, K.; Xia, J. *Chem. Rev.* **2001**, *101*, 2921.
5. Fischer, H. *Chem. Rev.* **2001**, *101*, 3581.
6. (a)Fukuda, T.; Goto, A.; Tsujii, Y. in *Handbook of Radical Polymerization*; Matyjaszewski, K.; Davis, T. P., Eds. Wiley, New York, **2002**. (b) Fukuda, T.; Goto, A.; Tsujii, Y. *ACS Symp. Ser.* **2000**, *768*, 27.
7. (a) Fischer, H. *Macromolecules* **1997**, *30*, 5666. (b) Fischer, H. *J. Polym. Sci. Part A.: Polym. Chem.* **1999**, *37*, 1885.
8. (a) Souaille, M; Fischer, H. *Macromolecules* **2000**, *33*, 7378. (b) Souaille, M; Fischer, H. *Macromolecules* **2001**, *34*, 2830. (c) Souaille, M; Fischer, H. *Macromolecules* **2002**, *35*, 248.

9. (a) Ohno, K. Tsujii, Y.; Miyamoto, T. ; Fukuda, T.; Goto, M.; Kobayashi, K. Akaike, T. *Macromolecules* **1998**, *31*, 1064. (b) Lutz, J.-F.; Lacroix-Desmazes, P.; Boutevin, B. *Macromol Rapid Commun* **2001**, *22*, 189. (c) Le Mercier, C.; Lutz, J.-F.; Marque, S.; Le Moigne, F.; Tordo, P., Lacroix-Desmazes, P.; Boutevin, B.; Couturier, J.-L.; Guerret, O.; Martschke, R.; Sobek, J.; Fischer, H. *ACS Symp. Ser.* **2000**, *768*, 108.

10. (a) Yoshikawa, C.; Goto, A.; Fukuda, T. *Macromolecules* **2002**, *353*, 5081. (b) Fukuda, T.; Yoshikawa, C.; Kwak, Y.; Goto, A.; Tsujii, Y. *ACS Symp. Ser.* This Volume.

11. (a) Ananchenko, G. S.; Fischer, H. *J. Polym. Sci. Part A.: Polym. Chem.* **2001**, *39*, 3604. (b) Ananchenko, G. S.; Souaille, M.; Fischer, H.; Le Mercier, C.; Tordo, P. *J. Polym. Sci. Part A.: Polym. Chem.* **2002**, *40*, 3264, and references therein.

12. Using MATLAB Version 6.1, The MathWorks Inc. **2001**.

13. Chauvin, F., PhD Thesis University of Marseille III, **2002**.

14. Goto, A.; Fukuda, T.; *Macromol. Chem. Phys.* **2000**, *201*, 2138.

15. From simulations of data kindly provided by D. Bertin and O. Guerret, Marseille.

16. Greszta, D.; Matyjaszewski, K. *Macromolecules* **1996**, *29*, 7661

17. Van Herk, A. M. *Macromol. Chem. Phys.* **1997**, *37*, 633.

18. Buback, M.; Gilbert, R. G.; Hutchinson, R. A.; Klumperman, B.; Kuchta, F. D.; Manders, B. G.; O'Driscoll, K. F.; Russell, G. T.; Schweer, J. *Macromol. Chem. Phys.* **1995**, *196*, 3267

19. Beuermann, S.; Buback, M.; Davies, T. P.; Gilbert, R. G.; Hutchinson, R. A.; Olaj, O. F.; Russell, G. T.; Schweer, J.; van Herk, A. M. *Macromol. Chem. Phys.* **1997**, *198*, 1545.

20. Listigover, N. A.; Georges, M. K.; Odell, P. G.; Keoshkerian, B. *Macromolecules* **1996**, *29*, 8992.

21. Marque, S.; Le Mercier, C.; Tordo, P.; Fischer, H. *Macromolecules* **2000**, *33*, 4403.

22. Sobek, J.; Martschke, R.; Fischer, H. *J. Amer. Chem. Soc.* **2001**, *123*, 2849, and references therein.

23. This was not recognized in our earlier work[11b] inspite of a thoughtful remark of a referee.

24. D. Bertin, O. Guerret, private communication.

25. Yoshikawa, C.; Goto, A.; Fukuda, T. *Macromolecules* **2003**, *36*, 908.

26. Lutz, J.-F.; Matyjaszewski, K. *Maromol. Chem. Phys.* **2002**, *203*, 1385.

Chapter 3

Mechanisms and Kinetics of Living Radical Polymerization: Absolute Comparison of Theory and Experiment

Takeshi Fukuda, Chiaki Yoshikawa, Yungwan Kwak, Atsushi Goto, and Yoshinobu Tsujii

Institute for Chemical Research, Kyoto University, Uji, Kyoto 611–0011, Japan

Theories that have been proposed to describe the polymerization rates and polydispersities in living radical polymerization (LRP) were experimentally tested for main branches of LRP including nitroxide-mediated polymerization (NMP), atom transfer radical polymerization (ATRP), iodide-mediated polymerization, and reversible addition-fragmentation chain transfer (RAFT) polymerization. The theories were verified on an absolute scale in both the absence and presence of conventional initiation. The cause for the marked retardation in polymerization rate observed in a dithiobenzoate-mediated RAFT polymerization of styrene was eluciated.

Introduction

The recent development of living radical polymerization (LRP) has·
achieved the combination of the robustness of radical chemistry with the
feasibility of living polymerization to finely control polymer structure (*I*). Main
reactions in LRP include the reversible activation shown in Scheme 1a, where
the dormant species P-X is activated to the alkyl radical P·, which, in the
presence of monomer, will propagate until it is deactivated back to P-X. Three
main mechanisms of reversible activation are currently considered important,
which are dissociation–combination (DC), atom transfer (AT), and degenerative
chain transfer (DT) processes (Scheme 1b–1d). In the DC process, the P-X bond
is cleaved by a thermal or photochemical stimulus to produce the stable or
persistent radical X· and the polymer radical P·. The AT process is kinetically
akin to the DC process, except that the activation is catalyzed by the activator A,
and the complex AX plays the role of a persistent radical. The DT process is a
chain transfer reaction in which activation and deactivation occur at the same
time, and is kinetically different from the former two. The reversible addition-
fragmentation chain transfer (RAFT: Scheme 2a) belongs to the DT category.

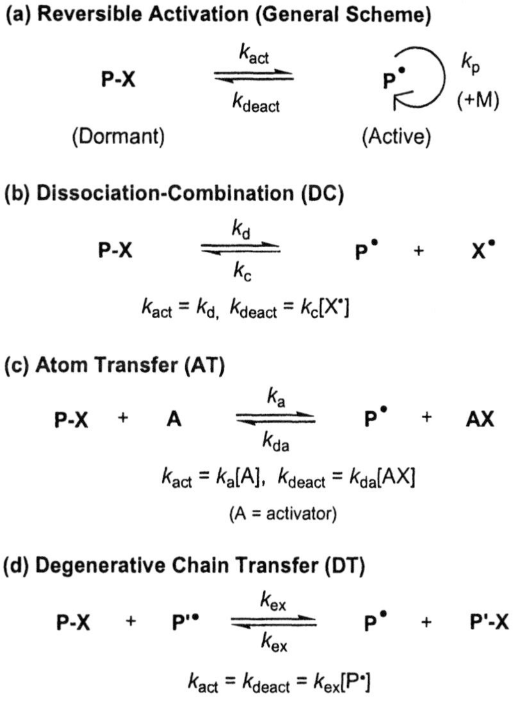

(a) Reversible Activation (General Scheme)

$$\text{P-X} \underset{k_{deact}}{\overset{k_{act}}{\rightleftharpoons}} \text{P}^\bullet \quad \begin{array}{c} k_p \\ (+M) \end{array}$$

(Dormant)   (Active)

(b) Dissociation-Combination (DC)

$$\text{P-X} \underset{k_c}{\overset{k_d}{\rightleftharpoons}} \text{P}^\bullet + \text{X}^\bullet$$

$$k_{act} = k_d, \ k_{deact} = k_c[\text{X}^\bullet]$$

(c) Atom Transfer (AT)

$$\text{P-X} + \text{A} \underset{k_{da}}{\overset{k_a}{\rightleftharpoons}} \text{P}^\bullet + \text{AX}$$

$$k_{act} = k_a[\text{A}], \ k_{deact} = k_{da}[\text{AX}]$$

(A = activator)

(d) Degenerative Chain Transfer (DT)

$$\text{P-X} + \text{P'}^\bullet \underset{k_{ex}}{\overset{k_{ex}}{\rightleftharpoons}} \text{P}^\bullet + \text{P'-X}$$

$$k_{act} = k_{deact} = k_{ex}[\text{P}^\bullet]$$

Scheme 1. Reversible activation processes.

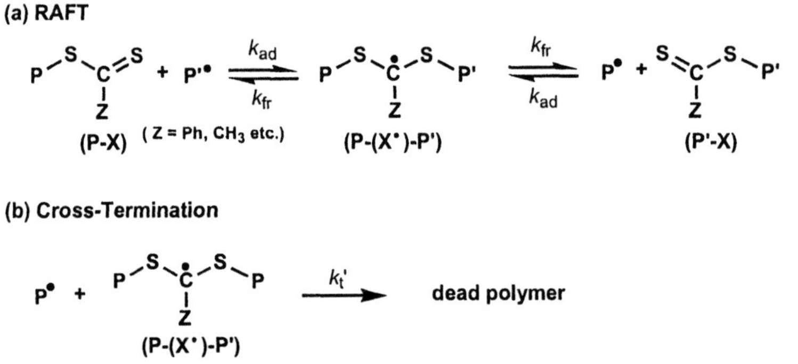

(a) RAFT

$$P-S-C(=S)-Z \;(P\text{-}X) \quad + \quad P'^{\bullet} \underset{k_{fr}}{\overset{k_{ad}}{\rightleftharpoons}} \quad P-S-C(^{\bullet})(S-P')-Z \;(P\text{-}(X^{\bullet})\text{-}P') \underset{k_{ad}}{\overset{k_{fr}}{\rightleftharpoons}} \quad P^{\bullet} + S=C(S-P')-Z \;(P'\text{-}X)$$

(Z = Ph, CH$_3$ etc.)

(b) Cross-Termination

$$P^{\bullet} + P-S-C(^{\bullet})(S-P)-Z \;(P\text{-}(X^{\bullet})\text{-}P') \xrightarrow{k_t'} \text{dead polymer}$$

Scheme 2. (a) RAFT and (b) cross–termination.

Theories have been developed describing the time evolutions of monomer consumption rate and chain length distribution in these systems. Qualitative aspects of those theories have been experimentally confirmed in part (*2*), but until recently, they have never been experimentally tested quantitatively. In this publication, we will review our results on absolute comparison of theory and experiment for the three main branches (Scheme 1b–1d) of LRP. It is *absolute* comparison, because the parameters appearing in the theoretical formula have all been determined by independent experiments. We will also discuss another important (and somewhat controversial) topic, the cause for the rate retardation observed in some RAFT polymerizations, by summarizing our published and unpublished results.

Theory of LRP

In a typical LRP run, the living chain is active for, say, 10^{-3} s and becomes dormant for about 10^2 s, which is followed by many such active-dormant cycles, thus the chain growing in an intermittent fashion. This gives characteristic kinetic features to LRP. As the number of active-dormant cycles increases, the chain length distribution becomes narrower. Thisn arrowing process and the related polymerization rate are different for two typical cases: the first case is the complete absence of conventional initiation. The existence of radical–radical termination will increase the persistent species ($X^{\bullet}$ in Scheme 1b or AX in Scheme 1c) with time, which will make the equilibrium incline towards the dormant state, or make the transient lifetime (active time per activation-deactivation cycle) shorter and shorter with increasing polymerization time t. The second case is the presence of conventional initiation. The conventional initiation will keep the concentration of the persistent species stationary, thus making the polymerization rate higher and the polydispersity index (PDI) smaller than in the absence of it.

Putting these remarks in theory, the following relations are obtained for LRP with a DC process: when $R_i = 0$ (3,4),

$$\ln([M]_0/[M]) = (3/2)k_p(K[P-X]_0/3k_t)^{1/3}t^{2/3} \tag{1}$$
$$[Y_B - (1/x_{n,B})]^{-1} = (3/8)k_d t \qquad \text{(small } t) \tag{2}$$

and when $R_i \neq 0$ and is sufficiently large (5,6),

$$\ln([M]_0/[M]) = k_p(R_i/k_t)^{1/2}t \tag{3}$$
$$[Y_B - (1/x_{n,B})]^{-1} = (1/2)k_d t \qquad \text{(small } t) \tag{4}$$

where R_i is the rate of conventional initiation, M is the monomer, the subscript "0" denotes the initial state, $K = k_d/k_c$, and k_p, k_t, k_d, and k_c are the rate constants of propagation, termination, dissociation, and combination, respectively; $Y_B + 1 = x_{w,B}/x_{n,B}$ with $x_{w,B}$ and $x_{n,B}$ being the weight- and number-average degrees of polymerization, respectively, of the grown portion (B) of the polymer. Given the PDI of the whole polymer ($Y + 1$), and that of the initiating portion (A) or the initiatingd ormant (low-mass or polymer) species, ($Y_A + 1$), Y_B can be calculated with

$$Y = w_A^2 Y_A + w_B^2 Y_B \tag{5}$$

where $w_A = (1 - w_B)$ is the weight fraction of the initiator. Equations 1–5 applies to the time range in which the quasi-equilibrium $k_d[P-X] = k_c[P^\bullet][X^\bullet]$ holds and in which the cumulative number of dead chains is sufficiently small compared to the number of dormant chains, i.e., $[P-X] \cong [P-X]_0$. (More specific conditions are given in refs (3), and it may easily be confirmed that the systems studied below meet these conditions.) It is also assumed that reactions other than disso-ciation (activation), combination (deactivation), termination, and conventional initiation are absent. For more general equations of rate and PDI, see refs (7).

For LRP with an AT process (ATRP), all the above equations are valid by the reinterpretations of $k_d = k_a[A]$, $k_c = k_{da}$, and $X^\bullet = AX$.

In LRP with a DT process, radicals must be generated by a conventional initiation to start and maintain polymerization. If the DT process is not a retarding or degrading one, or accompanies no side reaction, the stationary-state kinetics eq 3 will hold. For this type of LRP, the PDI is more conveniently represented by (8)

$$[Y_B - (1/x_{n,B})]^{-1} = (k_{ex}/k_p)[c/(2 - c)] \tag{6}$$

where c is the conversion.

Absolute Comparison of Theory and Experiment

In what follows, we will give a brief summary of our reported results for a nitroxide-mediated (DC–type) polymerization (9), a copper complex–mediated polymerization (ATRP) (10), an iodide-mediated (DT–type) polymerization (8), and a dithioester-mediated (DT–type or RAFT) polymerization (11). They are all styrene systems.

Nitroxide-Mediated Polymerization (NMP)

A polystyrene (PS)-DEPN adduct (M_n = 2200 and M_w/M_n = 1.13) was used as an initiating polymer, where DEPN is N-tert-butyl-1-diethylphosphono-2,2-dimethylpropyl nitroxide. Figure 1 shows the time evolution of the GPC curves for the bulk polymerization of styrene with this initiating polymer at 80°C without (Figure 1a) and with (1b) the conventional initiator BPO. These curves

Table 1. Kinetic Parameters Used in This Work

parameter	value	ref
a. NMP (PS-DEPN)		
$k_{i,th}$ (M^{-2}s^{-1})^a	4.5×10^{-12}	(16)
$k_{i,BPO}$ (s^{-1})^a	6.7×10^{-5}	(17)
k_p (M^{-1}s^{-1})	650	(18)
k_t (M^{-1}s^{-1})	3.0×10^8	(9)
k_d (s^{-1})	1.16×10^{-4}	(19)
$K = k_d/k_c$ (M)	1.7×10^{-10}	(9)
b. ATRP		
$k_{i,th}$ (M^{-2}s^{-1})^b	9.6×10^{-11}	(16)
$k_{i,VR110}$ (s^{-1})^b	3.0×10^{-5}	(10)
k_p (M^{-1}s^{-1})	1560	(18)
k_t (M^{-1}s^{-1})	3.6×10^8	(10)
k_a (M^{-1}s^{-1})	0.45	(20)
$K/[A] = k_a/k_{da}$ (–)	2.4×10^{-8}	(10)
c. DT (iodide)		
k_p (M^{-1}s^{-1})	650	(18)
$C_{ex} = k_{ex}/k_p$ (–)	3.6	(8)
d. RAFT (dithioacetate)		
k_p (M^{-1}s^{-1})	340	(18)
$C_{ex} = k_{ex}/k_p$ (–)	180	(11)
e. RAFT (dithiobenzoate)		
k_p (M^{-1}s^{-1})	340	(18)
k_{ad} (M^{-1}s^{-1})	4×10^6	(11)

$^a R_i = k_{i,th}[M]_0^3 + k_{i,BPO}[BPO]_0$ $^b R_i = k_{i,th}[M]_0^3 + k_{i,VR110}[VR110]_0$

give the conversion and the PDI of the product polymer at time t as well as the PDI of the initiating polymer (cf. the curve for $t = 0$), with which we can compute the polydispersity factor Y_B according to eq 5. (Note that these experiments were so designed as to yield accurate kinetic data, not to give polymers with a low polydispersity.)

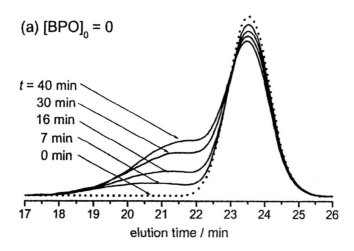

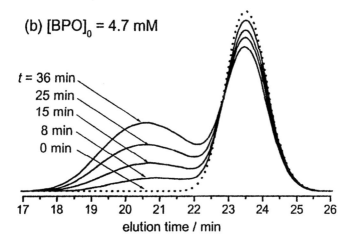

Figure 1. GPC chromatograms for the styrene/PS-DEPN/(BPO) systems (80 °C): [PS-DEPN]₀ = 25 mM; [BPO]₀ = (a) 0 and (b) 4.7 mM.

Table 1a lists the kinetic parameters used for the theoretical equations 1–4. They were either taken from the literature or, when literature data were unavailable, determined by independent experiments. The solid curve in Figure 2a shows eq 1 calculated with these data, which reproduces the experimental data (open and filled circles) very well. The solid and broken curves in Figure 2b show eq 2 and eq 4, respectively, which also reproduce the experimental polydispersity data satisfactorily. These are the first examples to experimentally justify the power-law equation 1 and the relevant PDI equation 2 on an absolute scale. The theoretical prediction that conventional initiation makes R_p larger (Figure 2a) and PDI smaller (Figure 2b) than in the absence of it has been thus confirmed. The dotted line is the best-fit representation of the plots (squares) according to eq 3 (from which we have obtained k_t (Table 1a) in this specific system (9) and used it in the above analysis of eq 1).

ATRP

An ω-polystyryl bromide (PS-Br) ($M_n = 1200$ and $M_w/M_n = 1.08$) was used as an initiating polymer. Polymerization of styrene in t-butylbenzene (50/50 v/v) was carried out at 110°C with the Cu(I)Br/dHbipy complex catalysis in the presence and absence of the conventional initiator VR110, where dHbipy is 4,4'-di-n-heptyl-2,2'-bipyridine and VR110 is 2,2'-azobis(2,4,4-trimethylpentane). Particular care was taken to avoid oxidation of Cu(I)Br prior to polymerization runs.

Table 1b lists the kinetic parameters used in this study. Again, parameters unavailable in literature were determined by independent experiments. To test eq 1 with $K = (k_a/k_{da})[Cu(I)Br]_0$, we examined the system with $[PS-Br]_0 = 13$mM and $[Cu(I)Br]_0 = 10$mM. It was confirmed that thermal initiation of the diluted styrene was entirely negligible in the examined time range (< 35 min). The duplicated experimental points (circles in Figure 3a) are well reproduced by the $t^{2/3}$-dependent linear line predicted by the theory (solid line), confirming the theory on an absolute scale. The addition of 40mM of the conventional initiator VR110 to this system increased R_p by a factor up to about 3, as shown by the squares in Figure 3a. Figure 3b shows the comparison of the PDI equations 2 and 4 ($k_d = k_a[Cu(I)Br]_0$) with the experiments with $[VR110]_0 = 0$ (circles) and $[VR110]_0 = 40$mM (square), respectively. In both systems, satisfactory agreement of theory and experiment was found. It was confirmed for the ATRP system, too, that conventional initiation not only increases the polymerization rate but lowers the polydispersity (at least at an initial stage of polymerization where the contribution of terminated chains to polydispersity is insignificant).

31

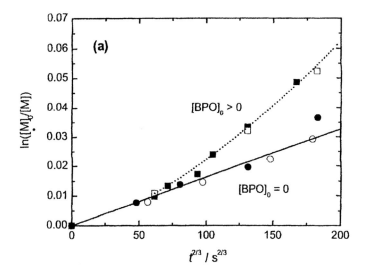

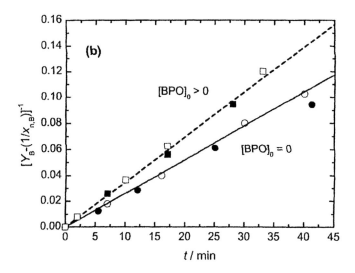

Figure 2. Plot of (a) ln([M]₀/[M]) vs t²ᐟ³ and (b) [Yᵦ−(1/xₙ,ᵦ)]⁻¹ vs t for the styrene/PS-DEPN/(BPO) systems (80 °C): [PS-DEPN]₀ = 25 mM; [BPO]₀ = 0 (• and •) and 4.7 mM (• and •). The experiments were duplicated. For (a), the solid line shows eq 1 with the independently determined rate constants (Table 1a), and the dotted line is the best-fit representation of the plots (quares) according to eq 3 (see also the text). For (b), the solid and broken lines show eqs 2 and 4, respectively, with the independently determined kd value (Table 1a).

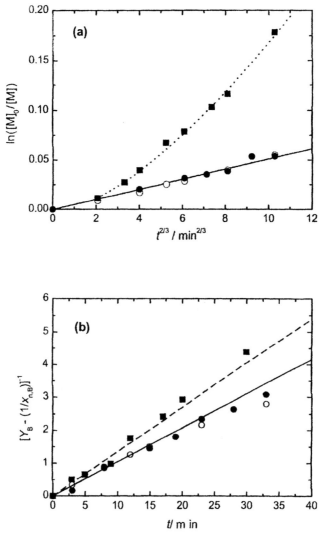

Figure 3. Plot of (a) ln([M]$_0$/[M]) vs t$^{2/3}$ and (b) [Y$_B$–(1/x$_{n,B}$)]$^{-1}$ vs t for the styrene/t-butylbenzene/PS-Br/Cu(I)Br/dHbipy/(VR110) system (110 °C): [PS-Br]$_0$ = 13 mM; [Cu(I)Br]$_0$ = 10 mM; [dHbipy]$_0$ = 30 mM; [VR110]$_0$ = 0 (• • and •) and 40 mM (•); [styrene]/[t-butylbenzene] = 50/50 v/v. The experiments were duplicated for the system without VR110. For (a), the solid line shows eq 1 with the independently determined rate constants (Table 1b), and the dotted line is the best-fit representation of the plots (quares) according to eq 3. For (b), the solid and broken lines show eqs 2 and 4, respectively, with the independently determined k$_a$ value (Table 1b).

Polymerization with a DT Process. 1. Iodide-Mediated Polymerization

The first detailed kinetic study on polymerization with a DT process was presented some time ago for an iodide-mediated polymerization of styrene. In brief, the bulk polymerization of styrene was carried out at 80°C with an ω-polystyryl iodide PS-I (M_n = 2000 and M_w/M_n = 1.26) as a DT agent and BPO as a conventional initiator. The polymerization followed the first-order kinetics and gave the same R_p as in the absence of the DT agent, thus confirming eq 3.

The PDI equation 6 with the transfer constant C_{ex} = k_{ex}/k_p determined independently (Table 1c) is given by the dotted line in Figure 4a. The experimental data for varying BPO concentrations fall on the same line, thus confirming the theory on an absolute scale.

Polymerization with a DT Process. 2. Dithioester-Mediated (RAFT) Polymerization

When a RAFT process is sufficiently fast with no side reaction accompanied, it provides an ideal DT process. Such an example is the polymerization of styrene mediated by polystyryl dithioacetate PS-SCSCH$_3$ (M_n = 1940 and M_w/M_n = 1.17). The bulk polymerization of styrene was carried out at 60°C with this DT (RAFT) agent and varying concentrations of BPO, confirming the first-order kinetics eq 3 to hold. The absolute value of R_p was approximately thes ame as in the absence of the RAFT agent. (Actually, it showed a weak dependence on the RAFT agent concentration: the larger was the RAFT agent concentration, the smaller was R_p. This matter will be discussed in the next section for the styrene polymerization with polystyryl dithiobenzoate, in which the rate retardation is highly significant.)

In Figure 4b, the PDI equation 6 with the independently determined transfer constant (Table 1d) is compared with the experimental data. Again, satisfactory agreement of theory and experiment is confirmed in Figure 4b.

Rate Retardation in RAFT Polymerization

As suggested above, significant retardation in polymerization rate has been noted in some ditioester-mediated (RAFT) polymerizations. Two opposing opinions, among others, have been proposed to explain this phenomenon. One is the slow fragmentation of the intermediate P-(X$^\bullet$)-P, which is assumed to be stable enough to cause no polymerization or no termination. This assumption

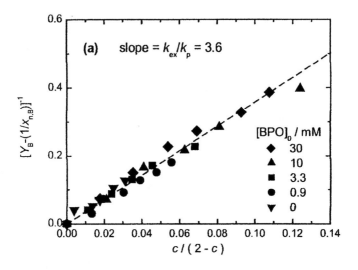

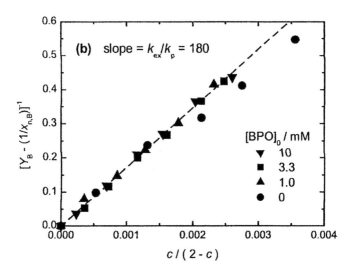

Figure 4. Plot of $[Y_B-(1/x_{n,B})]^{-1}$ vs $c/(2-c)$ (a) for the styrene/PS-I/(BPO) systems (80 °C): $[PS-I]_0 = 17$ mM; $[BPO]_0$ as indicated in the figure, and (b) for the styrene/PS-SCSCH$_3$/(BPO) systems (60 °C): $[PS-SCSCH_3]_0 = 0.45$ mM; $[BPO]_0$ as indicated in the figure.

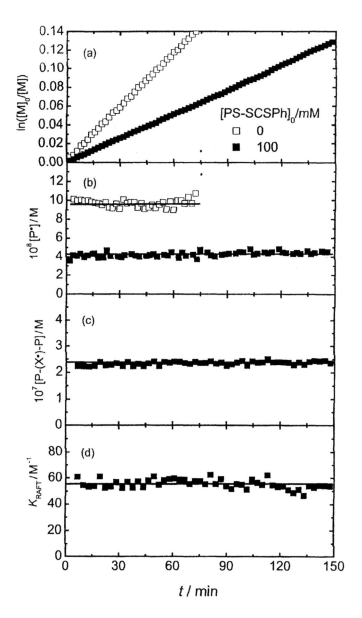

Figure 5. Plot of (a) ln([M]₀/[M]), (b) [P•], (c) [P-(X•)-P], and (d) K_{RAFT} (= [P-(X•)-P]/([P•][P-X])) vs t for the styrene/AIBN/(PS-SCSPh) systems (60 °C): [AIBN]₀ = 300 mM; [PS-SCSPh]₀ as indicated in the figure.

led to an estimate of the related rate constant of fragmentation k_{fr} of about $10^{-2} \cdot$ s^{-1}, typically (12). The other is the cross–termination between the intermediate and the linear radical $P^{\bullet}$ with a rate constant k_t' comparable in magnitude to k_t (cf. Scheme 2b). This assumption led to an estimate of k_{fr} of about 10^5 s^{-1} (13). Between these two estimates of k_{fr} is a difference of 7 orders of magnitude.

To elucidate the cause for the rate retardation, we studied the bulk polymerization of styrene at 60°C mediated by polystyryl dithiobenzoate PS-SCSPh ($M_n = 1100$ and $M_w/M_n = 1.08$) and initiated by the conventional initiator AIBN (14). Figure 5 shows the result of a series of kinetic analyses. The first-order plot of monomer concentration was linear in both the presence and absence of 100mM of the RAFT agent PS-SCSPh (Figure 5a), and the polymer radical concentration $[P^{\bullet}]$ estimated from the slope of the first–order plot and the known value of k_p (Table 1e) was smaller by a factor about 0.45 in the presence of the RAFT agent than in its absence (Figure 5b). The concentration of the intermediate radical P-($X^{\bullet}$)-P was determinable by ESR. It stayed nearly constant from an early stage of polymerization (Figure 5c), giving an equilibrium constant $K_{RAFT} = [P-(X^{\bullet})-P]/([P^{\bullet}][P-X])$ of 55 M^{-1} (Figure 5d). The fragmentation rate constant $k_{fr} = k_{ad}/K_{RAFT}$ evaluated with the known addition rate constant k_{ad} (Table 1e) was about 7×10^4 s^{-1}. This means that the fragmentation in this system is a fast process.

The existence of the addition-fragmentation equilibrium allows us to calculate the stationary concentration of $P^{\bullet}$ from the equality of the initiation and termination rates:

$$R_i = k_t [P^{\bullet}]^2 + k_t'[P^{\bullet}][P-(X^{\bullet})-P] \tag{7}$$

Combination of eq 7 with the equilibrium relation gives

$$R_p = R_{p,0}\{1 + K_{RAFT}(k_t'/k_t)[P-X]_0\}^{-1/2} \tag{8}$$

where $R_{p,0}$ is the polymerization rate for the RAFT agent-free ($[P-X]_0 = 0$) system. From eq 8 with the observed values of K_{RAFT} and $R_p/R_{p,0}$, k_t'/k_t is estimated to be about 0.8 (14). Namely, the cross-termination rate constant k_t' is similar to k_t.

If the cross-termination is recombination rather than disproportionation, it will produce a 3-arm star chain. The production of the 3-arm star polymer was confirmed in a model experiment: namely, PS-SCSPh and PS-Br model polymers of nearly the same chain length dissolved in t-butylbenzene were heated at 60°C in the presence of aC u(I)Br/Me$_6$TREN complex and Cu(0), where Me$_6$TREN is tris[2-(dimethyl amino)ethyl]amine (14,15). A high concentration of $P^{\bullet}$ produced by the activation of PS-Br will add to PS-SCSPh to produce the intermediate radical, which subsequently will be combined with another P$^\bullet$. The GPC chart of the product clearly showed the existence of the 3-arm chain produced by the cross–termination as well as the linear chain of doubled chain length produced by the conventional (homo-) termination. The concentration of the 3-arm chain relative to that of the linear chain was proportional to the concentration of the RAFT agent, as the theory demanded. This analysis also indicated that k_t' is similar to k_t in magnitude (15). The GPC chart of the 3-arm chain, taken after having been kept at 60°C for 25h, showed no detectable change, indicating that the chain is stable, and the cross-termination is a virtually irreversible process, at least at and below 60°C (15).

The importance of the cross-termination for a given RAFT system depends essentially on the magnitude of the equilibrium constant K_{RAFT} in that system, as eq 8 indicates. In this regard, the rate retardation observed for the dithiobenzoate/styrene system can be exceptionally large, for this system is characterized by an exceptionally large k_{ad} and hence K_{RAFT}, if k_{fr} is assumed to be less system dependent.

Conclusions

The analytical equations representing the polymerization rate and PDI in the three main branches of LRP (NMP, ATRP, and RAFT and iodide-mediated polymerizations) have been experimentally verified on an absolute scale in both the absence and presence of conventional initiation.

In the polystyryl ditiobenzoate-mediated RAFT polymerization of styrene, a marked retardation in polymerization rate was observed, which was ascribed to the cross-termination between the intermediate and propagating radicals to form a 3-arm star polymer.

Abbreviations

A	activator in ATRP (Scheme 1)
AX	deactivator in ATRP (Scheme 1)
c	monomer conversion
C_{ex}	degenerative chain transfer constant ($= k_{ex}/k_p$)
K	equilibrium constant in reversible activation ($= k_d/k_c = k_a[A]/k_{da}$) (Scheme 1)
k_a	activation rate constant in ATRP (Scheme 1)
k_{ad}	addition rate constant (Scheme 2)
k_{act}	pseudo-first-order activation rate constant (Scheme 1)
k_c	combination rate constant (Scheme 1)
k_d	dissociation rate constant (Scheme 1)

k_{da}	deactivation rate constant in ATRP (Scheme 1)
k_{deact}	pseudo-first-order deactivation rate constant (Scheme 1)
k_{ex}	degenerative chain transfer rate constant (Scheme 1)
k_{fr}	fragmentation rate constant (Scheme 2)
k_p	propagation rate constant
K_{RAFT}	equilibrium constant in the RAFT process ($= k_{ad}/k_{fr}$) (Scheme 2)
k_t	termination rate constant
k_t'	cross-termination rate constant (Scheme 2)
M	monomer
M_n	number-average molecular weight
M_w	weight-average molecular weight
$P^\bullet$	polymer radical
P-X	dormant species
R_i	(conventional) initiation rate
R_p	propagation rate
w_K	weight fraction of the subchain K (K= A or B, where A is the initiating portion and B is the grown portion of the chain) (eq 1)
$X^\bullet$	stable free radical
x_n	number-average degree of polymerization
x_w	weight-average degree of polymerization
Y	polydispersity factor ($Y = M_w/M_n - 1$)

References

1. (a) *Controlled Radical Polymerization;* Matyjaszewski, K., Ed.; ACS Symp. Ser. 685; American Chemical Society: Washington, DC, 1998. (b) *Controlled/Living Radical Polymerization;* Matyjaszewski, K., Ed.; ACS Symp. Ser. 768; American Chemical Society: Washington, DC, 2000. (c) *Handbook of Radical Polymerization;* Matyjaszewski, K.; Davis, T. P., Eds.; John Wiley & Sons, New York, NY, 2002.
2. (a) Kothe, T.; Marque, S.; Martschke, R.; Popov, M.; Fischer, H. *J. Chem. Soc., Perkin Trans.* **1998**, *2*, 1553. (b) Ohno, K.; Tsujii, Y.; Miyamoto, T.; Fukuda, T.; Goto, M.; Kobayashi, K.; Akaike, T. *Macromolecules* **1998**, *31*, 1064. (c) Lutz, J. –F.; Lacroix–Desmazes, P.; Boutevin, B. *Macromol. Rapid Commun.* **2001**, *22*, 189. (d) Zhang, H.; Klumperman, B.; Ming, W.; Fischer H.; van der Linde, R. *Macromolecules* **2001**, *34*, 6169. (e) Zhang, H.; Klumperman, B.; van der Linde, R. *Macromolecules* **2002**, *35*, 2261. (f) Chambard, G.; Klumperman, B.; German, A. L. *Macromolecules* **2002**, *35*, 3420. (g) Pintauer, T.; Zhou, P.; Matyjaszewski,K. *J. Am. Chem. Soc.* **2002**, *124*, 8196. (h) Shipp, D. A.; Yu, X. *Polym. Prepr. (Am. Chem. Soc., Div. Polym. Chem.)* **2002**, *43(2)*, 7.

3. (a) Fischer, H. *Macromolecules* **1997**, *30*, 5666. (b) Fischer, H. *J. Polym. Sci. Part A.: Polym. Chem.* **1999**, *37*, 1885. (c) Souaille, M.; Fischer, H. *Macromolecules* **2000**, *33*, 7378. (d) Fischer, H. *Chem. Rev.* **2001**, *101*, 3581.

4. Ohno, K.; Tsujii, Y.; Miyamoto, T.; Fukuda, T.; Goto, M.; Kobayashi, K.; Akaike, T. *Macromolecules* **1998**,*31* , 1064.

5. (a) Fukuda, T.; Terauchi, T.; Goto, A.; Ohno, K.; Tsujii, Y.; Miyamoto, T.; Kobatake, S.; Yamada, B. *Macromolecules* **1996**, *29*, 6396. (b) Goto, A.; Fukuda, T. *Macromolecules* **1997**,*30* , 4272.

6. Souaille, M.; Fischer, H. *Macromolecules* **2002**,*35* , 248.

7. (a) Fukuda, T.; Goto, A.; Tsujii, Y. In *Handbook of Radical Polymerization*; Matyjaszewski, K.; Davis, T. P., Eds. John Wiley & Sons, New York, NY, 2002; Chapter 9, p 407. (b) Fukuda, T.; Goto, A.; Tsujii, Y. In *Controlled/Living Radical Polymerization;* Matyjaszewski, K., Ed.; ACS Symp. Ser. 768; American Chemical Society: Washington, DC, 2000; p 27.

8. Goto, A.; Ohno, K.; Fukuda, T. *Macromolecules* **1998**, *31*, 2809.

9. Yoshikawa, C.; Goto, A.; Fukuda, T. *Macromolecules* **2002**,*35* , 5801.

10. Yoshikawa, C.; Goto, A.; Fukuda, T. *Macromolecules* **2003**,*36 (4)*, in press.

11. Goto, A.; Sato, K.; Tsujii, Y.; Fukuda,T.; Moad, G.; Rizzardo, E.; Thang, S. H. *Macromolecules* **2001**,*34* , 402.

12. Barner-Kowollik, C.; Quinn, J. F.; Morsley, D. R.; Davis, T. P. *J. Polym. Sci. Part A.: Polym. Chem.* **2001**,*39* , 1353.

13. Monteiro, M. J.; de Brouwer, H. *Macromolecules* **2001**, *34*, 349.

14. Kwak, Y.; Goto, A.; Tsujii, Y.; Murata, Y.; Komatsu, K.; Fukuda, T. *Macromolecules* **2002**, *35*, 3026.

15. Kwak, Y. et al., to be published.

16. Hui, A. W.; Hamielec, A. E. *J. Appl. Polym. Sci.*19 72, *16*, 749.

17. Molnar, S. *J. Polym. Sci., Part A-1* **1972**,*10* , 2245.

18. Gilbert, R. G. *Pure Appl. Chem.* **1996**,*68* , 1491.

19. Goto, A.; Fukuda, T. *Macromol. Chem. Phys.* **2000**, *201*, 2138.

20. Ohno, K.; Goto, A.; Fukuda, T.; Xia, J.; Matyjaszewski, K. *Macromolecules* **1998**, *31*, 2699.

Conventional Radical
Polymerization

Chapter 4

Termination Kinetics of Acrylate and Methacrylate Homo- and Copolymerizations

Michael Buback[1,2], Mark Egorov[1], and Achim Feldermann[1]

[1]Institut für Physikalische Chemie der Georg-August-Universität,
Tammannstrasse 6, D–37077 Goettingen, Germany
[2]Corresponding author: email: mbuback@gwdg.de

Termination kinetics of acrylate and methacrylate homo- and copolymerizations at 40⁰C and 1000 bar have been measured up to high conversion via the SP–PLP technique. DMPA and α-methyl-4-(methylmercapto)-α-morpholino propiophenone have been used as photoinitiators. An initial plateau region of constant k_t is seen which, with MMA, extends up to about 20 % monomer conversion and is followed by a reduction in k_t by about three orders of magnitude up to 50 % conversion. Dodecyl acrylate (DA) shows a distinctly different behavior in that constant (plateau) k_t is observed up to 75 % conversion. Copolymerization k_t in the plateau region is adequately described by a penultimate unit effect model. The chain-length dependence of k_t is investigated for several homopolymerizations using the expression: $k_t = k_t^0 \cdot i^{-\alpha}$, where i is chain length. With the exception of DA, where α is about 0.4 in the entire conversion range, α is close to 0.16 in the plateau region and increases once the gel effect region is reached.

Introduction

During recent years accurate propagation rate coefficients of free-radical homo- and copolymerizations have become available by applying the pulsed laser polymerization–size-exclusion chromatography (PLP–SEC) technique (*1, 2, 3*). To take full advantage of these data for polymerization modeling and optimization, also other rate coefficients, such as the termination rate coefficient, k_t, need to be precisely known. k_t's are also required for describing controlled living polymerizations in which both conversion and free-radical size increase during reaction. Modeling termination rate is by far more difficult as compared to propagation rate because of diffusion control of k_t. In addition to the dependence on temperature and pressure, k_t is affected by the diffusivity of macroradicals and thus depends on free-radical size and on the viscosity of the polymerizing medium. The latter quantity is determined by polymer content and by structural and dynamic properties of polymer molecules. The single pulse–pulsed laser polymerization (SP–PLP) method (*3, 4*) allows for a point-wise probing of termination kinetics during the course of polymerization up to high conversion. In addition to providing chain-length averaged termination rate coefficients, the technique allows for measuring the chain-length dependence of k_t. Within a perfect SP–PLP experiment, the laser pulse instantaneously creates a significant concentration of small photoinitiator-derived radicals which immediately start to propagate and thus, unless transfer reactions come into play, provide a narrow (Poisson-type) free-radical distribution with chain length i linearly increasing with time t after applying the laser pulse. Termination under SP–PLP conditions thus occurs between free radicals of approximately identical size.

Actually, k_t/k_p is the important primary kinetic parameter from SP–PLP. With k_p from PLP–SEC, chain-length averaged $<k_t>$ is immediately obtained from k_t/k_p. First such SP–PLP experiments have been carried out for a series of homopolymerizations (*3, 4*) and for intra-family acrylate or methacrylate systems (*5, 6*). These studies were extended to binary copolymerizations of inter-family acrylate–methacrylate systems. The first such inter-family system under investigation via SP–PLP was DA–dodecyl methacrylate (DMA) (*6*). This system is special in that the two homo-k_t's are almost identical. More complex situations occur with copolymerizations where both the homo-k_p's and homo-k_t's are significantly different, as is the case with the systems MMA–DA and DMA–methyl acrylate (MA) (*7*). Data for these systems measured at 40°C and high pressure, mostly 1000 bar, but for MMA also at 2000 bar will be presented. High-pressure conditions were selected because of the improved signal-to-noise quality of the spectroscopically measured SP–PLP conversion vs. time traces. High p is associated with a larger conversion per pulse due to pressure-induced propagation and pressure-retarded termination. With the weakly compressible monomer systems no reason is seen why mechanistic evidence deduced from studies at high p should not apply to ambient polymerization conditions. In addition to studying chain-length averaged copolymerization $<k_t>$ as a function of monomer conversion, the chain-length dependence of k_t at (almost) constant degree of monomer conversion is investigated for a series of homopolymerizations.

Experimental Part

The SP–PLP experiments have been carried out as described elsewhere (*3, 8*). The photoinitiators α-methyl-4-(methylmercapto)-α-morpholino propiophenone (MMMP, 98 %, Aldrich Chemie) and 2,2-dimethoxy-2-phenylacetophenone (DMPA, 99 %, Aldrich Chemie) were used as supplied at initial concentrations close to $5 \cdot 10^{-3}$ mol·L^{-1}. Methyl acrylate (>99 %, stabilized with 0.005 wt.% hydroquinone monomethylether, Fluka Chemie), butyl acrylate (>99 %, stabilized with 0.005 wt.% hydroquinone monomethylether, Fluka Chemie), dodecyl acrylate (which actually is a mixture of 55 wt.% DA and 45 wt.% tetradecyl acrylate, Fluka Chemie), methyl methacrylate (>99 %, stabilized with 0.005 wt.% hydroquinone monomethylether, Fluka Chemie), and dodecyl methacrylate (DMA, ≈96 %, Aldrich Chemie) were purified by distillation under reduced pressure in the presence of K$_2$CO$_3$ and treated by several freeze-pump-thaw cycles to remove dissolved oxygen. The samples were irradiated with XeF excimer laser pulses (at 351nm) of 2 to 3 mJ energy per pulse. Laser-induced monomer conversion is monitored via online NIR spectroscopy of the C–H modes (at the C=C double bond) at around 6170 cm^{-1}. After applying a series of excimer laser pulses, each being followed by microsecond time-resolved near-infrared spectroscopic measurement of pulse-induced polymerization, the reaction cell is introduced into the sample chamber of an IFS 88 Fourier transform IR/NIR spectrometer (Bruker) where absolute (overall) monomer concentration is checked. During each polymerization experiment, SP–PLP measurements are carried out until the reacting system becomes inhomogeneous or the photoinitiator is consumed.

Copolymerization

Before addressing the chain-length dependence of the termination rate coefficient, the conversion dependence of overall (chain-length averaged) $<k_t>$ will be briefly discussed taking the inter-family copolymerization systems MMA–DA and DMA–MA as examples. The pulsed laser induced monomer conversion measured (with a time resolution of microseconds) as a function of time t after applying the laser pulse at $t = 0$ is fitted to eq 1 which assumes ideal kinetics in that k_t is constant:

$$\frac{c_M(t)}{c_M^0} = \left(1 + 2 \cdot \langle k_t \rangle \cdot c_R^0 \cdot t\right)^{-k_p/(2 \cdot k_t)} \tag{1}$$

where c_M is monomer concentration, c_M^0 and c_R^0 are monomer and radical concentrations at time $t = 0$, respectively, and k_p is the propagation rate coefficient. The fitting procedure yields the two parameters: $\langle k_t \rangle / k_p$ and $\langle k_t \rangle \cdot c_R^0$. It should be noted that the chain-length averaged $\langle k_t \rangle$ value deduced from eq 1 may vary with the extension of the time region t used for the fitting procedure. This point will be stressed again in the subsequent section. It needs further to be mentioned that k_t refers to the IUPAC-preferred notation:

$$\frac{dc_R}{dt} = -2 \cdot k_t \cdot (c_R)^2 \tag{2}$$

The primary focus in the fitting procedure via eq 1 is on $\langle k_t \rangle / k_p$. With k_p being available from PLP–SEC experiments (1, 2, 3), $\langle k_t \rangle$ is immediately obtained from the primary experimental parameter $\langle k_t \rangle / k_p$. As c_R^0 is less easily accessible from independent experiments, the second parameter from fitting conversion vs. time data, $\langle k_t \rangle \cdot c_R^0$, less useful for estimates of $\langle k_t \rangle$.

Figure 1 shows a plot of $\langle k_t \rangle / k_p$ vs. overall monomer conversion, X, for MMA and DA homopolymerizations and for MMA–DA copolymerizations with initial MMA mole fractions, f_{MMA}^0 of 0.50, 0.25, and 0.06, respectively.

During copolymerization to high X, MMA mole fraction, f_{MMA}, varies because polymerization of MMA is preferred over DA polymerization. This variation may be quantitatively estimated from the known reactivity ratios (9). The SP–PLP experiments have been carried out at 40°C and 1000 bar with DMPA as the photoinitiator. Low conversion $\langle k_t \rangle / k_p$ of MMA exceeds the corresponding DA value by almost three orders of magnitude. This large difference is due to MMA k_t being by more than one order of magnitude above and MMA k_p being by close to two orders of magnitude below the corresponding DA values. The significant scatter on the MMA $\langle k_t \rangle / k_p$ data is due to high k_t and low k_p which is associated with unfavorable SP–PLP signal-to-noise quality. $\langle k_t \rangle / k_p$ is more or less constant at low X. As the propagation rate coefficient for both MMA and DA homopolymerization will not vary with conversion, the change of $\langle k_t \rangle / k_p$ with X in both homopolymerizations is entirely due to changes of $\langle k_t \rangle$. A remarkable observation from Figure 1 is that $\langle k_t \rangle$ of DA remains constant over the experimental conversion range up to about 80 % whereas $\langle k_t \rangle$

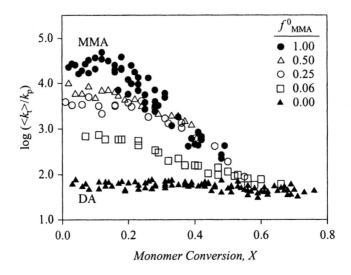

Figure 1. Log <k_t>/k_p of MMA–DA bulk copolymerizations at 40 °C and 1000 bar plotted against X, the overall monomer conversion. f_{MMA}^0 is the initial mole fraction of MMA in the comonomer mixture.
(Modified from reference 7. Copyright 2002.)

of MMA, after an initial plateau region up to about 20 % monomer conversion, enormously drops toward higher X.

The plateau region also of the DA-rich copolymerization systems extends to fairly large X. This effect has already been seen with MA–DA copolymerization (6), where it is particularly pronounced. The reason why the plateau region for DA-rich systems is less pronounced with MMA–DA systems than with MA–DA systems, see the $f_{MMA}^0 = 0.06$ data in Figure 1 as compared to the $f_{MMA}^0 = 0$ (DA) ones, is due to the fact that MMA polymerizes preferentially. Upon increasing X, this results in a relative enhancement of the DA content of the monomer mixture and thus in an enhancement of copolymerization k_p. As a consequence, <k_t>/k_p is lowered toward increasing X when data are plotted for the same initial monomer mixture composition, f_{MMA}^0. In the (intra-family) MA–DA copolymerization, f_{MA} stays close to f_{MA}^0 during the entire course of a polymerization up to high conversion (6).

The <k_t>/k_p values of MMA-rich and DA-rich systems are very different at low X, but approach each other at high degrees of monomer conversion. This remarkable transition is strongly indicative of different types of diffusion control of k_t being operative. At low conversion, intra-coil processes of entangled macroradicals or of short-lived contact pairs of such radicals control termination rate. This type of control is referred to as segmental diffusion. At high conversion, termination rate may run into reaction diffusion control (*10*). According to this latter mechanism, free-radical sites under conditions of severe hindrance of center-of-mass diffusion approach each other by propagation steps in conjunction with segmental mobility of the chain ends (*11*). The rate coefficient of termination by reaction diffusion, $k_{t,RD}$, may be expressed by eq 3:

$$<k_t>_{RD} = C_{RD} \cdot k_p \cdot (1-X) \tag{3}$$

where C_{RD} is a reaction-diffusion coefficient. That reaction diffusion is operative may be demonstrated by plotting $<k_t>/(k_p \cdot (1-X))$ vs. monomer conversion (*12*).

In what follows, $<k_t>$ will be analyzed within the plateau region, which is relatively narrow with MMA-rich systems but extends up to fairly high conversions with DA-rich systems. Copolymerization termination rate coefficients, $<k_t>_{copo}$, are deduced from experimental ($<k_t>/k_p)_{copo}$ via the copolymerization propagation rate coefficients, $k_{p,copo}$, known from independent PLP–SEC experiments (*13*). $<k_t>_{copo}$ data of DMA–MA at 40°C and 1000 bar and for overall monomer conversion up to 20 % are presented in Figure 2.

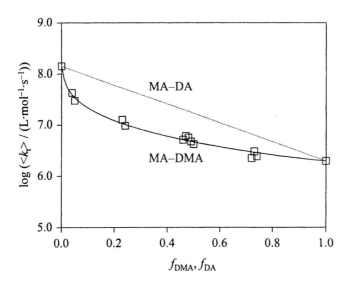

Figure 2. Chain-length averaged termination rate coefficients for the plateau region of DMA–MA copolymerizations at 40°C and 1000 bar plotted vs. the actual DMA mole fraction. The full line represents the fit of the data to eq 6. The dotted line illustrates MA–DA data for the same p and T conditions. (Modified from reference 7. Copyright 2002.)

The data points exhibit some cluster-type structure which is due to the slight variation of f_{DMA} with X that allows for deducing $k_{t,copo}$ at (slightly) different f_{DMA} from successive SP–PLP experiments carried out during one copolymerization experiment at increasing X. The log $<k_t>$ vs. f_{DMA} correlation is curved for DMA–MA and thus differs from the corresponding close-to-linear relation observed with the intra-family DA–MA system (dotted line in Figure 2) (6, 7). The full line in Figure 2 is a fit of the experimental $<k_t>$ data to the same penultimate unit effect (PUE) model that has been successfully applied toward representing the linear log $<k_t>$ vs. f dependence seen with the intra-family systems. The PUE model, which traces back to Russo and Munari (14, 15) has been thoroughly investigated and applied by Fukuda et al. (16) and has recently been used and further explored in our group (6, 7). The relevant equations will only briefly be reiterated here. The copolymerization termination rate coefficient, $<k_t>_{copo}$, is given by eq 4:

$$k_{t,copo} = \sum_{l=1}^{2} \sum_{k=1}^{2} \sum_{i=1}^{2} \sum_{j=1}^{2} P_{ij} \cdot P_{kl} \cdot k_{tij,kl} \tag{4}$$

where P_{ij} and P_{kl} are the populations of free radicals with terminal units j and l and with penultimate units i and k, respectively. The P_{ij}'s are estimated from the homo-propagation rate coefficients, k_{ii}, from terminal and penultimate reactivity ratios, r_i and r_{ii}, respectively, and from the radical reactivity ratios, s_i and s_j, according to the expressions provided by the Fukuda group (17). Ten PUE termination rate coefficients $k_{tij,kl}$ are contained in eq 4. To reduce this number, rate coefficients for termination between "unlike" free radicals, $k_{tij,kl}$, are replaced by the associated "like" coefficients $k_{tij,ij}$ and $k_{tkl,kl}$, via the so-called geometric mean model illustrated by eq 5 (6, 16):

$$k_{tij,kl} = (k_{tij,ij} \cdot k_{tkl,kl})^{0.5} \tag{5}$$

Inserting $k_{tij,kl}$ from eq 5 into eq 4 yields eq 6:

$$k_{t,copo}{}^{0.5} = k_{t11,11}{}^{0.5} \cdot P_{11} + k_{t21,21}{}^{0.5} \cdot P_{21} + k_{t22,22}{}^{0.5} \cdot P_{22} + k_{t12,12}{}^{0.5} \cdot P_{12} \tag{6}$$

where $k_{t11,11}$ and $k_{t22,22}$ are the associated homo-termination rate coefficients. Eq 6 has been successfully used for fitting $k_{t,copo}$ of numerous copolymerization systems (6, 16) including rather special systems such as styrene–diethyl fumarate (18).

With the acrylate and methacrylate intra-family systems studied so far, very reasonable fits of experimental $k_{t,copo}$ data to eq 6 have been obtained even with a simplified version of the PUE model in which the two "unlike" termination rate coefficients, $k_{t12,12}$ and $k_{t21,21}$, were assumed to be identical. The full line in

Figure 2 is also fitted by using this simplified version of the PUE k_t model. The P_{ij}'s are estimated from reactivity ratios $r_{mac} = 2.55$, $r_{ac} = 0.29$, deduced via composition analysis of a series of alkyl acrylate–alkyl methacrylate systems studied under high-pressure conditions (13). The radical reactivity ratios, which also go into the estimate of P_{ij}, were assumed to be $s_{mac} = 2$ and $s_{ac} = 0.5$. These values are close to the ones reported for the butyl acrylate–methyl methacrylate system (19). The coefficients resulting for 1000 bar and 40°C are $k_{t12,12} = k_{t21,21} = 3.08 \cdot 10^7$ L·mol^{-1}·s^{-1} which is in-between the associated homo-termination rate coefficients, $k_{tDMA,DMA} = 1.66 \cdot 10^6$ L·mol^{-1}·s^{-1} and $k_{tMA,MA} = 1.35 \cdot 10^8$ L·mol^{-1}·s^{-1}. Fitting of $k_{t,copo}$ to eq 6 with $k_{t12,12}$ being allowed to differ from $k_{t21,21}$ yields some minor improvement of the fit, but results in $k_{t12,12}$ and $k_{t21,21}$ values which are not both in-between $k_{tDMA,DMA}$ and $k_{tMA,MA}$. The important point to note from Figure 2 is that irrespective of the clear difference in log $<k_t>$ vs. monomer mole fraction, curved vs. linear, the same penultimate unit model is capable of representing the dependence of log $<k_t>_{copo}$ on monomer feed composition. This finding allows for estimates of $<k_t>_{copo}$ (within the initial plateau region) on the basis of a very small number of parameters which, with the exception of the unlike termination rate coefficient, $k_{t12,12}$ or $k_{t21,21}$, are mostly known or are available from independent experiments.

Chain-length Dependence of Homopolymerization Termination Rate Coefficients

Because of the special feature of a narrow size distribution of free radicals with chain length i increasing linearily with time t after applying the pulse (at $t = 0$), SP–PLP experiments provide access to measuring the chain-length dependence of k_t. The narrow Poisson-type distribution is, of course, restricted to the early time region in which chain-transfer processes play no major role. The perfect SP–PLP experiment further requires that both free-radical species from photoinitiator decomposition readily add to monomer and thus start macromolecular growth. The experiments described above have been carried out using DMPA (2,2-dimethoxy-2-phenylacetophenone) as the photoinitiator.

As is shown in Scheme 1, DMPA decomposes into an acetal radical 1 and a benzoyl radical 2. Whereas 2 is highly efficient in adding to monomer, 1 does not noticeably add to monomer in the dark time period after the pulse, but may react with radicals and thus behaves like an inhibitor species. The poor propagation activity of 1 has first been described by the Fischer group (20) and has recently been demonstrated through MALDI experiments by the Davis group (21). This low reactivity is understood as resulting from resonance stabilization of 1, which is illustrated by the series of resonance structures depicted in the lower part of Scheme 1. The simultaneous initiation and inhibition activity of the

DMPA-derived species results in a rather peculiar SP–PLP behavior: The monomer conversion vs. time traces measured at different initial DMPA contents, but otherwise identical reaction conditions, intersect each other (22). This crossing behavior provides access to measuring the chain-length dependence of the termination rate coefficient, $k_t(i)$ (22). Measuring several SP–PLP traces under conditions where all reaction conditions except DMPA concentration are identical, however, becomes increasingly difficult toward moderate and high degrees of monomer conversion. For this reason, it is highly desirable to use an ideal photoinitiator which decomposes into two free-radical fragments both of which easily add to a monomer molecule. Such a photoinitiator allows for deducing $k_t(i)$ from a single SP–PLP trace.

Before describing experiments with such an "ideal" initiator, it should be pointed out that the inhibition activity of **1** does not affect k_t determination via SP–PLP at low degrees of DMPA concentration. The SP–PLP experiments presented in the previous section have been carried out under such conditions. The $\langle k_t \rangle$ data presented above thus may be considered as "true" chain-length averaged termination rate coefficients.

Investigations by Kuelpmann into several photoinitiators revealed that α-methyl-4-(methylmercapto)-α-morpholino-propiophenone (MMMP), see Scheme 2, is a close-to-ideal photoinitiator which rapidly decomposes into two propagating free-radical species (23). MMMP was used in SP–PLP experiments at 40°C on BA, DA, and DMA (at 1000 bar) and on MMA (at 2000 bar).The measurements have been carried out within extended ranges of monomer conversion, partly up to 70 %. The time (interval) t, after applying the laser pulse at $t = 0$, in which chain transfer to monomer may be neglected at 40°C is about 0.1 s for acrylates and about 1.0 s for methacrylates.
The analysis of chain-length dependent k_t has been restricted to this initial time region in which free-radical size distribution after each pulse is of Poisson type and is linearly correlated with t.

Figure 3 shows the dependence of $\langle k_t \rangle$ on monomer conversion for bulk homopolymerizations at 1000 bar and 40°C of MA (circles) and DMA (squares).

The $\langle k_t \rangle$ values were obtained by fitting the SP–PLP traces to eq 1 for shorter and longer time intervals (see legend to Figure 3). Application of eq 1 yields an average termination coefficient $\langle k_t \rangle$, which refers to a smaller average chain length when the fitting procedure is carried out for a smaller initial time region of the SP–PLP trace. As can be seen from Figure 3, the $\langle k_t \rangle$ values for shorter time t and thus shorter chains (open symbols) are above the values deduced from fitting the same signal over a more extended time range (associated full symbols). The differences are minor at low and moderate degrees of monomer conversion and are more pronounced at higher conversion where data points from the same signal, but taken for different time intervals, are clearly separated. The data in Figure 3 suggests that k_t decreases with increasing free-radical chain length.

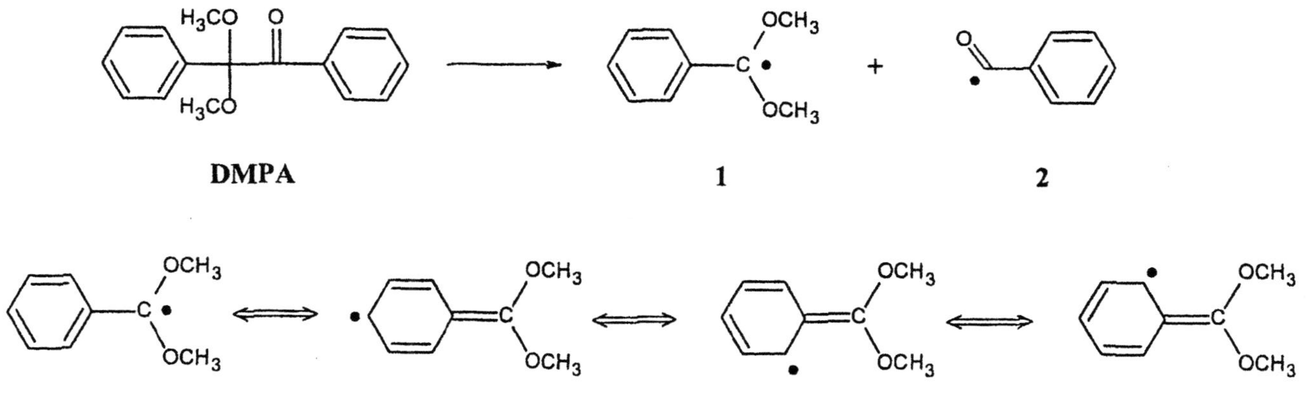

Scheme 1. Decomposition of the photoinitiator DMPA (2,2-dimethoxy-2-phenylacetophenone) (upper part) and resonance structures of the methoxy-type fragment 1 (lower part).

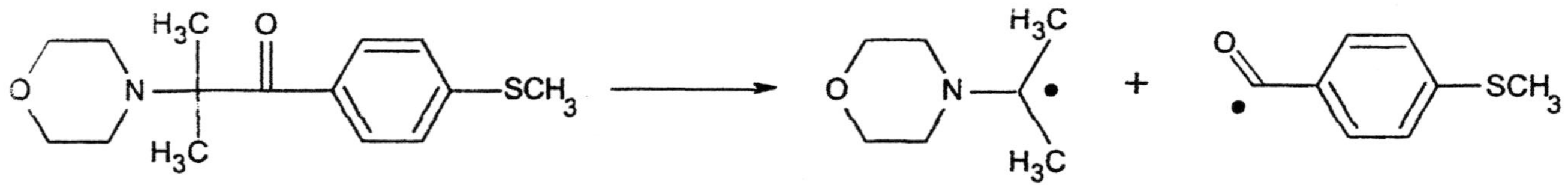

Scheme 2. Decomposition of the photoinitiator MMMP (α-methyl-4-(methylmercapto)-α-morpholino propiophenone).

MMMP

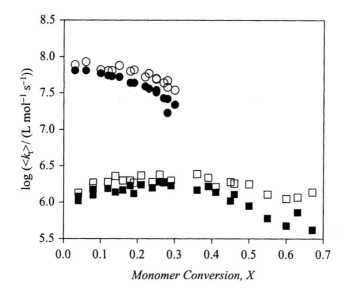

Figure 3. Log <k_t> calculated for different time ranges after a laser pulse at t = 0 for MA (circles) and DMA (squares) at 1000 bar and 40°C. The time intervals are as follows: 0 to 0.02 s (open symbols) and 0 to 0.1 s (full symbols) for MA, and 0 to 0.1 s (open symbols) and 0 to 0.5 s (full symbols) for DMA.

In Figure 4, a plot similar to the one in Figure 3 is shown for MMA (circles) and DA (squares) homopolymerizations at 40°C. With MMA, the difference between the two associated <k_t> values is small at low degrees of monomer conversion but becomes significant above 20 % conversion. The difference in <k_t> of smaller and larger free radicals of DA is larger than with the other monomers, but does not markedly change with conversion. The data in Figure 4 is also indicative of k_t decreasing toward larger chain length (CL).

A more detailed quantitative analysis of SP–PLP signals requires fitting to an expression which considers CL-dependent k_t. Among the most widely used termination models is the so-called power-law model, which assumes that k_t varies with radical chain length i according to:

$$k_t(i) = k_t^0 \cdot i^{-\alpha} \tag{7}$$

where k_t^0 is the termination rate coefficient for very small free radicals. With the CL dependence of k_t being represented by eq 7, the change in relative monomer concentration after applying a laser pulse at $t = 0$ is given by the integral expression in eq 8, see Ref. (*24*):

$$\frac{c_M(t)}{c_{M,0}} = 1 - \frac{b}{c} \cdot \int_0^{t'} \left(b \int_0^{t'} (t_p / t'')^\alpha \, dt'' + 1 \right)^{-1} dt' \tag{8}$$

where $t_p = (k_p \cdot c_M^0)^{-1}$ is the time required for the first propagation step (and for each subsequent propagation step of the particular SP–PLP experiment at $c_M \cong$ constant), $b \equiv 2 \cdot c_R^0 \cdot k_t^0$ and $c \equiv 2k_t^0 / k_p$. Within eq 7 it is assumed that chain-transfer processes may be ignored, which is a good approximation in the initial period of the SP–PLP trace at times $t < t_{tr}$, where $t_{tr} = (k_{tr} \cdot c_M)^{-1}$ is the time interval after which chain transfer comes into play. Fitting of SP–PLP data to eq 8 yields k_t^0 / k_p, $c_R^0 \cdot k_t^0$ and α.

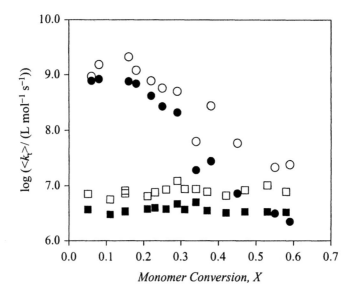

Figure 4. Log $<k_t>$ calculated for different time ranges after a laser pulse at $t = 0$ for MMA (circles) and DA (squares) at 1000 bar and 40 °C. The time intervals are as follows: 0 to 0.02 s (open symbols) and 0 to 0.1 s (full symbols) for DA, and 0 to 0.1 s (open symbols) and 0 to 0.5 s (full symbols) for MMA.

Plotted in Figure 5 are values of α deduced, via eq 8, for MA, BA, MMA and DMA homopolymerizations at monomer conversions up to about 20 %. Previously reported numbers for methyl acrylate are: $\alpha = 0.32$ (*22*) and, for chain lengths between $i = 50$ and 100, $\alpha = 0.35$-0.36 (*25*). The latter value may not be directly compared to the data of the present study as α was not determined for SP–PLP time regions corresponding to such small chain lengths.

The discrepancy to the value reported in Ref. (22) is probably due to the fact that in Ref. (22) k_t^0 has been fixed to $7 \cdot 10^8$ L·mol^{-1}·s^{-1}, whereas the fitting procedure of the present study resulted in a value of k_t^0 slightly above 10^8 L·mol^{-1}·s^{-1}. It should be noted, that the two quantities, α and k_t^0, are correlated to each other. If k_t^0 from Ref. (22) is used to deduce α from the present SP–PLP data, the resulting value of α would be in agreement with the literature value of 0.32. The previously reported values for MMA, $\alpha = 0.15$ (26) and $\alpha = 0.16$-0.17 (27) are close to the data in Figure 5, as are the literature values of α for styrene: $\alpha = 0.24$ (26) and $\alpha = 0.17$ (27).

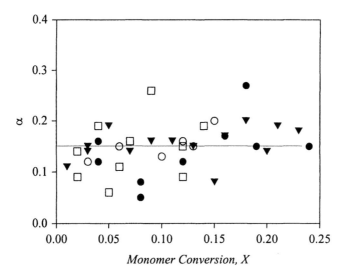

Figure 5. Exponent α obtained by fitting eq 8 to SP–PLP traces for MA (open circles), BA (triangles), DMA (full circles) at 1000 bar, MMA (squares) at 2000 bar, all at 40°C. The straight line refers to the mean value α = 0.15.

The arithmetic mean of the entire data set in Figure 5 for MA, BA, MMA, and DMA is $\alpha = 0.15$, represented by the straight line in the picture. This number is in close agreement with the theoretical value of $\alpha = 0.16$ predicted by Friedman and O'Shaughnessy (28, 29) for chemically controlled and for diffusion controlled termination reactions in dilute polymer solution of sufficiently large free radicals with radical functionality at (or close to) terminal positions of the macromolecular chain. For reaction between two free radicals with the active site of one radical being located somewhere on the backbone and the functionality of the second radical sitting at the chain end (end-interior reaction), α is predicted to be 0.27. For the interior-interior reaction of two free radicals, both with radical functionality on the backbone, α should be 0.43 (28).

Macroradicals with the radical site being located somewhere on the backbone should primarily occur at high degrees of monomer conversion and, within SP–PLP experiments, at large observation times, where t exceeds the time interval estimated for the transfer-to-polymer reaction. At lower conversions and at shorter observation times t, termination should proceed primarily via chain-end encounters. It is thus gratifying to note that the α values presented in Figure 5 are close to the predicted low-conversion value of $\alpha = 0.16$. The exponent α stays below 0.2 up to 15 or even 20 % monomer conversion, which is well above the range in which the polymer solution is considered to be dilute.

As was indicated by the data for MA and DMA in Figure 3, α increases toward higher conversion. This effect is particularly pronounced for MMA (see Figure 6) in the gel effect region at monomer conversions above 20 %. The significant increase of α with monomer conversion strongly suggests a different mechanism of diffusion control of termination rate. Whereas segmental diffusion controls the plateau-type behavior of k_t and the low value α characterizes the reduced probability (with increasing i) of chain-end encounters in a contact pair of macroradicals, the larger α under gel effect conditions is associated with termination rate control by center-of-mass diffusion of macroradicals in the entangled polymeric system.

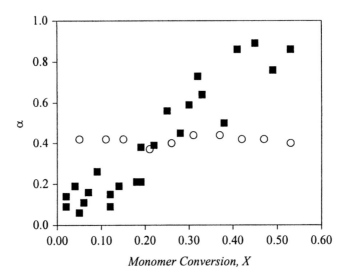

Figure 6. Conversion dependence of the exponent α for bulk homopolymerizations of MMA at 2000 bar / 40 °C (full squares) and DA at 1000 bar / 40 °C (open circles).

Also given in Figure 6 are the α values for DA up to about 50 % monomer conversion. The exponent α remains close to $\alpha = 0.4$ in the entire conversion range. It is not unexpected that α does not change under conditions where also chain-length averaged overall $<k_t>$ remains constant (see Figure 1). The relatively high value of α is not easily understood. Termination of mid-chain radicals may provide an explanation. There is however no indication from independent sources, e.g. from EPR, that such mid-chain radicals play a major role in termination during DA polymerization, but are not important during homopolymerization of the other monomers. A large fraction of mid-chain radicals would also affect k_p. Further studies are required to understand the CL dependence observed in DA polymerization.

It goes without saying that termination models other than the power-law expression may be used for fitting experimental SP–PLP traces. Eq 7 has been applied primarily because of its simplicity and because of the availability of a theoretically derived power-law expression for the CL dependence of termination rate in dilute polymer solution. The SP–PLP method thus clearly demonstrates that k_t decreases with chain length but does not allow for elucidation of the detailed termination mechanism so far.

Conclusions

The SP–PLP technique allows for probing the conversion and the CL dependence of termination kinetics. With DA and with DMA, chain-length averaged $<k_t>$ remains constant up to about 70 % conversion. $<k_t>$ of MMA, after an initial plateau region up to about 20 %, decreases by about three orders of magnitude. Copolymerization $<k_t>$ for acrylate–methacrylate systems in the plateau region is well described by a PUE model. Plateau k_t is assigned to segmental diffusion, whereas $<k_t>$ at higher conversion runs under control by center-of-mass diffusion and by reaction diffusion.

The CL dependence of k_t is described by a simple power-law model. The exponent α is close to 0.16, which is the theoretically predicted value for „chain end – chain end reactions" of large radicals in dilute polymer solutions. α increases once conversion exceeds the plateau region of k_t. This increase is very pronounced in the gel effect of MMA. α appears to be sensitive toward the type of termination-controlling mechanism. Among the monomers studied so far, only DA shows an unusual behavior in that the power-law exponent is rather high, $\alpha \cong 0.4$, and remains constant up to fairly large conversion.

Acknowledgement

Financial support by the *Deutsche Forschungsgemeinschaft (DFG)* through the *European Graduate School* "Microstructural Control in Free-Radical Polymerization" is gratefully acknowledged as is support by the *Fonds der Chemischen Industrie.*

References

1. Olaj, O. F.; Bitai, I.; Hinkelmann F. *Makromol. Chem.* **1987**, *188*, 1689.
2. Olaj, O. F.; Schnöll-Bitai I. *Eur. Polym. J.* **1989**, *25*, 635.
3. Beuermann, S.; Buback, M. *Prog. Polym Sci.* **2002**, *27*, 191.
4. Buback, M.; Hippler, H.; Schweer, J.; Vögele, H.-P. *Makromol. Chem. Rapid Commun.* **1986**, *7*, 261.
5. O'Driscoll, K. F.; Ito, K. *J. Polym. Sci.* **1979**, *17*, 3913.
6. Buback, M.; Kowollik, C. *Macromolecules* **1999**, *32*, 1445.
7. Buback, M.; Feldermann, A. *Aust. J. Chem.* **2002**, *55*, 475.
8. Buback, M.; Kowollik, C. *Macromolecules* **1998**, *31*, 3221.
9. Odian, G. *Principles of Polymerization*, 3rd edition; Wiley, N. Y. 1991.
10. Schulz, G. V. *Z. Phys. Chem. (Munich)* **1956**, *8*, 290.
11. Benson, S. W.; North, A. M. *J. Am. Chem. Soc.* **1962**, *84*, 935.
12. Buback, M.; Huckestein, B.; Russell, G. T. *Macromol. Chem. Phys.* **1994**, *195*, 539.
13. Buback, M.; Feldermann, A.; Barner-Kowollik, C.; Lacík, I. *Macromolecules* **2001**, *34*, 5439.
14. Russo, S.;Munari, S. *J. Macromol. Sci. Chem.* **1968**, *2*, 1321.
15. Bonta, G.; Gallo, B. M.; Russo, S. *J. Chem. Soc., Faraday Trans. 1* **1975**, *71*, 1721.
16. Fukuda, T.; Ide, N.; Ma, Y.-D. *Macromol. Symp.* **1996**, *111*, 305.
17. Fukuda, T.; Ma, Y.-D.; Inagaki, H. *Macromolecules* **1985**, *18*, 17.
18. Ma, Y.-D.; Sung, K.-S.; Tsujii, Y.; Fukuda, T. *Macromolecules* **2001**, *34*, 4749.
19. Hutchinson, R. A.; McMinn, J. H.; Paquet D. A., Jr., Beuermann, S.; Jackson, C. *Ind. Eng. Chem. Res.* **1997** , *36*, 1103.
20. Fischer, H.; Baer, R.; Hany, R.; Verhoolen, I.; Walbiner, M. *J. Chem. Soc., Perkin Trans.* **1990**, *2*, 787.
21. Vana, P.; Davis, T. P.; Barner-Kowollik, C. *J. Polym. Sci., Part A: Polym. Chem.* **2002**, *40*, 675.
22. Buback, M.; Busch, M.; Kowollik, C. *Macromol. Theory Simul.* **2000**, *9*, 442.
23. Buback, M.; Kuelpmann, A. submitted to *Macromol. Chem. Phys.*
24. Buback, M.; Kowollik, C.; Egorov, M.; Kaminsky, V. *Macromol. Theory Simul.* **2001**, *10,* 209
25. de Kock, J. B. L. Ph.D Thesis, Eindhoven 1999.
26. Mahabadi, H. K. *Macromolecules* **1985**, *18*, 1319.
27. Olaj, O. F.; Vana, P. *Macromol. Rapid Commun.* **1998**, *19*, 433 and 533.
28. Friedman, B.; O'Shaughnessy, B. *Macromolecules* **1993**, *26,* 5726.
29. Karatekin, E.; O'Shaughnessy, B.; Turro, N. J. *Macromol. Symp.* **2002**, *182*, 81.

Chapter 5

Lewis Acid-Catalyzed Tacticity Control during Radical Polymerization of (Meth)acrylamides

Yoshio Okamoto, Shigeki Habaue, and Yutaka Isobe

Department of Applied Chemistry, Graduate School of Engineering, Nagoya University, Furo-cho, Chikusa-ku, Nagoya 464–8603, Japan

The tacticity control during the radical polymerization of (meth)acrylamides was achieved in the presence of a catalytic amount of Lewis acids such as $Y(OTf)_3$ and $Yb(OTf)_3$. The conventional polymerization of N-isopropylacrylamide without Lewis acids produced an atactic polymer (meso diad = 42 – 46 %), whereas the polymer prepared using the Lewis acids in methanol had the meso diad of more than 90%. An analogous effect of the Lewis acids was observed during the polymerization of other (meth)acrylamides, including acrylamide, N,N-dimethylacrylamide, methacrylamide, N-methylmethacrylamide, N-isopropylmethacrylamide, N-phenylmethacrylamide, and (R)-N-[(methoxycarboxyl)-phenylmethyl]methacrylamide.

The tacticity control of a polymer is an important goal for polymer science and industry, because the property and function of a vinyl polymer significantly depend on the stereostructure (1-8). The stereocontrol during radical polymerization cannot be readily attained (5-7), although it has been widely used in industry, and therefore, stereoregular polymers have been mainly produced by ionic and coordinate polymerization processes. A growing free-radical is often very active and electrically neutral, which prevents it from interaction with other agents and makes the control during the propagation difficult. However, in the past decade, a remarkable advance has been made in the living radical polymerization, which enables the synthesis of block and graft copolymers with a defined structure by a radical process (9-12). On the other hand, the control of stereochemistry during radical polymerization has slowly advanced (5). The monomers such as dienes (13) and acrylonitrile (14) included in the host crystals afford the stereocontrolled polymers. The solid state polymerization of crystalline diene monomers produces stereoregular polymers (15). The radical polymerizations of the bulky methacrylates (16,17) and acrylamides (18,19) proceed in an isotactic-specific manner. Fluoroalcohols, such as perfluoro-tert-butyl alcohol ((CF_3)$_3$COH) and 1,1,1,3,3,3-hexafluoro-2-propanol ((CF_3)$_2$CHOH), can clearly influence the stereochemistry during the radical polymerization of vinyl esters (7,20,21) and methacrylates (7,22,23) through the hydrogen-bonding interaction between the monomers and the fluoroalcohols.

Recently, we examined the influence of various Lewis acids on the stereochemistry of the radical polymerization of polar monomers including α-substituted acrylates and (meth)acrylamides, and a clear effect has been observed. Poly[(meth)acrylamide]s are important materials due to their unique properties as gels or hydrophilic polymers (24-26). Among the many Lewis acids, rare earth metal trifluoromethanesulfonates (triflate: OTf) were particularly effective in increasing the isotacticity of various poly(acrylamide)s (27-29) and poly(methacrylamide)s (30,31). Lewis acids have been widely used in organic syntheses. In polymer chemistry, Lewis acids have been used to change the reactivity of monomers during copolymerization, and several alternating copolymers have been prepared (32-34). During the copolymerization of methyl methacrylate (MMA) and styrene, the addition of BCl_3 resulted in a highly coheterotactic alternating copolymer (35,36). On the other hand, during the homopolymerization, although Lewis acids affected monomer reactivity (37), their influence on the tacticity of the polymers had rarely been reported (5). Matsumoto et al. reported that $MgBr_2$ was effective in enhancing slightly the isotactic selectivity during the radical polymerization of MMA (38). We also found that Sc(OTf)$_3$ increased the isotacticity of the poly(methacrylate)s (7,29,39). For α-alkoxymethyl acrylates, the conventional polymerization without Lewis acids produced atactic polymers, but that in the presence of $ZnBr_2$ and Sc(OTf)$_3$ resulted in syndiotactic and isotactic polymers, respectively (40,41). Porter et al. reported that cyclic acrylamides produce isotactic polymers in the presence of rare earth metal triflates (42).

N,N-Disubstituted acrylamides can be polymerized by anionic initiators to produce stereoregular polymers (43,44), but (meth)acrylamides bearing an amide NH proton on the side chain cannot be anionically polymerized, and therefore, the stereocontrol during their polymerization has hardly been studied. In this chapter, we describes the isotactic-specific radical polymerization of (meth)acrylamides, including N-isopropylacrylamide (NIPAM), acrylamide (AM), N,N-dimethylacrylamide (DMAM), methacrylamide (MAM), N-methylmethacrylamide (MMAM), N-isopropylmethacrylamide (IPMAM), N-phenylmethacrylamide (PMAM), and (R)-N-[(methoxycarboxyl)phenylmethyl]-methacrylamide (MCPMMAM), in the presence of Lewis acids.

N-Isopropylacrylamide (NIPAM)

Table I shows the results of the polymerizations of NIPAM in the presence of various Lewis acids in methanol at 60°C. The polymerizations were carried out using Lewis acids in dry nitrogen. The polymerization in the absence of Lewis acids produced an atactic polymer (m / r = 44 / 56), whereas that in the presence of a catalytic amount of Lewis acids proceeded in an isotactic-specific manner. Most rare earth metal triflates except for $Sc(OTf)_3$ exhibited a similar isotacticity-enhancing effect and produced a polymer having a meso (m) value of about 80%. Rare earth metal chlorides were less effective than the corresponding triflates.

The effect of solvents was studied in the presence of $Y(OTf)_3$ (**Table II**). In the absence of Lewis acids, the influence of the solvents on the tacticity was very small and atactic polymers (m = 42~46%) were produced. The effect of Lewis acids significantly depended on the polymerization solvent, and the highest isotacticity-enhancing effect was observed in methanol. The effect of Lewis acids in alcoholic solvents decreased in the order of methanol > ethanol > isopropanol with increasing bulkiness. $Y(OTf)_3$ was also effective in tetrahydrofuran (THF), N,N-dimethylformamide (DMF), chloroform, toluene, 1,4-dioxane and even in water. However, the effect disappeared in DMSO.

Figure 1 illustrates the relationship between the $Y(OTf)_3$ concentration and the m value of the obtained polymers for the polymerization of NIPAM at 60°C, –20°C, and –78°C. A small amount of $Y(OTf)_3$ was enough to affect the stereostructure of the obtained polymers, and 8 mol% of $Y(OTf)_3$ to NIPAM resulted in a polymer having an m value of 90% at –20°C. The highest isotacticity was attained at $[Y(OTf)_3]_0 = 0.5$ mol/l ($[Y(OTf)_3]_0$ / $[NIPAM]_0 \approx 0.2$) at 60°C and –20°C (m = 92%), while the further addition of $Y(OTf)_3$ reduced the m value. Lewis acids were more effective at –20°C than at 60°C and –78°C. At –78°C, the Lewis acids may not catalytically function. **Figure 2** shows the ^1H NMR spectra of the atactic (A) and isotactic poly(NIPAM)s (B) measured at 170°C in DMSO-d_6. The diad tacticities of these polymers were determined on the basis of the peaks (1.3 ppm-1.8 ppm) of the methylene protons (45).

Table I. Effect of Lewis Acids in the Polymerization of NIPAM in Methanol[a]

Run	Lewis acid	Yield[b] %	Tacticity[c] m / r	Run	Lewis acid	Yield[b] %	Tacticity[c] m / r
1	None	82	45 / 55	10	$Gd(OTf)_3$	99	81 / 19
2	$Sc(OTf)_3$	86	62 / 38	11	$Tb(OTf)_3$	97	82 / 18
3	$Y(OTf)_3$	94	80 / 20	12	$Ho(OTf)_3$	98	83 / 17
4	$La(OTf)_3$	98	76 / 24	13	$Er(OTf)_3$	94	82 / 18
5	$Ce(OTf)_3$	89	79 / 21	14	$Tm(OTf)_3$	97	84 / 16
6	$Pr(OTf)_3$	96	81 / 19	15	$Yb(OTf)_3$	89	82 / 18
7	$Nd(OTf)_3$	99	80 / 20	16	$Lu(OTf)_3$	97	84 / 16
8	$Sm(OTf)_3$	87	81 / 19	17	$ScCl_3$	85	57 / 43
9	$Eu(OTf)_3$	94	82 / 18	18	$YbCl_3$	95	67 / 33

[a] $[NIPAM]_0 = 2.40$ mol/l. Initiator: AIBN (0.02 mol/l). Temp. = 60°C. Time = 3hr.
[b] Hot water-insoluble part.
[c] Determined by ^1H NMR (400MHz) in DMSO-d_6 at 170°C.

Table II. Effect of Solvent in the Polymerization of NIPAM in the Presence of $Y(OTf)_3$[a]

Run	Solvent	Yield[b] %	Tacticity[c] m / r
1	Methanol[d]	82	45 / 55
2	Methanol	94	80 / 20
3	Ethanol	90	73 / 27
4	Isopropanol	92	64 / 36
5	H_2O	94	57 / 43
6	THF	66	67 / 33
7	$CHCl_3$	75	62 / 38
8	1,4-Dioxane	96	53 / 47
9	Toluene	88	52 / 48
10	DMSO	96	47 / 53
11	DMF	>99	60 / 40

[a] $[NIPAM]_0 = 2.40$ mol/l. $[Y(OTf)_3]_0 = 0.20$ mol/l. Initiator: AIBN (runs 1-4, 6-11), $Na_2SO_3 + K_2S_2O_8$ (run 5) ([initiator]$_0 = 0.02$ mol/l). Temp.=60°C. Time = 3hr.
[b] Hot water-insoluble part.
[c] Determined by ^1H NMR (400MHz) in DMSO-d_6 at 170°C.
[d] Without $Y(OTf)_3$.

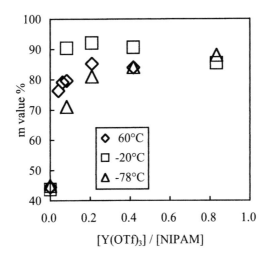

Figure 1. *Relationship between the [Y(OTf)₃] / [NIPAM] ratio and m value of the obtained poly(NIPAM). Conditions: [NIPAM]ₒ = 2.40 mol/l. Initiator: AIBN (0.02 mol/l) (60°C), AIBN with UV irradiation (-20°C), (n-Bu)₃B (0.10 mol/l) with air (-40°C). Time = 3hr (60°C), 24hr (-78°C, -20°C).*

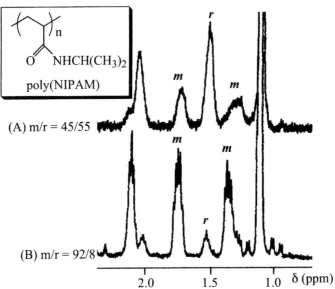

Figure 2. *¹H NMR spectra of poly(NIPAM)s prepared (A) in the absence and (B) presence of Y(OTf)₃ (0.50 mol/l) in methanol at –20°C [400MHz, DMSO-d₆, 170°C].*

Acrylamide (AM)

Poly(AM) and its copolymer are widely used for various applications including the clarification of waste water, oil recovery, and paper manufacture (26). **Table III** shows the results of the polymerization of AM under various conditions. Although atactic polymers of similar tacticities were obtained in the absence of Lewis acids regardless of temperature, in the presence of rare earth metal triflates, the isotactic polymers with different regularities were obtained depending on the polymerization conditions. The isotacticity increased in the order of $Sc(OTf)_3$ < $Y(OTf)_3$ < $Yb(OTf)_3$. The highest isotacticity (mm = 71) was observed for the polymerization with $Yb(OTf)_3$ at –40°C. However, the effect of the Lewis acids became negligible at –78°C. The tacticities of the poly(AM)s were determined by the ^{13}C NMR spectra in D_2O at 80°C on the basis of the peaks of the α-carbon according to the references (46,47). **Figure 3** shows the ^{13}C NMR spectra in the main chain region of the atactic (A) and isotactic poly(AM)s (B).

N,N-Dimethylacrylamide (DMAM)

The effects of Lewis acids during the radical polymerization of DMAM in methanol are shown in **Table IV**. The stereocontrol of this monomer has been examined by anionic (43,44) and radical processes (48). The m content of the polymers obtained without Lewis acids slightly increased when the temperature

Table III. Effect of Lewis Acids in the Polymerization of AM in Methanol[a]

Run	Lewis acid	[LA]$_o$ mol/l	Temp °C	Yield[b] %	Tacticity[c] mm/mr/rr
1	None	-	60	60	24 / 52 / 25
2	$Yb(OTf)_3$	0.10	60	84	47 / 40 / 13
3	None	-	0	60	22 / 49 / 29
4	$Sc(OTf)_3$	0.10	0	70	40 / 43 / 17
5	$Yb(OTf)_3$	0.10	0	50	65 / 29 / 6
6	$Y(OTf)_3$	0.10	0	90	58 / 33 / 9
7	None	-	-40	81	25 / 47 / 29
8	$Yb(OTf)_3$	0.20	-40	95	71 / 23 / 6
9	$Y(OTf)_3$	0.20	-40	93	61 / 30 / 8
10	$Yb(OTf)_3$	0.20	-78	>99	32 / 47 / 21

[a][AM]$_o$ = 1.00 mol/l. Initiator: AIBN (60°C), or AIBN with UV irradiation (0°C) ([AIBN]$_o$ = 0.02 mol/l), (n-Bu)$_3$B with air ([(n-Bu)$_3$B]$_o$=0.10 mol/l) (-78~-40°C). Time = 24hr.

[b]Polymers were purified by dialysis with a cellophane tube in water.

[c]Determined by ^{13}C NMR (125MHz) in D_2O at 80°C.

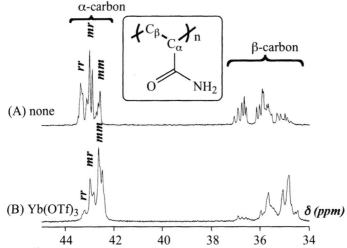

Figure 3. ^{13}C NMR spectra of poly(AM)s prepared (A) in the absence and (B) presence of Y(OTf)$_3$ (0.1mol/l) in methanol at 0°C. [125MHz, D$_2$O, 80°C]

changed from 60°C to –78°C. Similar to the polymerization of NIPAM, isotactic polymers were obtained during the polymerization in the presence of rare earth metal triflates, such as Yb(OTf)$_3$, Y(OTf)$_3$, and Lu(OTf)$_3$. Sc(OTf)$_3$ was less effective as well as for NIPAM, and the effect of MgBr$_2$ and ZnBr$_2$ was not obviously observed. During the polymerization in the presence of Yb(OTf)$_3$ at 0°C, the m value of the polymer reached 88%.

In **Table V**, the effects of the solvents are summarized for the polymerization in the presence of Yb(OTf)$_3$ at 60°C. The tacticity of the obtained polymer strongly depended on the solvents. Although the isotacticity of the polymers prepared in the absence of Lewis acids slightly increased in the order of methanol < ethanol < isopropanol < THF < toluene, the isotacticity-enhancing effect of Yb(OTf)$_3$ increased in the reverse order. This suggests that the polarity in an important factor for controlling the stereochemistry. For the Lewis acid to work catalytically, the exchange of the acid between polymer chain and monomer must be essential. In nonpolar solvents, this may be slowed down.

Methacrylamides

The radical polymerizations of methacrylamides also proceeded in an isotactic-specific manner by using a catalytic amount of Lewis acids (**Table VI**), although the conventional radical polymerization without Lewis acids produced syndiotactic-rich polymers. During the polymerizations of MAM with Yb(OTf)$_3$, the isotacticity and heterotacticity of the polymer increased. This

66

Table IV. Effect of Lewis Acids in the Polymerization of DMAM in Methanol[a]

Run	Lewis acid	Temp. °C	Yield[b] %	Tacticity[c] m / r
1	None	60	73	46 / 54
2	Sc(OTf)$_3$	60	76	78 / 22
3	Yb(OTf)$_3$	60	86	84 / 16
4	Y(OTf)$_3$	60	90	84 / 16
5	Lu(OTf)$_3$	60	85	85 / 15
6	MgBr$_2$	60	68	47 / 53
7	ZnBr$_2$	60	89	45 / 55
8	None	0	81	49 / 51
9	Yb(OTf)$_3$	0	76	88 / 12
10	None	-78	62	55 / 45
11	Yb(OTf)$_3$	-78	76	65 / 35

[a][DMAM]$_o$ = 1.00 mol/l. [LA]$_o$ = 0.10 mol/l. Initiator: AIBN ([AIBN]$_o$ = 0.02 mol/l) (temp. = 0~60°C), (n-Bu)$_3$B with air ([(n-Bu)$_3$B]$_o$ = 0.10 mol/l) (temp. = -78°C). Time = 24hr.
[b]Polymers were purified by dialysis with a cellophane tube in water.
[c]Determined by ^{1}H NMR (400MHz) in DMSO-d$_6$ at 170°C.

Table V. Effect of Solvent in the Polymerization of DMAM in the Presence of Yb(OTf)$_3$[a]

Run	Solvent	Lewis acid	[LA]$_o$ mol/l	Yield[b] %	Tacticity[c] m / r
1	Methanol	None	-	73	46 / 54
2	Methanol	Yb(OTf)$_3$	0.10	76	84 / 16
3	Toluene	None	-	96	53 / 47
4	Toluene	Yb(OTf)$_3$	0.10	84	55 / 45
5	THF	None	-	80	52 / 48
6	THF	Yb(OTf)$_3$	0.10	71	65 / 35
7	Ethanol	None	-	73	47 / 53
8	Ethanol	Yb(OTf)$_3$	0.10	60	80 / 20
9	Isopropanol	None	-	21	49 / 51
10	Isopropanol	Yb(OTf)$_3$	0.10	66	70 / 30

[a][DMAM]$_o$ = 1.00 mol/l. Initiator: AIBN ([AIBN]$_o$ = 0.02 mol/l). Temp. = 60°C. Time = 24hr.
[b]Polymers were purified by dialysis with a cellophane tube in water.
[c]Determined by ^{1}H NMR (400MHz) in DMSO-d$_6$ at 170°C.

polymer was insoluble in the common solvents and soluble only in strong acids such as concentrated sulfuric acid. The polymerization of (R)-MCPMMAM gave the polymers with a variety of tacticities. The polymerization of (R)-MCPMMAM without a Lewis acid at low temperature (-78°C) produced a syndiotactic polymer (rr = 85%), whereas the polymerization in the presence of Yb(OTf)$_3$ and MgBr$_2$ gave the isotactic (run 13, mm = 82%) and heterotactic polymers (run 14, mr = 63%), respectively. The former isotactic polymer was obtained in the presence of equimolar Yb(OTf)$_3$ to the monomer. The effects of the Lewis acids significantly depended on the monomers and solvents. For the polymerization of MMAM and IPMAM, methanol was a better solvent to attain a higher isotacticity than THF, while for PMAM and (R)-MCPMMAM, THF was a better solvent.

Table VI. Effect of Lewis Acids in the Polymerization of Various Methacrylamides[a]

Run	Monomer	Lewis acid	Solvent	Temp. °C	Yield[b] %	Tacticity[c] mm/mr/rr
1	MAM	None	Methanol	60	34	7 / 39 / 54
2	MAM	Yb(OTf)$_3$	Methanol	60	84	36 / 50 / 14
3	MMAM	None	Methanol	60	97	2 / 29 / 69
4	MMAM	Sc(OTf)$_3$	Methanol	60	92	28 / 55 / 17
5	MMAM	Y(OTf)$_3$	Methanol	60	91	46 / 40 / 14
6	MMAM	Yb(OTf)$_3$	Methanol	60	89	46 / 44 / 10
7	IPMAM	None	Methanol	20	16	~0 / 20 / 80
8	IPMAM	Yb(OTf)$_3$	Methanol	20	54	63 / 33 / 4
9	PMAM	None	THF	0	20	4 / 30 / 66
10	PMAM	Yb(OTf)$_3$	THF	0	55	55 / 38 / 7
11	(R)-MCPMMAM	None	THF	-20	33	~0 / 15 / 85
12	(R)-MCPMMAM	Yb(OTf)$_3$	THF	-20	59	68 / 22 / 10
13	(R)-MCPMMAM	Yb(OTf)$_3$	THF	-20	34	82 / 16 / 2
14	(R)-MCPMMAM	MgBr$_2$	CHCl$_3$	60	70	24 / 63 / 13

[a][Monomer]$_0$ = 2.4 mol/l (runs 1, 2), 2.0 mol/l (runs 3-6), 1.0 mol/l (runs 7-10), 1.1 mol/l (runs 11-13), 0.54 mol/l (run 14). [LA]$_0$ = 0.20 mol/l (runs 2, 6-8, 12), 0.10 mol/l (runs 8, 10, 14), 1.10 mol/l (run 13). Initiator: AIBN (0.02 mol/l) (runs 1-10, 14), (n-Bu)$_3$B (0.1 mol/l) (runs 11-13). Time = 24 hr (runs 1-10, 14), 48 hr (runs 11-13).

[b]THF-insoluble part (runs 1-6), hot water-insoluble part (runs 7, 8), water-methanol-insoluble part (runs 9, 10), methanol insoluble part (runs 11-14).

[c]Determined by ^{13}C NMR in D$_2$SO$_4$ at 60°C (runs 1, 2), ^{1}H NMR in DMSO-d$_6$ at 170°C (runs 3-6), ^{13}C NMR in DMSO-d$_6$ at 80°C (runs 7-14).

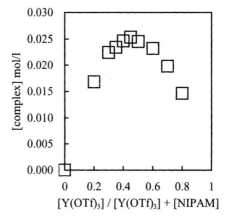

Figure 4. Job's plot based on the chemical shifts of NCH proton of NIPAM in methanol-d₄, where [complex] was calculated with the following equation: [complex] = [NIPAM] x Δδ / Δδ_f [Δδ_f = δ(NIPAM saturated by Y(OTf)₃) − δ(free NIPAM)]. [NIPAM] + [Y(OTf)₃] = 0.1 mol/l.

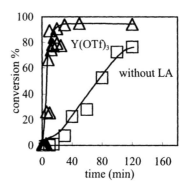

Figure 5. Time-conversion plots of the polymerization of NIPAM in the absence or presence of Y(OTf)₃ in methanol at 60°C. Conditions: [NIPAM]ₒ = 0.5mol/l. [AIBN]ₒ = 0.02mol/l. [Y(OTf)₃]ₒ = 0.20mol/l.

Mechanistic Study of Catalytic Function of Lewis Acid

As mentioned above, a small amount of Lewis acids resulted in a sufficient effect in increasing the isotactic selectivity during the polymerization of the (meth)acrylamides. In the ¹H NMR measurement of the NIPAM solution in methanol-d₄, the signal of the NCH proton shifted downfield with the addition of Y(OTf)₃. In order to determine the stoichiometry of the interaction between the Lewis acids and the monomers by Job's method (49), the ¹H NMR titration experiment was carried out for the NIPAM-Y(OTf)₃ system (**Figure 4**). These results suggest that a nearly 1:1 complex is formed between NIPAM and Y(OTf)₃.

Figure 5 shows the time-conversion plots for the polymerization of NIPAM in the absence and presence of Y(OTf)₃ in methanol at 60°C. The acceleration of the polymerization by Y(OTf)₃ was clearly observed. The accelerating effects of other Lewis acids during the radical polymerization have already been reported (37). This monomer activation effect of the Lewis acid appears to play a very important role in the catalytic action of the Lewis acids. **Figure 6** illustrates a plausible mechanism for the stereocontrol during the polymerization by Lewis acids. A monomer must be activated by the coordination of a Lewis acid (A). The activated monomer is preferentially polymerized and therefore, the Lewis acid is incorporated into the propagating chain end. The Lewis acid controls the conformation of the chain end for the polymerization to proceed in an isotactic-specific manner (B). The Lewis acids inside the polymer chain must weakly interact with the chain to readily be transferred to the monomer and activate it again (C).

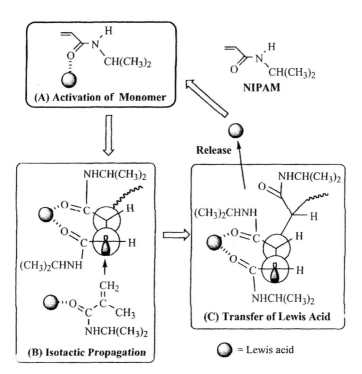

Figure 6. *Plausible mechanism of the stereocontrol in the polymerization of acrylamides by a catalytic amount of Lewis acids.*

70

Acknowledgements

This work was supported in part by the New Energy and Industrial Technology Development Organization (NEDO) under the Ministry of Economy, Trade and Industry, Japan, through the grant for the "Nanostructure Polymer Project" in the "Nanotechnology Materials Program" (2001-).

Literature Cited

1. Nakano, T.; Okamoto, Y. In *ACS Symp. Ser.* 685; Matyjazsewski, K., Ed.; ACS: Washington D.C., 1998, pp 451-462.
2. Pino, P.; Suter, U. W. *Polymer* **1976**, *17*, 977.
3. Hatada, K.; Kitayama, T.; Ute, K. *Prog. Polym. Sci.* **1988**, *13*, 189.
4. Yuki, H.; Hatada, K. *Adv. Polym. Sci.* **1979**, *31*, 1.
5. Matsumoto, A. In *Handbook of Radical Polymerization;* Matyjazsewski, K.; Davis, T. P., Eds; Wiley: New York, 2002, pp 691-773.
6. Nakano, T.; Okamoto, Y. *Macromol. Rapid Commun.* **2000**, *21*, 603.
7. Habaue, S.; Okamoto, Y. *Chem. Rec.* **2001**, *1*, 46.
8. Nagara, Y.; Nakano, T.; Okamoto, Y.; Gotoh, Y.; Nagura, M. *Polymer* **2001**, *42*, 9679.
9. Hawker, C.; Bosman, A. W.; Harth, E. *Chem. Rev.* **2001**, *101*, 3661.
10. Matyjaszewski, K.; Xai, J. *Chem. Rev.* **2001**, *101*, 2921.
11. Kamigaito, M.; Ando, T.; Sawamoto, M. *Chem. Rev.* **2001**, *101*, 3989.
12. Rizzardo, R.; Chiefari, J.; Mayadunne, R. T. T.; Moad, G.; Thang, S. H. In *ACS Symp. Ser.* 768; Matyjazsewski, K., Ed.; ACS: Washington D.C., 2000, pp 278-296.
13. Miyata, M. In *Inclusion Polymerization, Polymeric Materials Encyclopedia: Synthesis, Properties and Applications, Vol. 5;* Salamone, J. C., Ed.; CRC Press: New York, 1996, pp 3226-3233.
14. Kamide, K.; Yamazaki, H.; Okajima, K. *Polym. J.* **1985**, *17*, 1291.
15. Matsumono, A.; Matsumura, T.; Aoki, S. *Macromolecules* **1996**, *29*, 423.
16. Nakano, T.; Mori, M.; Okamoto, Y. *Macromolecules* **1993**, *26*, 867.
17. Nakano, T.; Matsuda, A.; Okamoto, Y. *Polym. J.* **1996**, *28*, 556.
18. Porter, N.A.; Allen, T. R.; Breyer, R. A. *J. Am. Chem. Soc.* **1992**, *114*, 7676.
19. Wu, W. X.; McPhail, A. T.; Porter, N. A. *J. Org. Chem.* **1994**, *59*, 1302.
20. Yamada, K.; Nakano, T.; Okamoto, Y. *Macromolecules* **1998**, *31*, 7598.
21. Yamada, K.; Nakano, T.; Okamoto, Y. *Proc. Jpn. Acad. B* **1998**, *74*, 46.
22. Isobe, Y.; Yamada, K.; Nakano, T.; Okamoto, Y. *Macromolecules* **1999**, *32*, 5979.

23. Isobe, Y.; Yamada, K.; Nakano, T.; Okamoto, Y. *J. Polym. Sci.: Part A: Polym. Chem.* **2000**, *38*, 4693.
24. Schild, H. G. *Prog. Polym. Sci.* **1992**, *17*, 163.
25. Pelton, R. *Adv. Colloid Interface Sci.* **2000**, *85*, 1.
26. Caulfield, M. J.; Qiao, G. G.; Solomon, D. H. *Chem. Rev.* **2002**, *102*, 3067.
27. Isobe. Y.; Fujioka, D.; Habaue, S.; Okamoto, Y. *J. Am. Chem. Soc.* **2001**, *123*, 7180.
28. Okamoto, Y.; Habaue, S.; Isobe, Y. *Polym. Prepr.* **2002**, *43(2)*, 134.
29. Okamoto, Y.; Habaue, S.; Isobe, Y.; Nakano, T. *Macromol. Symp.* **2002**, *183*, 83.
30. Suito, Y.; Isobe, Y.; Habaue, S.; Okamoto, Y. *J. Polym. Sci.: Part A: Polym. Chem.* **2002**, *40*, 2496.
31. Isobe, Y.; Suito, Y.; Habaue, S.; Okamoto, Y. *Polym. Prepr.* **2002**, *43(2)*, 148.
32. Hirooka, M.; Yabuuchi, H.; Morita, S.; Kawasumi, S.; Nakaguchi, K. *J. Polym. Sci. Polym. Lett.* **1967**, *5*, 47.
33. Patnaik, B. K.; Gaylord, N. G. *J. Polym. Sci. Polym. Chem. Ed.* **1971**, *9*, 347.
34. Wu, G. Y.; Qi, Y. C.; Lu, G. J.; Wei, Y. K. *Polym. Bull.* **1989**, *22*, 393.
35. Gotoh, Y.; Yamashita, M.; Nakamura, M.; Toshima, N.; Hirai, H. *Chem. Lett.* **1991**, 53.
36. Gotoh, Y.; Iihara, T.; Nakai, N.; Toshima, N.; Hirai, H. *Chem. Lett.* **1990**, 2157.
37. Seno, M.; Matsumura, N.; Nakamura, H.; Sato, T. *J. Appl. Polym. Sci.* **1997**, *73*, 1361.
38. Matsumoto, A.; Nakamura, S. *J. Appl. Polym. Sci.* **1999**, *74*, 290.
39. Isobe, Y.; Nakano, T.; Okamoto, Y. *J. Polym. Sci.: Part A: Polym. Chem.* **2001**, *39*, 1463.
40. Habaue, S.; Baraki, H.; Okamoto, Y. *Polym. J.* **2000**, *32*, 1022.
41. Baraki, H.; Habaue, S.; Okamoto, Y. *Macromolecules* **2001**, *34*, 4724.
42. Mero, C. L.; Porter, N. A. *J. Org. Chem.* **2000**, *65*, 775.
43. Kobayashi, M.; Okuyama, S.; Ishizone, T.; Nakahama, S. *Macromolecules* **1999**, *32*, 6466.
44. Kobayashi, M.; Ishizone, T.; Nakahama, S. *Macromolecules* **2000**, *33*, 4411.
45. Kitayama, T.; Shibuya, W.; Katsukawa, K. *Polym. J.* **2002**, *34*, 405.
46. Lancaster, J. E.; O'Connor, M. N. *J. Polym. Sci. Polym. Lett. Ed.* **1982**, *20*, 547.
47. Hikichi, K.; Ikura, M.; Yasuda, M. *Polym. J.* **1988**, *10*, 851.
48. Liu, W.; Nakano, T.; Okamoto, Y. *Polym. J.* **2000**, *32*, 771.
49. Cannors, K. A. In *Binding Constants: The Measurement of Molecular Complex Stability;* Wiley: New York, 1987, pp 24-28.

Chapter 6

Propagation and Termination in Free Radical Polymerization of Styrene to High Conversion Investigated by Electron Spin Resonance Spectroscopy

Per B. Zetterlund*, Hirotomo Yamazoe, Satoru Yamauchi, and Bunichiro Yamada*

Department of Applied and Bioapplied Chemistry, Graduate School of Engineering, Osaka City University, 3-3-138 Sugimoto, Sumiyoshi-ku, Osaka 558-8585, Japan

Abstract: The propagation and termination processes in the bulk free radical polymerization of styrene to high conversion have been investigated by electron spin resonance (ESR) spectroscopy. The focus has been the very highest conversion range where the propagation step is diffusion-controlled; at 70 °C, both the values of the propagation rate coefficient (k_p) and the termination rate coefficient ($<k_t>$) increased with increasing initiator concentration. At 120 °C, the same trend with initiator concentration was observed for k_p, although $<k_t>$ remained unchanged. Both k_p and f fall dramatically at approximately 80% conversion at 120 °C even though the temperature is above the glass transition temperature. The addition of a small amount of divinylbenzene has no major effect on k_p, which still remains under chemical control up to approximately 80% conversion at 70 °C. The effect on $<k_t>$ is more dramatic, especially at intermediate conversion.

Introduction

Major efforts have over the years been directed towards elucidating and quantifying the details of free radical bulk polymerization. Although significant progress has been made in recent years with improvements in the electron spin resonance (ESR) method *(1)* and the development of pulsed laser-based techniques (PLP) *(2,3)*, several aspects of the process are not fully understood, especially at high conversion where PLP methods are not suitable. The ESR technique is extremely valuable because it is in principle possible to estimate both the propagation rate coefficient (k_p) and the chain-length averaged termination rate coefficient ($<k_t>$) even as the conversion approaches its limiting value by directly measuring the free radical concentrations as functions of time.

PLP methods allow the determination of k_p under conditions of low conversion levels of a few percent. The value of $<k_t>$ can be estimated as a function of conversion up to high conversion levels, although current PLP methodology does not allow access to $<k_t>$ in the region where k_p falls due to diffusion control of the propagation step setting in. The current ESR study deals with a very wide conversion range, with specific attention being paid to the very highest conversion regime, which is currently the least understood. Determination of k_p by ESR is in principle straightforward, although caution is warranted as there exist several pitfalls *(4)*. Determination of $<k_t>$ is more complicated, and involves either steady-state or non-steady state measurements. In the current study we have taken the former approach, which allows direct access to the ratio $<k_t>/f$, where f denotes the initiator efficiency. It follows that f needs to be estimated independently to enable estimation of $<k_t>$. We have investigated the propagation and termination kinetics in the bulk free radical polymerization of styrene at 70 and 120 °C, the significance of these two temperatures being that at the latter temperature, vitrification will not occur even at the very highest conversion level. Furthermore, the effects of adding a small amount of the crosslinker divinylbenzene (DVB) have been investigated, considering the introduction of crosslinks on monomer and macroradical mobility.

Experimental

Materials

Commercially available St and DVB were purified by distillation under reduced pressure before use. DVB consisted of the *p*- and *m*-isomers of DVB (50.8 %) and ethylstyrene (49.2 %) according to ^1H NMR analysis. Dimethyl

2,2'-azobisisobutyrate (MAIB) was recrystallized from hexane, and *t*-butyl peroxide (TBP) was used as received (Wako Pure Chemicals).

Polymerizations

Bulk polymerizations of St were initiated by 0.05, 0.10 and 0.20 M MAIB at 70 °C, and by 0.05, 0.10 and 0.15 M TBP at 120 °C. Bulk polymerizations of St/DVB (0.05, 0.10 and 0.20 M DVB) at 70 °C were initiated by 0.10 M MAIB. FT-NIR measurements were carried out in a 5 mm o.d. Pyrex tube in an aluminum furnace at 70 and 120 °C. The consumption of St was monitored by the absorbance at 6150 cm^{-1} (assigned to the overtone absorption of $\nu_{C=C-H}$), employing a Jasco INT-400 Spectrometer with a MCT detector. Polymerizations for molecular weight measurement were conducted in glass ampules sealed under vacuum. The polymer was precipitated in methanol, and the conversion was determined by gravimetry. Molecular weights were measured with a Tosoh-800 series HPLC with GPC columns calibrated with polystyrene standards. The conversion range within which gelation occurred for St/DVB was determined by observing formation of insoluble polymer.

ESR

ESR spectra were recorded on a Bruker ESP300 during polymerization in the cavity in a 5 mm o.d. quartz tube sealed under vacuum. The line width and apparent splitting pattern change with conversion as a result of increasing viscosity (Figure 1). The values of the hyperfine coupling constants for the poly(St) radical, recorded under conditions of high resolution, have been reported previously *(1)*. 2,2,6,6-Tetramethylpiperidinyl-1-oxyl (TEMPO) in benzene was used for calibration of the relationship between ESR spectral signal intensity and radical concentration. The sensitivity of the ESR instrument is a function of conversion in the case of polar monomers such as methyl methacrylate (MMA) in bulk *(5)*. The sensitivity remains constant over the entire conversion range for the non-polar St *(1,5)*.

Methods

Propagation Rate Coefficient

The value of k_p as a function of conversion was determined from the FT-NIR and ESR data by use of eq 1:

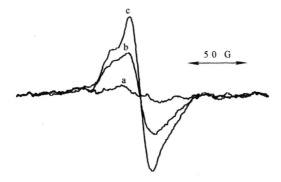

Figure 1. ESR spectra at different conversions during bulk free radical polymerization of St initiated with 0.20 M MAIB at 70 °C. (a) 24 % conversion (60 min), (b) 91 % (250 min), (c) 95 % (410 min).
(Reproduced from reference 8. Copyright 2001 American Chemical Society.)

$$\frac{d[St]}{dt} = -k_p[St][St\bullet] \qquad (1)$$

where [St•] is the propagating radical concentration.

Degree of Polymerization

The cumulative number average degrees of polymerization from GPC ($\overline{\overline{P_n}}$) were converted into the instantaneous values ($\overline{P_n}$) by use of eq 2 *(6,7)*:

$$\overline{P_n} = \overline{\overline{P_n}}\left(1 - \frac{x}{\overline{\overline{P_n}}}\frac{d\overline{\overline{P_n}}}{dx}\right)^{-1} \qquad (2)$$

where x is conversion. Semi-instantaneous values of the number average degrees of polymerization were also obtained by GPC curve subtraction *(5,8)*.

Initiator Efficiency

The value of f, and its conversion-dependence, was estimated from the relationship between the rate of polymerization (R_p), the rate of initiator decomposition *(9)* and $\overline{P_n}$ according to eq 3:

$$f = \frac{R_p}{\overline{P_n} k_d [I]} \tag{3}$$

where k_d is the rate constant for thermal decomposition of the initiator and [I] is the initiator concentration (8). Eq 3 refers to the case where termination occurs by combination, which is the dominant mode of termination for St (10). Furthermore, it rests on the assumption that there is no chain transfer in the system. It cannot be excluded that chain transfer to monomer may result in a slight overestimation of f (and thus also $<k_t>$), especially at high conversion where f falls dramatically. Chain transfer to polymer, which also may become significant at high levels of conversion, was not considered. The rate of thermal initiation of St at 120 °C was accounted for when applying eq 2 (11,12). The method employed to access $\overline{P_n}$ is error prone over narrow conversion ranges, limiting the accuracy of f thus obtained.

Termination Rate Coefficient

The value of $<k_t>$ was estimated over a very wide conversion range by first computing the ratio $<k_t>/f$ from eq 4:

$$\frac{<k_t>}{f} = \frac{k_d [I]_0 \exp(-k_d t)}{[St\bullet]^2} \tag{4}$$

Eq 4 is valid since $2 f k_d [I]_0 \exp(-k_d t) >> d[St\bullet]/dt$ under the current experimental conditions for the systems investigated. Subsequently, the value of f as a function of conversion was used to obtain $<k_t>$.

Results and Discussion

Conversion and Radical Concentrations

The conversions and radical concentrations for the bulk polymerization of St at 70 °C initiated by MAIB are displayed in Figure 2. The most interesting and hitherto not experimentally observed feature is the pronounced maxima in the radical concentrations that coincide with the maximum R_p's. Similar maxima have been detected in the bulk polymerization of MMA and for crosslinking systems such as MMA/ethylene glycol dimethacrylate (13,14). The free radical profiles at 120 °C also exhibited maxima, but the peaks were sharper and had shifted to considerably higher conversion levels (close to 90% conversion).

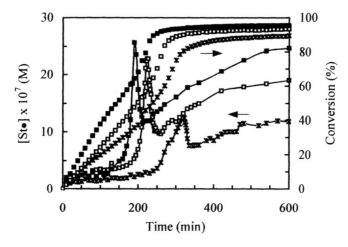

Figure 2. Conversion and free radical concentrations for bulk polymerization of St at 70 °C. [MAIB] = 0.20 (■), 0.10 (□), 0.05 M (∗).
(Reproduced with permission from reference 20. Copyright 2001 Wiley-VCH.)

The maxima were also detected for the St/DVB system at 70 °C *(15)*. The free radical concentration profiles are governed by the conversion dependences of $<k_t>$ and f according to eq 4. The increase in free radical concentration up to the maximum is due to the gel-effect, whereas the decrease is thought to be related to a dramatic fall in f *(8,13)*.

Initiator Efficiency

The instantaneous degree of polymerization obtained from eq 2 and the GPC subtraction technique as a function of conversion for the bulk polymerization of St initiated by 0.10 M MAIB at 70 °C are depicted in Figure 3. The agreement between the two methods is surprisingly good. The corresponding data at 120 °C (not shown) are qualitatively similar, although the increase in $\overline{P_n}$ with conversion is much less dramatic *(11)*. This is a result of the strong effect of temperature on the conversion dependence of $<k_t>$, in particular in the gel-effect region. The f values over the a wide conversion ranges for polymerization at 70 °C initiated by MAIB, and 120 °C initiated by TBP, are displayed in Figure 4. The values of f remain close to constant until approximately 80% conversion, where a dramatic fall is observed *(6,8,14,16)*. This drastic drop occurs at a somewhat higher conversion level at 120 °C. These drastic drops in f and k_p occur at very similar conversion levels. This is reasonable considering that the diffusion control of both processes depend on the mobility of species of similar sizes. It is noteworthy that at 120 °C, f falls dramatically even though the reaction temperature is above the glass transition temperature. It has however been pointed out that the drop in k_p at high

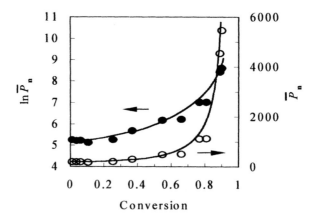

Figure 3. Instantaneous number average degree of polymerization for bulk polymerization of St at 70 ℃ initiated by 0.10 M MAIB. Points from GPC using "subtraction technique", lines are from eq 2.
(Reproduced from reference 8. Copyright 2001 American Chemical Society.)

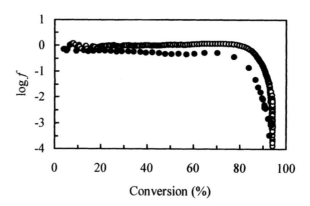

Figure 4. f vs. conversion in bulk polymerization of St for 0.10 M MAIB at 70 ℃ (full circles) and 0.10 M TBP at 120 ℃ (open circles). Data from refs. 8 and 11.

conversion in bulk polymerization does not coincide exactly with the conversion level at which the systems turns into a glass *(17)*. Extrapolation to zero conversion yields an initial value of 0.65 for MAIB, and close to 1 for TBP, in agreement with previous work and expectation since TBP undergoes no cage reaction, unlike MAIB *(10)*.

Propagation Rate Coefficient

Figure 5 shows the conversion dependence of k_p at 70 and 120 °C for bulk polymerization of St, and for St/DVB at 70 °C. The values of k_p at low conversion for St polymerization are somewhat lower than those given by the Arrhenius parameters reported by IUPAC *(18)* for unknown reasons. The value of k_p remains relatively constant up to approximately 80% conversion, where it falls by orders of magnitude over a narrow conversion range. The value obtained for k_p at 120 °C increases slightly at high conversion – this may be related to a temperature increase due to the reaction exotherm. It is interesting to note that k_p drops dramatically despite the temperature being above the glass transition temperature. Furthermore, even in the presence of the crosslinker DVB, albeit in rather low concentration but still high enough for gelation to occur at

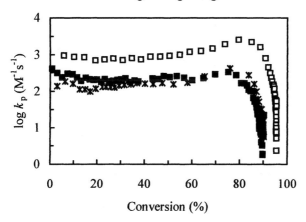

Figure 5. k_p vs. conversion for bulk polymerization of St and St/DVB. [MAIB] = 0.05 M at 70 °C (■); [TBP] = 0.15 M at 120 °C (□); [MAIB] = 0.10 M, [DVB] = 0.20 M, 70 °C (✶). Data from refs. 8, 11 and 15.

approximately 8-15% conversion, the propagation step remains chemically controlled up until roughly the same conversion level as for the homopolymerization of St. Thus, the dramatic fall in k_p at high conversion appears to occur at similar conversion levels in spite of significant changes of the

80

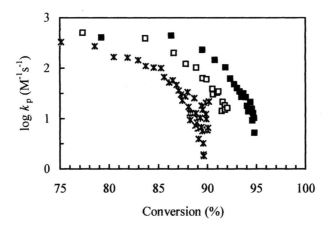

Figure 6. k_p vs. conversion at high conversion for bulk polymerization of St at 70 °C. [MAIB] = 0.20 (■), 0.10 (□), 0.05 M (✳).
(Reproduced with permission from reference 20. Copyright 2001 Wiley-VCH.)

polymer matrix as induced by temperature and crosslinker content. At conversion levels beyond approximately 80%, where propagation is diffusion controlled, there is a significant trend with initiator concentration at 70 °C; higher initiator concentration leads to a higher k_p at a given conversion (Figure 6). This is thought to be related to changes in the nature of the polymer matrix of dead chains caused by altering the initiator concentration. A higher initiator concentration leads to shorter chains being formed, and this may in turn result in lower monomer diffusion resistance. The diffusion coefficients are expected to be independent of matrix polymer molecular weights over approximately 10^4 g mol^{-1} *(19)*. The cumulative molecular weights of the sample at 70 °C with [MAIB] = 0.10 M are of the order 10^4 g mol^{-1}, and thus such a dependence may seem plausible. The same qualitative trend in k_p with initiator concentration was detected at 120 °C. However, the effect was weaker than at 70 °C, possibly related to the fact that the molecular weights were higher (of order 10^5 g mol^{-1} for [TBP] = 0.10 M) *(11,20)*.

Termination Rate Coefficient

The assumption was made that f and its conversion dependence is independent of the initiator concentrations for the St homopolymerizations at 70 and 120 °C *(6,16)*. If f were to increase with increasing initiator concentration in a manner similar to k_p, this would not alter the resulting qualitative trend in $<k_t>$ *(8)*. The values of $<k_t>$ over a wide conversion range are displayed for three

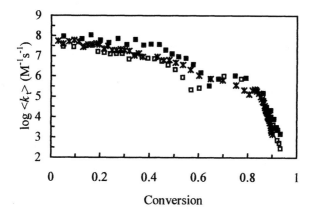

Figure 7. $<k_t>$ vs. conversion at 70 °C for bulk polymerization of St; [MAIB] = 0.20 (■), 0.10 (□), 0.05 M (✳).
(Reproduced from reference 8. Copyright 2001 American Chemical Society.)

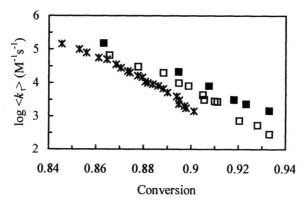

Figure 8. $<k_t>$ vs. conversion at high conversion for bulk polymerization of St at 70 °C; [MAIB] = 0.20 (■), 0.10 (□), 0.05 M (✳).
(Reproduced from reference 8. Copyright 2001 American Chemical Society.)

initiator concentrations for St at 70 °C (Figure 7). The initial plateau region has been assigned to the termination rate being controlled by segmental diffusion *(3,21,22)*. As the viscosity increases, center of mass diffusion becomes the rate determining step at approximately 20% conversion, where a significant decrease in $<k_t>$ is observed until another plateau region appears to be reached at approximately 60% conversion (this plateau may not be real, considering the scatter in the data). From this point onwards, the predominant termination process may be reaction diffusion *(3,21,23,24)*, with the dramatic drop in the very highest conversion regime being caused by the propagation step becoming diffusion controlled (Figure 6). The dips in the log k_t vs. conversion curves are believed to be artefacts caused by different heat transfer characteristics, resulting in slightly different conversions of the FT-NIR and ESR samples at the same reaction times *(8)*. The initiator concentration affects the chain length of both propagating radicals and dead polymer. Due to experimental scatter, it is difficult to conclude if chain-length dependent termination is manifested in the low conversion region (< 20%). However, in the region controlled by translational diffusion, the highest initiator concentration (corresponding to the shortest living and dead polymer chains) results in significantly higher values of $<k_t>$, probably as a result of chain-length dependent termination *(22,25,26)*. Similar effects have been observed in recent PLP studies *(22)*. In the conversion range where propagation is diffusion controlled (above approx. 80%), there is a marked dependence of $<k_t>$ on the initiator concentration; at a given conversion level, $<k_t>$ increases with the initiator concentration (Figure 8).

Figure 9 shows the values of $<k_t>$ for three different initiator concentrations for the bulk homopolymerization of St at 120 °C. The situation is entirely different from what was observed at 70 °C. The value of $<k_t>$ remains close to constant up to 80% conversion, where it falls many orders of magnitude over a narrow conversion range. It appears as if the conversion range where translational diffusion control limits $<k_t>$ is shifted to much higher conversion levels. No dependence on initiator concentration was observed at 120 °C. Both

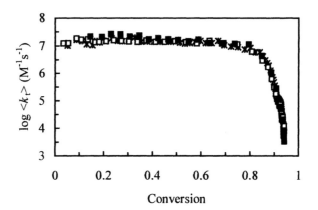

Figure 9. $<k_t>$ vs. conversion for bulk polymerization of St at 120 °C; [TBP] = *0.15 (■), 0.10 (□), 0.05 M (✳). Data from ref. 11.*

of these observations may be at least partly related to a relative increase in the rate of chain transfer to monomer with increasing temperature. This would increase the number of shorter chains, which terminated more rapidly due to chain-length dependent termination *(22,25,26)*. The drastic fall in k_t would then be controlled by reduced mobility of smaller species, and therefore occur at roughly the same conversion level as for k_p (governed by monomer mobility).

As the viscosity of the system gradually increases, the macroradicals eventually become essentially immobile, and the predominant termination mechanism becomes reaction diffusion *(3,8,21,23,24)*. The radical center moves as a result of propagation, and the value of $<k_t>$ can be related to k_p via the conversion and the reaction diffusion constant (C_{RD}):

$$k_{t,RD} = C_{RD} k_p (1 - x) \tag{5}$$

where $k_{t,RD}$ is the termination rate coefficient for termination by reaction diffusion and x is fractional monomer conversion. At 70 °C, the quantity $<k_t>/(k_p(1-x))$ remains close to constant at approximately 10^4 in the highest conversion range, but decreases somewhat above 80% conversion. The fact that both k_p and $<k_t>$ depend on the initiator concentration in a qualitatively similar fashion at 70 °C is what would be expected if reaction diffusion contributes significantly to the termination process, although it does not constitute proof that reaction diffusion is dominant. It is interesting to recall that at 120 °C, no initiator concentration dependence of $<k_t>$ could be detected in this conversion range (Figure 9). This may suggest that reaction diffusion is of less importance at this high temperature, as expected if termination is transfer-dominated as speculated above.

The effect of a small amount of the crosslinker DVB on the bulk polymerization of St at 70 °C was investigated. When decoupling $<k_t>$ from *f*, it was assumed that *f* remains unaffected by these small amounts of crosslinker. The value of $<k_t>$ normalized to the low conversion value ($<k_{t0}>$) for a series of DVB concentrations are shown in Figure 10. The low conversion plateau region decreases in length as the crosslinker content is increased, and the conversion levels where the initial drops occur approximately correlate with the gel-points. The greatest effect of the DVB content on $<k_t>$ can be seen in the conversion range 20-80%, where the formation of branched and eventually crosslinked structures imposes constraints on the diffusion processes. A plateau value in $<k_t>$ appears to be reached at around 60% conversion for the sample with the highest DVB content, which is followed by a sharp decrease of more than two orders of magnitude from 85% conversion up to the final conversion. This plateau may be related to reaction-diffusion, or it could be an artefact caused by the assumption that *f* is independent of DVB content (thus overestimating *f* for

84

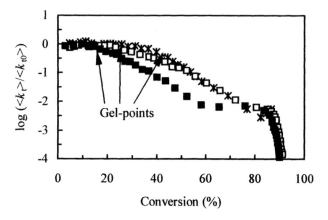

Figure 10. Relative values of <k_t> vs. conversion for bulk polymerization of St initiated by 0.10 M MAIB at 70 °C; [DVB] = 0.20 (■), 0.10 (□), 0.05 M (*). <k_{t0}> = <k_t> at low conversion. Data from ref. 15.

the samples with high DVB content). The absolute value of <k_t> at the plateau is approximately 8.7 x 10^4 M^{-1} s^{-1}.

Conclusions

The free radical bulk polymerization of St has been investigated to high conversion by ESR. Both k_p and <k_t> exhibited similar initiator concentration dependencies in the very highest conversion range for the polymerization of St at 70 °C. However, at 120 °C no effect of initiator concentration could be detected for <k_t>, although k_p did show a dependence similar to that at 70 °C. The values of k_p and f fall dramatically at approximately 80% conversion at 120 °C in absence of vitrification. Low levels of crosslinker have a very minor effect on k_p at 70 °C, which remained under chemical control up to conversion levels similar to during homopolymerization of St, despite gelation occurring before 20% conversion. The strongest effect of the crosslinker content on <k_t> was observed at intermediate conversion levels (20 – 80%). The end of the initial <k_t> plateau region correlated roughly with the gel-points.

References

1.   Yamada, B.; Westmoreland, D. G.; Kobatake, S.; Konosu, O. *Prog. Polym. Sci.* **1999**, *24*, 565.

2. Olaj, O. F.; Bitai, I.; Hinkelmann, F. *Makromol. Chem.* **1987**, *188*, 1689.
3. Beuermann, S.; Buback, M. *Prog. Polym. Sci.* **2002**, *27*, 191.
4. Tonge, M. P.; Kajiwara, A.; Kamachi, M.; Gilbert, R. G. *Polymer* **1998**, *39*, 2305.
5. Yamada, B.; Kageoka, M. *Polym. Bull.* **1992**, *28*, 75.
6. Sack, R.; Schulz, G. V.; Meyerhoff, G. *Macromolecules* **1988**, *21*, 3345.
7. Schulz, G. V.; Harborth, G. *Makromol. Chem.* **1947**, *1*, 106.
8. Zetterlund, P. B.; Yamazoe, H.; Yamada, B.; Hill, D. J. T.; Pomery, P. J. *Macromolecules* **2001**, *34*, 7686.
9. Dixon, K.W. *Decomposition Rates of Organic Free Radical Initiators*, In: *Polymer Handbook*, 4th ed, Brandrup, J.; Immergut, E.H.; Grulke, E.A., Eds., Wiley, New York 1999, p. II/1.
10. Moad, G.; Solomon, D. H. *The Chemistry of Free Radical Polymerization*; Pergamon: Oxford, 1995. Chapter 5, p. 229.
11. Zetterlund, P. B.; Yamauchi, S.; Yamada, B. *to be published*.
12. Hui, A. W.; Hamielec, A. E. *J. Appl. Polym. Sci.* **1972**, *16*, 749.
13. Zhu, S.; Tian, Y.; Hamielec, A. E.; Eaton, D. R. *Polymer* **1990**, *31*, 154.
14. Shen, J.; Tian, Y.; Wang, G.; Yang, M. *Makromol. Chem.* **1991**, *192*, 2669.
15. Zetterlund, P. B.; Yamazoe, H.; Yamada, B. *Polymer* **2002**, *43*, 7027.
16. Russell, G. T.; Napper, D. H.; Gilbert, R. G. *Macromolecules* **1988**, *21*, 2141.
17. Gilbert, R. G. *Emulsion Polymerization: A Mechanistic Approach*; Academic Press: London, 1995.
18. Buback, M.; Gilbert, R. G.; Hutchinson, R. A.; Klumperman, B.; Kuchta, F.-D.; Manders, B. G.; O'Driscoll, K. F.; Russell, G. T.; Schweer, J. *Macromol. Chem. Phys.* **1995**, *196*, 3267.
19. Faldi, A.; Tirrell, M.; Lodge, T. P.; von Meerwall, E. *Macromolecules* **1994**, *27*, 4184.
20. Yamazoe, H.; Yamada, B.; Zetterlund, P. B.; Hill, D. J. T.; Pomery, P. J. *Macromol. Chem. Phys.* **2001**, *202*, 824.
21. Buback, M. *Macromol. Chem. Phys.* **1990**, *191*, 1575.
22. Buback, M.; Kuchta, F.-D. *Macromol. Chem. Phys.* **1997**, *198*, 1455.
23. Schulz, G. V. *Z. Phys. Chem.* **1956**, *8*, 290.
24. Buback, M.; Huckestein, B.; Russell, G. T. *Macromol. Chem. Phys.* **1994**, *195*, 539.
25. Scheren, P. A. G. M.; Russell, G. T.; Sangster, D. F.; Gilbert, R. G.; German, A. L. *Macromolecules* **1995**, *28*, 3637.
26. Olaj, O. F.; Zoder, M.; Vana, P. *Macromolecules* **2001**, *34*, 441.

Chapter 7

Electron Spin Resonance Study of Conventional Radical Polymerization of *tert*-Butyl Methacrylates Using Radical Precursors Prepared by Atom Transfer Radical Polymerization

Atsushi Kajiwara[1] and Mikiharu Kamachi[2]

[1]Department of Materials Science, Nara University of Education, Takabatake-cho, Nara 630–8528, Japan
[2]Department of Applied Physics and Chemistry, Fukui University of Technology, Fukui 910–8505, Japan

Abstract: Radical precursors of poly(meth)acrylates with given chain lengths were prepared by atom transfer radical polymerization (ATRP) technique. Model radicals with given chain lengths were generated by reaction of the precursors with organotin compound. The radicals were observed by electron spin resonance (ESR) spectroscopy. ESR spectra of methacrylate model radicals with their degrees of polymerizations (P_n) of 30, 50, and 100 were measured. In comparison with these spectra, clear chain length dependence was observed and a detection of polymeric propagating radicals was clearly shown in the conventional radical polymerization. ESR spectra of model radicals of acrylates did not display chain-length dependence. The spectra suggested a formation of significant amount of mid-chain radicals from acrylate propagating radicals via 1,5-hydrogen shift. ESR spectra of dimeric model radical of acrylate clearly provided experimental evidence of the 1,5-hydrogen shift.

Introduction

Electron spin resonance (ESR) spectroscopy can theoretically provide direct information on the structures, properties, and concentrations of propagating radicals (1,2). Accordingly, ESR spectroscopy is a promising technique for obtaining information on the propagating radicals in the radical polymerization, if well-resolved ESR spectra of propagating radicals are observed (2-4).

We have detected well-resolved ESR spectra in the radical polymerizations of styrene and its derivatives, diene compounds, methacrylates, and vinyl esters in benzene or toluene solution (3-10). All these spectra are reasonably interpreted as propagating radicals.

As an extension of the previous works, in the present work, conventional radical polymerizations of (meth)acrylates were investigated with the aid of controlled radical polymerization technique. ESR spectra of model radicals with given chain lengths generated from radical precursors prepared by atom transfer radical polymerization (ATRP) were measured, and chain length dependent phenomena could be examined in the experiments.

Information on the chain lengths of propagating radicals is very important for kinetic study of radical polymerizations. We have estimated propagation rate constants (k_p) of radical polymerizations of several kinds of monomers by an ESR method (4-10). As part of the method, ESR spectra of propagating radicals have been detected and used for both of investigation of structures and estimation of steady-state concentrations of propagating radicals. The spectra have been considered to be due to propagating radicals whose chain lengths are long enough for estimation of k_p. However, there has been no clear experimental information on the actual chain length of the propagating radicals observed by ESR. The chain length was just speculated to be "long enough". In general, higher initiator concentrations than in usual radical polymerizations were employed in the ESR experiments for detection of well-resolved spectra. Under such conditions, the chain length of the propagating radicals will be shorter than that under the usual conditions. The chain length of the resulting polymers in the ESR samples can be estimated by measuring size exclusion chromatography (SEC) after ESR experiments. However, the resulting molecular weights, even if high enough, do not guarantee ESR detection of radicals with the high molecular weights, because detection of shorter radicals is much easier than that of longer radicals. Therefore, the chain length of the propagating radicals in the polymerizations should be examined carefully. In this work, ESR experimental evidence for the detection of polymeric propagating radicals in methacrylate polymerizations will be shown.

Historically, the interpretations of ESR spectra of radial polymerizations of acrylates have been much more complicated than for methacrylates. There have been some unsolved problems with the ESR spectra of propagating radicals of acrylates. Several kinds of ESR spectra under various conditions were reported as spectra of propagating radicals (11-17). Presence of mid-chain radicals in the course of radical polymerizations of acrylates has been discussed (14-18). In the present work, we investigated ESR spectra of model radicals of propagating radicals of acrylates with given chain lengths at several temperatures. As a

88

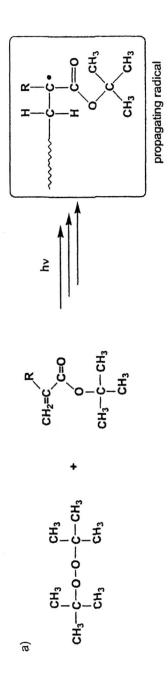

a)

propagating radical

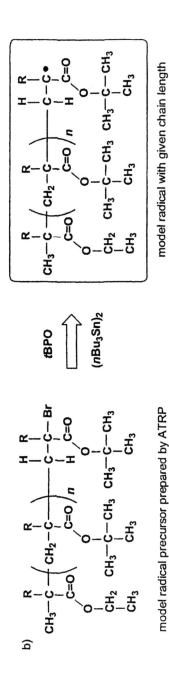

model radical precursor prepared by ATRP

model radical with given chain length

R = CH₃; methacrylate
R = H; acrylate

Scheme 1 Generation of propagating radicals (a) and model radicals (b). ESR spectra of methacrylate showed chain length dependence. ESR spectra of acrylate clearly exhibited occurrence of 1,5-hydrogen shift.

result, ESR spectra of propagating and mid-chain radicals were observed separately and were clearly assigned.

Results and Discussion

tert-Butyl Methacrylates

When a mixture of monomer and a radical initiator was heated or photo-irradiated in an ESR sample cell, propagating radicals were formed and polymerization proceeded (scheme 1a, R = CH₃). Well-resolved spectra of propagating radicals of *tert*-butyl methacrylate (*t*BMA) have been clearly detected at various temperatures as shown in Figure 1. The 16-line spectra were different from those of methacrylates (13-line or 9-line spectra), but is reasonably assigned to propagating radical as follows. Since the long-ranged hyperfine structures due to the methyl protons of the *tert*-butyl groups in the ester side chain did not appear, splitting lines due to three α-methyl protons and two β-methylene protons were clearly observed. Thus the values of their hyperfine coupling constants can be determined precisely. The spectroscopic feature of the spectra showed clear temperature dependence and it can be interpreted by hindered rotation model with presence of two stable conformations *(2)*. The intensity of the inner 8 lines increased with raising temperature, indicating that there are two exchangeable conformations whose existence have been shown in the elucidation of ESR spectra of methacrylates *(2)*. At 150 °C, the intensity of the inner 8 lines increased and the ESR spectrum can be interpreted as a single conformation, indicating that the energy difference between the two conformers is small.

The observed ESR spectrum of propagating radicals of *t*BMA at 150 °C is shown in Figure 2a along with simulated spectrum. The spectrum is completely simulated using hyperfine splitting constants of 1.40 mT for one proton, 1.16 mT for one proton, and 2.17 mT for three protons (Fig. 2b). A characteristic point of this result is estimation of different hyperfine splitting constants for the two methylene protons. This means that the rate of rotation of the end radical is not so fast as to be detected as equivalent methylene protons on the time scale of the ESR measurement. It is a 16-line spectrum. If the two β-methylene protons were equivalent, the total number of splitting lines would be 12 (4 (CH₃-) x 3 (-CH₂-)). This finding suggests the presence of a long propagated chain that hindered the rotation of the terminal bond for the propagating radical of 16-line spectrum.

If we can observe the ESR spectra of radicals with given chain length, chain length dependent phenomena could be examined. In order to clarify the phenomena, model radical precursors were prepared by the ATRP technique. ATRP can provide polymers with controlled molecular weights and the resulting polymers have terminal carbon-halogen bonds *(19-24)*. When the

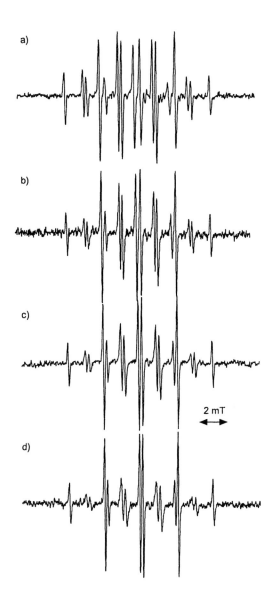

Figure 1 ESR spectra of propagating radicals of tBMA at various temperatures: a) 150, b) 120, c) 90, and d) 60 °C.

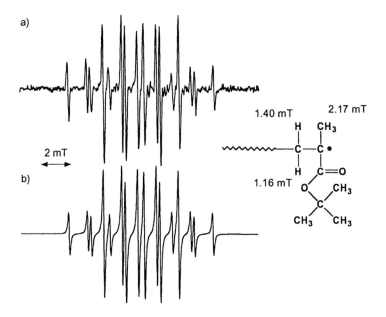

Figure 2 Structure of propagating radical of tBMA, ESR spectra of the propagating radical of tBMA at 150 °C (a), and simulated spectrum (b).

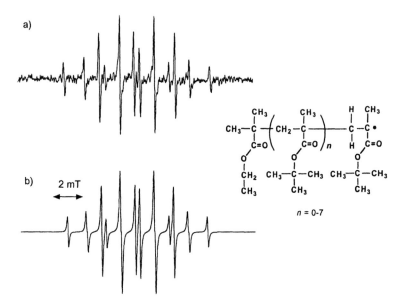

Figure 3 Structure of oligomeric model radical, ESR spectra of the oligomeric model radicals of tBMA (P_n = 2-7) at 150 °C (a), and simulated spectrum (b).

carbon-halogen bonds would be cleaved homolytically by reaction with organotin compounds, model radicals of propagating chains with given length could be generated (Scheme 1b, R = CH₃) *(25)*.

First, a mixture of oligomers containing 2-7 monomer units (P_n = 2-7) was prepared by ATRP and model radicals with short chain lengths were generated. The ESR spectrum of the radicals observed at 150 °C showed clear 12-line spectrum as shown in Figure 3a, which shows that the two β-methylene protons are almost equivalent in such small radicals at such high temperature. This finding indicates that rotation of the radical end is too fast to be detected as different methylene protons on the time scale of ESR spectroscopy.

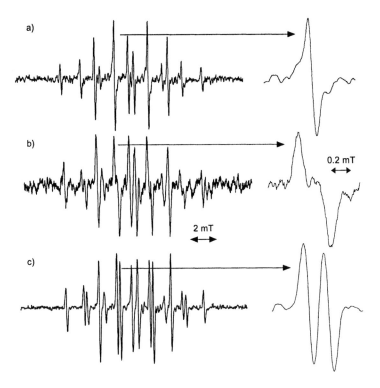

Figure 4 Comparison of ESR spectra of radicals with various chain length at 150 °C. a) oligomeric model radical, b) model radical with P_n = 100, and c) radicals in a radical polymerization (propagating radical). Characteristic lines have been enlarged on the right hand side.

In order to estimate the critical chain length which would show splitting of 16-line spectrum, model radical precursors with degrees of polymerization (P_n) of 30, 50, and 100 were prepared by ATRP. ESR spectra of radicals generated from these precursors were observed at various temperatures. These spectra showed similar temperature dependence as shown in Figure 1. In the case of P_n = 100, a 16-line spectrum was clearly observed at temperatures lower than 120 °C. The intensity of the inner 8 lines increased with raising temperature, and seems to coalesce into a single line at 150 °C (Fig. 4b). Similar ESR spectra were observed in radicals from polymeric precursors with P_n = 50 and 30. The intensity of the inner 8 lines seems to coalesce more clearly to a single line at 150 °C. The ESR spectra seemed to be 12-line spectrum, but the 4 lines coalesced insufficiently, indicating that the rate of the rotation of the end radical is not sufficiently fast for the methylene protons to be detected as equivalent on the time scale of the ESR experiment. The inner 4 lines of 12-line spectrum begin to separate into two lines at P_n = 30, and the separation becomes larger with increasing P_n owing to the lowering of the rate of the rotation. The separation was more clearly observed in the propagating radical, indicating that mobility of the end radical is restricted.

Comparison of the ESR spectra of oligomeric radicals (Fig. 3a), model radicals with P_n = 100, and radicals in polymerization system (Fig. 2a) at 150 °C in one figure is shown in Figure 4. From the comparison of the separation of inner lines, P_n of the propagating radical is considered to be higher than 100. Molecular weight (M_n) of the isolated polymers from polymerization system (Fig. 4c) was determined to be 30000 (P_n = 210) by SEC. The interpretation of the ESR spectra suggests that they correspond to "long" propagating radicals, and it is in agreement with what SEC tells us. It follows from what has been said that kinetics data on radical polymerizations determined on ESR spectra would have some more reality and reliability than before.

We can conclude that the 16-line spectrum in our ESR measurements is ascribable to polymeric radicals. ESR spectroscopy has provided structural information of the propagating radicals at their chain ends. Direct information on the chain length of the radicals has not been obtained from ESR measurements. Data shown in this research provide this information for the first time.

Further information on dynamic behavior of the propagating radicals can be obtained from these ESR spectra. The temperature dependence of these spectra can be simulated in consideration with dynamics of the radicals (26). Averaged exchange time between the two conformers was calculated from the simulation of the spectra in Figure 1. The activation energy of a rotation of the terminal C_α-C_β bond was estimated to be 21.2 kJ/mol (27).

tert-Butyl Acrylate

Interpretations of ESR spectra observed in acrylate polymerizations have been very complicated *(11-18)*. ESR spectra observed in radical polymerization systems of acrylates are not agreement with a spectrum of propagating radicals even under almost the same conditions as for methacrylates (Fig. 5d). Accordingly, it is difficult to interpret the spectrum to be that of propagating radicals. Spectroscopic change was observed in ESR spectra in the solution polymerization of *tert*-butyl acrylate (*t*BA) as shown in Figure 5. In the initial stage, a 6-line spectrum with narrow line width (Fig. 5a) was observed. This spectrum is reasonably assigned to be a propagating radical with two β-methylene protons and one α-proton (Fig. 5b). The spectrum had very short lifetime and readily changed to a totally different, long lived 7-line spectrum with broader line width (Fig. 5d). Observation of the latter spectrum was much easier than the former one. It was very difficult to detect the spectrum of the first species. Overlapped spectra of the first and latter spectra were observed in the intermediate stage (Fig. 5c). Two potential explanations for this change have been considered. One is a chain-length dependence of the spectra and the other is the occurrence of some kind of chemical reactions. These possibilities were examined by analysis of ESR spectra of model radicals with various chain lengths generated from polymeric radical precursors prepared by ATRP *(19-24)*.

Precursors of P_n = 20, 50, and 100 were prepared. The model radicals were generated by reaction with organotin compounds and the radicals were observed by ESR spectroscopy *(25)*. The three kinds of model radicals showed very similar ESR spectra and spectroscopic change. The ESR spectra observed for P_n = 100 are shown in Figure 6 as an example. These radicals showed almost the same 6-line ESR spectra of propagating radicals at low temperature (-30 °C) in the initial stage just like as shown in Figure 6a. This result indicates that there is no chain length dependence in the ESR spectra as the propagating radicals of acrylates. Furthermore, the observed 6-line spectra changed with raising temperature to 60 °C, and their change in the spectrum is in agreement with the change in the ESR spectra observed in the radical polymerization. All these things make it clear that some kind of chemical reaction is thus concluded to have occurred.

Next step is to clarify what kind of reaction takes place. In the case of radical polymerization of acrylates, 1,5-hydrogen shift to form mid-chain radical has been discussed for some time *(14-18)*. Yamada *et al*. observed ESR spectra like Figures. 5d and 6b, and reported that the spectra were due to mixture of propagating and mid-chain radicals *(16, 17)*. We show clear interpretations of ESR spectra of propagating and mid-chain radicals in this work.

96

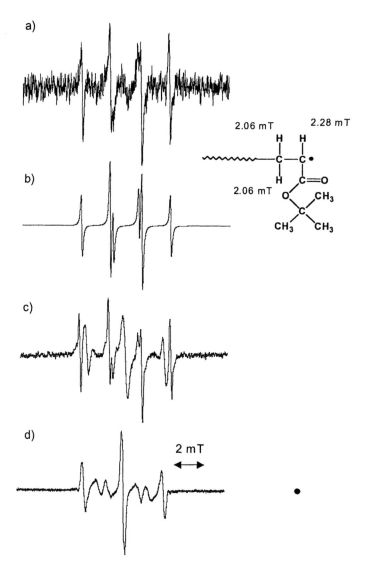

*Figure 5 ESR spectra of radicals in radical polymerization of tBA at -30 °C (a),
its simulated spectrum (b), at -10 °C (c), and at 60 °C (d). The initially
observed spectrum (a) readily changed to the final one (d) irreversibly with
raising temperature or with time at constant temperature.*

97

a)

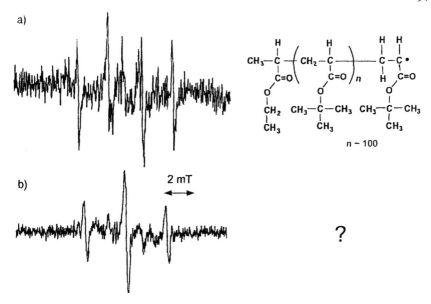

b)

2 mT

?

*Figure 6 ESR spectra of model radical (P_n = 100) at -30 °C (a) and at 60 °C
(b). Chain length was the same and spectrum showed irreversible change from
(a) to (b).*

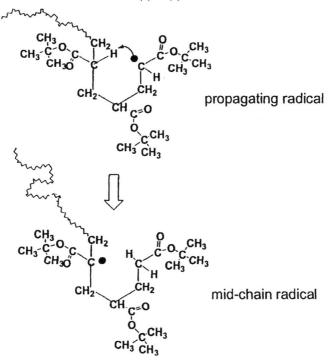

propagating radical

mid-chain radical

Figure 7 Schematic diagram of 1,5-hydrogen shift of propagating radical of tBA.

98

The proposed mechanism of 1,5-hydrogen shift is shown in Figure 7. Formation of six-membered ring enables the radical to migrate to the monomer unit two units before the terminal unit. Since this shift needs at least three monomer units, it is speculated that mid-chain radicals cannot be formed in a dimeric radical. A dimeric radical precursor was isolated from a mixture of oligomeric radical precursors, and the ESR spectrum of the dimeric radical was measured.

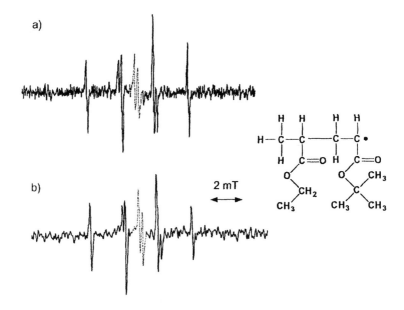

Figure 8   ESR spectra of dimeric radical of acrylate at 40 °C for 2 min (a) and 4 min (b) after generation. Spectra at center of the spectra indicated by dashed lines are probably due to tin radical.

The ESR spectra of dimeric model radical showed no change even at relatively high temperature (40 °C). Accordingly, formation of the mid-chain radicals took place through a 1,5-hydrogen shift reaction. As a result, the radical observed later (Fig. 5b) was due to mid-chain radicals. This mid-chain radical can also be formed by hydrogen abstraction from polyacrylates by oxygen centered radicals in the presence of poly*t*BA. The spectrum was very similar to both of those observed in the polymerization system (Fig. 5d) and reported by Westmoreland *et al. (18)*. Furthermore, it was reasonably simulated in consideration with the structure of the mid-chain radical.

It is well known that polyacrylate prepared by conventional radical polymerizations has many branches. Direct observations of the origin of the branching were provided by the present research work. On the basis of the results, we may say that estimation of k_p for acrylates is difficult by ESR method.

Conclusions

In the case of *tert*-butyl methacrylate, chain length dependent spectroscopic change was clearly observed in ESR spectra at 150 °C. At 150 °C, the ESR spectrum can be interpreted as a single conformation. Difference among various chain lengths was revealed apparently on the hyperfine splitting patterns in the spectra at the temperature. Clear experimental evidence for ESR detection of polymeric propagating radicals in radical polymerization was shown. In the case of *tert*-butyl acrylate, experimental evidence for formation of significant amount of mid-chain radicals from propagating radicals via 1,5-hydrogen shift was shown. Progress in controlled radical polymerization enables us to investigate conventional radical polymerization systems in detail.

Experimental

Materials: Monomers, initiators, copper salts, and ligands were purified in a usual manner.

Polymerization: The general procedures for the polymerization reactions can be obtained from ATRP literature *(19-24)*.

Generation of radicals: Model radicals were generated by cleavage of terminal C-Br bonds using organotin compounds *(25)*.

ESR measurements: ESR spectra of radicals were recorded on a JEOL JES RE-2X spectrometer operating in the X-band, utilizing a 100 kHz field modulation, and a microwave power of 1 mW. A TE_{011} mode cavity was used. ESR measurements were performed in mesitylene at 150 and 120 °C and in toluene at 90, 60, and 30 °C.

Characterization: Molecular weights and molecular weight distributions were measured using a TOSOH CCP&8020 series GPC system with TSK-gel columns. Polystyrene standards were used to calibrate the columns. Presence of terminal bromine atom was confirmed by ESI-MS spectroscopy.

Acknowledgements. The authors are grateful to Professor Krzysztof Matyjaszewski and Dr. Kelly A. Davis, Carnegie-Mellon University, for their kind advice for preparation of radical precursors by ATRP technique. The authors wish to thank Professor Antal Rockenbauer, Hungarian Academy of Sciences, for his kind permission to use his simulation program in consideration with dynamics behavior. Thanks are due to Professor Norikazu Ueyama, Dr. Taka-aki Okumura, and Mr. Hiroshi Adachi, Graduate School of Science, Osaka University, for measurements of ESI-mass spectra of radical precursors. This research work was partly supported by Grant-in-Aid (No. 13650937) from Ministry of Education, Culture, Sports, Science, and Technology of Japan.

100

References

1 Fischer, H. *Z.Naturforsch.*, **1964**,*19 a*, 267, 866.
2 Kamachi, M. *Adv. Polym. Sci.*,**19 87**, *82*, 207.
3 Kamachi, M. *J. Polym. Sci., Part A:*. *Polym. Chem.*,**2002**,*40* , 269.
4 Kamachi, M.; Kajiwara, A. *Macromol. Symp.*, **2002**, *179*, 53.
5 Kamachi, M.; Kajiwara, A. *Macromolecules*,**19 96**,*2 9*, 2378.
6 Kamachi, M.; Kajiwara, A. *Macromol. Chem. Phys*,. **1997**,*19 8*, 787.
7 Kamachi, M. In *Controlled Radical Polymerization*, ACS Symposium series 685; Matyjaszewski, K. ed., American Chemical Society: Washington, DC., 1998; Chapter 9, pp. 145-168.
8 Kajiwara, A.; Matyjaszewski, K.; Kamachi, M. In *Controlled Radical Polymerization*, ACS Symposium series 768; Matyjaszewski, K. ed., American Chemical Society, Washington, DC., 2000; Chapter 5, pp. 68-81.
9 Kamachi, M.; Kajiwara, A. *Macromol. Chem. Phys.*, **2000**,*20 1*, 2165.
10 Kajiwara, A.; Kamachi, M. *Macromol. Chem. Phys.*, **2000**,*20 1*, 2160.
11 Fischer, H.; Giacometti, G. *J. Polym. Sci.: C*,**19 67**,*16* , 2763.
12 Best, M. E.; Kasai, P. H. *Macromolecules*,**19 89**, *22*, 2622.
13 Gilbert, B. G.; Lindsay Smith, J. R.; Milne, E. C.; Whitwood, A. C.; Taylor, P. *J. Chem. Soc., Perkin Trans.* 2, **1993**, 2025.
14 Doetschman, D. C.; Mehlenbacher, R. C.; Cywar, D. *Macromolecules*, **1996**,*29* , 1807.
15 Sugiyama, Y. *Bull. Chem. Soc., Jpn.*,**19 97**, *70*, 1827.
16 Azukizawa, M.; Yamada, B.; Hill, D. J. T.; Pomery, P. J. *Macromol. Chem. Phys.*, **2000**, *201*, 774.
17 Yamada, B.; Azukizawa, M.; Yamazoe, H.; Hill, D. J. T.; Pomery, P. J. *Polymer*, **2000**,*41* , 5611.
18 Chang, H. R.; Lau, W.; Parker, H. -Y.; Westmoreland, D. G. *Macromol. Symp.*, **1996**,*11 1*, 253.
19 Wan, J. -S.; Matyjaszewski, K. *J. Am. Chem. Soc.*, **1995**, *117*, 5614.
20 Wan, J. -S.; Matyjaszewski, K. *Macromolecules*,**19 95**, *28*, 7901
21 Kajiwara, A.; Matyjaszewski, K.; Kamachi, M. *Macromolecules* **1998**, *31*, 5695.
22 Kajiwara, A.; Matyjaszewski, K. *Polym. J.* **1999**, *31*, 70.
23 *Controlled Radical Polymerization*, ACS Symposium series 685; Matyjaszewski, K. ed., American Chemical Society, Washington, DC., 1998.
24 *Controlled Radical Polymerization*, ACS Symposium series 768; Matyjaszewski, K. ed., American Chemical Society, Washington, DC., 2000.
25 Giese, B.; Damm, W.; Wetterich, F.; Zeitz, H.-G. *Tetrahedron Lett.*, **1992**, *33*, 1863.
26 Rockenbauer, A. *Mol. Phys. Rep.*,**19 99**, *26*, 117.
27 Kajiwara, A.; Kamachi, M.; Rockenbauer, A. to be published.

Atom Transfer Radical Polymerization

Mechanisms

Chapter 8

Living Radical Polymerization with Designed Metal Complexes

Masami Kamigaito, Tsuyoshi Ando, and Mitsuo Sawamoto

Department of Polymer Chemistry, Graduate School of Engineering,
Kyoto University, Kyoto 606–8501, Japan

This paper discusses recent developments in the metal-catalyzed living radical polymerization with designed metal complexes in our laboratory. The design was directed to increasing the catalytic activity and to widening the applicability to a variety of monomers by several methods. They include ruthenium(II) complexes with an electron-donating aminoindenyl ligand, a neutral and labile ethylene ligand, a hemilabile bidentate P,N-ligand, and a dinuclear iron(I) complex. These novel complexes were characterized by NMR, cyclic voltammetry, and X-ray crystallography. Most of them had a lower redox potential than the previous ruthenium(II) and iron(II) complexes and proved effective in a fast living radical polymerization.

Introduction

Living or controlled radical polymerization is now achieved with a variety of initiating systems based on a common concept, i.e., reversible activation of covalent bond of the dormant species into the growing radical species (1). The systems with nitroxide, metal catalysts, and RAFT agents have been widely developing in various aspects; initiating systems that can control polymer molecular weights more precisely, kind of monomers that can be polymerized in controlled fashion, and controlled architectures of polymers that can be obtained.

Because of their relative easiness in the procedures and their wide applicability, these polymerizations are becoming common as synthetic methods for well-defined polymers. However, there seems still much room for improvement in the radical systems from the viewpoint of rates and versatility.

We first discovered ruthenium-catalyzed living radical polymerization (2) of methyl methacrylate (MMA) with $RuCl_2(PPh_3)_3$ (1) and have been developing the metal-catalyzed systems (Scheme 1) by using ruthenium and other metals such as iron, nickel, rhenium, etc. or by introducing various ligands (1a, 1f). For example, Figure 1 summarizes evolution of a series of ruthenium and iron complexes in our laboratory. The first ruthenium complex proved highly effective in giving well-controlled polymethacrylates with controlled molecular weights [M_n (number-average molecular weight) = ([M]$_0$/[I]$_0$) × (molecular weight of monomer)], narrow molecular weight distributions (MWDs) ($M_w/M_n \leq$ 1.1), and high chlorine end functionality ($F_n = 1.0$) up to high conversions (2–4). However, the catalytic activity was not high, and the controllability for acrylate and styrene polymerizations was moderate. We then introduced electron-donating cyclopentadiene (Cp)-based ligands into the complex to lower the redox potential or to increase the activity. The ruthenium indenyl complex (2) induced a faster living radical polymerization of methacrylates (5) while the pentamethylcyclopentadienyl (Cp*) complex (3) proved versatile to result in good molecular weight control for three classes of monomers, methacrylates, acrylates, and styrenes ($M_w/M_n \leq 1.1$) (6). Another modulation for 1 is the use of a hydrophilic and ionic phosphine ligand, leading to the complex (4) with which enabled homogeneous living radical polymerization of hydrophilic monomers like 2-hydroxyethyl methacrylate in methanol (7). For iron counterparts, a PPh$_3$-based complex (5) was employed for MMA to show a slightly higher activity than the ruthenium (1) (8). The iron-Cp complex (6) was also effective and gave polymers of styrene and acrylate with controlled molecular weights (9–11).

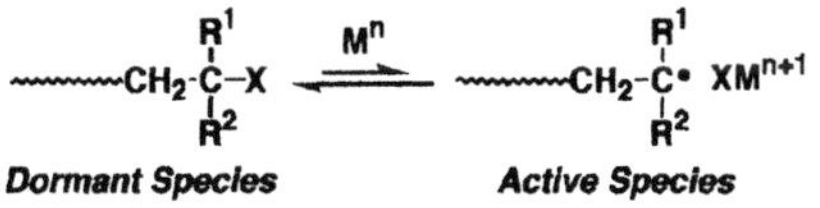

$$\text{\large\textasciitilde\textasciitilde\textasciitilde CH}_2\text{-}\overset{\displaystyle R^1}{\underset{\displaystyle R^2}{C}}\text{-X} \;\underset{M^n}{\overset{}{\rightleftharpoons}}\; \text{\large\textasciitilde\textasciitilde\textasciitilde CH}_2\text{-}\overset{\displaystyle R^1}{\underset{\displaystyle R^2}{C}}^{\bullet}\;\; XM^{n+1}$$

Dormant Species   **Active Species**

Scheme 1. Metal-catalyzed living radical polymerization.

This paper presents a brief overview on recent and further developments in ruthenium and iron catalysts (7–10) based on ligand or metal designs for faster and more versatile living radical polymerizations. The design is intended to increase the catalytic activity by several methods; i.e., introducing an electron-donating amino-group into the indenyl ligand (7) (12), replacing the chlorine atom in 1 with a neutral and labile ethylene ligand (8) (13), using a hemilabile bidentate P,N-ligand (9) (14), and employing iron(I) in place of iron(II) (15). These strategies would be hopeful as stated below.

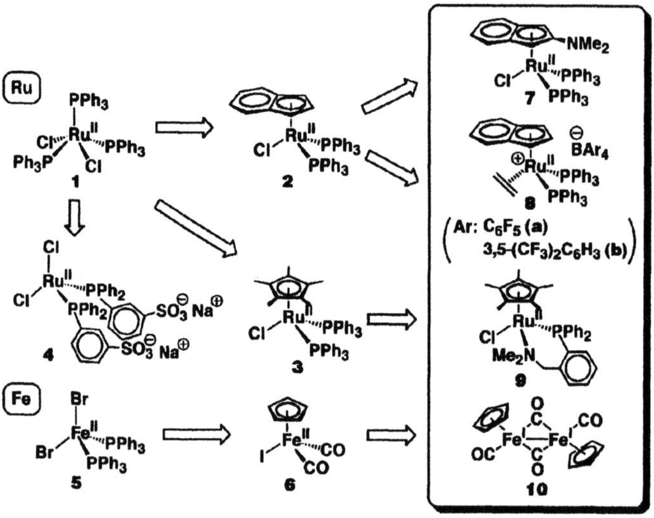

Figure 1. Evolution of ruthenium and iron catalysts for metal-catalyzed living radical polymerization in our laboratory.

Results and Discussion

Electron-Donating Aminoindenyl Ruthenium Complex (7)

One of the keys for increasing the catalytic activity in metal-catalyzed redox processes is to reduce the redox potential of metal complexes by introducing an electron-donating ligand, because the catalyst should give one electron to a carbon–halogen terminal on activation (Scheme 1). We thus prepared a new ruthenium complex with an aminoindenyl ligand [7, Ru(2-Me$_2$N-Ind)Cl(PPh$_3$)$_2$] and checked the catalytic activity, versatility, and reaction controllability (12).

The complex 7 was synthesized from the lithium salt of N-(1H-2-indenyl)-N,N-dimethylamine and RuCl$_2$(PPh$_3$)$_3$, and was confirmed by elemental analysis, NMR, and X-ray crystallography (Figure 2). X-Ray analysis showed that 7 is an 18-electron complex where the aminoindenyl ligand coordinates to the ruthenium center in η^5-fashion as indicated by a small slip parameter (Δ = 0.0320 Å) and hinge angle (HA = 12.6°) (16). Such a complex may release one of the phosphine ligands or may undergo slipping of the indenyl ring into a η^3-cooridination state to become active for abstraction of halogen from the dormant terminal. The bond length between the ruthenium and the phosphorous atoms was slightly longer in 7 than in 2, which suggests easier release of phosphine or

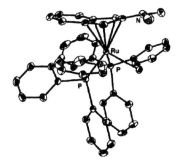

Figure 2. ORTEP diagram of 7 (thermal ellipsoids at the 50% level).

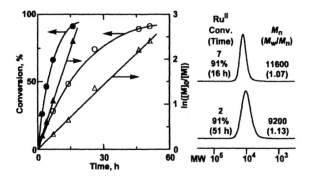

Figure 3. Polymerization of MMA with 7 (●, ▲) or 2 (O, △) coupled with (MMA)₂–Cl and Al(Oi-Pr)₃ in toluene at 80 °C: [MMA]₀ = 4.0 M; [(MMA)₂–Cl]₀ = 40 mM; [7 or 2]₀ = 4.0 mM; [Al(Oi-Pr)₃]₀ = 40 mM.

higher activity of 7. Its high activity was also suggested by cyclic voltammetry, where 7 had a lower redox potential than 2 ($E_{1/2}$ = 0.42 vs 0.55V vs Ag/AgCl).

The complex was employed for polymerization of MMA in conjunction with an MMA-dimer chloride [(CH₃)₂C(CO₂CH₃)CH₂C(CH₃)(CO₂CH₃)Cl; (MMA)₂–Cl] as an initiator in the presence of Al(Oi-Pr)₃. As shown in Figure 3, the new complex (7) induced a faster polymerization and gave polymers with narrower MWDs (M_w/M_n < 1.1) than the indenyl (2) does. The M_n increased in direct proportion to monomer conversion and agreed well with the calculated values assuming that one molecule of the initiator generates one living polymer chain. The fast polymerization accompanying with fine molecular weight control and narrow MWDs indicates that the complex 7 increases not only

106

polymerization rate but also the interconversion between the dormant and the radical species.

Figure 4 shows high molecular weight PMMA with $(MMA)_2-Cl/7/Al(Oi-Pr)_3$. The polymerization was fast irrespective of a low catalyst concentration (0.80 mM) in comparison to the initiator (8.0 mM). The M_n increased in direct proportion to monomer conversion and agreed well with the calculated values. The MWDs were narrower $(M_w/M_n < 1.1)$ than those of a similar M_n (~ 1×10^5) obtained with Cu(I) (17) although the highest M_n were set at lower in our case.

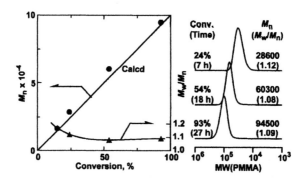

Figure 4. High molecular weight PMMA with $(MMA)_2-Cl/7/Al(Oi-Pr)_3$ in toluene at 80 °C: $[MMA]_0 = 8.0 M$; $[(MMA)_2-Cl]_0 = 8.0 mM$; $[7]_0 = 0.80 mM$; $[Al(Oi-Pr)_3]_0 = 40 mM$. The diagonal solid line indicates the calculated M_n assuming the formation of one living polymer per $(MMA)_2-Cl$ molecule.

The polymerization of styrene with 7 and $(MMA)_2-Cl$ was also examined in the presence of $Al(Oi-Pr)_3$. The molecular weights were in good agreement with the calculated values, where the MWDs were narrow even with the chloride initiator $(M_w/M_n = 1.26)$. However, for methyl acrylate (MA), the MWDs were broader $(M_w/M_n = 2.33)$ with $(MMA)_2-Cl$ while narrower with a bromide $[(CH_3)_2C(CO_2CH_2CH_3)Br$; EMA–Br] and an iodide $[(CH_3)_2C(CO_2CH_2CH_3)I$; EMA–I] initiator $(M_w/M_n = 1.47$ and 1.32, respectively). These are because of stronger secondary carbon–halogen bonds in the dormant species in polystyrene and polyacrylates, and because of weaker bond energy in the order of Cl > Br > I. A more active ruthenium catalyst may be needed for control of acrylate polymerizations by using the most stable carbon–chlorine species.

Cationic Indenyl Ruthenium Complex with Labile Ethylene Ligand (8)

Another approach to high catalytic activity is to promote formation of a vacant active site in complexes. The chlorine in 2 was thus substituted with a labile and neutral ethylene ligand to be converted into a cationic 18-electron ruthenium complex (8) with a weakly nucleophilic borate anion [BAr₄⁻; Ar = C₆F₅ (8a), 3,5-(CF₃)₂C₆H₃ (8b)], which is often employed for olefin polymerization. The catalysts were synthesized from the reaction between 2 and lithium [LiB(C₆F₅)₄] or sodium borate [NaB[3,5-(CF₃)₂C₆H₃]₄] under ethylene atmosphere. The elemental analysis showed that the observed contents of carbon and hydrogen were in good agreement with the calculated. ¹H NMR analysis indicated the presence of the borate anion and the ethylene that is coordinated to ruthenium, as shown in Figure 5. These results supported formation of the cationic ruthenium complexes (8).

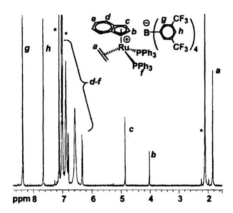

Figure 5. ¹H NMR spectrum of 8b in toluene-d₈ at room temperature. The signals marked with an asterisk are due to the solvent.

The catalysts were used for polymerization of MMA in conjunction with (MMA)₂-Cl and Al(Oi-Pr)₃ in toluene at 80 °C (13). The polymerizations with the cationic ruthenium complexes (8a and 8b) proceeded faster than that with the neutral counterpart (2) (Figure 6). The obtained polymers had unimodal MWDs and M_n that increased in direct proportion to monomer conversion. These results indicate that the cationic ruthenium complexes induce a fast living radical polymerization of MMA. However, slightly broader MWDs and slightly higher M_n may suggest some side reactions.

We further investigated copolymerization of MA and olefins such as 1-hexene with the cationic (8b) and the neutral complexes (2) in conjunction with an iodide initiator (MA–I) in the presence of Al(Oi-Pr)₃. Both catalysts led to copolymerizations where the hexene was consumed much slower than MA. The cationic complex showed a slightly higher activity. The molecular weights of the polymers increased with consumption of monomers while the MWDs were

108

broad (M_w/M_n = 2–3). ^{1}H and ^{13}C NMR analysis showed that the copolymers had about 10% hexene, which were distributed among MA units in the copolymers. There were almost no differences in the comonomer distributions between the polymers obtained with **2** and **8b**. Ethylene was also copolymerized with MA in place of 1-hexene. These results demonstrate that α-olefins, much less reactive than MA, may be copolymerized via a radical mechanism which is very similar to that in the metal-catalyzed living radical polymerization (Scheme 1). Further studies are now in progress.

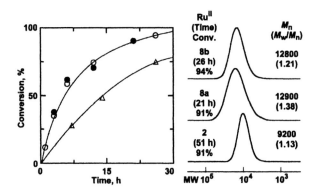

*Figure 6. Polymerization of MMA with **8a** (O) or **8b** (●) or **2** (△) coupled with (MMA)$_2$–Cl and Al(Oi-Pr)$_3$ in toluene at 80 °C: [MMA]$_0$ = 4.0 M; [(MMA)$_2$–Cl]$_0$ = 40 mM; [**8a** or **8b** or **2**]$_0$ = 4.0 mM; [Al(Oi-Pr)$_3$]$_0$ = 40 mM.*

Ruthenium Complex with Hemilabile Bidentate *P,N*-Ligand (9)

We have already found that the ruthenium-catalyzed radical polymerization was accelerated on addition of Al(Oi-Pr)$_3$ (*18*) or amine (*19, 20*), where the enhancement effects were larger for the latter. Furthermore, such additives often make the MWDs much narrower. This is most probably attributed to interaction between the ruthenium complex and the additives, which may result in a new ruthenium species with a higher activity. Although some interactions between the ruthenium complex and the additives were observed by NMR analysis of the mixtures (*19*), isolation of the amine-coordinated ruthenium species was difficult due to relatively weak coordination of amines to ruthenium.

We herein employed a chelating *P,N*-ligand in the ruthenium-catalyzed living radical polymerization for rate enhancement and more fine control of the polymerization. The nitrogen group in such a ligand can strongly interact or coordinate to the ruthenium center due to the chelating effect assisted by the strong coordination of the phosphine part to the ruthenium. One of the selected *P,N*-ligands is 2-(*N,N*-dimethylamino)benzyl diphenylphosphine (*21*), which possesses triphenylphosphine and trialkylamine moieties. The former is a good ligand for the ruthenium-catalyzed living radical polymerization while the latter is effective in rate enhancement and fine control in the polymerization.

The P,N-compound was added to polymerization of MMA with $(MMA)_2$–Cl/**3** in toluene at 80 °C (Figure 7) (*14*). The compound accelerated the polymerization (filled circles in Figure 7), where the rate enhancement was larger than those with n-Bu_2NH (filled triangles) and $Al(Oi\text{-}Pr)_3$ (open triangles), despite a lower concentration of the added P,N-compound. In contrast, a similar amine without the phosphine part, N,N-dimethylbenzylamine, led to little rate enhancement (open squares). Further addition of triphenylphosphine to the amine rather retarded the polymerization (filled squares). The chelating effect on the catalytic activity was thus shown clearly.

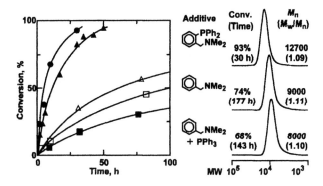

Figure 7. MMA polymerization with $(MMA)_2$–Cl/3 in the presence of additives [P,N-ligand (●), $Me_2NCH_2C_6H_5$ (□), $Me_2NCH_2C_6H_5 + PPh_3$ (■), $Al(Oi\text{-}Pr)_3$ (△), n-Bu_2NH (▲)] in toluene at 80 °C: $[MMA]_0 = 4.0$ M; $[(MMA)_2\text{-}Cl]_0 = 40$ mM; $[3]_0 = 4.0$ mM; $[additives]_0 = 16$ (●, □, ■) or 40 (△, ▲) mM.

The polymers obtained with the P,N-compound had narrow MWDs (M_w/M_n = 1.09) and controlled molecular weights that increased in direct proportion to monomer conversion. These results indicate that the P,N-compound containing alkylamine and triphenylphosphine moieties is effective in the RuCp*-mediated living radical polymerization. The P,N-ligand most probably coordinates to the ruthenium to modify the catalytic activity. The formation of a new complex was suggested by ^1H and ^{31}P NMR analysis of the mixture.

We thus synthesized the P,N-chelating Cp*-ruthenium complex from the ligand and $[RuCp*Cl]_4$, isolated it by recrystallization, and identified by elemental and NMR analyses. The ^1H and ^{31}P NMR spectra were the same as those of the in-situ formed compound above. The X-ray crystallographic analysis indicate that the P,N-ligand coordinates to the ruthenium by chelation

110

(Figure 8), where the bond length of Ru and P is almost the same as that of Ru and N (2.2982 vs 2.2611 Å) irrespective of larger radius of P than that N. This means that the phosphorous part strongly coordinates to the ruthenium and that the amine moiety is more labile. The redox potential of the P,N-chelated complex (9) was lower than the phosphine complex (3) (0.26 vs 0.46 V vs Ag/AgCl), which indicates a higher activity of 9. The complex actually induced a faster polymerization than 3, while the MWDs became slightly broader $(M_w/M_n = 1.22)$ than that obtained with a mixture of the P,N-ligand and 3. There seem to be some additional effects of excess ligands or the triphenylphosphine-complex on the deactivation process.

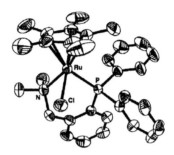

Figure 8. ORTEP diagram of 9 (thermal ellipsoids at the 50% level).

Dinuclear Iron(I) Complex (10)

Catalytic activity or redox potential can be modulated by the oxidation state of a metal center in addition to ligand design. We have recently found that dinuclear iron(I) complex (10) not only induced a faster polymerization of styrene than mononuclear iron(II) (6) (*10*) but also enabled polymerization of less reactive monomers like vinyl acetate to result in control of molecular weights (*22*). This is most probably attributed to the lower oxidation state or redox potential of the iron(I) species.

The iron(I) complex, 10, was thus employed for polymerization of MA in conjunction with EMA–I in toluene at 60 °C (*15*). The polymerization proceeded very fast to reach about 90% conversion in 1 h (Figure 9). The high activity is in a remarkable contrast to no activity of the mononuclear version 6 under the same conditions while the latter became moderately active (93% conversion in 80 h) on addition of Al(Oi-Pr)$_3$. However, the molecular weight control with EMA–I/10 was inferior to that with EMA–I/6/Al(Oi-Pr)$_3$ probably due to formation of excess radical species and/or the slow transformation of the resulting radical species into the dormant. For controlling the polymerization, molecular iodine (I$_2$) was added to the reaction mixture, where the additive may

111

not only serve as a radical scavenger to trap the growing radicals but may regenerate the C–I dormant terminal more rapidly. On addition of iodine (20 mM), the polymerization with the EMA–I/10 system was slightly retarded but reached a high conversion (93%) in 5 h. The MWDs of the polymers became much narrower ($M_w/M_n = 1.28$).

The polymers obtained in the presence of iodine had M_n that increased in direct proportion to monomer conversion (filled circles in Figure 10) and had

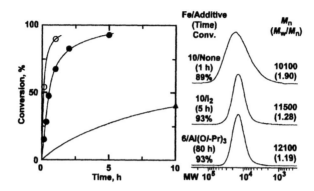

Figure 9. Polymerization of MA with EMA–I/10 (O), or EMA–I/10/I₂ (●), or EMA–I/6/Al(Oi-Pr)₃ (▲) in toluene at 60 °C: [MA]₀ = 4.0 M; [EMA–I]₀ = 40 mM; [10 or 6]₀ = 40 mM; [Al(Oi-Pr)₃]₀ = 40 mM; [I₂]₀ = 20 mM.

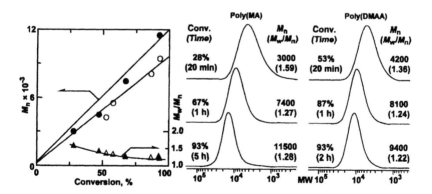

Figure 10. Polymerization of MA (●, ▲) or DMAA (O, △) with EMA–I/10/I₂ in toluene at 60 °C: [MA]₀ = 4.0 M; [EMA–I]₀ = 40 mM; [10]₀ = 40 mM; [I₂]₀ = 20 mM.

112

one initiating moiety per one polymer chain. This system can be also applicable for N,N-dimethylacrylamide (DMAA) to induce a similarly fast and controlled radical polymerization (open circles in Figure 10). The MWDs were similarly narrow (M_w/M_n = 1.22).

Although the working mechanism of the Fe(I) complex and the added iodine, as well as the possibility of degenerative iodide-transfer mechanism, should be further discussed, the Fe(I)-based initiating system is one of the most active and effective from the viewpoint of its wide applicability to various monomers such as acrylamides and vinyl acetate.

Experimental

General

All experiments involving air-sensitive metal compounds were performed under a moisture- and oxygen-free argon atmosphere (H_2O < 1 ppm; O_2 < 1 ppm) in a glove box (M.Braun Labmaster 130) or under argon or nitrogen with standard Schlenk line techniques. Toluene was dried overnight over calcium chloride and distilled from sodium/benzophenone ketyl. n-Pentane, n-hexane, n-octane, and tetralin were dried overnight over calcium chloride, and distilled twice over calcium hydride. Methylene chloride was dried overnight over calcium chloride, distilled from phosphorous pentoxide and then from calcium chloride. All solvents were degassed before use by freeze-thaw-pumping technique or by bubbling with dry nitrogen over 15 min. Water (Wako, distilled) was degassed under reduced pressure at 50 °C and saturated with dry nitrogen for several times. Methyl methacrylate (MMA) (Tokyo Kasei, >99%), styrene (Wako, >99%), and methyl acrylate (MA) (Wako, >98%) were dried overnight over calcium chloride and distilled twice over calcium hydride under reduced pressure before use. RuCl$_2$(PPh$_3$)$_3$ (1; Merck, >99%), Ru(Ind)Cl(PPh$_3$)$_2$ (2; Strem, >98%), Fe(Cp)I(CO)$_2$ (6; Aldrich, >97%), Fe$_2$Cp$_2$(CO)$_4$ (10; Aldrich, >99%), and Al(Oi-Pr)$_3$ (Aldrich, >99.99%) were used as received. The chloride initiator [(CH$_3$)$_2$C(CO$_2$CH$_3$)CH$_2$C(CH$_3$)(CO$_2$CH$_3$)Cl; (MMA)$_2$–Cl] was prepared as reported (23). The bromide initiator [(CH$_3$)$_2$C(CO$_2$Et)Br, EMA–Br] (Tokyo Kasei; >98%) was distilled twice over calcium hydride under reduced pressure. The iodide initiator [(CH$_3$)$_2$C(CO$_2$Et)I; EMA–I] was prepared as reported (10, 24).

Synthesis of Complexes

The aminoindenyl ruthenium complex (7) (12) was prepared from the lithium salt of the ligand [N-(1H-2-indenyl)-N,N-dimethylamine] (25, 26) and

113

$RuCl_2(PPh_3)_3$ in toluene at 90 °C. After removal of the solvent followed by extraction with CH_2Cl_2, the solid was washed with deionized water and n-hexane for several times. The complex was recrystallized by layering of n-hexane on the CH_2Cl_2 solution. ^1H NMR and X-ray crystallography indicated that one molecule of CH_2Cl_2 was contained per one molecule of the ruthenium compound. Anal. Calcd for $C_{48}H_{44}Cl_3NP_2Ru$: C, 63.76; H, 4.90; N, 1.55; Cl, 11.76. Found: C, 64.21; H, 4.96; N, 1.55, Cl, 11.63.

The cationic ruthenium complexes (8a and 8b) (13) were obtained from $Ru(Ind)Cl(PPh_3)_2$ and lithium $[LiB(C_6F_5)_4 \cdot Et_2O]$ and sodium borate $[NaB[3,5-(CF_3)_2C_6H_3]_4 \cdot 2H_2O]$, respectively, in CH_2Cl_2 at room temperature under an ethylene atmosphere. After filtration of the solution followed by removal of the solvent under vacuum, the complexes were precipitated by addition of n-hexane into the CH_2Cl_2 solution. Anal. Calcd for $C_{71}H_{41}BF_{20}P_2Ru$ (8a): C, 58.90; H, 2.85. Found (8a): C, 59.04; H, 3.14. Anal. Calcd for $C_{79}H_{53}BF_{24}P_2Ru$ (8b): C, 58.14; H, 3.27. Found (8b): C, 58.13; H, 3.21.

The P,N-chelated ruthenium complex (9) (14) was synthesized from $[Ru(Cp^*)Cl]_4$ (27) and the P,N-ligand (21) in toluene at 80 °C. After removal of the solvent followed by extraction with CH_2Cl_2, the complex was recrystallized by layering of pentane on the CH_2Cl_2 solution. Anal. Calcd for $C_{31}H_{37}ClNPRu$: C, 62.99; H, 6.31; N, 2.37; Cl, 6.00. Found: C, 62.80; H, 6.20; N, 2.29, Cl, 6.00.

Polymerization Procedures

Polymerization was carried out by the syringe technique under dry nitrogen in glass tubes equipped with a three-way stopcock or in baked and sealed glass vials. A typical example for the polymerization of MMA with 7/(MMA)$_2$–Cl/Al(Oi-Pr)$_3$ is given below: In a 50-mL round-bottomed flask was placed 7 (14.1 mg, 0.0156 mmol), toluene (0.40 mL), n-octane (0.33 mL), MMA (1.67 mL, 1.56 mmol), a solution of Al(Oi-Pr)$_3$ (1.25 mL of 125 mM in toluene, 0.156 mmol), and a solution of (MMA)$_2$–Cl (0.25 mL of 620 mM in toluene, 0.155 mmol) at room temperature. The total volume of reaction mixture was 3.90 mL. Immediately after mixing, four aliquots (0.50 mL each) of the solutions were injected into backed glass tubes. The reaction vials were sealed and placed in an oil bath kept at 80 °C under vigorous stirring. In predetermined intervals, the polymerization was terminated by cooling the reaction mixtures to −78 °C. Monomer conversion was determined from the residual monomer concentration by gas chromatography with n-octane as an internal standard.

The quenched reaction solutions were diluted with toluene (ca. 20 mL) and rigorously shaken with an absorbent [KYOWAAD-2000G-7 (Mg$_{0.7}$Al$_{0.3}$O$_{1.15}$); Kyowa Chemical Industry] (ca. 5 g) to remove the metal-containing residues. After the absorbent was separated by filtration (Whatman 113V), the filtrate was washed with water and evaporated to dryness to give the products, which were subsequently dried overnight under vacuum at room temperature.

114

Measurements

The MWD, M_n, and M_w/M_n ratios of the polymers were measured by size-exclusion chromatography (SEC) in chloroform at 40 °C on three polystyrene gel columns (Shodex K-805L × 3) that were connected to a Jasco PU-980 precision pump and a Jasco RI-930 refractive index detector. The columns were calibrated against twelve standard poly(MMA) samples (Polymer Laboratories; M_n = 630–1200000; M_w/M_n = 1.06–1.22) as well as the monomer. ^1H NMR spectra were recorded in $CDCl_3$ at 25 °C on a JEOL JNM-LA500 spectrometer, operating at 500.16 MHz. Polymers for NMR analysis were fractionated by preparative SEC (column: Shodex K-2002) to be freed from low molecular impurities originated from the catalysts. Cyclic voltammograms were recorded on a Hokuto Denko HZ-3000 apparatus. Measurements were carried out under argon at 0.10 Vs^{-1} in a CH_2ClCH_2Cl solution (5.0 mM) with n-Bu_4NPF_6 (100 mM) as the supporting electrolyte. A three-electrode cell, equipped with a platinum disk as a working electrode, a platinum wire as a counter electrode, and an Ag/AgCl electrode as a reference, was used. X-Ray data were collected at 130 K for **7** and at 293 K for **9** with a Bruker SAMRT-CCD area detector diffractometer using the MoKα radiation (λ = 0.71073 Å). Cell parameters were refined with 16868 reflections for **7** and 21848 reflections for **9**. The diffraction frames were integrated on the SAINT package. The structures were solved and refined with SHELXL-97.

Acknowledgment

This work was supported by the New Energy and Industrial Technology Development Organization (NEDO) under the Ministry of Economy, Trade and Industry (METI), Japan. We thank Professor Tamejiro Hiyama and Dr. Masaki Shimizu at Kyoto University and Professor Kyoko Nozaki and Mr. Koji Nakano at the University of Tokyo for the X-ray crystallographic analysis.

References

1. For recent reviews on living radical polymerization, see: (a) Kamigaito, M.; Ando, T.; Sawamoto, M. *Chem. Rev.* **2001**, *101*, 3689. (b) Matyjaszewski, K.; Xia, J. *Chem. Rev.* **2001**, *101*, 2921. (c) Hawker, C. J.; Bosma, A. W.; Harth, E. *Chem. Rev.* **2001**, *101*, 3661. (d) Matyjaszewski, K., Ed.; ACS Symposium Series 768; American Chemical Society: Washington, DC, 2000. (e) Sawamoto, M.; Kamigaito, M. In *Synthesis of Polymers*; Schlüter, A.-D., Ed.; Materials Science and Technology Series; Wiley-VCH: Weinheim, Germany, 1999, Chapter 6; pp 163–194. (f) Sawamoto, M.; Kamigaito, M. *CHEMTECH* **1999**, *29* (6), 30.

115

2. Kato, M.; Kamigaito, M.; Sawamoto, M.; Higashimura, T. *Macromolecules* **1995**, *28*, 1721.
3. Ando, T.; Kamigaito, M.; Sawamoto, M. *Macromolecules* **1996**, *29*, 1070.
4. Nonaka, H.; Ouchi, M.; Kamigaito, M.; Sawamoto, M. *Macromolecules* **2001**, *34*, 2083.
5. Takahashi, H.; Ando, T.; Kamigaito, M.; Sawamoto, M. *Macromolecules* **1999**, *32*, 3820.
6. Watanabe, Y.; Ando, T.; Kamigaito, M.; Sawamoto, M. *Macromolecules* **2001**, *34*, 4370.
7. Fuji, Y.; Watanabe, K.; Baek, K.-Y.; Ando, T.; Kamigaito, M.; Sawamoto, M. *J. Polym. Sci., Part A, Polym. Chem.* **2002**, *40*, 2055.
8. Ando, T.; Kamigaito, M.; Sawamoto, M. *Macromolecules* **1997**, *30*, 4507.
9. Kotani, Y.; Kamigaito, M.; Sawamoto, M. *Macromolecules* **1999**, *32*, 6877.
10. Kotani, Y.; Kamigaito, M.; Sawamoto, M. *Macromolecules* **2000**, *33*, 3543.
11. Onishi, I.; Baek, K.-Y.; Kotani, Y.; Kamigaito, M.; Sawamoto, M. *J. Polym. Sci., Part A, Polym. Chem.* **2002**, *40*, 2033.
12. Kamigaito, M.; Watanabe, Y.; Ando, T.; Sawamoto, M. *J. Am. Chem. Soc.* **2002**, *124*, 9994.
13. Nakano, H.; Sawauchi, C.; Ando, T.; Kamigaito, M.; Sawamoto, M. *Polym. Prepr. Jpn.* **2002**, *51*, 360.
14. Kobayashi, S.; Ando, T.; Kamigaito, M.; Sawamoto, M. *Polym. Prepr. Jpn.* **2002**, *51*, 126.
15. Kamigaito, M.; Onishi, I.; Kimura, S.; Kotani, Y.; Sawamoto, M. *Chem. Commun.* **2002**, 2694.
16. Cadierno, V.; Díez, J.; Gamasa, M. P.; Gimeno, J.; Lastra, E. *Coord. Chem. Rev.* **1999**, *193–195*, 147.
17. Xue, L.; Agarwal, U. S.; Lemstra, P. *Macromolecules* **2002**, *35*, 8650
18. Ando, T.; Kamigaito, M.; Sawamoto, M. *Macromolecules* **2000**, *33*, 6732.
19. Hamasaki, S.; Kamigaito, M.; Sawamoto, M. *Macromolecules* **2002**, *35*, 2934.
20. Hamasaki, S.; Sawauchi, C.; Kamigaito, M.; Sawamoto, M. *J. Polym. Sci., Part A, Polym. Chem.* **2002**, *40*, 617.
21. Rauchfuss, T. B.; Patino, F. T.; Roundhill, D. A. *Inorg. Chem.* **1975**, *14*, 652.
22. Wakioka, M.; Baek, K.-Y.; Ando, T.; Kamigaito, M.; Sawamoto, M. *Macromolecules* **2002** *35*, 330.
23. Ando, T.; Kamigaito, M.; Sawamoto, M. *Macromolecules* **2000**, *33*, 2819.
24. Curran, D. P.; Bosch, E.; Kaplan, J.; Newcomb, M. *J. Org. Chem.* **1989**, *54*, 1826.
25. Edlund, U. *Acta. Chem. Scand.* **1973**, *27*, 4027.
26. Klosin, J.; Kruper, W. J.; Nikias, P. N.; Roof, G. R.; De Waele, P.; Abboud, K. A. *Organometallics* **2001**, *20*, 2663.
27. Fagan, P. J.; Ward, M. D.; Calabrese, J. C. *J. Am. Chem. Soc.* **1989**, *111*, 1698.

Chapter 9

Controlled Radical Polymerization Catalyzed by Ruthenium Complexes: Variations on Ru-Cp#

Sébastien Delfosse[1], Aurore Richel[1], François Simal[1,3], Albert Demonceau[1,*], Alfred F. Noels[1], Oscar Tutusaus[2], Rosario Núñez[2], Clara Viñas[2], and Francesc Teixidor[2]

[1]Laboratory of Macromolecular Chemistry and Organic Catalysis, University of Liége, Sart-Tilman (B.6a), B–4000 Liége, Belgium
[2]Institut de Ciéncia de Materials, CSIC-UAB, Campus de Bellaterra, Cerdanyola, 08193 Barcelona, Spain
[3]Current address: UCB Chemicals, Research and Technology, Anderlechtstraat 33, B–1620 Drogenbos, Belgium

A series of isoelectronic ruthenium-based complexes of the general formula $[RuX(Cp^{\#})L_2]$ ($Cp^{\#}$ = cyclopentadienyl or cyclopentadienyl derivatives) were synthesized, and their relative catalytic activities were determined by monitoring the atom transfer radical polymerization of methyl methacrylate, n-butyl acrylate, and styrene. $[RuCl(Cp^*)(PPh_3)_2]$ and $[RuCl(Ind)(PPh_3)_2]$ were found to be highly efficient catalysts for ATRP, producing polymers with narrow molecular weight distributions ($M_w/M_n < 1.2$). The following order of increasing efficiency was determined : $[RuCl(Cp)(PPh_3)_2]$ $<<$ $[RuCl(Ind)(PPh_3)_2]$ $<$ $[RuCl(Cp^*)(PPh_3)_2]$. In sharp contrast, ruthenacarboranes were inefficient in ATRP, demonstrating therefore the prominent role of the $Cp^{\#}$ ligand. The effect of the phosphine ligands was also investigated, and additional studies indicated that the release of a phosphine ligand occurred prior to the activation of the carbon-halogen bond of both the initiator and polymer growing chain end by the unsaturated ruthenium center.

It was in 1995 that Sawamoto (1) and Matyjaszewski (2) independently reported their seminal papers on atom transfer radical polymerization (ATRP). Among the numerous catalytic systems developed for ATRP, ruthenium plays a prominent role (3). $[RuCl_2(PPh_3)_3]$ (**A**, Scheme 1) was the first complex employed for the metal-catalyzed controlled radical polymerization of methyl methacrylate in conjunction with CCl_4 as an initiator in the presence of a Lewis acid as an additive.

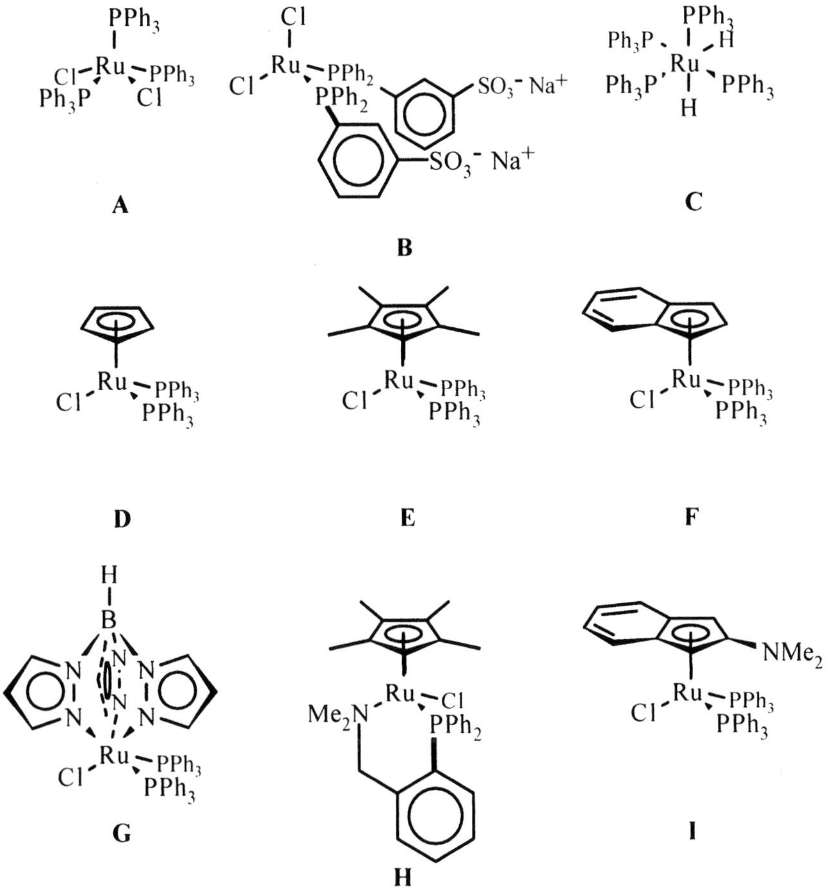

PPh₃
│
Cl‑Ru‑PPh₃
Ph₃P   Cl

A

Cl
│
Cl‑Ru‑PPh₂
      PPh₂ ⟨benzene⟩ SO₃⁻ Na⁺
      ⟨benzene⟩‑SO₃⁻ Na⁺

B

PPh₃
Ph₃P   │   H
      Ru
Ph₃P   │   PPh₃
      H

C

Cl‑Ru‑PPh₃
      PPh₃ (cyclopentadienyl)

D

Cl‑Ru‑PPh₃
      PPh₃ (pentamethylcyclopentadienyl)

E

Cl‑Ru‑PPh₃
      PPh₃ (indenyl)

F

H
│
B
(tris-pyrazolylborate)
Cl‑Ru‑PPh₃
      PPh₃

G

Me₂N‑Ru‑Cl
        PPh₂ (Cp*, P,N-chelating)

H

Cl‑Ru‑PPh₃
      PPh₃ (aminoindenyl, NMe₂)

I

Scheme 1. Representative Sawamoto's ruthenium complexes for ATRP

Since then, the development of new ruthenium catalysts for ATRP has been a dramatic success (Scheme 1). Interestingly, half-sandwich ruthenium complexes (**D-F**) proved very successful, with the indenyl derivative (**F**) providing the fastest controlled radical polymerization of methyl methacrylate *without* Lewis acid activation. Furthermore, addition of an amine such as n-Bu₂NH dramatically increased the rate so that completion of the polymerization was attained in 5 h at 100 °C without broadening of the molecular weight distributions (4). Very recently, Sawamoto disclosed preliminary results obtained using half-metallocene ruthenium complexes with either P,N-chelating (**H**) or aminoindenyl ligands (**I**) (5).

Back in 1999, we found that the 18-electron complex [RuCl₂(p-cymene)(PCy₃)] (p-cymene = 4-isopropyltoluene) (**1**) was a versatile and efficient catalyst precursor for promoting ATRP of vinyl monomers *without* cocatalyst activation (6). We then moved to [RuCl₂(=CHPh)(PCy₃)₂], **2**, the Grubbs' ruthenium benzylidene complex which has had a tremendous impact on olefin metathesis (7). Since initial results showed that complex **2** proved to be highly efficient also for ATRP (6), we expanded our investigations (8) towards N-heterocyclic carbene-containing ruthenium benzylidene complexes, **3** and **4** (9). Later on, we also discovered (independently of Sawamoto) that half-sandwich

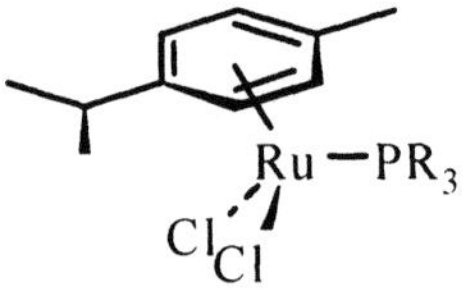

Ru—PR$_3$

Cl
Cl

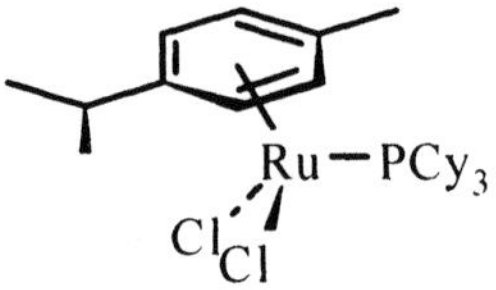

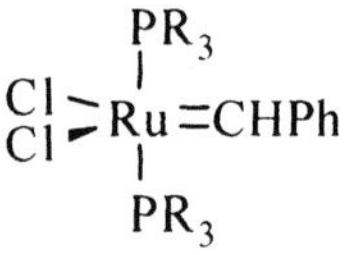

1

Ru—PCy$_3$

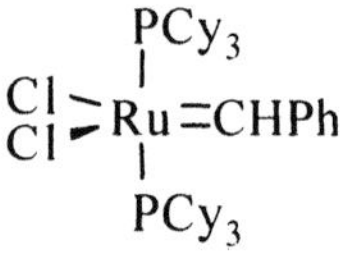

Cl
Cl

PR$_3$
$\begin{array}{c} Cl \\ Cl \end{array}$ Ru=CHPh
PR$_3$

2

PCy$_3$
$\begin{array}{c} Cl \\ Cl \end{array}$ Ru=CHPh
PCy$_3$

3

PCy$_3$
$\begin{array}{c} Cl \\ Cl \end{array}$ Ru=CHPh
R—N$\quad$N—R

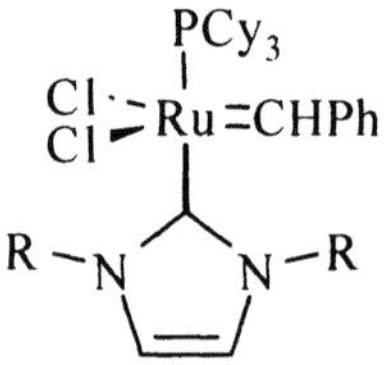

4

R—N$\quad$N—R
$\begin{array}{c} Cl \\ Cl \end{array}$ Ru=CHPh
R—N$\quad$N—R

Activation

k_a

wwC—X $\quad+\quad$ L$_n$Ru^{II} $\quad\rightleftharpoons\quad$ wwC• $\quad+\quad$ L$_n$Ru^{III}X

k_d

Deactivation

dormant
species

active
species

k_p M

polymer

polymer

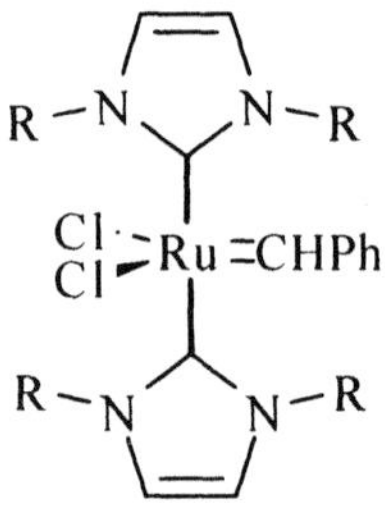

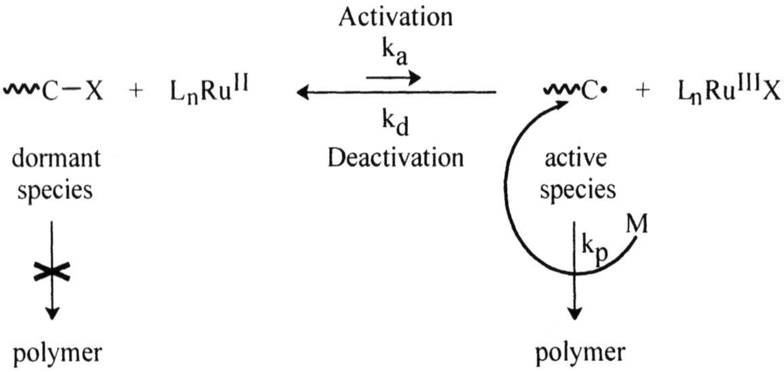

Scheme 2. Generally accepted mechanism for ATRP

ruthenium complexes, [RuCl(Cp#)(PPh₃)₂], were efficient and markedly active catalysts for ATRP (*10*).

Since ATRP is based on a dynamic equilibration between active propagating radicals and dormant species (Scheme 2), it is anticipated that catalytic engineering at the metal center should shift this equilibrium to the most suitable position, so as to maintain a low concentration of propagating radicals while keeping a useful rate of polymerization for polymers to be obtained on a sensible time-scale. To further improve the catalyst efficiency in the ATRP process, we have launched a detailed investigation on the role of the ligands in complexes of the general formula [RuX(Cp#)LL'] (Scheme 3). The present contribution is aiming at illustrating how variation of the cyclopentadienyl-based ligand (Cp#), ancillary ligands LL' (phosphine or carbon monoxide), and X (chloride, hydride, or allyl) influences ATRP of vinyl monomers, such as methyl methacrylate, *n*-butyl acrylate, and styrene.

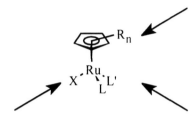

Scheme 3. Tuning of ruthenium-cyclopentadienyl derivatives

Molecular Engineering of [RuX(Cp#)LL'] Complexes

Influence of the Cp# ligand

Nowadays, the factors governing catalyst activity and the mechanism by which half-sandwich ruthenium catalysts perform radical reactions are not known. Investigations were undertaken to address both of these topics in the following way : By varying the ligand sphere around the ruthenium catalyst, we wished to determine how the electronic and steric properties of the ligands affect catalyst activity.

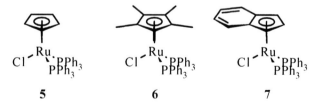

Ruthenium complexes **5–7** and ruthenacarboranes **8** were compared for two reasons : (1) The cyclopentadienyl anion (Cp), its pentamethylcyclopentadienyl (Cp*) and indenyl (Ind) derivatives, and the carboranyl anion present in complexes **8** are isolobal and uninegative ligands. Upon Cp# substitution, it is expected to modify the electronic contributions in these systems. The higher electron donating ability of Cp* compared to Cp is well-established, and the capacity of carboranyl ligands ([C₂B₉H₁₁]²⁻) to stabilize uncommon and high oxidation states of the metals as well. (2) On the other hand, Cp# substitution also results in changing the steric properties of the ligands, which are expressed by the cone angle, θ (Scheme 4). In this way, Cp* is obviously bulkier than Cp and, most probably, than the carboranyl ligand $[C_2B_9H_{11}]^{2-}$, although the relative size of the latter compared to Cp and Cp* is still a question under debate.

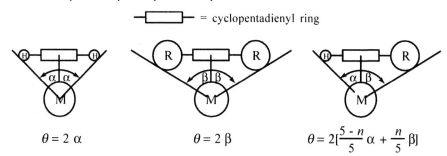

$$\theta = 2\,\alpha \qquad \theta = 2\,\beta \qquad \theta = 2[\frac{5-n}{5}\,\alpha + \frac{n}{5}\,\beta]$$

Scheme 4. Cone angle (θ) for the cyclopentadienyl ring, C_5H_5, a pentasubstituted cyclopentadienyl ring, C_5R_5, and measurement of the average cone angle for an n substituted cyclopentadienyl ring, $C_5H_{5-n}R_n$ (*11*)

The indenyl ligand poses a more complex problem since it is known to undergo a facile metal ring slippage from η^5- to η^3-coordination, leading to the creation of a vacant coordination site on the metal to host an entering ligand or substrate (Scheme 5) (*12*).

η^5-, 18-electron
complex

η^3-, 16-electron
complex

Scheme 5. The indenyl effect

First, half-sandwich ruthenium complexes **5-7** were employed as catalysts to survey the scope of the radical polymerization of vinyl monomers and to better apprehend the role of the Cp$^{\#}$ ligand. Methyl methacrylate (MMA), *n*-butyl acrylate (BA), and styrene (S) were selected as model vinyl monomers. Ethyl 2-bromo-2-methylpropionate, ethyl 2-bromopropionate, and (1-bromoethyl)benzene were used as initiators for the polymerization of MMA, BA, and S, respectively, because of a structure quite comparable to that of the dormant species of PMMA, PBA, and PS chains. Polymerization reactions were carried out *without* cocatalyst at 85 or 110 °C (according to the monomer) for 16 h.

As illustrated in Table 1, [RuCl(Cp*)(PPh₃)₂] (**6**) provided an efficiency that generally surpassed that of related complexes, **5** and **7**. For instance, **6** led to a rather slow polymerization of styrene (27 % yield), but the polymer thus obtained had a very narrow molecular weight distribution (M_w/M_n = 1.10). Complex **6** also induced the controlled polymerization of *n*-butyl acrylate in high yield (91 %) and with a high degree of control (M_w/M_n = 1.20) (*13*). In addition, in both cases, the initiation efficiency (*f*) was close to

Table 1. ATRP of Methyl Methacrylate, n-Butyl Acrylate, and Styrene
Catalyzed by Half-Sandwich Ruthenium Complexes 5-7

Complex	Polymer yield (%)	M_n^d	M_w/M_n^d	f^e
Methyl methacrylate[a]				
5	28	12 000	1.27	0.95
6	19	9 600	1.45	0.80
7	45	21 000	1.30	0.85
n-Butyl acrylate[b]				
5	4	1 500	1.35	0.99
6	91	36 000	1.20	0.95
7	67	46 000	2.15	0.55
Styrene[c]				
5	10	9 000	1.7	0.45
6	27	11 000	1.10	0.95
7	57	35 000	1.55	0.65

[a] $[MMA]_0:[initiator]_0:[Ru]_0 = 800:2:1$ (initiator, ethyl 2-bromo-2-methyl-propionate; temperature, 85 °C; reaction time, 16 h. [b] $[n$-Butyl acrylate$]_0:[initiator]_0:[Ru]_0 = 600:2:1$ (initiator, ethyl 2-bromopropionate; temperature, 85 °C; reaction time, 16 h. [c] $[Styrene]_0:[initiator]_0:[Ru]_0 = 750:2:1$ (initiator, (1-bromoethyl)benzene; temperature, 110 °C; reaction time, 16 h. [d] Determined by size-exclusion chromatography (SEC) with PMMA and polystyrene calibration, respectively. [e] f (initiation efficiency) $= M_{n.th}/M_{n.exp}$ with $M_{n.th} = ([monomer]_0/[initiator]_0) \times MW_{monomer} \times conversion.$

unity as indicated by the observation that the number-average molecular weights (M_n) agreed very well with the calculated values, assuming that one molecule of the initiator generates one living polymer chain. With methyl methacrylate, the molecular weight distribution of the obtained polymer was broader ($M_w/M_n = 1.45$); the GPC curve was bimodal, and also showed additional shoulders.

[RuCl(Cp)(PPh$_3$)$_2$] (5) and [RuCl(Ind)(PPh$_3$)$_2$] (7) were more active than their Cp* analogue (6) for MMA polymerization, and gave narrower molecular weight distributions. The reverse was observed for the polymerization of n-butyl acrylate and styrene, as exemplified by the broadening of the molecular weight distributions for PBA ($M_w/M_n = 2.15$) and PS ($M_w/M_n = 1.55$) obtained using [RuCl(Ind)(PPh$_3$)$_2$] as the catalyst.

For several polymerization reactions, the semi logarithmic plots of monomer conversion against time were linear. The linear plots suggest a first-order kinetic with respect to the monomer concentration and that constant radical concentrations were maintained during the reactions. In other cases, a significant deviation from linearity was observed. As shown in Figure 1, in the early stage of MMA polymerization catalyzed by [RuCl(Ind)(PPh$_3$)$_2$] the $\ln([M]_0/[M])$ vs. time plot is linear ($y = 0.016087 + 0.077936\ x$; $r^2 = 0.981$, for the first five hours of reaction), indicating that the radical concentration remained constant. However, the plot begins to deviate from the first-order kinetic and slows down at approximately 30 % conversion. This suggests that there is a higher contribution of termination reactions, which results from an increase in [Ru(III)] and radical concentration.

The molecular weight and polydispersity of PMMA as the functions of monomer conversion are shown in Figure 1. The molecular weight increases linearly with conversion ($y = 418.54 + 454.18\ x$; $r^2 = 0.992$), while M_w/M_n are around 1.2-1.3, reaching a minimum of 1.17 at approximately 25 % conversion.

122

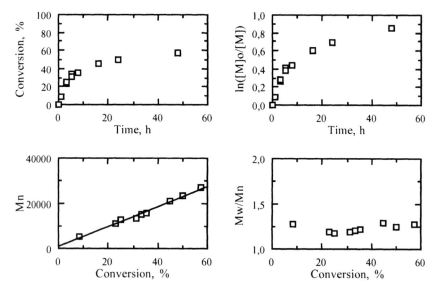

Figure 1. *Polymerization of methyl methacrylate initiated by ethyl 2-bromo-2-methylpropionate and catalyzed by [RuCl(Ind)(PPh₃)₂] (7). at 85 °C (Reaction conditions same as in Table 1).*

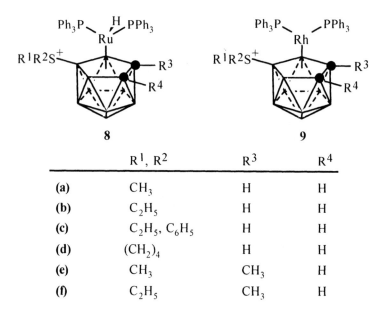

	R^1, R^2	R^3	R^4
(a)	CH_3	H	H
(b)	C_2H_5	H	H
(c)	C_2H_5, C_6H_5	H	H
(d)	$(CH_2)_4$	H	H
(e)	CH_3	CH_3	H
(f)	C_2H_5	CH_3	H

Table 2. ATRP of Methyl Methacrylate, n-Butyl Acrylate, and Styrene Catalyzed by Ruthena- (8) and Rhodacarborane (9) Complexes

Complex	Polymer yield (%)	M_n^d	M_w/M_n	Complex	Polymer yield (%)	M_n^d	M_w/M_n
Methyl methacrylate[a]							
8a	93	25 700	1.65	9a	93	18 300	1.95
8b	93	25 200	1.6	9b	93	14 000	1.85
8c	88	18 100	1.65	9c	99	25 700	1.9
8d	88	23 400	1.7	9d	84	14 800	1.9
8e	94	25 300	1.75	9e	86	20 300	1.85
8f	90	28 400	1.65	9f	90	20 800	1.85
n-Butyl acrylate[b]							
8a	99	23 100	6.7	9a	99	23 600	4.0
8b	99	56 000	7.1	9b	97	34 000	1.9
8c	97	22 800	1.7	9c	97	36 500	3.0
8d	97	22 500	7.3	9d	99	21 800	3.9
8e	99	78 000	6.1	9e	95	56 000	5.6
8f	99	23 600	5.3	9f	98	40 000	5.1
Styrene[c]							
8a	88	21 200	3.05	9a	65	6 700	2.8
8b	25	8 200	1.5	9b	51	11 500	2.15
8c	40	3 900	1.8	9c	69	18 200	1.85
8d	22	9 900	2.1	9d	45	14 800	2.05
8e	25	10 300	2.1	9e	77	16 000	1.9
8f	54	21 000	2.7	9f	83	25 600	1.6

[a,b,c,d] Reaction conditions same as in Table 1, except for n-butyl acrylate : [n-butyl acrylate]$_0$: [initiator]$_0$:[complex]$_0$ = 1200:4:1, instead of 600:2:1 (Table 1).

To further enlarge the set of $Cp^{\#}$ ligands and to better apprehend their impact on the catalytic process, ruthenacarboranes (8) were prepared and tested in ATRP (Table 2). With MMA and BA, complexes 8 were more active than complexes 5-7. With styrene, however, they displayed similar activities, and polymerizations proceeded smoothly. In addition, ruthenacarboranes generally gave polymers with uncontrolled molecular weights and broad (or very broad) molecular weight distributions. In some cases, the GPC curves of the polymers clearly showed some shoulder(s) at higher or lower elution times compared to the main peak, or both (Figure 2). In few cases, especially with PBA, a bimodal distribution was evidenced.

Rhodacarboranes (9) also served as catalysts for ATRP. They behaved similarly to ruthenacarboranes (Table 2). They were very active, even at a temperature as low as 30 °C. However, whatever the temperature, polymerizations were never controlled and polydispersities were broad. Quite interesting was the influence of the temperature on both M_n and M_w/M_n (Table 3). At 30 °C, M_n was quite high ($\sim$ 350 000), indicating a low initiation efficiency. As the temperature increased, M_n decreased and M_w/M_n became broader, reaching 3.8 at 75 °C.

Rhodacarboranes also displayed unexpected reactivity profiles. As shown in Figure 3, styrene polymerization started immediately, then reached a plateau for 6-7 h corresponding to approximately 17 % conversion, after which the reaction rate was first-order with respect to the monomer ($y = -0.01124 + 0.054674\ x$; $r^2 = 0.989$).

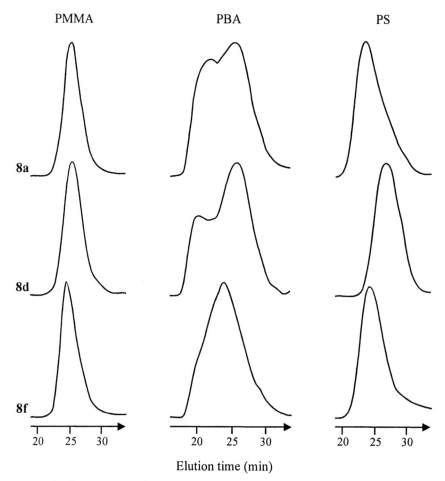

Figure 2. GPC curves of poly(methyl methacrylate)s, poly(n-butyl acrylate)s, and polystyrenes obtained using complexes **8a, 8d,** and **8f** (Reaction conditions same as in Table 2).

Furthermore, the polymerization was not controlled in terms of molecular weights and molecular weight distributions. M_n slightly decreased, whereas M_w/M_n remained constant (2.5-2.9) with monomer conversion (Figure 3).

Influence of the Phosphine Ligand

To further expand the potentials of the original $[RuCl(Cp^*)(PPh_3)_2]$ catalyst (**6**), we anticipated that the dynamic equilibration between active propagating radicals and dormant species (as sketched in Scheme 2) should also be fine-tuned through modification of the phosphine ligands. Thus, we replaced the PPh_3 ligands in complex **6** by isosteric p-substituted triarylphosphines so as to modify the electronic properties of the phosphine while maintaining the cone angle constant at 145°.

Table 3. ATRP of *n*-Butyl Acrylate Catalyzed by Rhodacarborane 9d[a]

Temperature (°C)	Polymer yield (%)	M_n[b]	M_w/M_n
30	72	340 000	2.45
45	90	135 000	2.75
60	96	55 000	3.4
75	99	23 800	3.75
85	99	21 800	3.9

[a,b] Reaction conditions same as in Table 2.

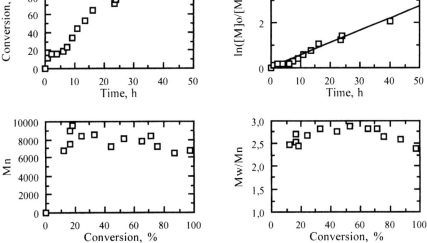

Figure 3. *Polymerization of styrene initiated by (1-bromoethyl)benzene and catalyzed by rhodacarborane* ***9a****, at 110 °C (Reaction conditions same as in Table 1).*

Two phosphines were selected on the basis of their electron-donating ability : tris-(4-trifluoromethylphenyl)phosphine and tris(4-methoxyphenyl)phosphine. The phosphine ability to donate σ electrons to the metal may be estimated by the basicity of the free ligand, expressed as the pK_a value for the conjugate acid (HPR_3^+) (*14*):

$$P(p\text{-}CF_3C_6H_4)_3 \ (-1.75) \ < \ PPh_3 \ (2.73) \ < \ P(p\text{-}CH_3OC_6H_4)_3 \ (4.59)$$

An examination of the results collected in Table 4 reveals that the ruthenium-tris-(4-methoxyphenyl)phosphine complex (**11**) was more active than the parent complex (**6**). With *n*-butyl acrylate and styrene, this was at the expense of the polydispersity which was slightly higher. With methyl methacrylate, however, the molecular weight distribution was narrower when using complex **11** (M_w/M_n = 1.32) as the catalyst instead of **6** (M_w/M_n = 1.45). In addition, the GPC curve of PMMA obtained using complex **11** was monomodal, whereas it was bimodal when using **6**. The ruthenium-tris(4-trifluoromethyl-phenyl)phosphine complex (**10**) displayed a reactivity similar to that of **6**.

$$6 \quad (Ar = C_6H_5)$$
$$10 \quad (Ar = p\text{-}CF_3\text{-}C_6H_4)$$
$$11 \quad (Ar = p\text{-}CH_3O\text{-}C_6H_4)$$

Table 4. ATRP of Methyl Methacrylate, n-Butyl Acrylate, and Styrene Catalyzed by Ruthenium Pentamethylcyclopentadienyl Complexes 6, 10, and 11

Complex	Polymer yield (%)	M_n^d	M_w/M_n^d	f^e
Methyl methacrylate[a]				
6	19	9 600	1.45	0.80
10	25	13 500	1.19	0.75
11	72	30 000	1.32	0.95
n-Butyl acrylate[b]				
6	91	36 000	1.20	0.95
10	96	37 500	1.20	0.99
11	96	37 500	1.28	0.99
Styrene[c]				
6	27	11 000	1.10	0.95
10	21	10 000	1.09	0.80
11	52	21 000	1.14	0.95

[a-c] Reaction conditions same as in Table 1.

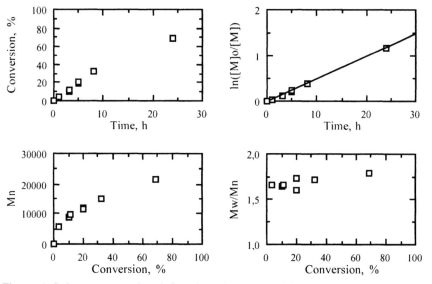

Figure 4. *Polymerization of methyl methacrylate initiated by ethyl 2-bromo-2-methyl-propionate and catalyzed by [RuCl(Ind)(PPh₃)₂] (7), at 85 °C, in the presence of 1 equivalent of PPh₃ (Reaction conditions same as in Table 1).*

The only exception was observed with methyl methacrylate : The molecular weight distribution of the obtained PMMA was narrower with **10** (M_w/M_n = 1.19) than with **6** (M_w/M_n = 1.45).

These results indicate that electron-donating phosphines such as P(p-CH$_3$OC$_6$H$_4$)$_3$ exert a significant role in stabilizing the Ru(III) intermediate species (Scheme 2). They also suggest that catalyst's engineering and fine-tuning can easily be achieved by using suitable and readily available ligands.

Previous studies in Kharasch chemistry (*15*) suggested that the release of a phosphine ligand from [RuCl(Cp*)(PAr$_3$)$_2$] was of utmost importance and most likely occurred prior to the activation of the dormant species by the unsaturated ruthenium center. With this in mind, kinetics of the polymerization of MMA involving [RuCl(Ind)(PPh$_3$)$_2$] in the presence of an excess of PPh$_3$ were undertaken. In the presence of 1 equivalent of PPh$_3$ with respect to ruthenium (Figure 4), the polymerization was slower (k_{app} = 1.38 x 10^{-5} s^{-1}) than without added PPh$_3$ (k_{app} = 2.16 x 10^{-5} s^{-1}). The semi logarithmic plot of MMA conversion against time was linear (y = - 0.018232 + 0.049556 x; r^2 = 0.999) throughout the reaction. Interestingly, the M_n vs. conversion plot showed a significant deviation from linearity, and the molecular weight distribution became broader (~ 1.7 instead of 1.3 without added PPh$_3$).

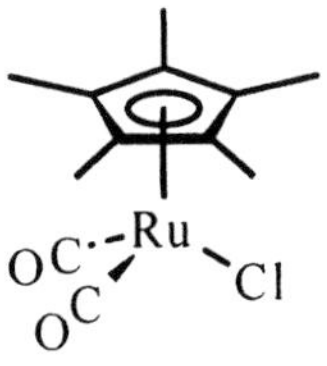

12

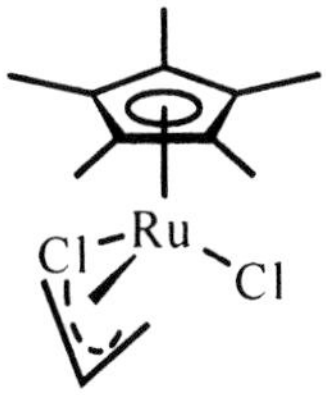

16

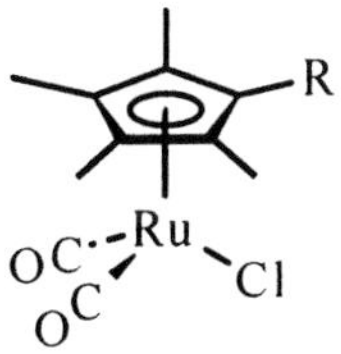

13 (R = C(O)-C$_6$H$_5$)
14 (R = CH=CH-C(O)-CH$_3$)

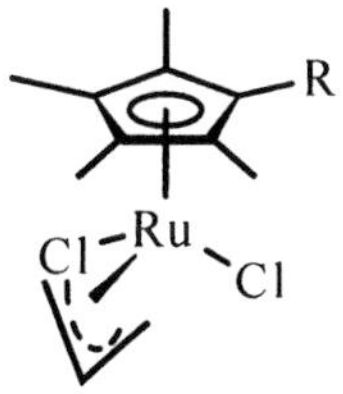

17 (R = C(O)-C$_6$H$_5$)
18 (R = CH=CH-C(O)-C$_6$H$_5$)

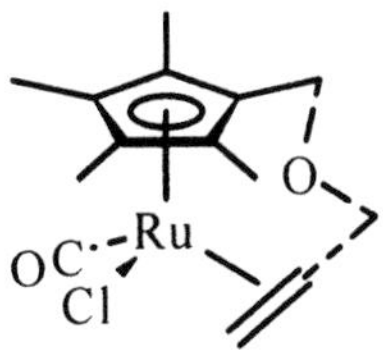

15

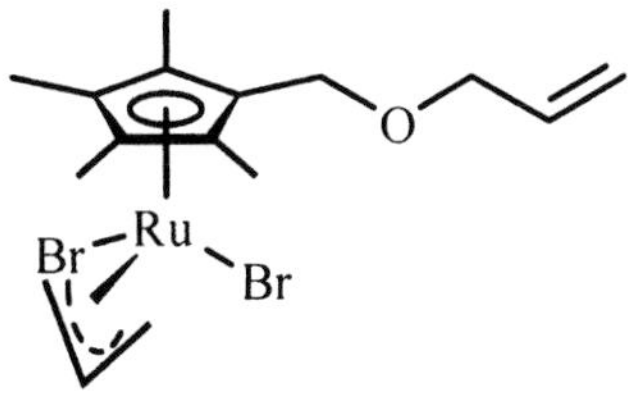

19

20

Table 5. ATRP of Methyl Methacrylate and Styrene Catalyzed by Ruthenium Complexes 12-20[a]

Complex	Methyl methacrylate			Styrene		
	Polymer yield (%)	M_n^b	M_w/M_n	Polymer yield (%)	M_n^b	M_w/M_n
12	1	-	-	50	25 000	1.85
13	1	-	-	40	56 000	3.4
14	3	145 000	2.2	21	29 000	2.05
15	16	70 000	2.05	33	20 000	1.95
16	1	9 500	1.75	30	40 000	1.95
17	4	13 000	1.7	5	2 600	2.45
18	1	31 000	1.7	1	-	-
19	1	-	-	5	12 500	1.75
20	25	480 000	2.15	38	55 000	2.1

[a, b] Same as in Table 1.

In order to assess the importance of the release of the phosphine ligand, we replaced PPh_3 by the carbonyl ligand, which forms stronger bonds to metal centers than most phosphines. Thus, we investigated a new class of ruthenium(II) complexes, $[RuCl(Cp^\#)(CO)_2]$, bearing two carbon monoxide ligands and Cp* or various substituents tethered $Cp^\#$ ligands (Table 5). Complexes **12-15** and their cationic derivative (**20**) proved to be poorly active, and molecular weight distributions were broad. As expected, related ruthenium(IV) complexes (**16-19**) were practically inactive under ATRP conditions.

Conclusions

In exploring the reactivity of many analogous $[RuCl(Cp^\#)(PR_3)_2]$ complexes for ATRP of methyl methacrylate, *n*-butyl acrylate, and styrene, it was found that $[RuCl(Cp^*)(PPh_3)_2]$ and $[RuCl(Ind)(PPh_3)_2]$ were highly efficient catalysts. Highly electron-donating ligands such as carboranes led to inefficient catalysts. As indicated by several experimental facts, the mechanism would involve the dissociation of a phosphine ligand, giving rise to a 16-electron species at which activation of the carbon-halogen bond of the initiator and of the dormant species could take place.

Acknowledgments

This work was financially supported by a grant-in-aid (O.T.) joined to the project MAT98-0921 from CICYT (Spain). Part of this work has also been carried out in the

129

framework of the COST D17/0006/00 and TMR-HPRN CT 2000-10 'Polycat' programs (European Union). The Belgian team is also grateful to the 'Fonds National de la Recherche Scientifique' (F.N.R.S.), Brussels, for the purchase of major instrumentation, and the 'Région wallonne' (Programme FIRST Europe) for a fellowship to A.R.

References

1. Kato, M.; Kamigaito, M.; Sawamoto, M.; Higashimura, T. *Macromolecules* **1995**, *28*, 1721.
2. Wang, J.-S.; Matyjaszewski, K. *J. Am. Chem. Soc.* **1995**, *117*, 5614.
3. (a) Sawamoto, M.; Kamigaito, M. *Chemtech* **1999**, *29(6)*, 30. (b) Kamigaito, M.; Ando, T.; Sawamoto, M. *Chem. Rev.* **2001**, *101*, 3689.
4. (a) Hamasaki, S.; Kamigaito, M.; Sawamoto, M. *Macromolecules* **2002**, *35*, 2934. (b) Hamasaki, S.; Sawauchi, C.; Kamigaito, M.; Sawamoto, M. *J. Polym. Sci., Part A: Polym. Chem.* **2002**, *40*, 617.
5. (a) Sawamoto, M.; Kamigaito, M.; Ando, T. *Macro Group UK*, University of Warwick, July 29th - August 1st, 2002, IL42. (b) Kamigaito, M.; Ando, T.; Sawamoto, M. *224th ACS National Meeting*, August 18-22, 2002, POLY 34. (c) Kamigaito, M.; Ando, T.; Sawamoto, M. *Polym. Prepr.* **2002**, *43(2)*, 3.
6. (a) Simal, F.; Demonceau, A.; Noels, A. F. *Angew. Chem.* **1999**, *111*, 559; *Angew. Chem. Int. Ed.* **1999**, *38*, 538. (b) Simal, F.; Jan, D.; Delaude, L.; Demonceau, A.; Spirlet, M.-R.; Noels, A. F. *Can. J. Chem.* **2001**, *79*, 529.
7. Trnka, T. M.; Grubbs, R. H. *Acc. Chem. Res.* **2001**, *34*, 18.
8. Simal, F.; Delfosse, S.; Demonceau, A.; Noels, A. F.; Denk, K.; Kohl, F. J.; Weskamp, T.; Herrmann, W. A. *Chem. Eur. J.* **2002**, *8*, 3047.
9. Weskamp, T.; Schattenmann, W. C.; Spiegler, M.; Herrmann, W. A. *Angew. Chem.* **1998**, *110*, 2631; *Angew. Chem. Int. Ed.* **1998**, *37*, 2490.
10. Simal, F. PhD Thesis, University of Liège, **2000**.
11. White, D.; Coville, N. J. *Adv. Organomet. Chem.* **1994**, *36*, 95.
12. Calhorda, M. J.; Romão, C. C.; Veiros, L. F. *Chem. Eur. J.* **2002**, *8*, 868.
13. Watanabe, Y.; Ando, T.; Kamigaito, M.; Sawamoto, M. *Macromolecules* **2001**, *34*, 4370.
14. Rahman, Md. M.; Liu, H.-Y.; Eriks, K.; Prock, A.; Giering, W. P. *Organometallics* **1989**, *8*, 1.
15. (a) Simal, F.; Wlodarczak, L.; Demonceau, A.; Noels, A. F. *Tetrahedron Lett.* **2000**, *41*, 6071. (b) Simal, F.; Wlodarczak, L.; Demonceau, A.; Noels, A. F. *Eur. J. Org. Chem.* **2001**, 2689.

Chapter 10

Toward Structural and Mechanistic Understanding of Transition Metal-Catalyzed Atom Transfer Radical Processes

Tomislav Pintauer, Blayne McKenzie, and Krzysztof Matyjaszewski

Department of Chemistry, Carnegie Mellon University, 4400 Fifth Avenue, Pittsburgh, PA 15213

Summary: Structural and mechanistic aspects of transition metal catalyzed atom transfer radical polymerization (ATRP) were discussed. Structures of Cu^I and Cu^{II} active ATRP catalysts were investigated using various spectroscopic techniques and were found to be dependent on the nature of the complexing ligand, solvent and temperature. The overall equilibrium constant for ATRP was correlated with the redox potentials of Cu^I complexes with nitrogen based ligands and equilibrium constant for halogen dissociation from the corresponding Cu^{II}-X complexes. Additionally, various techniques to measure the kinetics of elementary reactions in the ATRP such as activation, deactivation and initiation rate constants were presented.

Introduction and Background

The synthesis of macromolecules with well-defined compositions, architectures and functionalities represents an ongoing effort in the field of polymer chemistry. Over the past few years, atom transfer radical polymerization (ATRP) has emerged as a very powerful and robust technique to meet these goals (*1-3*). The basic working mechanism of ATRP (Scheme 1) involves a reversible switching between two oxidation states of a transition metal complex (*4,5*). It originates from Atom Transfer Radical Addition (ATRA), which is a well known and widely used reaction in organic synthesis (*6,7*).

Scheme 1. Proposed Mechanism for ATRP.

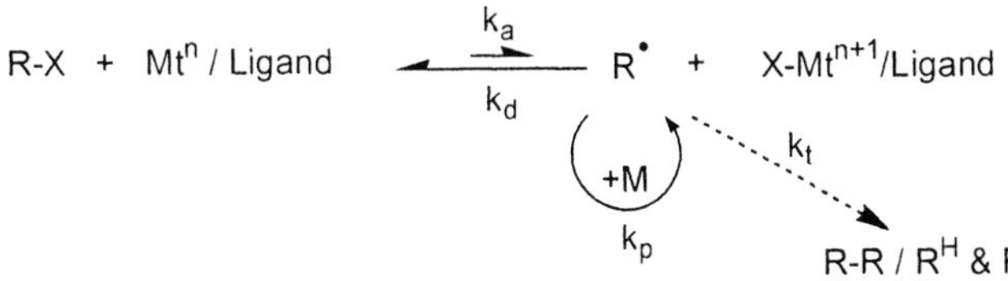

$$R\text{-}X \; + \; Mt^n / Ligand \; \underset{k_d}{\overset{k_a}{\rightleftharpoons}} \; R^\bullet \; + \; X\text{-}Mt^{n+1}/Ligand$$

$$\left(\;_{+M}\;\right) \overset{k_p}{\underset{}{}} \qquad \overset{k_t}{\searrow} \qquad R\text{-}R \,/\, R^H \; \& \; R^=$$

Homolytic cleavage of the alkyl halogen bond (R-X) by a transition metal complex in the lower oxidation state generates an alkyl radical and a transition metal complex in the higher oxidation state. The formed radicals can initiate the polymerization by adding across the double bond of a vinyl monomer, propagate, terminate by either coupling or disproportionation, or be reversibly deactivated by the transition metal complex in the higher oxidation state. The thermodynamics and kinetics of the atom transfer step defines the control over final polymer structure, namely molecular weight, polydispersity and end functionality (*8*).

Structural and mechanistic studies are crucial in further understanding of this mechanism and are inherently part of the future developments in the ATRP. The important factors that need to be considered in the structural aspects of the ATRP include the structures of the catalysts in solution and their solvent and temperature dependence, the role of complexing ligand and how does it effect the catalyst properties (e.g. redox potential), participation of the catalyst in side reactions other than atom transfer, and characterization of other active intermediates (e.g. radicals). The mechanistic studies, on the other hand, should aim at determining the rate constants for elementary reactions occurring in the ATRP such as activation, deactivation and initiation, and more importantly correlate them with reaction parameters such as catalyst, alkyl halide and monomer structure, temperature and solvent. Such studies can lead to a development of optimal catalysts and generally improve the overall catalytic process.

In this article, we present on overview of the structural and mechanistic aspects of transition metal catalyzed ATRP, with an emphasis on copper catalyzed ATRP.

Structural Aspects of Transition Metal Catalyzed ATRP

General Structural Features of the ATRP Catalysts

Structural characterization of the ATRP active transition metal complexes plays an important role in the overall catalytic process and still remains a very challenging task. The catalyst consists of a transition metal center accompanied by a complexing ligand and counterion which can form a covalent or ionic bond with the metal center. The efficient catalyst should be able to expand its coordination sphere and oxidation number upon halogen abstraction from alkyl halide or dormant polymer chains. Additionally, the catalyst should not participate in any side reactions which would lower its activity or change the radical nature of the ATRP process. The concurrent reactions which can occur in the ATRP include: (a) monomer, solvent or radical coordination, (b) oxidation/reduction of radicals to radical cations/anions, respectively, (c) β-hydrogen abstraction, (d) disproportionation, etc. So far, a variety of transition metal complexes have been successfully used in the ATRP. They include compounds from Group 6 (Mo (9)), 7 (Re (10)), 8 (Fe (11-13), Ru (14)), 9 (Rh (15)), 10 (Ni (16), Pd (17)) and 11 (Cu (4,18,19)). The following discussion will concentrate on the copper complexes with nitrogen based ligands commonly used for ATRP in our laboratories.

Structural Features of Cu^I Complexes in ATRP

A variety of structural techniques have been used to study ATRP active Cu^I and Cu^{II} complexes in both solid state and solution. They included solid state X-ray crystallography (20,21), extended X-ray absorption fine structure (EXAFS) (22,23), cyclic voltammetry (24), electrospray ionization mass spectrometry (ESI-MS) (25), EPR (26,27), and UV-Vis spectroscopy (21). The catalysts typically consist of a copper(I) halide accompanied by a nitrogen based complexing ligand. Variety of bidentate, tridentate and tetradentate nitrogen based ligands have been utilized in our laboratories and are summarized in

Scheme 2. In addition, ligands such as pyridineimines (*18*) and phenanthrolines (*28*) have also been used. Cu^I complexes with bpy based ligands have been characterized by solid state X-ray crystallography and typically consist of distorted tetrahedral $[Cu^I(bpy)_2]^+$ cations (Figure 1) accompanied by a non coordinating anions (BF_4^-, ClO_4^-, Br^-, $Cu^IBr_2^-$) (*29-32*). The structure of the complex is strongly dependent on the solvent polarity and temperature. For

Scheme 2. Nitrogen based ligands typically used in copper mediated ATRP.

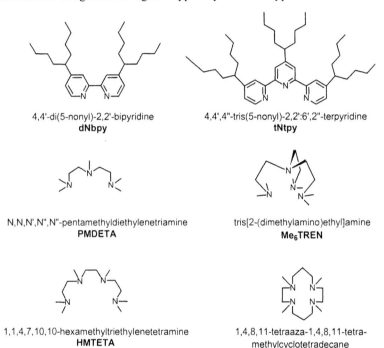

4,4'-di(5-nonyl)-2,2'-bipyridine
dNbpy

4,4',4"-tris(5-nonyl)-2,2':6',2"-terpyridine
tNtpy

N,N,N',N",N"-pentamethyldiethylenetriamine
PMDETA

tris[2-(dimethylamino)ethyl]amine
Me₆TREN

1,1,4,7,10,10-hexamethyltriethylenetetramine
HMTETA

1,4,8,11-tetraaza-1,4,8,11-tetra-methylcyclotetradecane
Me₄CYCLAM

example, Cu^IBr complex with 2 eq. of dNbpy in non polar medium such as toluene, methyl acrylate or styrene predominantly exists as $[Cu^I(dNbpy)_2][Cu^IBr_2]$ (*22*). In polar medium such as MeOH, $[Cu^I(dNbpy)_2][Br]$ is preferred (Table 1) (*23*). Additional structures have also been proposed in literature and include neutral $[Cu^I(dNbpy)(Br)]_2$ complex (*33*). Tetradentate Me₆TREN, HMTETA and Me₄CYCLAM ligands are also expected to form four coordinated complexes with Cu^I (Figure 1).(*34,35*) The situation with tridentate PMDETA and tNtpy ligands is different because the 18 electron

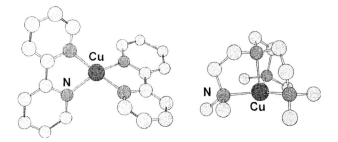

Figure 1. Structures of [Cu^I(bpy)$_2$] and [Cu^I(Me$_6$TREN)] cations in [Cu^I(bpy)$_2$] [ClO$_4$] (Ref. (31)) and [Cu^I(Me$_6$TREN)] [ClO$_4$] (Ref. (35)), respectively..

Table 1. Structural parameters of Cu^{I}Br/2dNbpy complex, determined by EXAFS measurements under ambient conditions at the Cu K- and Br K-edge.[a]

Solvent	Backscatt.	N	r [Å]	Proposed Structure
Toluene	Cu-N	3.2(5)	2.02(2)	[Cu^I(dNbpy)$_2$]$^+$[Cu^{I}Br$_2$]$^-$
	Cu-Br	1.8(3)	2.25(2)	
	Br-Cu	1.0(2)	2.23(2)	
Sty	Cu-N	2.9(4)	2.03(2)	[Cu^I(dNbpy)$_2$]$^+$[Cu^{I}Br$_2$]$^-$
	Cu-Br	1.4(2)	2.25(2)	
	Br-Cu	0.8(1)	2.24(2)	
MA	Cu-N	2.8(4)	2.01(2)	[Cu^I(dNbpy)$_2$]$^+$[Cu^{I}Br$_2$]$^-$
	Cu-Br	1.3(2)	2.26(2)	
	Br-Cu	0.8(1)	2.25(2)	
MeOH	Cu-N	3.3(5)	2.02(2)	[Cu^I(dNbpy)$_2$]$^+$[Br]$^-$
	Cu-Br	0.0	/	
	Br-Cu	0.0	/	

[a]N = coordination number, r = absorber-backscatterer distance.

rule would leave an open coordination sphere which can be occupied by coordinating anions such as Br$^-$ or Cl$^-$. On the other hand, in the case of noncoordinating anions such as BF$_4^-$, BPh$_4^-$ or ClO$_4^-$, the empty coordination sphere can be occupied by the solvent or monomer molecules. The monomer coordination was demonstrated recently in characterization of [Cu^I(PMDETA)(π-MA)][BPh$_4$] and [Cu^I(PMDETA)(π-Sty)][BPh$_4$] complexes (36). ^{1}H NMR studies (29) of some Cu^I complexes with nitrogen based ligands also indicated a rapid ligand exchange reactions, which in the presence of coordinating solvent or monomer can give rise to other active species that can coexist in the ATRP medium.

Structural Features of Cu[II] Complexes in ATRP

Copper(II) complexes that are active in the ATRP have been characterized using a variety of spectroscopic techniques (20,21,23). The Cu[II] complexes showed either a trigonal bipyramidal structure in the case of dNbpy ligand ([Cu[II](dNbpy)$_2$Br][Br]), or a distorted square pyramidal coordination in the case of triamines and tetramines ([Cu[II](PMDETA)Br$_2$], [Cu[II](tNtpy)Br$_2$], [Cu[II](Me$_6$TREN)][Br], Cu[II](HMTETA)Br][Br] and [Cu[II](Me$_4$CYCLAM)Br][Br]) (20). Depending on the type of the amine ligand, the complexes were either neutral (triamines) or ionic (bpy and tetramines). The counterions in the case of the ionic complexes were either bromide (Me$_4$CYCLAM and HMTETA) or the linear [Cu[I]Br$_2$]⁻ (dNbpy). No direct correlation was found between the Cu[II]-Br bond length and the deactivation rate constant in the ATRP, which suggested that other parameters such as the entropy for the structural reorganization between the Cu[I] and Cu[II] complexes might play an important role in determining the overall activity of the catalyst in the ATRP.

Additionally, EXAFS (23) and UV (37) measurements indicated bromide dissociation in polar medium such as MeOH or H$_2$O. The EXAFS data at the Cu and Br K-edges for copper(II) complexes are summarized in Table 2. The dissociation of bromide anions from copper(II) centers is evident from a decrease in the Br/Cu coordination numbers. This reaction might be responsible for the fast ATRP in polar medium since the concentration of the deactivator is significantly lower than in nonpolar medium.

Correlating Redox Potential with Catalyst Activity

The coordination of the ligand is inherently related to the ability of metal complexes to be oxidized/reduced by the corresponding R-X (38). The overall equilibrium constant for ATRP (Scheme 3) can be expressed as the product of the equilibrium constants for electron transfer between metal complex (K_{ET}), electron affinity of the halogen (K_{EA}), bond homolysis of the alkyl halide (K_{BH}) and heterolytic cleavage of the Mt^{n+1}-X bond or "halogenophilicity" (K_{HP}). Therefore, for a given alkyl halide R-X, the activity of the catalyst in the ATRP depends not only on the redox potential, but also on the halogenophilicity of the transition metal complex. Both parameters are affected by the nature of the transition metal and ligand, including the binding constants, basicity, back-

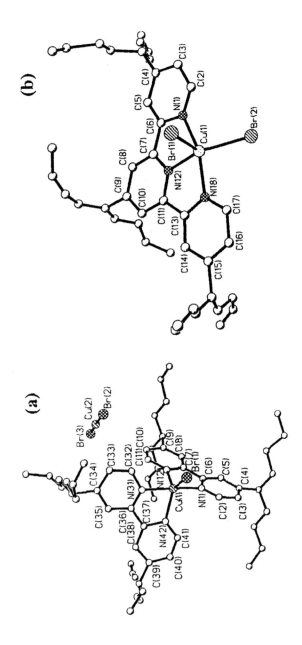

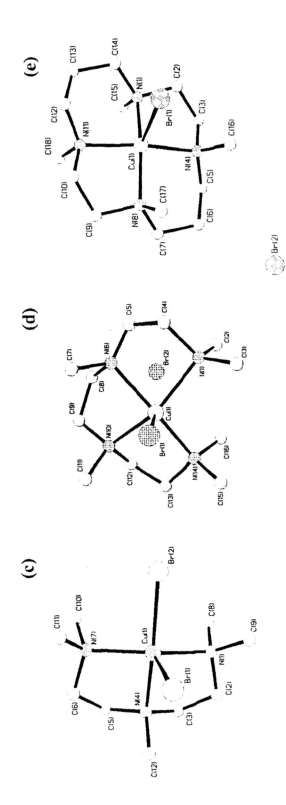

Figure 2. Molecular structures of copper(II) complexes precipitated from ATRP solutions. $[Cu^{II}(dNbpy)_2Br]^+[Cu^IBr_2]^-$ (a), $[Cu^{II}(tNtpy)Br_2]$ (b), $[Cu^{II}(PMDETA)Br_2]$ (c), $[Cu^{II}(HMTETA)Br]^+[Br]^-$ (d) and $[Cu^{II}(Me_4CYCLAM)Br]^+[Br]^-$ (e).

138

Table 2.   Room Temperature EXAFS Measurements of Cu^{II}Br$_2$ Complexes with bpy, PMDETA, and Me$_6$TREN Ligands.[a]

Complex	Solvent	Backscatt.	N	r [Å]
Cu^{II}Br$_2$/2dNbpy	Solid[b]	Cu-N	4.0	2.02(2)
		Cu-Br	1.0	2.41(2)
		Br-Cu	1.0	2.41(2)
	MeOH	Cu-N	2.8(4)	2.03(2)
		Cu-Br	0.6(1)	2.43(2)
		Br-Cu	0.3(0)	2.42(2)
Cu^{II}Br$_2$/PMDETA	Solid[b]	Cu-N	3.0	2.09(2)
		Cu-Br	2.0	2.44(2), 2.64(2)
		Br-Cu	1.0	2.42(2), 2.63(2)
	MeOH	Cu-N	3.1(5)	2.09(2)
		Cu-Br	1.1(2)	2.41(2)
		Br-Cu	0.4(1)	2.39(2)
	H$_2$O	Cu-N	4.0(6)	2.06(2)
		Cu-Br	0.0	/
Cu^{II}Br$_2$/Me$_6$TREN	Solid[b]	Cu-N	4.0	2.09(2)
		Cu-Br	1.0	2.41(2)
		Br-Cu	1.0	2.40(2)
	MeOH	Cu-N	3.2(5)	2.13(2)
		Cu-Br	1.4(2)	2.38(2)
		Br-Cu	0.9(1)	2.38(2)
	H$_2$O	Cu-N	4.1(6)	2.06(2)
		Cu-Br	0.3(0)	2.42(2)
		Br-Cu	0.0	/

[a]N = coordination number, r = absorber-backscatterer distance.
[b]Coordination numbers were fixed to the crystallographically known values for [Cu^{II}(bpy)$_2$Br]$^+$[Br]$^-$, [Cu^{II}(PMDETA)Br$_2$] and [Cu^{II}(Me$_6$TREN)Br]$^+$[Br]$^-$.

Scheme 3. Representation of atom transfer equilibrium by redox processes, homolytic dissociation of alkyl halide and heterolytic cleavage of Mt^{n+1}-X bond.

Atom Transfer (Overall Equilibrium)

$$R\text{-}X \;+\; Cu^I\text{-}Y\,/\,Ligand \;\underset{k_d}{\overset{k_a}{\rightleftharpoons}}\; R^{\bullet} \;+\; X\text{-}Cu^{II}\text{-}Y/Ligand$$

Contributing Reactions

$$Cu^I\text{-}Y\,/\,Ligand \;\underset{}{\overset{K_{ET}}{\rightleftharpoons}}\; Cu^{II}\text{-}Y/Ligand \;+\; e^{\ominus}$$

$$X^{\bullet} \;+\; e^{\ominus} \;\underset{}{\overset{K_{EA}}{\rightleftharpoons}}\; X^{\ominus}$$

$$R\text{-}X \;\underset{}{\overset{K_{BH}}{\rightleftharpoons}}\; R^{\bullet} \;+\; X^{\bullet}$$

$$X^{\ominus} \;+\; Cu^{II}\text{-}Y/Ligand \;\underset{}{\overset{K_{HP}}{\rightleftharpoons}}\; X\text{-}Cu^{II}\text{-}Y/Ligand$$

$$K_{ATRP} = \frac{k_a}{k_d} = K_{EA}K_{BH}K_{HP}K_{ET} \quad \text{or} \quad \frac{K_{ATRP}}{K_{EA}K_{BH}} = K_{ET}K_{HP} \quad (1)$$

bonding, steric effects, etc. For a complexes that have similar halogenophilicities, the redox potential can be used as a measure of catalyst activity in the ATRP. This was demonstrated in the linear correlation between K_{ATRP} and $E_{1/2}$ (Figure 2) for a series of Cu^I complexes with nitrogen based ligands (24,37). However, redox potential might not be sufficient to compare

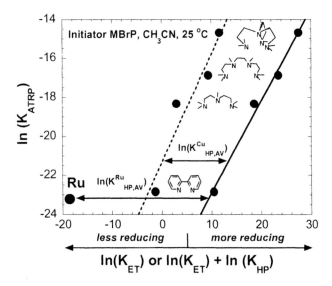

Figure 2. A plot of ln(K_{ATRP}) vs. ln (K_{ET}) (broken line) and ln(K_{ET}) + ln(K_{HP}) (solid line) for methyl 2-bromopropionate in CH_3CN at RT.

K_{ATRP} for different transition metals since they are expected to have different halogenophilicities (eg. Ru (39) vs. Cu).

Characterization of Radical Intermediates in the ATRP

Characterization of radical intermediates in the ATRA/ATRP still remains a challenging task. The principal problem lies in the fact that radicals in the ATRP are present at very small concentrations. Nonetheless, the presence of radical intermediates in the ATRP has been deduced indirectly based on several experimental observations: (1) radical inhibitors, such as galvinoxyl and TEMPO, inhibit the polymerization, (2) ATRP can be carried out in the presence of H_2O (40) and is tolerant to a variety of functionalities (8), (3) the reactivity ratios are similar to those reported for the conventional free radical polymerizations ($41,42$), (4) reverse ATRP (43), and (5) similar rates of racemization, halogen exchange and trapping reactions (44). Recently, radicals have been detected by EPR in a network-forming ATRP of poly(ethylene glycol) dimethylacrylate (45). The study of the reactivities of active intermediates generated in the ATRA/ATRP could provide further evidence for the radical nature of the process (*vide infra*).

Mechanistic Aspects of Transition Metal Catalyzed ATRP

Measurements of Rate Constants in the ATRP

ATRP is a complex process based on several elementary reactions. Similarly to the conventional radical polymerizations, the elementary reactions in the ATRP consist of initiation, propagation and termination. The rate constants of radical propagation are systematically evaluated by pulse laser polymerization techniques ($46,47$). The rate constants of termination of small radicals are usually accessible through EPR measurements (48). Perhaps, the most important reaction parameters in the ATRP are activation (k_a), deactivation (k_d) and initiation rate (k_i) constants. They depend on the structure of the alkyl halide, monomer, and more importantly transition metal complex. Systematic evaluation of the elementary rate constants is crucial in further understanding of the ATRP mechanism. Absolute and relative values can provide an important information on the nature and reactivity of active intermediates in the ATRP.

Determination of the Activation Rate Constants

Activation rate constants (k_a) in the ATRP are typically determined from model studies in which the transition metal complex is reacted with alkyl halide in the presence of radical trapping agents such as TEMPO ($49,50$). The rates are typically determined by monitoring the rate of disappearance of alkyl halide in the presence of excess activator (L_nMt^z) and TEMPO. Under such conditions, $\ln([RX]_0/[RX]_t) = -k_a[L_nMt^z]_0t$. The values were also determined for some polymeric systems using GPC techniques (51).

Determination of the Deactivation Rate Constants

The deactivation rate constants (k_d) are much less studied in the ATRP. The principal reason is the lack of techniques in measuring relatively fast deactivation process. One of the methods includes the clock reaction in which the generated radicals are simultaneously trapped with TEMPO and the deactivator ($49,52$). Additionally, k_d for more active catalysts has been estimated from the initial degree of polymerization without the reactivation, end functionality and molecular weight distributions ($53-55$).

Determination of the Equilibrium Constant for the ATRP

The apparent equilibrium constant for ATRP, $K_{ATRP}^{app} = K_{ATRP}/[Cu^{II}]$, is typically determined directly from polymerization kinetics using $\ln([M]_0/[M]_t)$ vs. t plots. More precise values can be obtained from model studies using the analytical solution of the persistent radical effect (56):

$$[L_nMt^{z+1}X] = (3K_{EQ}^2 k_t[RX]_0^2[L_nMt^z]_0^2)^{1/3} t^{1/3}$$

Typically, the activator (L_nMt^z) is reacted with the excess alkyl halide (RX) and the concentration of the generated deactivator ($L_nMt^{z+1}X$) monitored as a function of time. The plot of $[L_nMt^{z+1}X]$ vs. $t^{1/3}$ is then used to calculate K_{EQ}, provided that the termination rate constant (k_t) is known (Figure 3).

Simultaneous Determination of the Activation, Deactivation and Initiation Rate Constants in the ATRP/ATRA

We have recently reported a method to determine activation (k_a), deactivation (k_d) and initiation (k_i) rate constants in ATRP/ATRA using monomer trapping techniques and analytical solution of the persistent radical

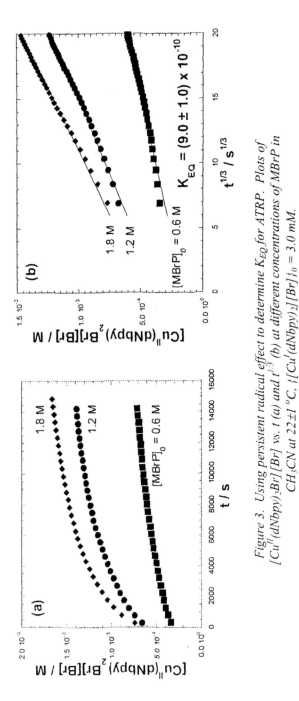

Figure 3. Using persistent radical effect to determine K_{EQ} for ATRP. Plots of $[Cu^{II}(dNbpy)_2Br][Br]$ vs. t (a) and $t^{1/3}$ (b) at different concentrations of MBrP in CH_3CN at $22 \pm 1\ ^\circ C$, $\{[Cu^I(dNbpy)_2][Br]\}_0 = 3.0\ mM$.

effect (57). The rate of disappearance of alkyl halide R-X, assuming steady state radical concentration and neglecting termination is given by:

$$\frac{1}{k_{app}} = -\frac{dt}{d\ln[RX]} = \frac{1}{k_a[L_nMt^z]} + \frac{k_d[L_nMt^{z+1}X]}{k_ak_i[M][L_nMt^z]}$$

From the plot of $1/k_{app}$ vs. $[L_nMt^{z+1}X]/[M]$ one can determine k_a and k_d/k_i. (Figure 4) The analytical solution of the persistent radical effect is then used to determine $K_{EQ} = k_a/k_d$, which enables simultaneous determination of k_a, k_d and k_i. The values of the activation and deactivation rate constants calculated using this methodology (Table 3) were independent of the monomer used. Furthermore, for tBuBrP, the initiation rate constants (k_i) are in excellent agreement with the literature values (58). Therefore, this system provides the

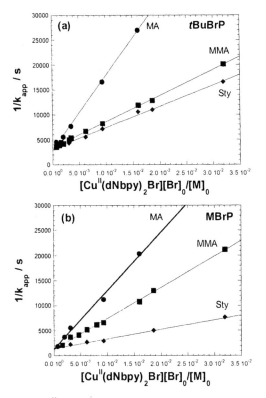

Figure 4. $1/k_{app}$ vs. $\{[Cu^{II}(dNbpy)_2Br][Br]\}_0/[M]_0$ for tBuBrP (a) and MBrP (b) in CH_3CN at 22 ± 1 °C; $[tBuBrP]_0 = [MBrP]_0 = 3.1$ mM, $\{[Cu^{I}(dNbpy)_2][Br]\}_0 = 31$ mM.

first evidence in which the reactivity of the radicals in ATRP was found to be identical to the conventional free radical addition reactions, which is a significant finding towards more comprehensive understanding of ATRP and ATRA mechanisms. This supports involvement of radical intermediates in

Table 3. Summary of Kinetic Parameters for tBuBrP and MBrP Using [Cu^I(dNbpy)$_2$][Br] Catalyst in CH$_3$CN at 22±1 °C.

Initiator	Monomer	k_ax10^2 [a]	k_dx10^{-7} [b]	k_i x10^{-4} [c]	
tBuBrP	MA	1.1	1.0	1.9±0.38	1.5[d]
	MMA	0.95	0.86	5.6±1.1	6.0[d]
	Sty	1.0	0.91	6.9±1.4	9.3[d]
MBrP	MA	2.5	2.8	3.1	/
	MMA	2.5	2.8	5.8	/
	Sty	2.2	2.4	18	/

[a] k_a, k_d and k_i in M^{-1}s^{-1}. [b] k_d for tBuBrP and MBrP was calculated using k_a and K_{EQ} = 1.1 x 10^{-9} and 9.0 x 10^{-10}, respectively. [c] k_i was calculated using k_d and k_d/k_i values obtained from the slopes in Fig. 4 (a and b). [d] Determined using time-resolved EPR spectroscopy (ref. (*58*)).

ATRP deduced earlier from similar rates of racemization, halogen exchange and trapping (*44*).

Conclusions

In conclusion, structural features of ATRP active Cu^I and Cu^{II} complexes were discussed based on several spectroscopic techniques. The structures were found to depend on the complexing ligand, solvent and temperature. Additionally, redox properties and halogenophilicity were correlated with the equilibrium constant for the ATRP. Lastly, various methods to study the elementary reactions in the ATRP were presented. They included measurements of activation, deactivation and initiation rate constants, and overall equilibrium constant for the ATRP. Structural and mechanistic studies are a continuous part of the future developments in the ATRP.

Acknowledgment. The financial support from the National Science Foundation through grant CHE 0096601 and the CMU CRP Consortium is acknowledged.

Literature Cited

1. Matyjaszewski, K.; Davis, T. P. *Handbook of Radical Polymerization*; John Wiley & Sons, Inc.: Hoboken, 2002.
2. Matyjaszewski, K.; Xia, J. *Chem. Rev.* **2001**, *101*, 2921-2990.
3. Kamigaito, M.; Ando, T.; Sawamoto, M. *Chem. Rev.* **2001**, *101*, 3689-3745.
4. Wang, J.-S.; Matyjaszewski, K. *J. Am. Chem. Soc.* **1995**, *117*, 5614-5615.
5. Patten, T. E.; Xia, J.; Abernathy, T.; Matyjaszewski, K. *Science* **1996**, *272*, 866-868.
6. De Campo, F.; Lastecoures, D.; Verlhac, J.-B. *J. Chem. Soc., Perkin Trans. 1* **2000**, *4*, 575-580.
7. Curran, D. P. In *Comprehensive Organic Synthesis*; Trost, B. M., Flemming, I., Eds.; Pergamon: NY, 1992; Vol. 4, p 715.
8. Coessens, V.; Pintauer, T.; Matyjaszewski, K. *Prog. Polym. Sci.* **2001**, *26*, 337-377.
9. Brandts, J. A. M.; van de Geijn, P.; van Faassen, E. E.; Boersma, J.; van Kotten, G. *J. Organomet. Chem.* **1999**, *584*, 246.
10. Kotani, Y.; Kamigaito, M.; Sawamoto, M. *Macromolecules* **1999**, *32*, 2420.
11. Matyjaszewski, K.; Wei, M.; Xia, J.; McDermott, N. E. *Macromolecules* **1997**, *30*, 8161-8164.
12. Ando, T.; Kamigaito, M.; Sawamoto, M. *Macromolecules* **1997**, *30*, 4507-4510.
13. Teodorescu, M.; Gaynor, S.; Matyjaszewski, K. *Macromolecules* **2000**, *33*, 2335.
14. Kato, M.; Kamigaito, M.; Sawamoto, M.; Higashimura, T. *Macromolecules* **1995**, *28*, 1721-1723.
15. Moineau, G.; Granel, C.; Dubois, P.; Jerome, R.; Teyssie, P. *Macromolecules* **1998**, *31*, 542.
16. Granel, C.; Dubois, P.; Jerome, R.; Teyssie, P. *Macromolecules* **1996**, *29*, 8576-8582.
17. Lecomte, P.; Drapier, I.; Dubois, P.; Teyssie, P.; Jerome, R. *Macromolecules* **1997**, *30*, 7631.
18. Haddleton, D. M.; Jasieczek, C. B.; Hannon, M. J.; Shooter, A. J. *Macromolecules* **1997**, *30*, 2190-2193.
19. Wang, J.-S.; Matyjaszewski, K. *Macromolecules* **1995**, *28*, 7901.
20. Kickelbick, G.; Pintauer, T.; Matyjaszewski, K. *New. J. Chem.* **2002**, *26*, 462-468.
21. Pintauer, T.; Qiu, J.; Kickelbick, G.; Matyjaszewski, K. *Inorg. Chem.* **2001**, *40*, 2818-2824.
22. Kickelbick, G.; Reinohl, U.; Ertel, T. S.; Weber, A.; Bertagnolli, H.; Matyjaszewski, K. *Inorg. Chem.* **2001**, *40*, 6-8.

146

23. Pintauer, T.; Reinohl, U.; Feth, M.; Bertagnolli, H.; Matyjaszewski, K. *Polym. Prepr. (Am. Chem. Soc., Div. Polym. Chem.)* **2002**, *43(2)*, 219-220.

24. Qiu, J.; Matyjaszewski, K.; Thouin, L.; Amatore, C. *Macromol. Chem. Phys.* **2000**, *201*, 1625-1631.

25. Pintauer, T.; Jasieczek, C. B.; Matyjaszewski, K. *J. Mass. Spectrom.* **2000**, *35*, 1295-1299.

26. Kajiwara, A.; Matyjaszewski, K.; Kamachi, M. *Macromolecules* **1998**, *31*, 5695-5701.

27. Kajiwara, A.; Matyjaszewski, K. *Macromol. Rapid Commun.* **1998**, *19*, 319-321.

28. Destarac, M.; Bessiere, J. M.; Boutevin, B. *Macromol. Rapid Commun.* **1997**, *18*, 967-974.

29. Levy, A. T.; Olmstead, M. M.; Patten, T. E. *Inorg. Chem.* **2000**, *39*, 1628-1634.

30. Burke, P. J.; McMillin, D. R.; Robinson, W. R. *Inorg. Chem.* **1980**, *19*, 1211-1214.

31. Munakata, M.; Kitagawa, S.; Asahara, A.; Masuda, H. *Bull. Chem. Soc. Jpn.* **1987**, *60*, 1927-1929.

32. Willett, R. D.; Pon, G.; Nagy, C. *Inorg. Chem.* **2001**, *40*, 4342-4352.

33. Skelton, B. W.; Waters, A. F.; White, A. H. *Aust. J. Chem.* **1991**, *44*, 1207-1215.

34. Becker, M.; Heinemann, F. W.; Knoch, F.; Donaubauer, W.; Liehr, G.; Schindler, S.; Golub, G.; Cohen, H.; Meyerstein, D. *Eur. J. Inorg. Chem.* **2000**, 719-726.

35. Becker, M.; Heinemann, F. W.; Schindler, S. *Chem. Eur. J.* **1999**, *5*, 3124-3129.

36. Pintauer, T.; Tsarevsky, N. V.; Kickelbick, G.; Matyjaszewski, K. *Polym. Prepr. (Am. Chem. Soc., Div. Polym. Chem.)* **2002**, *43(2)*, 221-222.

37. Pintauer, T.; McKenzie, B.; Matyjaszewski, K. *Polym. Prepr. (Am. Chem. Soc., Div. Polym. Chem.)* **2002**, 217-218.

38. Ambundo, E. A.; Deydier, M. V.; Grall, A. J.; Aguera-Vega, N.; Dressel, L. T.; Cooper, T. H.; Heeg, M. J.; Ochrymowycz, L. A.; Rorabacher, D. B. *Inorg. Chem.* **1999**, *38*, 4233-4242.

39. Ando, T.; Kamigaito, M.; Sawamoto, M. *Macromolecules* **2000**, *33*, 5825-5829.

40. Qiu, J.; Gaynor, S.; Matyjaszewski, K. *Macromolecules* **1999**, *32*, 2872.

41. Roos, S. G.; Muller, A. H. E. *Macromolecules* **1999**, *32*, 8331-8335.

42. Haddleton, D. M.; Crossman, M. C.; Hunt, K. H. *Macromolecules* **1997**, *30*, 3992-3995.

43. Xia, J.; Matyjaszewski, K. *Macromolecules* **1997**, *30*, 7692-7696.

44. Matyjaszewski, K.; Paik, H.-J.; Shipp, D. A.; Isobe, Y.; Okamoto, Y. *Macromolecules* **2001**, *34*, 3127-3129.

45. Yu, Q.; Zeng, F.; Zhu, S. *Macromolecules* **2001**, *34*, 1612-1618.
46. Beuermann, S.; Buback, M. *Prog. Polym. Sci.* **2002**, *27*, 191-255.
47. Fischer, H.; Radom, L. *Angew. Chem. Int. Ed.* **2001**, *40*, 1340-1371.
48. Fischer, H.; Paul, H. *Acc. Chem. Res.* **1987**, *20*, 200-206.
49. Matyjaszewski, K.; Paik, H.-j.; Zhou, P.; Diamanti, S. J. *Macromolecules* **2001**, *34*, 5125-5131.
50. Goto, A.; Fukuda, T. *Macromol. Rapid Commun.* **1999**, *20*, 633-636.
51. Ohno, K.; Goto, A.; Fukuda, T.; Xia, J.; Matyjaszewski, K. *Macromolecules* **1998**, *31*, 2699-2701.
52. Matyjaszewski, K.; Gobelt, B.; Paik, H.-j.; Horwitz, C. P. *Macromolecules* **2001**, *34*, 430-440.
53. Gromada, J.; Matyjaszewski, K. *Macromolecules* **2002**, *35*, 6167-6173.
54. Chambard, G.; Klumperman, B.; German, A. L. *Macromolecules* **2002**, *35*, 3420-3425.
55. Greszta, D.; Matyjaszewski, K. *Macromolecules* **1996**, *29*, 7661-7670.
56. Fischer, H. *Chem. Rev.* **2001**, *101*, 3581-3610.
57. Pintauer, T.; Zhou, P.; Matyjaszewski, K. *J. Am. Chem. Soc.* **2002**, *124*, 8196-8197.
58. Knuehl, B.; Marque, S.; Fischer, H. *Helv. Chim. Acta* **2001**, *84*, 2290-2300.

Chapter 11

Mechanistic Aspects of Copper-Mediated Living Radical Polymerization

Jeetan Lad, Simon Harrisson, and David M. Haddleton

Department of Chemistry, University of Warwick, Coventry CV4 7AL,
United Kingdom

A series of aminomethacrylate ((dimethylamino)ethyl methacrylate, (diethylamino)ethyl methacrylate and *t*-butylaminoethyl methacrylate) monomers and a series of methoxy[poly(ethylene glycol)] methacrylate macromonomers of different molecular weights have been copolymerized with methyl methacrylate under free radical and transition-metal mediated conditions. Significantly greater levels of comonomer incorporation were observed under transition-metal mediated conditions for all comonomers. This is attributed to complex formation between comonomer and catalyst, which is presumed to affect the reactivity of the monomer double bond.

Transition-metal mediated living radical polymerization, reported independently by Sawamoto (*1*) and Matyjaszewski (*2*) in 1995, has proved to be a remarkably efficient method of producing polymers with a wide range of functionalities and architectures (*3*). The polymerization is widely agreed to follow a free-radical mechanism, shown in Scheme 1, and containing as its key step the reversible abstraction of a (pseudo)halogen from the initiator or dormant polymer chain, creating a free radical which propagates via free-radical addition to monomer. Reactivity ratios, tacticities, and insensitivity to many functional groups are in most cases similar to those observed in free-radical polymerizations. Nevertheless, the presence of an additional step (activation/deactivation) and of an additional species (the metal complex) in the reaction may be expected to affect the mechanism to some extent. Certain monomers that contain donor atoms such as N or O are likely to coordinate to the catalyst, and with many catalysts, π-bonding of the monomer to the catalyst may be significant (*4*). Furthermore, a number of studies on similar transition-metal mediated additions (*5-7*) have implied the involvement of a "caged" radical, which is constrained within the coordination sphere of the catalyst.

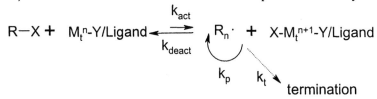

$$R{-}X + M_t^n{-}Y/Ligand \underset{k_{deact}}{\overset{k_{act}}{\rightleftharpoons}} R_n\cdot + X{-}M_t^{n+1}{-}Y/Ligand$$

$k_p \quad k_t$ termination

Scheme 1. Literature mechanism of transition-metal mediated living radical polymerization.

Under appropriate conditions, significant mechanistic differences can be observed between free-radical and transition metal-mediated polymerizations. For example, changing the polarity of the reaction medium has dramatic effects on the rate of polymerization, e.g. in the presence of oxyethylene groups (*8*), substituted phenols (*9*) and in aqueous (*10*) or ethylene carbonate (*11*) solution. More generally, while it is clear that transition-metal mediated polymerizations resemble conventional free-radical polymerizations, data obtained from free-radical polymerizations cannot be assumed to be applicable to transition-metal mediated polymerizations. In this contribution, we present the results of some recent experiments, which show significant mechanistic differences between the two types of polymerization.

In a previous study, the reactivity ratio of MMA with poly(lactic acid) methacrylate was measured and found to be significantly different between conventional free radical and atom transfer radical polymerizations (*12*). In that work and a previous study (*13*) the difference in the reactivity ratios between the two types of polymerization was attributed to the large size of the macromonomer: in conventional free radical polymerization, rapid growth of the polymer chain depletes the local concentration of macromonomer, which diffuses to the active chain end more slowly than the small comonomer. The much longer chain lifetimes associated with living polymerizations allow the macromonomer to diffuse to the active site, maintaining equality between local and bulk concentrations. However, monomers such as poly(lactic acid methacrylate) that contain donor atoms such as N or O are also potentially able to coordinate to the catalyst. Thus, the aim of this work was to determine whether the observed difference in reactivity ratios could be due to monomer coordination to the transition metal used in transition-metal mediated polymerization. This was investigated using low mass monomers such as (dimethylamino)ethyl methacrylate which should diffuse at approximately the same rate as the comonomer. In a further set of experiments, a series of methoxy[poly(ethylene glycol)] methacrylate monomers containing different numbers of repeat units were copolymerized with MMA under transition-metal mediated conditions to investigate the effect of increasing molecular weight on monomer reactivity.

Results and Discussion

There have been several reports on reactivity ratios in ATRP and other transition-metal mediated polymerizations (*14-16*). These have generally concluded that the reactivity ratios are very similar to those observed in conventional free radical polymerization. Any differences are generally small compared to the precision of the data. Recently, simulations have shown that large (order of magnitude) differences in the rates of activation of the two monomers can produce apparent reactivity ratios that differ from those observed in free radical polymerizations, even though the underlying rate coefficients of homo- and cross-propagation are unchanged (*17*). Once again, however, these differences are generally small except in the very unusual case where both reactivity ratios are significantly greater than 1 (*17*).

In the case of the methyl methacrylate-poly(lactic acid)methacrylate systems (*12*) the differences between conventional and transition-metal mediated polymerizations are large, and well outside experimental error. This report prompted us to investigate the reactivities of a range of small monomers that would be expected to coordinate with the catalyst. The monomers chosen were (dimethylamino)ethyl methacrylate (DMAEMA, **1**), (diethylamino)ethyl methacrylate (DEAEMA, **2**) and (*t*-butylamino)ethyl methacrylate (TBAEMA, **3**).

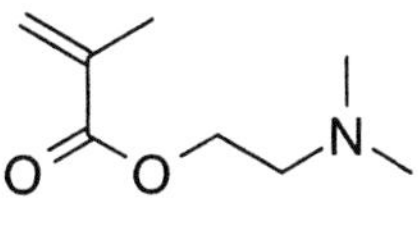

DMAEMA, 1 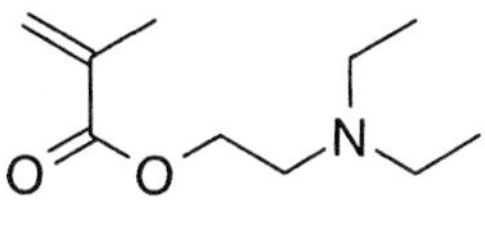   DEAEMA, 2

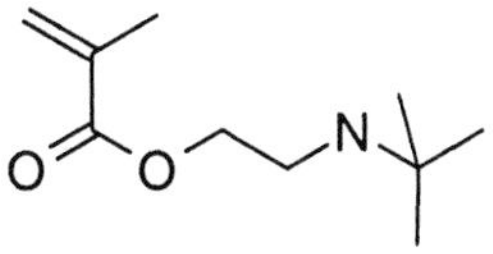

TBAEMA, 3

These monomers were copolymerized with a large excess of MMA, and the relative rates of consumption of the comonomers were monitored using NMR spectroscopy. Molecular weights of the resulting polymers were not measured, but ATRP polymerizations under similar conditions (100:1 monomer:catalyst ratio) give polymers with number-average degrees of polymerization very close to the theoretical value of $100.x$ (where x is conversion).

The use of a large excess of MMA allows the reactivity ratio of MMA (r_{MMA}) towards the comonomer to be evaluated using a simplified form of the copolymer composition equation, which can be readily integrated. The resulting, integrated equation is shown below (eq. 1):

$$[M_1]/[M_1]_0 = ([M_2]/[M_2]_0)^{r1} \tag{1}$$

in which M_1 is the monomer in excess (MMA in these experiments). This expression is equivalent to the more common, linear expression given in equation 2 (*20*).

$$\ln\{[M_1]_0/[M_1]\} = r_1.\ln\{[M_2]_0/[M_2]\} \tag{2}$$

Significant chain length dependence of the rate coefficients of propagation is generally observed in the first few propagation steps (*18*). This leads to problems in the application of the instantaneous form of copolymer composition equation to living polymerizations as at low conversions the polymer is dominated by oligomeric species. By using an integrated form of the copolymer composition equation, it is possible to obtain results at conversions of 0.1-0.95 ($M_n \sim 1000 - 9500$). Within this molecular weight range, the rate coefficients of propagation and hence the reactivity ratios are expected to be approximately constant and equal to their long-chain limits.

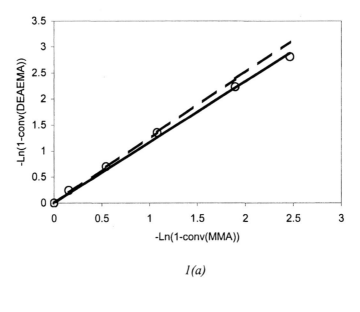

1(a)

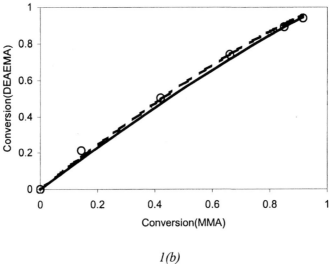

1(b)

Figure 1. Linear and non-linear fits to experimental data (MMA-TBAEMA copolymerization by copper-mediated polymerization) using (a) equation 2. (b) equation 1. Open circles are experimental data, solid lines show linear fit to equation 2, dotted lines non-linear fit to equation 1.

In free radical polymerizations high molecular weight polymer is produced from the start of the reaction and reactivity ratios should be constant across the entire conversion range.

While equation 2 is a convenient form in which to graph results, the transformation of experimental results ($[M]/[M]_0$) into logarithmic expressions produces severe distortions in the experimental error structure, with the effect that data points are effectively given greater weight as the conversion increases. This is illustrated in Figure 1, which shows experimental data from a copolymerization of MMA and DEAEMA under ATRP conditions, with lines of best fit calculated using both linear (eq. 2) and non-linear (eq. 1) methods. It is clear from Figure 1b that the linear method provides an excellent fit to the final data points, but at the expense of data at lower conversions. The non-linear method produces a line, which fits all data points equally well. This distortion also affects the estimation of errors in r_1, which will generally be underestimated using the linear method.

In this work the problem was avoided by the use of equation 1 in conjunction with non-linear least squares fitting (*20*). The results obtained are shown in Table I. It can be seen from the table that linear fitting can give erroneous results, particularly when the reactivity ratio to be measured is not close to unity. Figure 2 gives a graphical representation of the results, including 95% confidence intervals (obtained from the non-linear fitting procedure).

There is a clear difference between the free-radical and transition-metal mediated copolymerizations. Unlike the MMA-poly(lactic acid)methacrylate case (*12*), this cannot be explained by differences in hydrodynamic radius between the comonomers, as these differences are minimal. The increased incorporation of the aminoethyl methacrylate (AEMA) monomers into the transition-metal mediated copolymers could potentially be explained by AEMA-terminated radicals having an increased rate constant of activation, k_{act}, compared to MMA-terminated radicals. It is by no means clear, however, that this would be sufficient to produce the large deviations observed here, even if k_{act} of AEMA monomers were several orders of magnitude greater than that of

Table I. Reactivities of Coordinating Monomers (r_{MMA}) in Free Radical and Transition-Metal Mediated Copolymerizations

M_2	Free radical		Transition-metal mediated	
	Linear[a]	*NLLS*[b]	*Linear*[a]	*NLLS*[b]
DMAEMA	0.96	0.96(2)	0.77	0.74(3)
DEAEMA	0.99	0.98(1)	0.86	0.79(3)
TBAEMA	0.98	0.97(1)	0.74	0.69(3)

[a] Estimated by linear fitting of eq. 2. [b] Estimated by nonlinear least squares fitting of eq. 1. Figures in parentheses are standard errors in the final digit.

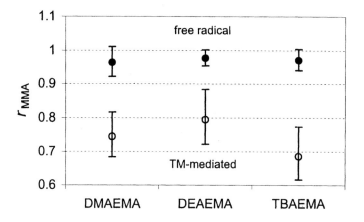

Figure 2. Reactivities of coordinating monomers (r_{MMA}) in free radical (filled) and transition-metal mediated polymerizations (open), showing 95% confidence intervals.

MMA, as deviations from free-radical behavior according to this model appear to be small except when both monomer reactivity ratios are substantially greater than 1 (e.g. examples 1C, 3B, 3C and 2D in reference *17*). Such a large k_{act} would make controlled polymerization of AEMAs by transition-metal mediated polymerization virtually impossible due to the high radical concentrations and termination rates that would ensue. Nevertheless, there are many reports of successful controlled polymerizations of these monomers (*21-24*).

Hence it appears that the two explanations previously advanced to support differences in reactivities between ATRP and free radical copolymerizations are untenable in this case. Many reactivity ratios show solvent effects (*25*), but the addition of catalytic amounts of copper complex is unlikely to have a significant effect on the solvent polarity. Such solvent effects are most often seen when charge-transfer structures play a significant role in stabilizing the transition state of the cross-propagation reaction as in styrene-MMA or styrene-acrylonitrile – this is unlikely to be the case for MMA-DMAEMA where the two double bonds are very similar.

NMR spectra of mixtures of aminoethyl methacrylates and copper(I) bromide in d_8-toluene show clear shifts in the absorptions of protons close to the nitrogen atom as well as the downfield vinylic proton (*cis*- to the ester moiety), even in the presence of equimolar amounts of pyridyl methanimine ligand (Figure 3). No change was observed in the position of the IR absorptions of the carbonyl groups of the aminomethyl methacrylates in the presence of copper(I) bromide.

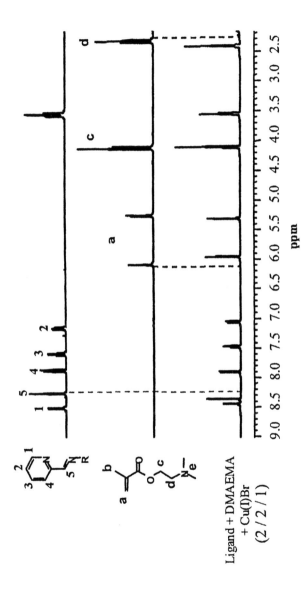

Figure 3. ^1H NMR spectra of N-propyl 2-pyridylmethanimine (top), DMAEMA (middle) and a mixture of both ligands with Cu(I)Br.

This is indicative of coordination to copper through the amine and ester groups of the aminoethyl methacrylate, as shown in Scheme 2. The structure shown is not definitive, and we are unable to explain why the vinylic proton *cis* to the ester group is shifted upfield, indicating increased shielding, rather than dowfield as might be expected from deshielding due to the copper. However, as this shift is only observed in the presence of copper (and not in mixtures of ligand and DMAEMA) it is logical to attribute it to interactions between the proton and either the copper itself or other ligands. Further evidence of the ability of DMAEMA to form complexes with copper is provided by the ability of this monomer to undergo living radical polymerization in the absence of additional ligand – presumably in this case the monomer itself acts as ligand (*26*). Coordination may affect the monomer reactivity by altering the electronic structure of the double bond, or simply through a mass effect (the mass of the monomer-copper complex will be much greater than that of the monomer alone, and this should cause an increase in the pre-exponential factor of the rate constant) (*27*). It is well known that the presence of Lewis acids may affect reactivity ratios, particularly in the case of STY-MMA polymerizations (*18, 28*); it is possible that a similar effect is observed in this case, with copper acting as the Lewis acid.

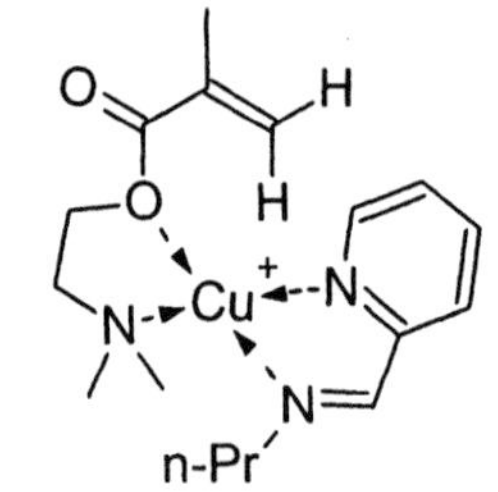

Scheme 2. Possible structure of a DMAEMA / N-propyl 2-pyridyl methanimine / copper(I) complex.

Trends in reactivities within the monomers appear to support this explanation, as the secondary amine, TBAEMA, shows the greatest deviation from free-radical reactivity ratios, while the most sterically hindered tertiary amine, DEAEMA, shows the least, suggesting that the extent of deviation is correlated with decreasing congestion around the nitrogen. This is difficult to explain except in the context of monomer-copper complex formation.

The study was extended to a series of methoxy[poly(ethylene glycol)] methacrylates (**4, 5**) in order to investigate the effect of increasing molecular weight on reactivity. Similar results were seen as for the amino methacrylates, with higher levels of incorporation into the copolymer (lower r_{MMA}) in transition-metal mediated polymerizations than in free radical polymerizations for all molecular weights (Figure 3, Table II).

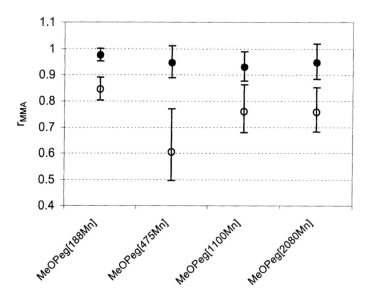

4: MW = 188

5a: M_n = 475, n ~ 6
5b: M_n = 1100, n ~ 21
5c: M_n = 2080, n ~ 43

Table II. Reactivities of Methoxy[Poly(ethylene glycol) Methacrylate Monomers (r_{MMA}) in Free Radical and Transition-Metal Mediated Copolymerizations

Macromonomer	Free radical	Transition-metal mediated
4	0.98(1)	0.85(3)
5a	0.95(3)	0.60(5)
5b	0.93(2)	0.76(4)
5c	0.95(3)	0.76(3)

Figure 3. Reactivities of methoxy[poly(ethylene glycol)] methacrylates as a function of number average molecular weight in free radical (closed circles) and transition-metal mediated polymerizations (open circles). Error bars show 95 % confidence intervals.

Within the transition-metal mediated results, there is a general trend towards lower r_{MMA} as molecular weight increases. This trend is not compatible with a diffusion control explanation, which predicts that increasing chain length will only affect reactivities in free radical polymerization. The results presumably reflect macromonomers' increased ability to coordinate to copper as the number of potential ligating groups are increased. This should be counterbalanced by the increased likelihood that the copper will be coordinated to donor atoms, which are too far away from the monomer double bond to influence its reactivity as chain length increases. The value of r_{MMA} decreases sharply from $M_n = 188$ (**4**) to $M_n = 475$ (**5a**) before increasing to what appears to be a stable value (**5b** and **c**) at longer chain lengths. This suggests that **5a**, containing on average 6 oxyethylene repeat units, is close to the optimum length for complexing the copper catalyst in such a way that it remains sufficiently close to the monomer double bond to affect its reactivity.

Conclusions

We have shown that for a series of aminoethyl methacrylate monomers and a series of poly(ethylene glycol) macromonomers of different weights, the monomer reactivities towards methyl methacrylate differ significantly from those observed in conventional free radical polymerizations. This is attributed to complex formation between monomer and catalyst. It is likely that monomer reactivities in transition-metal mediated copolymerizations will differ significantly from those measured in free radical copolymerizations for a wide range of functionalized monomers with the potential to undergo similar monomer-catalyst interactions.

Experimental

Materials.

Methyl methacrylate and aminomethacrylate monomers (99%, inhibited with monomethyl ether hydroquinone) were obtained from Aldrich and passed over a column of activated basic alumina before use. Methoxy[poly(ethylene glycol)] methacrylate monomers (Aldrich) were used as received. Copper(I) bromide (Aldrich, 98%) was purified by a modification of the method of Keller and Wycoff (*29*). N-propyl 2-pyridylmethanimine was synthesized according to the method of Haddleton et al. (*30*). Azobis(isobutyronitrile) (ACROS Chimica, 98%) was recrystallized from methanol. All other chemicals were purchased from Aldrich or ACROS Chimica and used as received.

159

Analysis.

NMR spectra were obtained on a Bruker DPX400 spectrometer using d_8-toluene as solvent. Infra-red absorption spectra were recorded on a Bruker VECTOR-22 FTIR spectrometer using a Golden Gate diamond attenuated total reflection cell.

Copolymerizations.

Mixtures of MMA and aminoethyl methacrylate monomers **1**, **2**, and **3** (96.5 wt% MMA, 3.5 wt% aminoethyl methacrylate) were copolymerized by both conventional free radical polymerization with AIBN initiation and by copper(I) bromide mediated living radical polymerization using N-propyl pyridyl-methanime as ligand (total monomer:ligand:copper = 100:2:1). All copolymerizations were carried out in solution in toluene or d_8-toluene (1:1 toluene:monomer) at 90 °C under nitrogen after freeze-pump-thaw degassing. For the transition metal mediated polymerizations, samples were taken regularly and conversion was determined by ^{1}H NMR spectroscopy (d_8-toluene, 400MHz). Free radical polymerizations were carried out in sealed NMR tubes within the NMR spectrometer (400MHz) using d_8-toluene as solvent, and scanned at regular intervals. Conversions for each monomer were calculated from the NMR spectra by comparing the combined areas of the methylene peaks (2H) with the area of the combined monomer and polymer methoxy (MMA and methoxy[poly(ethylene glycol)] methacrylates, 3H) or $-CH_2O-$ (aminoethyl methacrylates, 2H) peaks. Reactivity ratios were calculated from these results using a modified version of the Jaacks method (*19*) incorporating nonlinear least squares fitting.

Acknowledgements

We thank EPSRC (SH, GR/R16228) and Avecia (JL, studentship) for funding this work.

References

1. Kato, M.; Kamigato, M.; Sawamoto, M.; Higashimura, T. *Macromolecules* **1995**, *28*, 1721.
2. Wang, J. S.; Matyjaszewski, K. *J. Am. Chem. Soc.* **1995**, *117*, 5614.
3. Matyjaszewski, K.; Xia, J. *Chem. Rev.* **2001**, *101*, 2921-2990.

160

4. Kamigaito, M.; Ando, T.; Sawamoto, M. *Chem. Rev.* **2001**, *101*, 3689-3745.
5. Kameyama, M.; Kamigata, N.; Kobayashi, M. *J. Org. Chem.* **1987**, *52*, 3312-3316.
6. Bland, W. J.; Davis, R.; Durrant, J. L. A. *J. Organomet. Chem.* **1984**, *267*, C45-48.
7. Bland, W. J.; Davis, R.; Durrant, J. L. A. *J. Organomet. Chem.* **1985**, *280*, 397.
8. Haddleton, D. M.; Perrier, S.; Bon, S. A. F. *Macromolecules* **2000**, *33*, 8246-8251.
9. Haddleton, D. M.; Clark, A. J.; Crossman, M. C.; Duncalf, D. J.; Heming, A. M.; Morsley, S. R.; Shooter, A. J. *Chem. Commun.* **1997**, 1173-4.
10. Wang, X.-S.; Armes, S. P. *Macromolecules* **2000**, *33*, 6640-6647.
11. Matyjaszewski, K.; Nakagawa, Y.; Jasieczek, C. B. *Macromolecules* **1998**, *31*, 1535.
12. Shinoda, H.; Matyjaszewski, K. *Macromolecules* **2001**, *34*, 6243-6248.
13. Roos, S. G.; Mueller, A. H. E.; Matyjaszewski, K. *Macromolecules* **1999**, *32*, 8331.
14. Moineau, G.; Minet, M.; Dubois, P.; Teyssié, P.; Senninger, T.; Jérôme, R. *Macromolecules* **1999**, *32*, 27.
15. Haddleton, D. M.; Crossman, M. C.; Hunt, K. H. *Macromolecules* **1997**, *30*, 3992.
16. Ziegler, M. J.; Matjyaszewski, K. *Macromolecules* **2001**, *34*, 415.
17. Matyjaszewski, K. *Macromolecules* **2002**, *35*, 6773.
18. Moad, G.; Solomon, D. H. *The Chemistry of Free Radical Polymerisation*; Pergamon: Oxford, 1995.
19. Jaacks, V. *Makromol. Chem.* **1972**, *161*, 161.
20. Gans, P. *Data Fitting in the Chemical Sciences*; John Wiley & Sons: Chichester, 1992.
21. Zhang, Z.; Liu, G.; Bell, S. *Macromolecules* **2000**, *33*, 7877.
22. Huang, W.; Kim, J.-B.; Bruening, M. L.; Baker, G. L. *Macromolecules* **2002**, *35*, 1175.
23. Shen, Y.; Zhu, S.; Zeng, F.; Pelton, R. *Macromolecules* **2000**, *33*, 5399.
24. Narrainen, A. P.; Pascual, S.; Haddleton, D. M. *J. Pol. Sci. Part A: Polym. Chem.* **2002**, *40*, 439.
25. Coote, M. L.; Davis, T. P. In *Handbook of Radical Polymerization*; Matyjaszewski, K.; Davis, T. P., Eds.; John Wiley & Sons: Hoboken, NJ, 2002; pp 263-300.
26. Lad, J.; Haddleton, D. M., unpublished results.
27. Heuts, J. P. A. In *Handbook of Radical Polymerization*; Matyjaszewski, K.; Davis, T. P., Eds.; John Wiley & Sons: Hoboken, NJ, 2002; pp 1-76.
28. Kırcı, B.; Lutz, J.-F.; Matyjaszewski, K. *Macromolecules* **2002**, *35*, 2448.
29. Keller, R. N.; Wycoff, H. D. *Inorg. Synth.* **1947**, *2*, 1.
30. Haddleton, D. M.; Kukulj, D.; Duncalf, D. J.; Heming, A. M.; Shooter, A. J. *Macromolecules* **1998**, *31*, 5201.

Chapter 12

ESR Study and Radical Observation in Transition Metal-Mediated Polymerization: Unified View of Atom Transfer Radical Polymerization Mechanism

Aileen R. Wang[1], Shiping Zhu[1,*], and Krzysztof Matyjaszewski[2]

[1]Department of Chemical Engineering, McMaster University, 1280 Main Street West, Hamilton, Ontario, L8S 4L7, Canada
[2]Department of Chemistry, Carnegie Mellon University, 4400 Fifth Avenue, Pittsburgh, PA 15213

Using an on-line electron spin resonance (ESR) technique, we studied the electron paramagnetic species of oxidized metal centers and carbon radicals in three representative catalyst/ligand systems of transition metal-mediated polymerization of methyl methacrylate. These systems were: (1) CuBr/Bpy, (2) RuCl$_2$(PPh$_3$)/Al(OiPr)$_3$, and (3) CuBr/N-pentyl-2-pyridylmethanimine. Ethylene glycol dimethacrylate was used as crosslinker to trigger the diffusion-controlled radical deactivation so that the radical population was accumulated to an ESR detectable level. Methacrylate radicals were observed in all the systems. Their signal hyperfine structures were typical of nine line and were identical to those observed in conventional free radical polymerization processes. There were also peculiar signals observed at low dimethacrylate levels in (2) and (3). These paramagnetic species were present without the monomer addition. The nature of the propagating center type was analyzed. This work provided a unified view for the radical mechanisms of atom transfer radical polymerization.

162

Introduction

A polymerization system is considered to be living as long as there is no permanent radical termination or transfer reaction (*1,2*). Living polymerization provides good control over chain structure and functionality. Among the developed living processes, controlled radical polymerization enjoys further advantages including versatility of monomer type, tolerance of water and protonic impurities, and mild reaction conditions. Due to the active nature of radicals, living radical process is achieved by frequently and temporarily capping propagating radical centers so that equilibrium between the dormant and active species is established. The capping molecules can be transition metal halide (atom-transfer radical polymerization or ATRP) (*3,4*), nitroxide (stable free radical polymerization or SFRP) (*5*), or dithioester (reversible addition-fragmentation transfer polymerization or RAFT) (*6*). One of the key requirements for a process to be living is the high rate of radical capping (deactivation) that prevents the centers from experiencing permanent termination.

The transition metal-mediated atom transfer radical polymerization has attracted enormous attention since its discovery due to potential industrial applications in synthesizing well-defined polymer chains (see the recent review papers (*7,8*)). However, the mechanistic understanding of the process is still lacking and requires more effort. ATRP is considered to be an extension of atom-transfer radical addition (ATRA) (*9,10*). There are numerous evidences in supporting the proposed radical mechanism (*11*). However, there are also some peculiar experimental observations that lead to other suggestions (*12*).

An important aspect of the mechanistic study is the direct observation of radical intermediates during polymerization. For this purpose, Kajiwara et al. (*13*) carried out an ESR study on ATRP of styrene mediated by CuBr/4,4'-di(5-nonyl)-2,2'-bipyridine (dNbpy) initiated by 1-phenylethyl bromide. However, only Cu(II) signals were observed. The reason for this lack of radical detection is the low radical concentration due to the high radical deactivation rate in:

$$PBr + Cu(I)Br/L \leftrightarrow P^{\bullet} + Cu(II)Br_2/L$$

where P = polymer chain, L = ligand; and $\bullet$ = radical center. Normally, a minimal level of radical concentration $>10^{-7}$ mol/l is needed to be detected with a recognizable signal/noise ratio.

Recently, we used an on-line ESR technique to investigate the ATRP of poly(ethylene glycol) dimethacrylate mediated by CuBr/1,1,4,7,10,10-hexamethyltriethylenetetramine (HMTETA) initiated by methyl α-bromophenylacetate (MBP). For the first time, we observed the radical intermediates involved in an ATRP process (*14*). The ESR signal appeared to have a typical 9-line hyperfine structure arising from methacrylate radicals that experience a glass/solid environment in the polymer matrix (*15,16*). The success was due to network formation in the system that imposed diffusion limitations to the CuBr/L complex. The diffusion-controlled radical deactivation favored radical generation. Also because of the network structure, these trapped radicals had reduced mobility, and thus avoided bimolecular termination.

In a more recent paper (*17*), we measured the radical and Cu(II) concentration profiles in the ATRP of methyl methacrylate (MMA)/ethylene glycol dimethacrylate (EGDMA) with the MBP/CuBr/HMTETA system. The work provided a quantitative explanation for the change of the equilibrium constant $K_{eq} = k_{act}/k_{dea}$ during the polymerization. The polymerization process showed three definable stages. In the first stage, the Cu(II) concentration increased continuously but slowly. The methacrylate radical signal was not detectable because its concentration was below the sensitivity of the ESR machine. In the second stage, the Cu(II) concentration increased dramatically. The methacrylate radical signal started to appear and increased synchronously with the Cu(II) concentration. This autoacceleration was because the radical deactivation became diffusion-controlled. In the third stage, the Cu(II) and radical concentrations increased gradually and reached a steady state due to radical trapping in the network.

As a part of continuous effort in elucidating the ATRP radical mechanisms, in this work, we carried out ESR studies on different catalyst/ligand systems for MMA/EGDMA polymerization. Three representative systems were selected. System 1 was initiated by ethyl 2-bromoisobutyrate (EBIB) mediated by CuBr/2,2'-bipyridine (*3*). System 2 was initiated by CCl$_4$ mediated by dichlorotris(triphenylphosphine) ruthenium (RuCl$_2$(PPh$_3$)$_3$) activated by aluminum isopropoxide (Al(OiPr)$_3$) (*4*). System 3 was initiated by EBIB mediated by CuBr/*N*-pentyl-2-pyridylmethanimine (PPMI) (*18*). For comparison, we also included the MBP/CuBr/HMTETA system. The objective of this work was to provide a unified view about the radical mechanisms for these different systems.

Experimental

Materials. Methyl methacrylate (MMA), ethylene glycol dimethacrylate (EGDMA), carbon tetrachloride (CCl$_4$) from Aldrich were distilled over CaH$_2$ and stored at 0 °C prior to use. Ethyl 2-bromoisobutyrate (EBIB, 98%), methyl α-bromophenylacetate (MBP, 97%), copper bromide (CuBr, 98%), 2,2'-bipyridine (Bpy, 99%), 1,1,4,7,10,10-hexamethyl-triethylenetetramine (HMTETA, 97%), dichlorotris(trisphenylphosphine) ruthenium (RuCl$_2$(PPh$_3$)$_3$, 97%), aluminum isopropoxide (Al(OiPr$_3$)$_3$, 98%), 2-pyridinecarboxaldehyde (99%), amylamine (99%), magnesium sulphate (MgSO$_4$) were all purchased from Aldrich and used as received. *N*-pentyl-2-pyridylmethanimine (PPMI) was synthesized following the procedure reported in the literature (*18*). Scheme 1 shows the molecular structures for some of the chemicals used in the work.

MMA

$$CH_2{=}CH{-}\overset{\overset{\displaystyle CH_3}{|}}{C}{-}\overset{\overset{\displaystyle O}{\|}}{}{-}O{-}CH_3$$

EGDMA

$$CH_2{=}\overset{\overset{\displaystyle CH_3}{|}}{C}\quad\overset{\overset{\displaystyle CH_3}{|}}{C}{=}CH_2$$
$$O{=}C\quad C{=}O$$
$$\overset{|}{O}\quad\overset{|}{O}$$
$$CH_2{-}CH_2$$

Bpy

EBIB

$$CH_3{-}\overset{\overset{\displaystyle CH_3}{|}}{\underset{\underset{\displaystyle Br}{|}}{CH}}{-}\overset{\overset{\displaystyle O}{\|}}{C}{-}O{-}CH_2{-}CH_3$$

$RuCl_2(PPh_3)_3$

$$\begin{array}{c} PPh_3 \\ | \\ Cl{-}Ru{-}PPh_3 \\ Ph_3P^{\nearrow}\ \ Cl \end{array}$$

$Al(OiPr)_3$

$$\begin{array}{c} CH_3\quad\quad\quad\quad CH_3 \\ C{-}O{-}Al{-}O{-}C \\ CH_3\quad\ \ \overset{|}{O}\quad\ \ CH_3 \\ CH_3{-}C{-}CH_3 \end{array}$$

PPMI

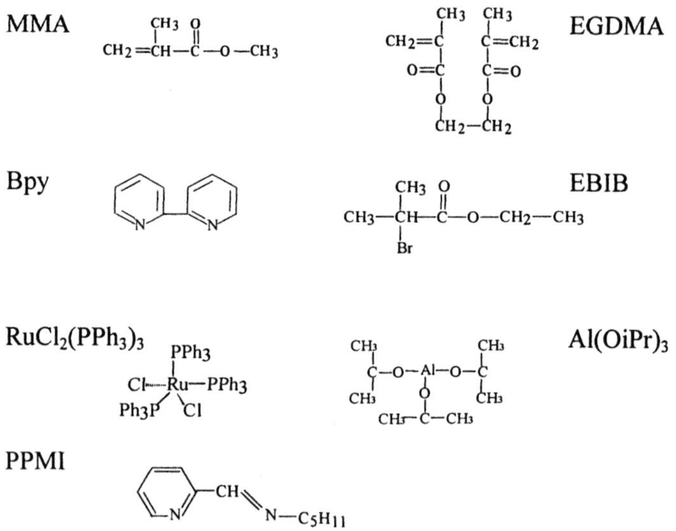

Scheme 1. Molecular structures for some chemicals used in the work.

Sample Preparation. For System 1 and 3, the ratio of monomer (vinyl basis)/catalyst /ligand/initiator = 100/1/2/1 (mol/mol). Take the MMA/EGDMA 80/20 (vinyl based mol/mol) sample as an example. 8.55 mg CuBr (0.0584 mmol), 18.44 mg Bpy (0.117 mmol) for System 1 or 22.1 μl of PPMI (0.117 mmol) for System 3, and 0.5 ml MMA (4.67 mmol of vinyl group), 0.11 ml EGDMA (1.17 mmol of vinyl group) were charged to a dry ampoule (5 mm o.d., 3.3 mm i.d.). The ampoule was then sealed with a rubber septum and the mixture was degassed by three cycles of ultrasonic wave - ultra purity nitrogen. 8.8 μl of EBIB (0.0584 mmol) was added by a degassed syringe and the ampoule was shaken for 1 min prior to the ESR measurement.

In System 2, the ratio of monomer/catalyst/additive/initiator = 100/0.5/2/2 (mol/mol). Take the MMA/EGDMA 80/20 (vinyl based mol/mol) as an example. 0.5 ml MMA (4.67 mmol of vinyl group), 0.11 ml EGDMA (1.17 mmol of vinyl group), and 11.3 μl of CCl_4 (0.117 mmol) were charged to a dry ampoule (5 mm o.d., 3.3 mm i.d.). 24.34 mg $Al(OiPr)_3$ and 28.88 mg $RuCl_2(PPh_3)_3$ (0.0292 mmol) were sequentially added to the system. The ampoule was then sealed with a rubber septum and the mixture was degassed by three cycles of ultrasonic wave - ultra purity nitrogen.

ESR measurement. All ESR measurements were carried out on-line with BRUKER EPR spectrometer (EP072). The ESR spectrometer was operated at

0.505 mW power, microwave frequency 9.42 GHz, 100 kHz modulation frequency and modulation amplitude 3.0 G, time constant 10.24 sec, gain 5.02×10^5,.unless otherwise indicated. The polymerization took place when the ampoule was inserted into the EPR cavity with its temperature maintained at the preset level by a BRUKER variable temperature unit. The EPR spectra were recorded at different time intervals during the polymerization.

Results and Discussion

1) EBIB/CuBr/Bpy

Figures 1a-d show the ESR spectra at the four EGDMA levels. In pure MMA experiments (Figure 1a), only Cu(II) signals were observed. We made every effort in optimizing ESR parameters for possible radical observation. The absence of radical signals should not be interpreted as no radicals present in the system. The radical concentration at this condition is most likely below the sensitivity of the ESR machine, even at the conversion when the system reached its glass state.

The Cu(II) signal appeared from the very beginning of the polymerization. It has the definable hyperfine structure characteristic of Cu(II) species with a reasonable signal/noise ratio. The [Cu(II)], estimated from the change in the signal intensity, increased continuously with time in the early stage. The intensity change after 30 min was not significant up to 365 min, indicating nearly constant Cu(II) concentration in the system. The absolute Cu(II) concentrations in pure MMA were low, evident from the noise background in the spectra.

Increasing the EGDMA level to 10% (vinyl basis, approximately equal to weight fraction) yielded higher signal/noise Cu(II) spectra as shown in Figure 1b, indicating higher Cu(II) concentrations than pure MMA. The signal intensity and thus Cu(II) concentration kept increasing till 180 min and then leveled off. However, there was still no radical signal detectable. Further increasing the EGDMA level to 20% resulted in further improvement in the signal/noise ratio (Figure 1c). More importantly, some hyperfine structure at the carbon radical field started to appear at 117 min. The intensity of this structure increased with time and became most visible at 247 min, but started to decay after that.

Figure 1d shows the spectra with 50% EGDMA. The signal intensity was very high. The Cu(II) showed from the beginning, and continuously increased. The high-field structure appeared at 20 min, and its intensity also increased continuously. At 50 min, it became definable. The signal had a hyperfine structure identical to the spectra observed in the conventional free radical polymerization of methacrylate monomers (15,16). This 9-line structure (outer 5 inner 4 - see 180 min) was attributed to methacrylate radicals trapped in a polymer matrix. The radical concentration kept increasing till 460 min. At this stage, its intensity was almost comparable to that of Cu(II) species.

166

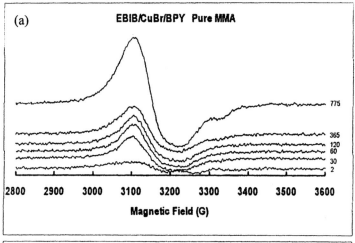

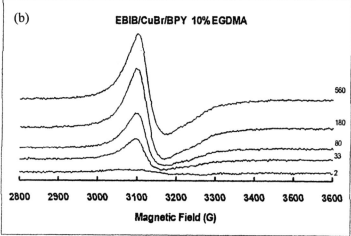

Figure 1. ESR spectra (5 scans) observed in bulk ATRP of (a) pure MMA, (b) 10% EGDMA, (c) 20% EGDMA, and (d) 50% EGDMA, initiated by ethyl 2-bromoisobutyrate (EBIB) and mediated by CuBr/2,2'-bipyridine, at 90 ^{0}C and time (min) as indicated. [M]/[CuBr]/[Bpy]/[EBIB] = 100/1/2/1. MMA/EGDMA in vinyl molar fraction (~ weight fraction).

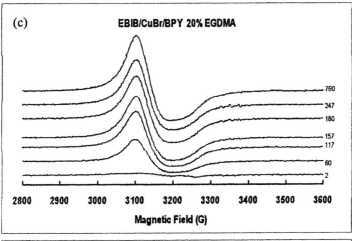

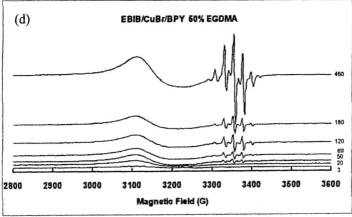

Figure 1. *Continued.*

The appearance of the methacrylate radical signal can be logically explained by the equilibrium:

$$PBr + Cu(I)Br/Bpy \leftrightarrow P^\bullet + Cu(II)Br_2/Bpy$$

This equilibrium was established after a short period of time (in a few minutes) by termination of initiating radicals that generated a required level of [Cu(II)]. The polymerization proceeded at a radical concentration lower than the ESR detection limit (about $10^{-7} \sim 10^{-6}$ mol/l). The deactivation rate constant k_{dea} ($\sim 10^7$ l/mol.sec) is many orders of magnitude higher than that of activation k_{act} ($\sim 10^{-1} - 10^1$ l/mol.sec). At higher conversions when significant amount of EGDMA reacted, the system gelled. The three-dimensional network structure imposed significant diffusion limitations to reacting species. Diffusion control affects first the fastest reaction, i.e. the bimolecular radical termination. Translational diffusion of chain species inside a network is virtually limited.

When the polymerization further proceeded, small molecules also experienced diffusion limitations. These molecules included monomers and catalyst/ligand complexes. However, the rate constant of deactivation that involved $P^\bullet$ and $CuBr_2$/Bpy was much higher than those of activation (PBr with CuBr/Bpy) and propagation ($P^\bullet$ with MMA or EGDMA, $k_p \sim 10^3$ l/mol.sec). Therefore, the deactivation reaction became diffusion-controlled after bimolecular radical termination. Decrease in k_{dea}, while k_{act} remained constant, favored the generation of methacrylate radicals. The radical signal finally became observable. Further increasing monomer conversion led to tighter network structure that further decreased k_{dea} and increased radical concentration.

We have also investigated other possible radical sources in the ATRP of MMA/EGDMA. One possibility was thermal initiation. A 50% EGDMA sample without initiator and catalyst/ligand addition was placed in an oil batch at 90 ^{0}C and measured ESR at 60, 120, 180, 480, and 1440 min. The sample was still in liquid, and no methacrylate radical signal was observed.

CuBr/Bpy complex has very low solubility in MMA, that limits its mediation ability in the polymerization. We also carried out the ATRP of MMA mediated by MBP/CuBr/HMTETA. Figure 2a shows the spectra at 90 ^{0}C with 18.2% EGDMA. Compared to the bipyridine system, the HMTETA system yielded much higher methacrylate radical concentrations. The effect of the ligand on the catalyst center was also evident from the change of the Cu(II) hyperfine structure, which were different from that in the EBIB/CuBr/Bpy system. The 9-line signal appeared at 33 min and its intensity increased dramatically after that. An extensive investigation on this system revealed that the time for the radical signal to appear, i.e., the onset of diffusion-controlled radical deactivation, was determined by a complete gelation of the system. The monomer conversion and gel fraction data were measured at the time of radical signal appearance for different EGDMA levels (17). The onset-point monomer conversion changed, but the gel fractions were all reached to almost 100%.

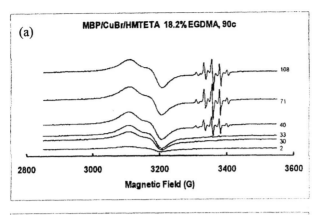

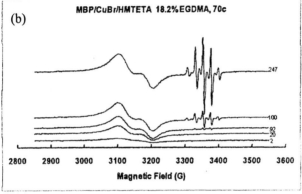

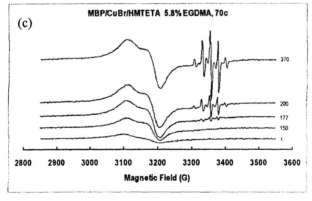

Figure 2. ESR spectra (single scan) observed in bulk ATRP of (a) 18.2% EGDMA, 90 ⁰C, (b) 18.2% EGDMA, 70 ⁰C, and (c) 5.8% EGDMA, 70 ⁰C initiated by methyl α-bromophenylacetate and mediated CuBr/1,1,4,7,10,10-hexamethyl triethylenetetramine, at time (min) as indicated. [M]/[CuBr] /[HMTETA]/[MBP] = 100/1/1/1.

We also examined the effect of temperature on the radical appearance. Figure 2b shows the result of 70 ^{0}C with 18.2% EGDMA. Compared to 90 ^{0}C (Figure 2a), the radical signal appeared later at 90 min, but the methacrylate radical spectra were as clear. We also measured pure MMA with MBP/CuBr/HMTETA and observed only Cu(II) signal. A small amount of EGDMA would help to elucidate the radical signal. Figure 2c shows the spectra with 5.8% EGDMA at 70 ^{0}C. The radical signal having the same hyperfine structure appeared at 170 min.

2) CCl$_4$/RuCl$_2$(PPh$_3$)$_3$/Al(OiPr)$_3$

Figures 3a-d show the ESR spectra measured with the CCl$_4$/RuCl$_2$(PPh$_3$)$_3$/Al(OiPr)$_3$ system. Other experimental conditions were the same as in System 1. In pure MMA, the signals at the early stage of polymerization were very weak. The Ru(III) signal intensity increased gradually. This ruthenium system appeared to be more complicated than those of CuBr/Bpy and CuBr/HMTETA. First, the Ru(III) spectrum experienced peak broadening and splitting. The Ru(III) signal was initially a single-lined spectrum and became double-lined The transition started at 30 min.

There was another signal at the shoulder of the Ru(III) spectrum. This signal was centered at 3350 G. Its intensity reached a maximum at 152 min and started to decrease after that. Based on the location information, it might be from some carbon radicals. We have thought about many possibilities and carried out tests accordingly, but have not reached a solid conclusion regarding the nature and function of this radical type. The identification work of this signal is underway.

Figure 3b shows the spectra measured at 10% EGDMA. The Ru(III) spectrum underwent broadening and splitting as in pure MMA. However, the unidentified signal at 3350 G remained till 80 min. Then, some hyperfine structures started to develop. The signal at 100 min showed a resemblance to the 9-line methacrylate radical. The central peak clearly overlapped with the unidentified signal. The unidentified signal survived for longer time. At 616 min, the methacrylate radical signal disappeared, but the unidentified signal remained.

The methacrylate radical signal became very clear at higher EGDMA levels. Figure 3c show the ESR spectra with 20% EGDMA. The spectra at 80 and 145 min were typical of the 9-line methacrylate radical. The spectra before 80 min and after 145 min had some influence from the unidentified signal. This became even more clear with 50% EGDMA as shown in Figure 3d. The signal starting from 26 min was very close to that of methacrylate radical. The radical concentration increased dramatically after that. The 9-line signal became dominant at 100 min (note the intensities of the spectra at 100, 570 and 850 min were divided by a factor of 5 in order to fit into the figure). The influence of the unidentified signal was still observable in the spectra collected at the early stage as well as the final ones.

An examination of all the data with the CCl$_4$/RuCl$_2$(PPh$_3$)$_3$/Al(OiPr)$_3$ system revealed a common feature. The methacrylate radical signal started to appear when the system gelled and imposed significant diffusion limitations to the radical deactivation reaction. The signal intensity, which is approximately proportional to the radical concentration, increased rapidly and reached a maximum followed by a decrease due to bimolecular radical termination. On the other side, the unidentified signal appeared earlier and disappeared later than that of methacrylate radical. The change in its intensity was much smaller during the polymerization.

3) EBIB/CuBr/PPMI

Figures 4a-d show the ESR spectra with the EBIB/CuBr/PPMI system. There was an interesting observation for this system. In pure MMA, the Cu(II) signal showed from the very beginning. Its intensity decreased with time and reached a minimum at 76 min before it came back and started to increase. The Cu(II) spectrum was also broader than those in the EBIB/CuBr/Bpy and MBP/CuBr/HMTETA systems. Starting at 74 min, a sharp single-line signal appeared in the carbon radical field. The appearance and growth of this single-line signal was accompanied with further broadening of the Cu(II) signal. The intensity of this single-line signal increased and appeared to be very stable. It is unknown if this single-line signal was from the same radical source as that in the $CCl_4/RuCl_2(PPh_3)_3/Al(OiPr)_3$ system. We tried many different experiments including different degassing methods, different ampoule reactor types, and various means of generating possible bromine radical. So far, we were not able to identify this species.

Figure 4b shows the spectra with 10% EGDMA. The 9-line methacrylate radical signal became very clear at 66 min. Its intensity increased with time, reached a maximum before decrease, and finally disappeared. However, the single-line signal persisted with its intensity continuously increased. The species became dominant at the final stage and appeared to be very stable in the polymer matrix. Figure 4c shows the spectra with 20% EGDMA. Figure 4d shows those with 50% EGDMA. The appearance of the methacrylate radical signal started earlier and the intensity became stronger. However, the trends in the signal development were the same as in 10% EGDMA. The methacrylate radical signal came and gone, while the single-line signal grew continuously.

4) Propagating center type

We also measured ESR for the systems without MMA/EGDMA addition. Toluene was used as solvent. The objective was to examine if the single-lined signals were generated from side reactions of initiator and catalyst/ligand molecules. Surprisingly, the single-line signals were present in EBIB/CuBr/PPMI (see Figure 5a) and $CCl_4/RuCl_2(PPh_3)_3/Al(OiPr)_3$ (Figure 5b), but not in EBIB/CuBr/Bpy and MBP/CuBr/HMTETA. The signal in the EBIB/CuBr/PPMI system was particularly strong, and its intensity increased with time. The signal in the $CCl_4/RuCl_2(PPh_3)_3/Al(OiPr)_3$ system was weak and did not change much with time. We also measured ESR for the combinations of EBIB+CuBr, EBIB+PPMI, and CuBr+PPMI in toluene at different time

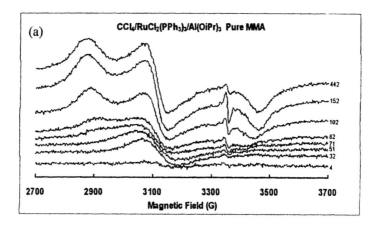

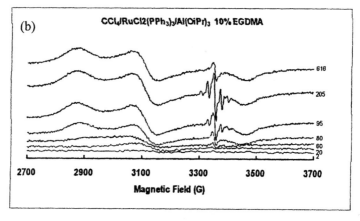

*Figure 3. ESR spectra (5 scans) observed in bulk ATRP of (a) pure MMA, (b) 10% EGDMA, (c) 20% EGDMA, and (d) 50% EGDMA, initiated by CCl_4, and dichlorotris(triphenylphosphine) ruthenium and activated by $Ai(OiPr)_3$, at 90 ^0C and time (min) indicated. $[M]/[RuCl_2(PPh_3)_3]/ [Al(OiPr)_3 /[CCl_4] = 100/0.5/2/2$. The signal intensities indicated by * were divided by a factor of 5. Modulation amplitude 10.0 G.*

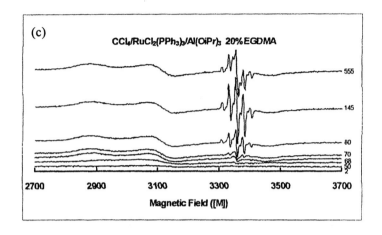

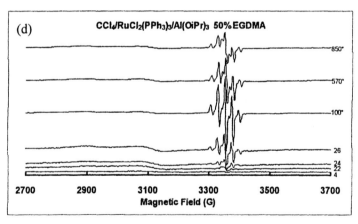

Figure 3. *Continued.*

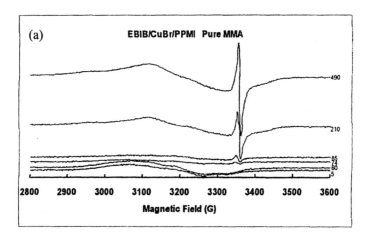

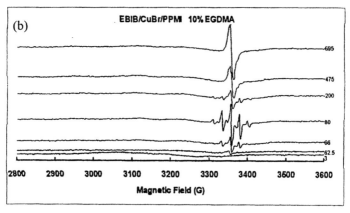

*Figure 4. ESR spectra (5 scans) observed in bulk ATRP of (a) pure MMA, (b) 10% EGDMA, (c) 20% EGDMA, and (d) 50% EGDMA, initiated by EBIB and mediated by CuBr/ N-pentyl-2-pyridylmethanimine, at 90 ^{0}C and time (min) as indicated. [Vinyl]/[EBIB]/[CuBr]/[PPMI] = 100/1/1/2. The signal intensities indicated by * were divided by a factor of 20.*

175

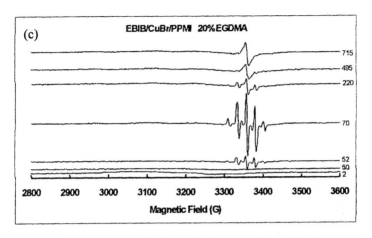

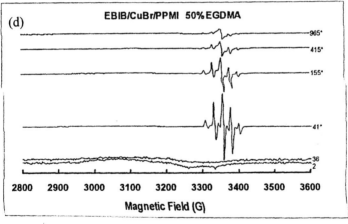

Figure 4. *Continued.*

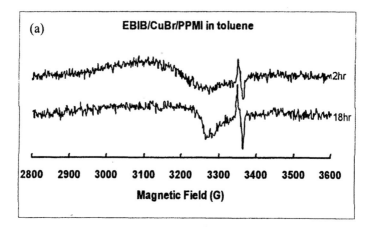

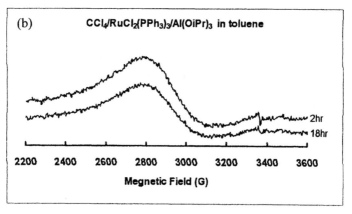

Figure 5. ESR spectra observed in the toluene solution of (a) EBIB/CuBr/N-pentyl-2-pyridylmethanimine, and (b) CCl₄/RuCl₂(PPh₃)₃/Al(OiPr)₃, at 90 ⁰C and time indicated. ESR operation parameters are the same as in Figures 4 and 3, respectively.

intervals and 90 ⁰C, and observed no signals other than the oxidized metal center when CuBr was included. These measurements revealed that the single-line signals were not associated with monomers or monomeric units in polymer chains. The lack of hyperfine structure also indicated that they were unlikely to be from initiating radicals. The possibilities for these single-line species being some types of propagating centers have therefore been ruled out.

The single-line paramagnetic species were long-lived in polymer networks. For example, we measured the EBIB/CuBr/PPMI sample of 10%EGDMA again after storage at room temperature for two months. There was no decay in the signal intensity. We broke the ampoule and exposed the sample to air for two hours at room temperature and for another 15 min at 90 ⁰C. No significant change was observed, as shown in Figure 6. This experiment proved that the species were trapped in the polymer matrix, but not in the ampoule wall, and that they were non-reactive with oxygen molecules.

This work demonstrated the presence of methacrylate radicals in all four ATRP systems. The hyperfine structures of their representative ESR spectra are identical. These signals origin from methacrylate radicals trapped in a glassy/solid matrix, and are the same as those observed in conventional free radical polymerization processes. Due to the sensitivity limit of ESR machine, we were not able to detect the methacrylate radicals in the pure MMA systems.

What is the origin of methacrylate radicals? It was demonstrated that they were not generated by thermal initiation. It was also demonstrated that the single-line species observed in Systems 2 and 3 were not propagating centers. On the other hand, what was the effect of the polymer network in the polymerization? It should not have any chemical influence on the MMA/EGDMA polymerization.

A most plausible physical function was the diffusion limitations imposed on the reacting species such as radical chains, catalyst/ligand complexes, and monomer molecules. The direct observation of methacrylate radical, as well as the oxidized transition metal species, revealed that the oxidized metal complex molecules ($CuBr_2$/Bpy in System 1, $RuCl_3$/$(PPh_3)_3$ in System 2, and $CuBr_2$/PPMI in System 3) diffused freely in the reaction mass, as suggested by the equilibrium

$$PX + Mt^{(n)}X_n/L \leftrightarrow P^{\bullet} + Mt^{(n+1)}X_{n+1}/L.$$

In pure MMA system or at low conversion of EGDMA, these complex species experienced no difficulty in diffusing back to the chain radical center $P^{\bullet}$ and efficiently capped the center. Upon network formation, the diffusion limitations

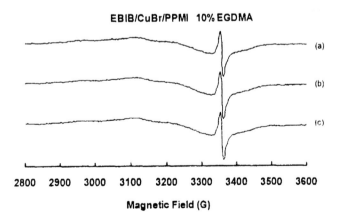

EBIB/CuBr/PPMI 10% EGDMA

Magnetic Field (G)

Figure 6. ESR spectra observed in 10%EGDMA EBIB/CuBr/PPMI sample (a) after stored at room temperature for two months, (b) exposed the sample to air for two hours at room temperature, and (c) for another 15 min at 90 ^{0}C. ESR operation parameters are the same as in Figure 4.

imposed on the species by the network slowed down their mobility, and thus increased the radical concentration to a level detectable by ESR. If the complex molecules never left the radical centers, they would not be caught by the network.

Another basic question is whether there existed some type of intermediate species such as

$$P^{\bullet}\text{--}X\text{--}Mt^{(n+1)}X_n/L \qquad \text{or} \qquad P^{\bullet}\text{--}Mt^{(n+1)}X_n/L$$

If these intermediates existed, were they responsible for monomer propagation? Both structures would require equimolar concentrations of $P^{\bullet}$ and Cu(II) species. However, all the reported here and earlier[ll] ESR data indicate much higher concentration of $Mt^{(n+1)}$ than $P^{\bullet}$. Even if a stable radical intermediates were present and capable of monomer insertion (this hypothesis needs to be proven), the species coexisted with the free methacrylate radicals. The ATRP propagating centers have reactivities typical for free radicals.[19] ATRP systems are more sensitive to solvents and/or other conditions than their conventional counterparts, because of possible reactions and interactions with catalyst/ligand molecules that alter their ability to mediate.

Conclusions

We carried out the ESR measurements for the ATRP of MMA/EGDMA mediated by three catalyst systems, i.e., System 1 initiated by ethyl 2-bromoisobutyrate (EBIB) mediated by CuBr/2,2'-bipyridine, System 2 initiated by CCl$_4$ mediated by dichlorotris(triphenylphosphine) ruthenium activated by aluminum isopropoxide, and System 3 initiated by EBIB mediated by CuBr/N-pentyl-2-pyridylmethanimine. The methacrylate radical concentration in pure MMA was lower than the limit of ESR sensitivity. The dimethacrylate was added for the purpose of network formation that triggered diffusion limitations to chain radicals as well as catalyst/ligand complex species. The fast reactions such as bimolecular radical termination and radical deactivation thus became diffusion controlled. The diffusion-controlled radical deactivation moved the equilibrium $PX + Mt^{(n)}X_n/L \leftrightarrow P^{\bullet} + Mt^{(n+1)}X_{n+1}/L$ to the right-hand-side and consequently increased the radical concentration.

Methacrylate radicals were observed in all the systems upon the network formation. Their signals showed the typical 9-line hyperfine structure that was identical to those observed in conventional free radical polymerization processes. We also observed peculiar signals in System 2 and System 3, but ruled out the possibility of these species being propagating centers, because these signals were also present without the monomer addition. The methacrylate radical appeared to be the center for monomer propagation. This work supports the radical mechanisms in the ATRP systems.

179

Acknowledgements. We would like to thank the Natural Sciences and Engineering (NSERC) of Canada for financial support of this research, and the Ministry of Education, Science, and Technology (MEST) for PREA award.

References

1   Szwarc, M. *Nature* **1956**, *178*, 1168. (b) Szwarc, M. *J. Am. Chem. Soc.* **1956**, *78*, 2656.
2   Webster, O.W. *Science* **1991**, *251*, 887
3   (a) Wang, J.S.; Matyjaszewski, K. *J. Am. Chem. Soc.* **1995**, *117*, 5614. (b) Wang, J.S.; Matyjaszewski, K. *Macromolecules* **1995**, *28*, 7901
4   (a) Kato M.; Kamigaito, M.; Sawamoto, M.; Higashimura, T. *Macromolecules* **1995**, *28*, 1721. (b) Kotani, Y.; Kato, M.; Kamigaito, M.; Sawamoto, M. *Macromolecules* **1996**, *29*, 6979.
5   Georges, M.K.; Veregin, R.P.N.; Kazmaier, P.M.; Hamer, G.K. *Macromolecules* **1993**, *26*, 2987.
6   Chiefari, J.; Chong, Y.K.; Ercole, F.; Krstina, J.; Jeffery, J.; Le. T.P.T.; Mayadunne, R.T.A.; Meijs, G.F.; Moad, C.L.; Moad, G.; Rizzardo, E.; Thang, S.H. *Macromolecules* **1998**, *31*, 5559.
7   Matyjaszewski, K. *Chem. Rev.* **2001**, *101*, 2921.
8   Kamigaito, M.; Ando, T.; Sawamoto, M. *Chem. Rev.* **2001**, *101*, 3689.
9   Bellus, D. *Pure Appl..Chem.* **1985**, *57*, 1827
10 (a) Curran, D.P. *Synthesis* **1988**, 417, (b) Curran, D.P. *Synthesis* **1988**, 489
11 Matyjaszewski, K. *Macromolecules* **1998**, *31*, 4710.
12 Haddleton, D.M; Lad, J.; Harrisson, S. *ACS Polym. Prepr.* **2002**, *43*, 13.
13 (a) Matyjaszewski, K.; Kajiwara, A. *Macromolecules* **1998**, *31*, 548. (b) Kajiwara, A.; Matyjaszewski, K.; Kamachi, M. *Macromolecules* **1998**, *31*, 5695.
14 Yu, Q.; Zeng, F.; Zhu, S. *Macromolecules* **2001**, *34*, 1612.
15 Kamachi, M. *Adv. Polym. Sci.* **1987**, *82*, 207. (b) Kamachi, M. *J. Polym. Sci. Polym. Chem.*, **2002**, *40*, 269.
16 Zhu, S.; Tian, Y.; Hamielec, A.E.; Eaton, D.R. *Macromolecules* **1990**, *23*, 1144. (b) Zhu, S.; Tian, Y.; Hamielec, A.E.; Eaton, D.R. *Polymer* **1990**, *31*, 154. (c) Tian, Y.; Zhu, S.; Hamielec, A.E.; Fulton, D.B.; Eaton, D.R. *Polymer* **1992**, *33*, 384.
17 Wang, A.R.; Zhu, S. *Macromolecules* **2002**, *35*, 9926
18 Haddleton, D.M.; Jasieczek, C.B.; Hannon, M.J.; Shooter, A.J. *Macromolecules*, **1997**, *30*, 2190.
19 Pintauer, T.; Zhou, P.; Matyjaszewski, K., *J. Am. Chem. Soc.* **2002**, *124*, 8196.

Chapter 13

Peculiarities in Atom Transfer Radical Copolymerization

Bert Klumperman[1], Grégory Chambard[2], and Richard H. G. Brinkhuis[3]

[1]Department of Polymer Chemistry (SPC), Eindhoven University of Technology, P.O. Box 513, 5600 MB Eindhoven, The Netherlands
[2]Dow Chemical, P.O. Box 48, 4530 AA Terneuzen, The Netherlands
[3]Akzo Nobel Chemicals Research, Dpt. CPR, P.O. Box 9300, 6800 SB Arnhem, The Netherlands

In recent literature it is quite common to treat living radical copolymerization completely analogous to its free radical counterpart. When doing so it is often observed that small deviations in the copolymerization behavior occur. In this work it is shown that the observed deviations can be interpreted on the basis of reactions specific for living radical polymerization, such as the activation – deactivation equilibrium in atom transfer radical polymerization. Reactivity ratios obtained from atom transfer radical copolymerization data, interpreted according to the conventional terminal model deviate from the true reactivity ratios of the propagating radicals. Furthermore, it is shown that the copolymerization of monomers that strongly deviate in equilibrium constant between active and dormant species may exhibit for low molar mass copolymers a significant correlation between molar mass distribution and chemical composition distribution. The rationale behind this correlation is that the chains that happen to add a relatively large fraction of the more dormant monomer lag behind in their growth. The more dormant monomer refers to the monomer creating a chain end with the lower equilibrium constant in the activation – deactivation process. This monomer will be preferentially present in the low molar mass fraction of the distribution, whereas the other monomer will be present to a larger extent in the high molar mass fraction.

Introduction

The majority of studies on living radical polymerization focuses either on homopolymerization or on block copolymerization. Some publications focus on statistical copolymerizations, and the possibility to synthesize gradient or tapered copolymers (*1*). In the majority of publications on statistical copolymers it is common practice to evaluate the reactivity ratios of the living radical copolymerization system, and compare the results with the free radical copolymerization analogue (*2*). In doing so, it is often found that reactivity ratios in LRcP deviate slightly from those in FRcP (*3*). There is also very recent theoretical evidence that copolymerizations with intermittent activation may show behavior deviating from conventional copolymerization (*4*). Usually the experimentally observed differences are small enough to safely interpret the values of LRcP as being in good agreement with the values of FRcP. Here we want to show that a good agreement is not always expected. Based on simulations as well as on experimental data it will be shown that reactivity ratios determined from LRcP are only apparent reactivity ratios. It needs to be stressed that the presently discussed phenomenon clearly differs from the preferential addition of the primary (initiator derived radical) to one of the two comonomers in conventional FRcP. The phenomenon of preferential addition has been investigated before (*5*).

Somewhat related to this seemingly theoretical exercise is the desire from e.g. coatings perspective to synthesize functional (co)polymers with a narrow functionality distribution. It is well known that if the aim of the synthesis is a relatively low molar mass copolymer, with an average functionality of two, a significant fraction of non-functional polymer will be present. This is the case when the method of incorporation is via conventional free radical copolymerization of e.g. a non-functional methacrylate and hydroxy-ethyl-methacrylate. For ATR copolymerizations, the large differences in equilibrium constant between acrylates and methacrylates turn out to be quite beneficial in the synthesis of copolymers with narrow functionality distribution. Also on this topic, simulations and experimental evidence will be presented to stress the scope of this phenomenon.

The effect of initiation on atom transfer radical copolymerization

The copolymerization of styrene (STY) and butyl acrylate (BA) has been studied by several groups via conventional free radical copolymerization (FRcP) as well as by atom transfer radical copolymerization (ATRcP) (*6*). In order to determine the reactivity ratios that govern a copolymerzation, two methods are commonly used:

Low conversion copolymerization, and subsequent fitting of the copolymer composition *versus* monomer feed composition to the well-known Mayo-Lewis equation (*7*).

Measurement of the individual rates of conversion of the two comonomers as a function of overall monomer conversion, and subsequent fitting of the data to the integrated version of the Mayo-Lewis equation, also known as the Skeist equation (8).

It is generally accepted that the first method is not applicable to ATRcP. The short chain lengths associated with low conversion in a living radical polymerization (LRP) violate the assumptions inherent to the Mayo-Lewis model. The effect of the initiator-derived (primary) radical, and the known deviating reactivities of short-chain radicals would yield results that are not applicable to long chain copolymerizations. Therefore, studies on ATRcP are usually conducted according to the second method.

In an attempt to shed some light on the small but significant differences between reactivity ratios from FRcP and ATRcP, a number of simulations were carried out. The whole set of differential equations for both FRcP as well as for ATRcP were numerically solved without any assumption regarding steady state of the radicals. In both systems, chain transfer was neglected. Bimolecular termination was considered to be independent of chain length. The terminal model was used in the simulations, which means that radical reactivities are assumed to be dependent only on the terminal monomer unit in a growing radical. It is known that this assumption results in a poor description of the rate of polymerization, but that it adequately describes the copolymer composition data as a function of monomer feed composition (9). In view of the purpose of this simulation, it is expected that the choice to use the terminal model does not influence the results. When carrying out the simulations it was confirmed that the results were independent on the step-size of integration. During the simulation, the concentrations of all relevant species were tracked as a function of time, *i.e.* both types of chain end radicals, both types of dormant species, and both monomers.

For the sake of availability, the simulations were carried out using the rate constants of primary radical addition of the ethyl-2-isobutyryl radical to styrene (STY) and butyl acrylate (BA). Also the other rate constants were taken from literature sources, and are listed in Table I.

On inspection of the results from various simulations it turned out that there is a significant difference in the ratio of the two active radicals between the FRcP and ATRcP scenarios. In FRcP, the ratio of concentrations of the two radicals is gradually, and monotonically changing to follow the well-known composition drift in copolymerization at a non-azeotropic comonomer ratio. In Figure 1 this is clearly demonstrated by the dashed curve. If one rearranges the steady state assumption, which is usually made in the derivation of the Mayo-Lewis equation, it follows that the ratio of the radical ratios over the comonomer ratios should be a constant. On the right-hand axis in Figure 1 this ratio is represented. The dashed line, which is associated with the right-hand axis clearly proves the applicability of this assumption for FRcP. In ATRcP the same behavior is observed at monomer conversions above approximately 15%. At lower conversions however it is clear that there is a significant deviation from the FRcP behavior. The deviation at high conversion, which is most prominent in the $(p_{STY}/p_{BA})/ (f_{STY}/f_{BA})$ curve of ATRcP, is most likely due to f_{STY} approaching zero. An artifact originating from the numerical approximation is held responsible for the deviation.

Table I. Initial concentrations and parameters used in the simulations of ATRcP and FRcP

Parameter / Concentration	ATRcP	FRcP
f_S^0	0.75	0.75
$[M_t^n]_0$	$5 \cdot 10^{-2}$ mol·L^{-1}	-
$[I-X]_0$	$5 \cdot 10^{-2}$ mol·L^{-1}	$5 \cdot 10^{-2}$ mol·L^{-1}
k_S^1 (10)	$5.50 \cdot 10^3$ L·mol^{-1}·s^{-1}	
k_B^1 (10)	$1.20 \cdot 10^3$ L·mol^{-1}·s^{-1}	
k_{SS} (11)	$1.58 \cdot 10^3$ L·mol^{-1}·s^{-1}	$1.58 \cdot 10^3$ L·mol^{-1}·s^{-1}
k_{BB} (12)	$7.84 \cdot 10^4$ L·mol^{-1}·s^{-1}	$7.84 \cdot 10^4$ L·mol^{-1}·s^{-1}
K_{act}^1 (13)	0.43 L·mol^{-1}·s^{-1}	-
K_{deact}^1 (14)	$6.8 \cdot 10^7$ L·mol^{-1}·s^{-1}	-
r_S (15)	0.89	0.89
r_B (15)	0.24	0.24
k_{act}^S (13)	0.43 L·mol^{-1}·s^{-1}	-
k_{act}^B (13)	$7.5 \cdot 10^{-2}$ L·mol^{-1}·s^{-1}	-
k_{deact}^S (14)	$6.8 \cdot 10^7$ L·mol^{-1}·s^{-1}	-
k_{deact}^B (14)	$1.0 \cdot 10^8$ L·mol^{-1}·s^{-1}	-
k_{dis} (16)	-	$2.0 \cdot 10^{-6}$ s^{-1}
k_t^0 (17)	$1.0 \cdot 10^8$ L·mol^{-1}·s^{-1}	$1.0 \cdot 10^8$ L·mol^{-1}·s^{-1}

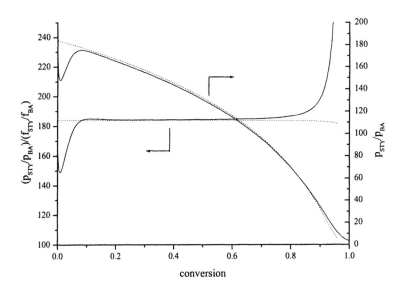

Figure 1. Ratio of active species (p_{STY}/p_{BA}) and of active species over comonomer ((p_{STY}/p_{BA})/ (f_{STY}/f_{BA})) as a function of conversion for a simulated ATRcP (drawn curves), and simulated FRcP (dotted curves). For parameters used in the simulations see Table I.

184

From earlier work in the area of LRP it is known that it may take some time for the pseudo-equilibrium between dormant and active species to establish. The build-up of persistent radicals is associated with termination reactions. Therefore, depending on the kinetic parameters of a particular polymerization, it may take a certain percentage of monomer conversion for the equilibrium to set in. It is expected that in the case of copolymerization an additional effect plays a role. If the primary radical has a certain preference for addition to one of the two monomers, the ratio of the dormant species may deviate from the expected ratio in the long chain limit. The dormant species that get activated gradually shift the ratio to the expected value, which is likely to be identical to the FRcP case. Obviously the consumption of initiator is not instantaneous. The consequence of this is that throughout the initial stages of the polymerization one may expect the creation of dormant species in the non-equilibrium ratio.

The shape of the drawn curves in Figure 1 can be explained according to these two opposing phenomena. On one hand the creation of a non-equilibrium ratio of dormant radicals, and on the other hand the tendency of activated chains

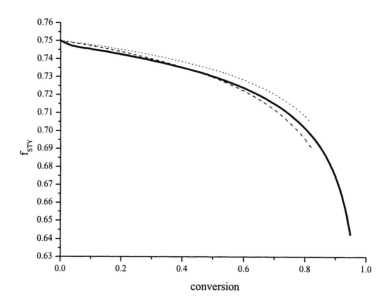

Figure 2. Fraction styrene in a STY/BA reaction mixture as predicted by several simulations. Bold curve: ATRcP according to the parameters in Table I; dotted curve: FRcP with reactivity ratios identical to the ATRcP; dashed curve: r_S = 0.89, r_B = 0.19.

to return to the equilibrium ratio. The two effects lead to the characteristic dip in the curves as seen in Figure 1.

The non-equilibrium ratio of the dormant species logically results in a different ratio of the two active chain ends during the first stages of the polymerization. With the probabilities of addition of the two monomers only being determined by the nature of the chain end radical, this leads to a deviation in fractional comonomer conversion compared to FRcP. When ATRcP is carried out to determine reactivity ratios as indicated above, this may lead to apparent values that do not reflect the true reactivity of the chain ends. In Figure 2 an attempt is illustrated to visualize the effect described above. The bold drawn curve is the fraction of STY in the residual monomer mixture as a function of overall monomer conversion for the simulation described above. The dashed and dotted curves are FRcP traces calculated for various reactivity ratios. It is clearly seen that when the same reactivity ratios are used in simulating ATRcP and FRcP, there is a systematic deviation between the two curves. Numerically the fit can be improved by adjusting the reactivity ratios in the FRcP simulation. However, this leads to a situation in which the two traces have a different shape. Unfortunately there is no experimental data yet that allows a critical evaluation of this simulated effect, but it is likely to be a realistic explanation of the small but significant deviations often observed between FRcP and ATRcP. Currently we investigate copolymerizations by on line techniques to follow individual comonomer conversions with a relatively high sampling frequency. The different shapes of the comonomer conversion curves should become apparent from these experiments.

One additional remark should be made in relation to the treatment of FRcP in this publication. The Mayo-Lewis model has been derived for the long chain limit in which initiation and termination effects can be neglected. In order to verify the effect of the long chain assumption on comonomer ratio *versus* conversion additional simulations were carried out. Simulations with high initiation rates may show some deviation from the equivalent ones for the long chain limit (results not shown here). However, the effect of reducing the kinetic chain length from around 1,800 to around 130 is much smaller than the effects described above resulting from variations in rate (and selectivity) of primary radical addition.

The effect of equilibrium constants in ATRcP on MMCCDs

In certain applications, *e.g.* functional acrylic oligomers in coatings, it is beneficial to combine low molar mass with a narrow functionality distribution. These requirements originate from the desire to have low viscosity during processing, and good network properties after crosslinking of the material. In traditional research on acrylic polyols the incorporation of functional materials as hydroxy ethyl (meth)acrylate served to introduce functionality in the chains, with molecular weights and average functionalities (number of OH-groups per chain) being lowered with the desire to use less solvent . Molecular weight of so-called high solids acrylic polyols can be less than $\overline{M}_n \approx 2000$. In conventional FRcP this inevitably leads to the introduction of mono- and non-functional polymer chains. These chains finally end up in the network as dangling ends or extractable material, respectively An obvious way to overcome this problem is the synthesis of telechelic polymers carrying a functionality on both chain ends. Synthetically easier is the statistical copolymerization of a functional monomer into a (meth)acrylic polymer chain in an LRP process. The narrow MMD obtainable can be translated into narrower functionality distributions. Acrylic polyol copolymers usually have both methacrylate as well as acrylate comonomers in their backbone. The hydroxy functional monomer can either be an acrylate or a methacrylate. . The question that we want to answer is that whether the choice to assign the functional group to either the acrylate or the methacrylate component in the copolymer would affect the functionality distribution of the final product.

If one looks at the rate parameters of the various acrylates and methacrylates it is obvious that propagation rate constants of acrylates are more than an order of magnitude larger than those of methacrylates. However, for the equilibrium constant between active and dormant species in ATRP exactly the reverse is true, acrylates are much more in the dormant state than methacrylates. When a composite reactivity is defined for the event in which a dormant chain is activated, adds a monomer, and is deactivated again, it turns out that methacrylates are approximately 40 times more reactive than acrylates, since this quantity is determined by the product of the rate constants of activation and that of propagation. For homopolymerizations this leads to differences in the rate of polymerization, but what happens when methacrylates are copolymerized with acrylates?

Figure 3 shows the result of a set of Monte Carlo simulations (based on parameters which were reported in literature (2)) in which acrylates and methacrylates are copolymerized in different ratios. The system is somewhat simplified, since it assumes an infinitely fast equilibrium between the dormant and active state of the chains. However, the general results presented here are not strongly influenced by this assumption. The simulations are carried ou in such a way that composition drift is avoided, i.e. a monomer feed strategy is invoked in the simulation. Figure 3 shows the polydispersity index as a function of the fraction of methacrylate in the comonomer mixture. As can be seen, the polydispersity index of a polymethacrylate significantly increases when small amounts of acrylate are incorporated. The effect is weaker when methacrylate is incorporated in polyacrylate. This asymmetry can be rationalized by simply imagining what happens to polymer chains upon incorporation of a comonomer. A methacrylate polymerizes relatively rapidly under ATRP conditions due to a favorable equilibrium constant for the activation/deactivation equilibrium. When a chain adds an acrylate comonomer, it will be inactive for a significantly longer time. Thus it lags behind in growth compared to the chains that add less of the acrylate.

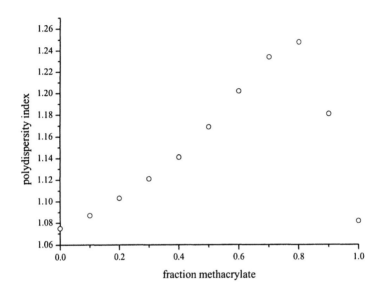

Figure 3. Polydispersity index as a function of fraction of methacrylate monomers, for a simulated ATRcP (monomer conversion 0.99 in all cases). Molar ratio monomer/initiator 10.

The result is an increased polydispersity index. When the similar action is considered for a polyacrylate chain that occasionally adds a methacrylate unit the effect will be less drastic. The majority of propagation steps are acrylate homopropagation steps. The occasional addition of a methacrylate results in a chain that, after deactivation, relatively quickly gets reactivated and propagates (most likely with an acrylate). Also this will result in a broadening of the MMD, but it is readily appreciated that this broadening is less severe than in the former case.

When a copolymer of acrylate and methacrylate is synthesized it is expected that the chemical composition as a function of molar mass will vary. Based on the same arguments as outlined above for the width of the molar mass distribution it is expected that low molar mass chains will be enriched in acrylate, whereas the higher molar mass will be enriched in methacrylate. Monte Carlo simulations are compared with experimental data on chemical composition as a function of molar mass. A copolymer of butyl acrylate and methyl methacrylate was synthesized via ATRcP with an overall fraction of MMA F_{MMA} = 0.55, and a number average degree of polymerization P_n = 8.5. The copolymer was subjected to MALDI-TOF-MS as well as to HPLC-ESI-MS. Part of the mass spectrum from the MALDI experiment is shown in Figure 4.

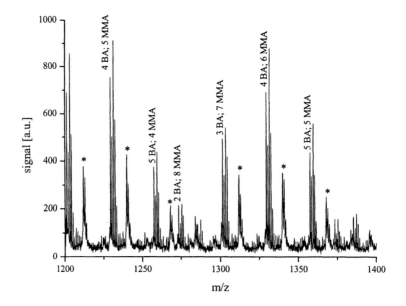

Figure 4. Experimental MALDI-TOF-MS spectrum of MMA-BA copolymer with F_{MMA} = 0.55, and $\overline{P}_n$ = 8.5; peaks indicated with an asterisk are the result of post source fragmentation.

The shortest chains of the distribution were obscured by matrix peaks so that we had to rely on the HPLC-ESI-MS data there. It is known that quantification of MALDI data can lead to erroneous results due to preferential ionization, or due to mass discrimination. In the present case we are dealing with a relatively narrow molar mass distribution. Furthermore, from calculations as well as from experimental data it seems that the very great majority of dormant chains carries a butyl acrylate unit as the terminal monomer. The published lactone formation from a PMMA chain with a terminal bromine atom is therefore completely absent (*18*). Peaks like the one indicated in Figure 4 were assigned to the various copolymer compositions and integrated to obtain a best estimate of the relative amounts of the individual chains. The peaks indicated with an asterisk are due to post source fragmentation and are not included in the further calculations. The justification for this procedure can be found in more detail elsewhere. (*19*)

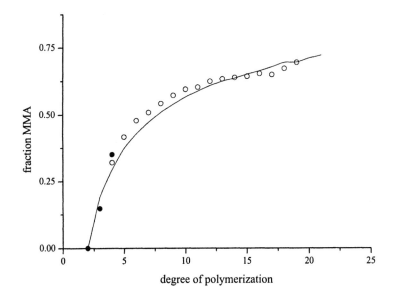

Figure 5. Mole fraction MMA (m/(m+n)) for chains with mass EBIB-MMA$_m$-BA$_n$, as a function of chain length. Open symbols derived form MALDI, solid symbols derived from ESI data. Solid curve based on simulations as discussed in the text.

In Figure 5, the fraction MMA is plotted as a function of chain length. Noteworthy is that the shortest chains (degree of polymerization 1 and 2) do not contain any MMA. As indicated above, the fraction acrylate at low molar mass is larger than the average value $F_{BA} = 0.45$. There is a fairly good agreement between experimental values and results from the Monte Carlo simulation (drawn curve).

Returning to the case of copolymers comprising a functional monomer, it is clear that this phenomenon is expected to have a direct impact on the functionality distribution. If we allow ourselves to generalize these MMA-BA results to acrylate-methacrylate copolymers in which one of the monomers contains a functional side group, we can predict based on similar simulations that a functional acrylate will be enriched in the lowest molar mass chains when copolymerized with a non-functional methacrylate, whereas it will be depleted in the highest molar mass chains. In comparison to a situation without molar mass – chemical composition dependency, this situation is favorable because the number of chains of extremely low functionality as well as that of extremely high functionality is lowered. Conversely, the presence of chains of such extreme functionality will be enhanced when a copolymer of a functional methacrylate and a non-functional acrylate is considered.

Experimental

Materials

Methyl methacrylate (MMA) (99%), n-butylacrylate (BA) (99+%), N,N,N',N'',N''-pentamethyldiethylene triamine (PMDETA) (99%), ethyl-2-bromo-isobutyrate (EBIB) (98%), ethylene glycol dimethyl ether (99%), $CuCl_2.2H_2O$ (99+%) and CuCl (99%) were supplied by Aldrich. The monomers were purified by passing them over a basic alumina column, the other reagents were used as received.

Procedure

The sample for mass spectrometry studies was prepared via atom transfer radical copolymerization of methyl methacrylate (MMA) and n-butyl acrylate (BA). The polymerization was initiated by ethyl-2-bromo-isobutyrate (EBIB), and mediated by a $CuCl/CuCl_2/PMDETA$ system, using ethylene glycol dimethyl ether (EGDME) as solvent. A reaction vessel was filled with 91.2 g EGDME, 0.37 g (3.7 mmol) CuCl, 0.03 g (0.02 mmol) $CuCl_2.2H_2O$, 1.97 g (11.2 mmol) PMDETA, 21.9 g (219 mmol) MMA and 51.9 g (405 mmol) BA, raising the temperature to 75 °C, and adding 19.5 g (100 mmol) EBIB to start the reaction. Simultaneously, a mixture of 43.8 g (438 mmol) MMA and 28.1 g (219 mmol) BA was slowly added over the course of the first two hours of the reaction. The reaction was stopped after 5 hours and 30 minutes, at an overall conversion level of 60 % by weight of the monomers. Based on this conversion level, a theoretical

$\overline{P}_n$ (assuming 100% initiator efficiency) of 8.5 is calculated (including the initiator unit). The MMA/BA molar ratio incorporated was 55/45 as determined via ^{1}H NMR (ester side chain resonances) on the polymer. SEC analysis yielded an $\overline{M}_n$ = 1065 (polystyrene standards), and a polydispersity index, PDI=1.39. The polymer was purified by passing it over a cation exchange column to remove Cu ions and the PMDETA ligands.

Analysis

SEC analysis was carried out on a styrene-divinylbenzene type mixed bed column (Mix C) from Polymer Labs, equipped with a RI detector. The eluens used was THF with 0.5% acetic acid, at a flow rate of 1.0 mL/min at ambient temperature.

Matrix-assisted laser-desorption ionization time-of-flight mass spectrometry (MALDI-TOF-MS) experiments were carried out on an Applied Biosystems Voyager DE-STR. The instrument was operated in reflector mode. 3,5-Dimethoxy-4-hydroxycinnamic acid (DMHCA, 98 %, Aldrich) was used as the matrix, and sodium trifluoro-acetate (NaTFA, 98%, Aldrich) was added to ensure cationization by sodium. Polymer sample, DMHCA, and NaTFA were dissolved in stabilized THF (HPLC grade, Biosolve Ltd.) at concentrations of 5 mg/mL, 23 mg/mL, and 5 mg/mL resp. A 6 : 6 : 1 mixture of the polymer, DMHCA, and NaTFA solutions (50 μL) was spotted on the target plate from a pipette and subsequently air-dried. Instrument settings during the MALDI measurement were as follows. Accelerating voltage 25000 V, grid voltage 70%, mirror voltage ratio 1.12, extraction delay time 100 ns. Laser intensity 2150, laser repetition rate 20 Hz, low molar mass gate 800 Da.

Electron spray ionization MS (ESI-MS) measurements were carried out on a Micromass model LCT mass spectrometer. The MS was operated in the positive ion mode at a resolution of 5000 (FWHM). The source temperature was 110 °C, desolvation temperature 150 °C, cone voltage set at 30V, the RF lens at 200V and the capillary voltage was 3000V. Nitrogen was used as desolvation and nebulizer gas.

Peak intensities were determined for all clusters assigned to dormant polymer chains with masses identifiable as Na-ethylisobutyrate-MMA$_n$BA$_m$-Br (*20*) (dominant in the MALDI pattern over any non Br-containing peaks)

MALDI data on the lowest chain lengths (less than three monomers added) were obscured by matrix peaks. For these oligomers, ESI data were used.

192

References

1. Matyjaszewski, K.; Ziegler, M. J.; Arehart, S. V.; Greszta, D.; Pakula, T., *J. Phys. Org. Chem.* **2000**, *13*, 775
2. Ziegler, M. J.; Matyjaszewski, K., *Macromolecules* **2001**, *34*, 415
3. Haddleton, D. M.; Crossman, M. C.; Hunt, K. H.; Topping, C.; Waterson, C.; Suddaby, K. G., *Macromolecules* **1997**, *30*, 3992
4. Matyjaszewski, K., *Macromolecules* **2002**, *35*, 6773
5. Galbraith, M.N.; Moad, G.; Solomon, D.H.; Spurling, T.H., *Macromolecules* **1987**, *20*, 675
6. Arehart, S. V.; Matyjaszewski, K., *Macromolecules* **1999**, *32*, 2221; Dubé, M.A.; Penlidis, A.; O'Driscoll, K.F., *Can. J. Chem. Eng.* **1990**, *68*, 974
7. Mayo, F.R.; Lewis, F.M., *J. Am. Chem. Soc.* **1944**, *66*, 1594
8. Skeist, I., *J. Am. Chem. Soc.* **1946**, *68*, 1781
9. Fukuda, T.; Ma, Y.D.; Inagaki, H., *Macromolecules* **1985**, *18*, 17
10. Zytowski, T.; Knühl, B.; Fischer, H. *Helv. Chim. Acta* **2000**, *83*, 658
11. Buback, M.; Gilbert, R.G.; Hutchinson, R.A.; Klumperman, B.; Kuchta, R.-D.; Manders, B.G.; O'Driscoll, K.F.; Russell, G.T.; Schweer, J. *Macromol. Chem. Phys.* **1995**, *196*, 3267
12. Manders, B.G. *Ph.D. thesis*; Technische Universiteit Eindhoven: Eindhoven, 1997; Chapter 5
13. Chambard, G.; Klumperman, B.; German, A.L., *Macromolecules* **2000**, *33*, 4417
14. Chambard, G. *Ph.D. thesis*; Technische Universiteit Eindhoven: Eindhoven, 2000 (ISBN 90-386-2622-3); Chapter 4
15. Chambard, G. *Ph.D. thesis*; Technische Universiteit Eindhoven: Eindhoven, 2000 (ISBN 90-386-2622-3); Chapter 5
16. Arbitrarily chosen value for a slowly decomposing initiator ($t_{1/2} \approx 100$ hrs).
17. Value based on work by de Kock, J.B.L. (*Ph.D. thesis*; Technische Universiteit Eindhoven: Eindhoven, 1999 (ISBN 90-386-2701-7)
18. Borman, C.D.; Jackson, A.T.; Bunn, A.; Cutter, A.L.; Irvine, D.J., *Polymer* **2000**, *41*, 6015
19. Brinkhuis, R.H.G.; Klumperman, B. – to be published
20. The Br end group from a methacrylate chain end is lost during the MALDI-experiment. The fact that we mainly see intact chains, with the Br end group still present is attributed to the large fraction of acrylate chain end dormant species. From the simulations we were able to estimate that around 93% of all dormant chains contain an acrylate as the terminal unit.

Chapter 14

Atom Transfer Radical Polymerization of Methyl Methacrylate Utilizing an Automated Synthesizer

Huiqi Zhang, Martin W. M. Fijten, Richard Hoogenboom, and Ulrich S. Schubert*

Macromolecular Chemistry and Nanoscience, Eindhoven University of Technology and Dutch Polymer Institute, P.O. Box 513, 5600 MB Eindhoven, The Netherlands

The homogeneous atom transfer radical polymerization of methyl methacrylate mediated by CuBr/N-(n-hexyl)-2-pyridylmethanimine was successfully carried out using an automated synthesizer. The effects of initiators, solvents and reactant ratios on the polymerization were investigated. Three different kinds of initiators, namely ethyl 2-bromoisobutyrate, (1-bromoethyl)-benzene, and p-toluenesulfonyl chloride, were utilized to initiate the polymerization and ethyl 2-bromoisobutyrate was proven to be the best initiator for the studied system in terms of molecular weight control and narrow polydispersities of the obtained polymers. The solvents used (i.e., toluene, p-xylene, and n-butyl-benzene) were found to have strong effects on the polymerization. The reactions in toluene and p-xylene were well-controlled and almost the same polymerization rates were observed. However, the reaction rate dramatically increased in the case of n-butylbenzene, resulting in radical termination during the polymerization. The initiator and Cu(I) concentrations had a positive effect on the polymerization rate. In the meantime, all the reactions were controlled and polymers with predetermined molecular weights and low polydispersity indices (< 1.3) were obtained.

Introduction

Controlled "living" radical polymerization (CRP) systems have attracted great attention from both the academic research and industry during the past years because of their versatility (1,2). Among the most successful CRPs are nitroxide-mediated polymerization (NMP) (3), atom transfer radical polymerization (ATRP) (4,5,6), and reversible addition-fragmentation chain transfer (RAFT) polymerization (7). All these techniques allow the preparation of well-defined polymers with predetermined molecular weights, low polydispersities, functional groups, and various architectures under relatively mild reaction conditions.

ATRP has been the most extensively studied CRP system (8,9). The success depends largely on a reversible dynamic equilibrium between the dormant species (P_n-X) and the active species (radicals, $P_n\cdot$) (Scheme 1).

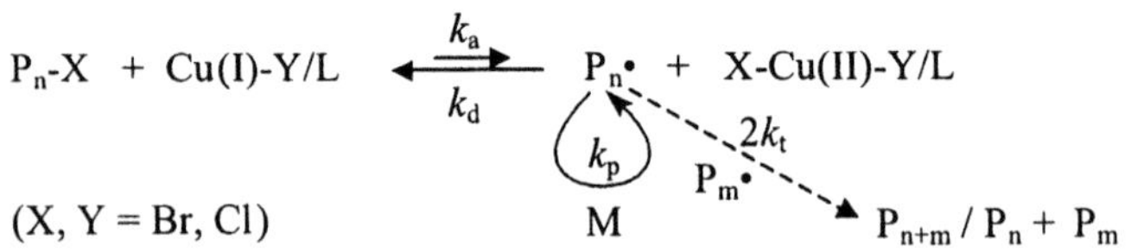

$$P_n\text{-}X \ + \ Cu(I)\text{-}Y/L \ \underset{k_d}{\overset{k_a}{\rightleftharpoons}} \ P_n\cdot \ + \ X\text{-}Cu(II)\text{-}Y/L$$

$$(X, Y = Br, Cl) \qquad\qquad M \qquad k_p \searrow \quad P_m\cdot \quad \overset{2k_t}{\longrightarrow} \ P_{n+m} \, / \, P_n + P_m$$

Scheme 1

The equilibrium determines the radical concentration and subsequently the rates of polymerization and termination. A successful ATRP system should have a very low equilibrium constant ($K_{eq} = k_a/k_d \sim 10^{-7}$) (10), which keeps the active species at a very low concentration (approximately 10^{-8} to 10^{-7} M) (11) and thus greatly minimizes the radical termination reactions. Many parameters in ATRP such as the structure and concentration of the utilized monomers, catalysts (metals and ligands), initiators, solvents, reactant ratios, and the reaction temperature can significantly influence the equilibrium, which makes the optimization of the reaction conditions very time-consuming, in particular when a new reaction system is investigated. Therefore, practical techniques for a rapid parallel and maybe even automated synthesis, which would allow an efficient high-throughput screening to obtain optimal reaction conditions and to understand the polymerization parameters, are highly suitable for this research direction. Recently, an automated parallel synthetic approach has been applied in ATRP, but only the screening of the molecular weights of the copolymers prepared via the ATRP of styrene and butyl acrylate was presented (12,13). In addition, no detailed information concerning the utilized process has been provided. In this paper, the investigation of the homogeneous ATRP of methyl methacrylate (MMA) mediated by Cubr/N-(n-hexyl)-2-pyridylmethanimine (NHPMI) utilizing an automated synthesizer is reported. The experimental set-up, the parallel synthetic procedure, the reproducibility of the parallel approach, and the comparison with the conventional laboratory experiments are described. Furthermore, the effects of initiators, solvents and reactant ratios on the polymerization are also presented.

Experimental Section

Materials

MMA (Aldrich, 99%) was washed twice with an aqueous solution of sodium hydroxide (5%) and twice with distilled water, dried with anhydrous magnesium sulfate overnight, and then distilled over calcium hydride under vacuum. The distillate was stored at –18 °C before use. CuBr (Aldrich, 98%) was stirred with acetic acid for 12 h, washed with ethanol and diethyl ether, and then dried under vacuum at 75 °C for 3 days. The purified CuBr was stored in an argon atmosphere. NHPMI was synthesized by condensation of pyridine-2-carboxaldehyde (Acros, 99%) and n-hexylamine (Acros, 99%) as described elsewhere (14). Toluene (Biosolve Ltd., AR) was distilled over calcium hydride. p-Xylene (Aldrich, 99+%, anhydrous), n-butylbenzene (Acros, 99+%), ethyl 2-bromoisobutyrate (EBIB, Aldrich, 98%), (1-bromoethyl)benzene (BEB, Aldrich, 97%), p-toluenesulfonyl chloride (TSC, Acros, 99+%), deactivated neutral aluminium oxide (Merck, for column chromatography), and all the other chemicals were used as received.

Instruments and Measurements

The reactions were carried out in a computer-controlled Chemspeed ASW 2000 automated synthesizer (Figure 1). Five reactor blocks could be used in parallel and each block had 4 to16 reaction vessels depending on their volumes (100 to 13 mL). Each reaction vessel was jacketed with an oil bath and was equipped with a cold-finger reflux condenser. The temperature of the oil bath was controlled by a Huber Unistat 390 W Cryostat and could vary from –90 to 150 °C and the temperature of the reflux liquid was controlled by a Huber ministat compatible control and could change from –10 to 50 °C. The reaction vessels were connected with a membrane pump, which could be utilized for inertization or evaporation processes. Mixing was performed by a vortex process

196

(0 to 1400 rpm). A glove box was available, which kept an argon atmosphere outside the reaction system. The automated synthesizer was connected to an online size exclusion chromatography (SEC) and an off-line gas chromatography (GC). A Gilson liquid handling system was used in the automated synthesizer.

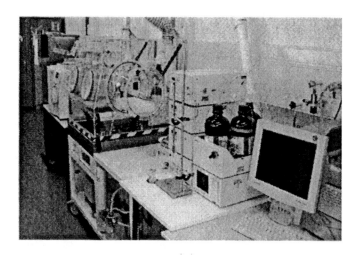

Figure 1. Visualization of the automated synthesizer set-up.

The monomer conversion was determined from the concentration of the residual monomer using an off-line Interscience Trace GC with an auto-sampler, equipped with a Rtx-5 (Crossbond 5% diphenyl-95% dimethyl polysiloxane) capillary column (30 m × 0.25 mm ID × 0.25 μm df) with polymerization solvents (toluene, *p*-xylene, and *n*-butylbenzene) as internal references. After optimization, the GC measurement of one sample required 5 min. Molecular weights and molecular weight distributions (MWDs) were measured with an online SEC set-up (Shimadzu gel permeation chromatography (GPC) equipped with a LC-10AD VP pump and a RID-6A differential refractometer) at ambient temperature. Tetrahydrofuran (THF) was used as the eluent at a flow rate of 1.0 mL/min. A linear column (PLgel 5 μm Mixed-D, Polymer Laboratories, 30 cm) was used. The calibration curve was prepared with poly(methyl meth-acrylate) (PMMA) standards. The GPC measurement of one sample required 15 min.

Polymerization Procedures

A typical ATRP was carried out in the automated synthesizer as follows: CuBr (0.0596 g, 0.42 mmol) was manually added to the reaction vessels (75 mL, three parallel reactions). Inertization (three cycles of vacuum (15 min)/argon filling) of these reaction vessels was conducted at 120 °C to remove the oxygen and moisture. The temperature of the oil bath was then lowered to 25 °C. A degassed stock solution of MMA (6.2407 g, 62.33 mmol) in p-xylene (5.7734 g) and a degassed stock solution of NHPMI (0.2372 g, 1.25 mmol) in p-xylene (2.8867 g) were added subsequently. The reaction temperature was increased to 90 °C and at the same time the reflux liquid was cooled to –5 °C. After the reaction mixtures were vortexed at 600 rpm for 15 min at 90 °C, a degassed solution of initiator EBIB (0.0811 g, 0.42 mmol) in p-xylene (2.8867 g) was added during 2 min. The reaction mixtures were then vortexed at 600 rpm, and the polymerizations were sampled at suitable time periods throughout the reactions. The samples were diluted with THF, and parts of them were used for GC measurements in order to determine the monomer conversion. The rest was purified automatically by passing through aluminium oxide columns in the solid phase extraction (SPE) set-up prior to the SEC measurements.

The ATRP in the conventional set-up was carried out according to the procedure as previously reported (15).

Results and Discussion

The ATRP of MMA mediated by CuBr/NHPMI (Scheme 2) was investigated with an automated synthesizer. The volume of the solvent used was always twice that of MMA in each ATRP system and a molar ratio of initiator to CuBr to NHPMI of 1:1:3 was utilized. A homogeneous dark brown solution was obtained when the reaction mixture was heated to 90 °C.

The fast screening of the reaction conditions for ATRP by using an automated synthesizer can significantly speed up the research. However, it should be evaluated in advance whether the automated synthesizer can provide reproducible results as well as comparable results with those obtained from the conventional laboratory experiments. Therefore, selected experiments were carried out both in the automated synthesizer and in a conventional set-up in the laboratory in order to allow a detailed comparison of the obtained results. The reproducibility of the ATRP carried out in the automated synthesizer as well as the comparability of the results from the automated synthesizer and the conventional experiments has been confirmed and described elsewhere (16). In addition, an automated purification procedure was developed to purify the polymers prepared via ATRP for the SEC measurements. Hand-made deactivated aluminium oxide columns (0.5 cm) in SPE cartridges (length = 5.6 cm, diameter = 0.6 cm) including porous polyethylene frits and ASPEC caps together with 2 mL of THF (as the eluent) were utilized to purify the polymers (16).

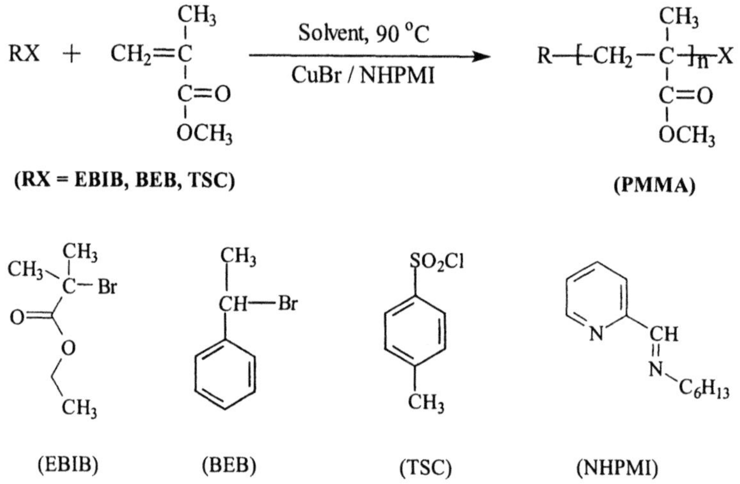

$$RX + CH_2=\underset{\underset{OCH_3}{\overset{\overset{CH_3}{|}}{\underset{|}{C}}}}{\overset{}{C}}=O \quad \xrightarrow[\text{Cubr / NHPMI}]{\text{Solvent, 90 °C}} \quad R\text{---}[CH_2\text{---}\underset{\underset{OCH_3}{\overset{\overset{CH_3}{|}}{\underset{|}{C}}}}{\overset{}{C}}\text{=}O]_n\text{X}$$

(RX = EBIB, BEB, TSC)   (PMMA)

(EBIB)   (BEB)   (TSC)   (NHPMI)

Scheme 2

ATRP of MMA with Different Initiators

Figure 2 shows the effect of the initiators used on the ATRP of MMA at 90 °C in *p*-xylene with CuBr/NHPMI as the catalyst and $[MMA]_0/[\text{initiator}]_0/$ $[CuBr]_0/[NHPMI]_0 = 150:1:1:3$. Three different kinds of initiators were used in this study, including ethyl 2-bromoisobutyrate (EBIB), (1-bromoethyl)-benzene(BEB), and *p*-toluenesulfonyl chloride (TSC). Two parallel reactions for each ATRP system were carried out in the automated synthesizer and good reproducibility of the reactions can be clearly seen in Figure 2. In all cases, almost a linear relationship between $\ln([M]_0/[M])$ and reaction time t was obtained, indicating that no significant radical termination was present in the polymerization processes. The initiators revealed only a little influence on the kinetics of the polymerization and the reaction rates. In addition, an induction period of 19 to 86 min was observed in these systems. The origin of this induction period is not very clear yet, and further investigation is going on at present.

The effects of the initiators on the molecular weights and polydispersities of the polymers were also studied (Figure 3). All the polymers were purified by using the above-mentioned deactivated aluminium oxide columns automatically before the SEC measurements. The ATRP initiated by EBIB was well-controlled in terms of the molecular weights and MWDs of the polymers. The number-average molecular weights determined by SEC, $M_{n,SEC}$, increased linearly with increasing monomer conversion and were close to the theoretical values (i.e., $M_{n,th}$) calculated according to eq 1, indicating that the polymerization was living and controlled.

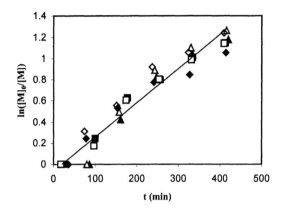

*Figure 2. Plots of ln([M]₀/[M]) versus reaction time t for the ATRP of
MMA in p-xylene at 90 °C with EBIB (■,□), BEB (▲,△), and TSC (◆,◇) as
initiators. [MMA]₀/[initiator]₀/[CuBr]₀/[NHPMI]₀ = 150:1:1:3.*

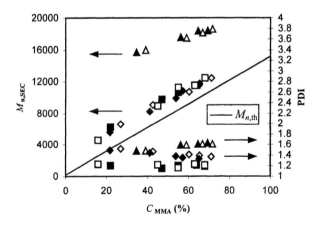

*Figure 3. Dependence of Mₙ,SEC and PDIs of the polymers on the monomer
conversion of the ATRP of MMA in p-xylene at 90 °C using EBIB (■,□), BEB
(▲,△), and TSC (◆,◇) as initiators. [MMA]₀/[initiator]₀/[CuBr]₀/[NHPMI]₀
= 150:1:1:3.*

$$M_{n,th} = ([M]_0/[RX]_0)M_M C_M + M_{RX} \qquad \text{(eq 1)}$$

where $[M]_0$ and $[RX]_0$ are the initial concentrations of monomer and initiator, respectively, M_M and M_{RX} are the molecular weights of monomer and initiator, respectively, and C_M is the monomer conversion. The obtained polymers revealed polydispersity indices (PDIs) around 1.2 throughout the reactions. The TSC-initiated polymerization also provided polymers with predetermined molecular weights and the PDI values of the polymers decreased with the monomer conversions down to 1.4 at high monomer conversion, which is a typical characteristic of the controlled polymerizations. However, an uncontrolled polymerization was obtained when BEB was used as the initiatior, which yielded polymers with much higher molecular weights than $M_{n,th}$ and PDI values close to 1.6, probably due to the inefficient initiation of BEB for the ATRP of MMA (17). Among all the initiators utilized, EBIB was proven to be the most efficient initiator in terms of the polymerization rate as well as the molecular weights and polydispersities of the polymers. Although TSC is a much faster initiator than EBIB (18), the polymers obtained in the former case revealed higher polydispersities than those obtained in the latter case. This might be ascribed to the fact that polymers containing a majority of chlorine end groups were formed in the case of TSC while the polymers obtained from the ATRP initiated by EBIB were end-capped with bromine (19). Since the Cu-Cl bond is more stable than Cu-Br bond, the deactivation process should be slower in the case of TSC, resulting in higher polydispersities of the obtained polymers. In the following studies, EBIB was utilized as the initiator for all the polymerization systems.

ATRP of MMA in Different Solvents

The ATRP of MMA with CuBr/NHPMI as the catalyst and EBIB as the initiator was carried out at 90 °C in a series of solvents such as toluene, p-xylene, and n-butylbenzene to study the effect of solvents on the polymerization (two parallel reactions for each ATRP system in p-xylene and n-butylbenzene). The kinetic results revealed that $\ln([M]_0/[M])$ increased almost linearly with reaction time t throughout the reactions when toluene and p-xylene were used as solvents, demonstrating that the radical concentrations remained constant during the reactions (Figure 4). However, a curved kinetic plot was found for the ATRP in n-butylbenzene, suggesting the occurrence of radical termination during the ATRP process (15). The polymerization rate in toluene was almost the same as that in p-xylene, while the reaction in n-butylbenzene proceeded much faster. The differences in the polymerization rates in different solvents were also observed from the different viscosities of the reactions. The reaction mixture in n-butylbenzene was so viscous at the end of the reaction (about 7 h) that stirring became impossible, while the reaction mixtures in toluene and p-xylene were still low viscous solutions.

The effect of the polarity of the solvents on the ATRP of MMA catalyzed by CuBr/N-alkyl-2-pyridylmethanimine has been investigated and faster reactions were observed in more polar solvents (20). However, the reactions in different apolar solvents have hardly been compared, largely due to the opinion that the

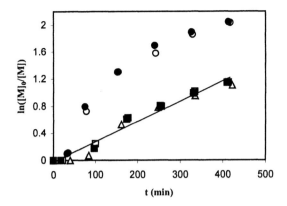

Figure 4. Plots of ln([M]₀/[M]) versus reaction time t for the ATRP of MMA at 90 °C with [MMA]₀/[EBIB]₀/[CuBr]₀/[NHPMI]₀ = 150:1:1:3 and different solvents such as n-butylbenzene (●,○), p-xylene (■,□), and toluene (△).

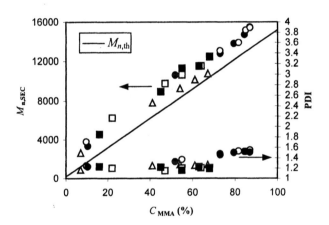

Figure 5. Dependence of $M_{n,SEC}$ and PDIs of the polymers on the monomer conversion of the ATRP of MMA at 90 °C in toluene (△), p-xylene (■,□), and n-butylbenzene (●,○). [MMA]₀/[EBIB]₀/[CuBr]₀/[NHPMI]₀ = 150:1:1:3.

different apolar solvents with close polarity, such as toluene and p-xylene, have no influence on the polymerization rate. This is consistent with our experiments in toluene and p-xylene but not in the case of n-butylbenzene. The cause of the remarkable increase of the polymerization rate in n-butylbenzene is not clear yet, and further investigations are ongoing at present. Figure 4 also revealed the slow initiation at the beginning of the reactions.

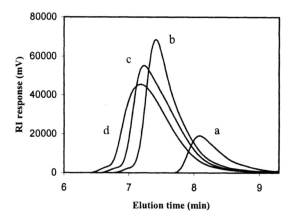

Figure 6. GPC traces of the polymers prepared via ATRP at 90 °C with n-butyl-benzene as the solvent. C_{MMA} = 10% (a), 55% (b), 73% (c), and 87% (d).

The solvents were found to have little influence on the molecular weights of the polymers (Figure 5). The $M_{n,SEC}$ increased linearly with increasing monomer conversion and were comparable to $M_{n,th}$ values. The ATRP systems in toluene and p-xylene resulted in polymers with unimodal GPC traces and PDI values < 1.3 throughout the reactions (Figure 5). The PDI values of the polymers prepared via the ATRP in n-butylbenzene, however, increased with monomer conversion and were larger than 1.5 when the monomer conversion was higher than 80%. This large increase of the PDI values of the polymers might in part be ascribed to the presence of radical termination during the reaction. In addition, the high viscosity of the reaction mixture at the end of the reaction, which was likely to influence the activation and deactivation processes of the equilibrium in ATRP (Scheme 1), could also be partially responsible for this phenomenon. It is known that the ATRP of MMA with a too high radical concentration will mainly result in radical disproportionation (21). Therefore, the $M_{n,SEC}$ of the polymers prepared via the ATRP of MMA with a too high radical concentration should be comparable with those of the polymers obtained from the well-controlled systems, but the PDI values of the polymers would be larger in the former case. This agrees well with our experimental results (Figure 5). The GPC traces of the polymers obtained from the ATRP in n-butylbenzene are shown in Figure 6. Small shoulders were observed on the high molecular weight sides of the GPC traces of the obtained polymers at relatively high monomer conversions, suggesting that radical coupling was also present in the ATRP with n-butylbenzene as the solvent. Besides, the tailing of the GPC traces was clearly observed, which could be attributed to the dead polymers resulted from the radical termination during the reactions.

Effect of Initiator and Cu(I) Concentrations

A series of reactions at different reactant ratios were carried out at 90 °C using the automated synthesizer in order to investigate the effect of the initiator and Cu(I) concentrations on the polymerization (Figure 7, two parallel reactions for each system). The results showed that $\ln([M]_0/[M])$ increased almost linearly with reaction time t for all the reactions with $[MMA]_0/[EBIB]_0/[CuBr]_0/[NHPMI]_0$ = 150:1:1:3, 100:1:1:3, and 50:1:1:3, revealing that all the polymerizations proceeded in a controlled way and no significant radical termination took place during the reactions. The polymerization rate increased with the increase of the initiator and Cu(I) concentrations, and the average apparent rate constants (k_{app} = slope of the kinetic plot) of the three reactions were 0.0054, 0.0038, 0.0030 min^{-1}, respectively.

Figure 8 shows that the $M_{n,SEC}$ increased linearly with monomer conversion for all the reactions with different reactant ratios. However, they were slightly higher than the $M_{n,th}$ values revealing that persistent radical effect took place at the beginning of the reactions and thus lowered the initiation efficiency of the systems (22). The PDI values of the polymers obtained from these systems were below 1.3 and almost identical, indicating the well-controlled reactions.

Conclusions

This paper describes the successful application of an automated synthesizer in the homogeneous ATRP of MMA mediated by CuBr/NHPMI. The polymerizations initiated by EBIB, BEB, and TSC resulted in nearly linear kinetic plots of $\ln([M]_0/[M])$ versus reaction time t and identical polymerization rates. However, the molecular weights and PDIs of the obtained polymers were significantly influenced by the utilized initiators. The ATRP with EBIB and TSC as initiators provided controlled reactions while an uncontrolled system was obtained for the ATRP with BEB as the initiator. EBIB was proven to be the best initiator for the studied system. The effect of the solvents, including toluene, p-xylene, and n-butylbenzene, on the polymerization was studied. Both the linear kinetic plots of $\ln([M]_0/[M])$ versus reaction time t and the linear dependence of the molecular weights on the monomer conversion were obtained for the ATRP systems in toluene and p-xylene. A curved kinetic plot was observed for the ATRP in n-butylbenzene although the molecular weights still increased linearly with the monomer conversion. The polymerization in toluene proceeded as fast as that in p-xylene, while the reaction rate dramatically increased in the case of n-butylbenzene, leading to radical termination reactions (radical disproportionation and coupling) and higher PDI values of the obtained polymers. The molecular weights of the polymers were almost not influenced by the solvents used and they were all comparable to the $M_{n,th}$ values. An increase of the initiator and Cu(I) concentrations resulted in an increase of the polymerization rate. All the studied reactions with different reactant ratios were controlled in terms of the molecular weights and PDI values of the obtained polymers. In addition, an induction period was present in most of the studied systems, which requires further investigation.

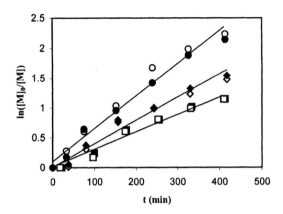

Figure 7. Plots of ln([M]₀/[M]) versus reaction time t for the ATRP of MMA in p-xylene at 90 °C using [MMA]₀/[EBIB]₀/[CuBr]₀/[NHPMI]₀ = 150:1:1:3 (■,□), 100:1:1:3 (◆,◇), and 50:1:1:3 (●,○), respectively.

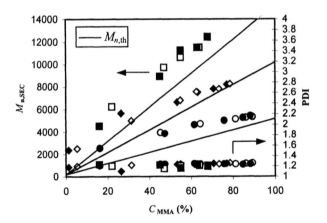

Figure 8. Plots of Mₙ,SEC and PDIs versus the monomer conversion for the ATRP of MMA in p-xylene at 90 °C. [MMA]₀/[EBIB]₀/[CuBr]₀/[NHPMI]₀ = 150:1:1:3 (■,□), 100:1:1:3 (◆,◇), and 50:1:1:3 (●,○).

Acknowledgement

We thank NWO, the Dutch Polymer Institute (DPI) and the Fonds der Chemischen Industrie for the financial support and Chemspeed Ltd. for the outstanding collaboration.

References

1. Matyjaszewski, K. *Controlled Radical Polymerization*; ACS Symposium Series 685; American Chemical Society: Washington, DC, 1998.
2. Matyjaszewski, K. *Controlled/Living Radical polymerization — Progress in ATRP, NMP, and RAFT*; ACS Symposium Series 768; American Chemical Society: Washington, DC, 2000.
3. Johnson, C. H. J.; Moad, G.; Solomon, D. H.; Spurling, T. H.; Vearing, D. *J. Aust. J. Chem.* **1990,** *4 3*, 1215.
4. Kato, M.; Kamigaito, M.; Sawamoto, M.; Higashimura, T. *Macromolecules* **1995,** *2 8*, 1721.
5. Wang, J. S.; Matyjaszewski, K. *J. Am. Chem. Soc.* **1995,** *117*, 5614;
6. Wang, J. S.; Matyjaszewski, K. *Macromolecules* **1995,** *28*, 7901.
7. Chiefari, J.; Chong, Y. K.; Ercole, F.; Krstina, J.; Jeffery, J.; Le, T. P. T.; Mayadunne, R. T. A.; Meijs, G. F.; Moad, C. L.; Moad, G.; Rizzardo, E.; Thang, S. H. *Macromolecules* **1998,** *31*, 5559.
8. Matyjaszewski, K.; Xia, J. *Chem. Rev.* **2001,** *101*, 2921.
9. Kamigaito, M.; Ando, T.; Sawamoto, M. *Chem. Rev.* **2001,** *101*, 3689.
10. Hong, S. C.; Matyjaszewski, K. *Macromolecules* **2002,** *35*, 7592.
11. Matyjaszewski, K.; Patten, T. E.; Xia, J. *J. Am. Chem. Soc.* **1997,** *1 19*, 674.
12. Nicolaou, K. C.; Hanko, R.; Hartwig, W. *Handbook of Combinatorial Chemistry-Drugs, Catalysts, Materials*; Wiley-VCH, Weinheim, 2002; *Vol. 2*, p1050.
13. Nielsen, R. B.; Safir, A. L.; Petro, M.; Lee, T. S.; Huefner, P. *Polym. Mater. Sci. Eng.* **1999,** *8 0*, 92.
14. Haddleton, D. M.; Crossman, M. C.; Dana, B. H.; Duncalf, D. J.; Heming, A. M.; Kukulj, D.; Shooter, A. J. *Macromolecules* **1999,** *32*, 2110.
15. Zhang, H.; Klumperman, B.; Ming, W.; Fischer, H.; van der Linde, R. *Macromolecules* **2001,** *3 4*, 6169.
16. Zhang, H.; Fijten, M. W. M.; Hoogenboom, R.; Reinierkens, R.; Schubert, U. S. *Macromol. Rapid Commun.* **2003,** *2 4*, 81.
17. Matyjaszewski, K.; Wang, J. L.; Grimaud, T.; Shipp, D. A. *Macromolecules* **1998,** *3 1*, 1527.
18. Percec, V.; Barboiu, B.; Kim, H. J. *J. Am. Chem. Soc.* **1998,** *1 20*, 305.
19. Matyjaszewski, K.; Shipp, D. A.; Wang, J. L.; Grimand, T.; Patten, T. E. *Macromolecules* **1998,** *3 1*, 6836.
20. Haddleton, D. M.; Kukulj, D.; Duncalf, D. J.; Heming, A. X.; Shooter, A. J. *Macromolecules* **1998,** *3 1*, 5201.
21. Snijder, A.; Klumperman, B.; van der Linde, R. *Macromolecules* **2002,** *35*, 4785.
22. Fischer, H. *Chem. Rev.* **2001,** *101*, 3581.

Chapter 15

Asymmetric Atom Transfer Radical Polymerization: Enantiomer-Selective Cyclopolymerization of *rac*-2,4-Pentanediyl Dimethacrylate Using Chiral ATRP Initiator

Toyoji Kakuchi, Masashi Tsuji, and Toshifumi Satoh

Division of Molecular Chemistry, Graduate School of Engineering, Hokkaido University, Sapporo 060–8628, Japan

The enantiomer-selective radical cyclopolymerization of the racemic mixture of (2*R*,4*R*)-2,4-pentanediyl dimethacryrate (*RR*-1) and (2*S*,4*S*)-2,4-pentanediyl dimethacrylate (*SS*-1) was achieved using chiral atom transfer radical polymerization (ATRP) initiating systems. *RR*-1 and *SS*-1 were predominantly polymerized using methyl 2-bromoisobutyrate /CuBr/(*S*,*S*)-2,6-bis(4-isopropyl-2-oxazolin-2-yl)pyridine and methyl 2-bromoisobutyrate/RuCl₂(PPh₃)₃/(*S*)-1,1'-bi-2-naphthol, respectively. The enantiomer selectivity ratio was 1.3 ~ 3.2 for the Cu-catalyzed ATRP initiating system and ca. 1.0 ~ 3.4 for the Ru-catalyzed ATRP initiating system.

The synthetic methodology of optically active polymers is of great interest from the viewpoint of the fine stereocontrol of polymerization. Asymmetric polymerization, such as the asymmetric synthesis polymerization, helix-sense-selective polymerization, and enantiomer-selective polymerization, has been achieved using chiral ionic and coordination initiating systems (*1-3*). For the asymmetric radical polymerization, Wulff et al. and we established the radical cyclocopolymerization of a divinyl monomer having a chiral template with an achiral vinyl monomer leading to the main chain chiral vinyl polymer after removal of the chiral template unit from the precursor polymer (*4,5*). Although Okamoto et al. reported the helix-sense-selective radical polymerization with enantiomer selectivity using the optically active phenyl-2-pyridyl-*o*-tolylmethyl methacrylate with various enantiomeric excesses, to our knowledge, little is known about the enantiomer-selective radical polymerization of a racemic monomer (6).

Atom transfer radical polymerization (ATRP), which has been developed from the Kharash addition and cyclization and used extensively in organic synthesis, has been a major step forward in the "living/controlled" polymerization by Matyjaszewski et al. and Sawamoto et al. (7,8). Although ATRP has been demonstrated to provide excellent control over the molecular weight and polydispersity and used for the synthesis of various macromolecular architectures such as block copolymers and star polymers, the stereochemical control of the polymer main chain is still insufficient (9). Thus, of great interest is the endeavor to finely control the stereochemistry of the radical polymerization based on ATRP. In this chapter, we describe the enantiomer-selective radical cyclopolymerization of a racemic monomer using a chiral ATRP initiating system, i.e., the polymerization of rac-2,4-pentanediyl dimethacrylate (rac-1) was carried out using the initiating system consisting of methyl 2-bromoisobutyrate (3), metal sources (Cu-1~2 and Ru-1), and chiral ligands (l-1~3) or additives (a-1~5).

Experimental Section

Measurements. The ^{1}H and ^{13}C NMR spectra were recorded using a JEOL JNM-400II spectrometer in deutero-chloroform at 25 °C. The optical rotatory measurements were performed in chloroform at 28 °C using a Jasco DIP 1000 digital polarimeter. The molecular weights were measured by gel permeation chromatography in THF using a Jasco GPC 900 system equipped with three polystyrene columns (Shodex KF-804L). The number-average molecular weights (M_ns) and molecular weight distributions (M_w/M_ns) were calculated on the basis of a polystyrene calibration. The chiral high-performance liquid chromatography (HPLC) analysis was performed using a Jasco HPLC system (PU-980 Intelligent HPLC pump and UV 975 Intelligent UV detector) equipped with a Daicel CHIRALCEL OB-H column (eluent, hexane/2-propanol (vol. ratio, 100/1); flow rate, 0.5 mL min^{-1}). The gas chromatography (GC) analysis was recorded using a Shimadzu GC-17A gas chromatograph equipped with J&W Scientific 30 m DB-1 column.

Monomer. rac-2,4-Pentanediyl dimethacrylate (rac-1) was obtained by esterification of rac-2,4-pentanediol with methacryloyl chloride in N-methyl-2-pyrrolidinone at room temperature (10).

208

Scheme 1

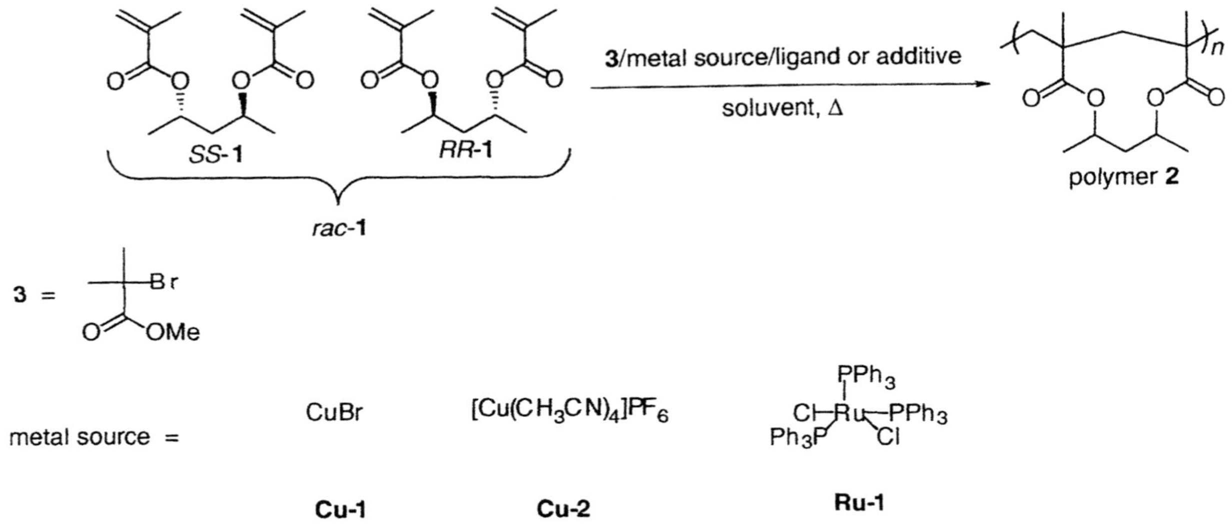

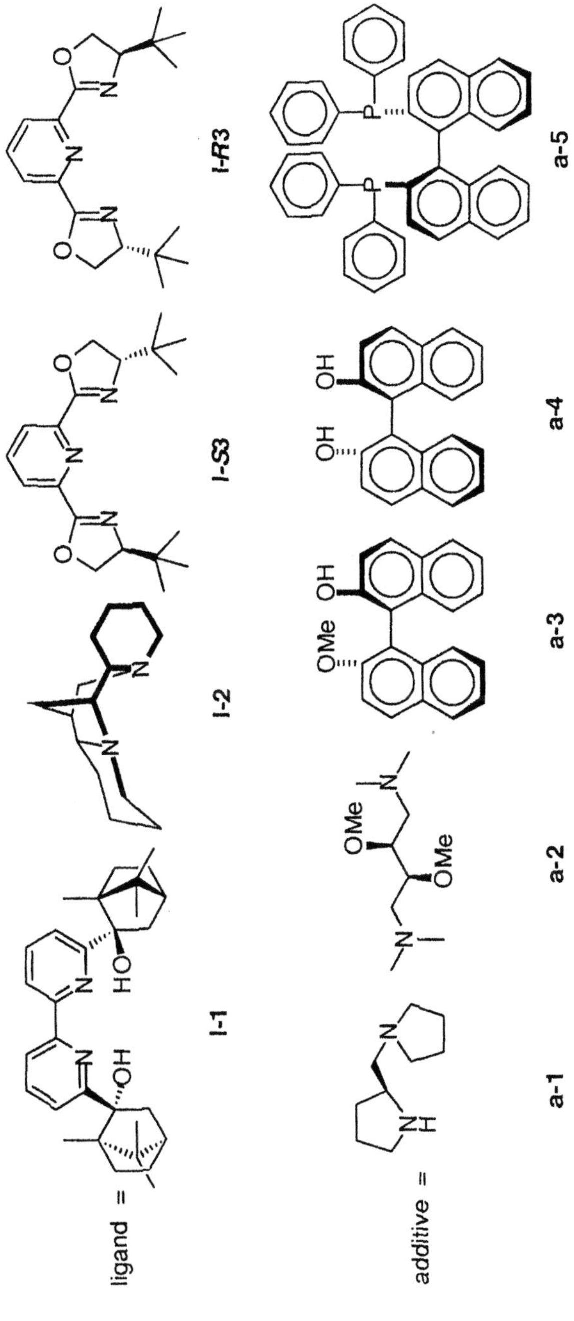

ligand =

I-1   I-2   I-S3   I-R3

additive =

a-1   a-2   a-3   a-4   a-5

Polymerizations. A typical example of the polymerization using **3/Cu-1/l-1** as an initiating system is given follows: In a glovebox under a moisture- and oxygen-free argon atmosphere (H_2O and O_2 < 1 ppm), *rac*-**1** (500 mg, 2.08 mmol), **3** (3.7 mg, 2.08 x 10^{-2} mmol), **Cu-1** (5.9 mg, 4.16 x 10^{-2} mmol), and **l-1** (38.3 mg, 8.32 x 10^{-2} mmol) were dissolved in anisole (20.8 mL). Two mL of the mixture was placed in a dry test tube, capped, and then taken out of the glovebox. The test tube content was stirred at 90 °C. At the end of the polymerization, an aliquot (20 μL) of the reaction mixture was added to hexane (0.5 mL) and filtered through a 0.25 μm pore membrane filter. The sample was analyzed for the monomer conversion and enantiomeric excess (*e.e.*) by HPLC equipped with a CHIRALCEL OB-H column and for the M_n and M_w/M_n by SEC. The residual polymerization mixture was passed through a short alumina column to remove the metal salts, and the solvent was removed under reduced pressure. The residue was poured into hexane and the precipitate was filtered. The obtained powder was purified by reprecipitation with chloroform-methanol and dried in *vacuo*.

Results and Discussion

Copper-catalyzed Cyclopolymerization of *rac*-1. The atom transfer radical polymerization of *rac*-**1** was carried out using an initiating system consisting of **3**, CuX, and chiral amine ligands (Table I). All the polymerizations homogeneously proceeded and the obtained polymers were soluble in chloroform and tetrahydrofuran. Because the characteristic absorption due to the methacrylate groups was not observed in the ^{13}C NMR spectrum of the resulting polymer, the polymerization of *rac*-**1** proceeded through a cyclopolymerization mechanism to afford the polymer essentially consisting of cyclic constitutional repeating units, *i.e.*, the extent of the cyclization was ca. 100 %.

The number average molecular weights (M_ns) of the resulting polymers ranged from 11,000 to 13,500 and the molecular weight distributions (M_w/M_ns) were relatively narrow in the range of 1.21 ~ 1.30. The chiral amine ligand affected the enantiomer selectivity, *i.e.*, the enantiomeric excess (*e.e.*) of the recovered monomer was 6.8 % ~ 15.3 %. In addition, the enantiomer selectivity changed with the chirality of the amine ligands used, *i.e.*, (2*S*,4*S*)-2,4-pentanediyl dimethacrylate (*SS*-1) was predominantly polymerized using **l-*R*3**, whereas (2*R*,4*R*)-2,4-pentanediyl dimethacrylate (*RR*-1) using was predominantly polymerized **l-*S*3**. The obtained polymers exhibited optical activity, and the absolute values of the specific rotation ($[\alpha]_{435}$) of the obtained polymers were from 18.9° to 38.6°. For polymer **2** prepared using **3/Cu-1/l-*S*3**, the SEC chromatograms using RI (lower) and polarimetric (upper) detectors are shown in Figure 1.

Table I. Enantiomer-selective polymerization of *rac*-1 using copper-catalyzed chiral ATRP initiating system [a]

CuX/ligand	time hr	recovered monomer conv.[b] %	e.e.[c] %	enriched isomer	polymer 2 M_n (M_w/M_n) [d]	$[\alpha]_{435}$ [e]
Cu-1/l-1	4	26.9	6.8	SS-1	13,500 (1.23)	-18.9°
Cu-1/l-2	12	18.5	9.4	SS-1	11,000 (1.30)	-35.5°
Cu-1/l-S3	6	24.6	15.3	SS-1	12,300 (1.25)	-36.5°
Cu-1/l-R3	6	22.8	13.4	RR-1	11,200 (1.23)	+38.6°
Cu-2/l-1	9	18.5	4.7	SS-1	10,500 (1.21)	-19.6°
Cu-2/l-2	24	21.3	3.4	SS-1	12,900 (1.25)	-12.3°

[a] $[rac\text{-}1]_0 = 0.1$ mol L⁻¹; $[rac\text{-}1]_0/[3]_0/[CuX]_0/[ligand]_0 = 200/1/2/4$; solvent, anisole; temperature, 90 °C.
[b] Determined by HPLC equipped with CHIRALCEL OB-H column.
[c] Enantiomeric excess of recovered monomer.
[d] Determined by SEC in THF using polystyrene standards.
[e] $c = 0.3$ in chloroform at 28 °C

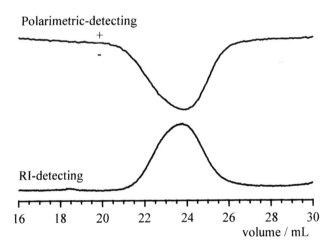

Figure 1. SEC chromatograms using RI (lower) and polarimetric (upper) detectors for polymer 2 prepared using 3/**Cu-1/l-S3**.

The peak shapes of the two chromatograms were almost similar, suggesting that the chirality of polymer **2** should be due to the excess amount of the *RR*-1 units in the copolymer composition. Figure 2 shows the relations between the M_n and M_w/M_n of the obtained polymers and the monomer conversion for the polymerization of *rac*-1 using **3/Cu-1/l-*S*3**. The M_n values increased with the increasing monomer conversion, and the M_w/M_n values were relatively narrow in the range of 1.26 ~ 1.29. Figure 3 shows the kinetic plots for the polymerization of *rac*-1 using **3/Cu-1/l-*S*3**. The apparent polymerization rates of *SS*-1 and *RR*-1 were both first order with respect to the monomer concentration, indicating that the polymerization of *rac*-1 using **3/Cu-1/l-*S*3** was living-like.

The consumption rate of *RR*-1 was faster than that of *SS*-1. The enantiomer-selectivity ratio (*r*) was calculated using eq. 1, the monomer conversion and the *e.e.* value of the recovered monomer (*12*),

$$r = \ln \frac{(1 - \text{conversion} / 100)(1 - e.e. / 100)}{(1 - \text{conversion}/100)(1 + e.e. / 100)} \tag{1}$$

According to eq. 1, for the polymerization of *rac*-1 using the **3/Cu-1/**ligand, the *r* value was calculated as 3.2 for l-*S*3, 1.6 for l-1, 1.3 for l-2, and 3.1 for l-*R*3. Figure 4 shows the change in the *e.e.* of the recovered monomer and the specific rotation of the obtained polymers with the monomer conversion for the polymerization of *rac*-1 using **3/Cu-1/l-*S*3**. The *e.e.* value of the recovered monomer increased with the increasing monomer conversion. On the other hand, the specific rotation ([α]$_{435}$) of the resulting polymers decreased with the increasing monomer conversion, which was caused by the optical purity (*o.p.*) of the resulting polymers decreasing with the increasing monomer conversion. In general, the *o.p.* value of the resulting polymer is given by eq. 2 (*13*),

$$o.p. = \frac{e.e.}{\text{conversion} - e.e.} \tag{2}$$

For example, the *o.p.* value of the resulting polymer was calculated to be 56.3 % at a 13.1 % monomer conversion and 8.0 % at a 80.8 % monomer conversion. This change in the *o.p.* was in good agreement with that of the specific rotation, indicating that the optical activity of the polymer was attributable to the excess amount of the *RR*-1 units in the obtained polymer. These results indicated that the chiral Cu-complex affected the addition of *rac*-1 to the growing end, in which the *SS*-1 enantiomer of *rac*-1 was predominantly polymerized, *i.e.*, the enantiomer-selective radical polymerization.

Copper-catalyzed Copolymerization and Homopolymerization of *RR*-1 and *SS*-1. The enantiomer-selective polymerization of *rac*-1 using the chiral ATRP initiating system can be treated as the usual cyclocopolymerization of *RR*-1 and *SS*-1 in the following reactions,

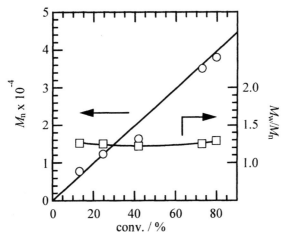

Figure 2. Dependence of M_n (open circle) and M_w/M_n (open square) on monomer conversion for polymerization of rac-1 using **3/Cu-1/l-S3**.

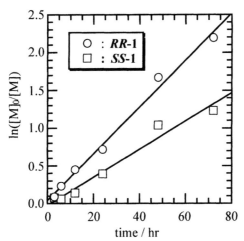

Figure 3. Kinetic plots for polymerization of rac-1 using **3/Cu-1/l-S3**.

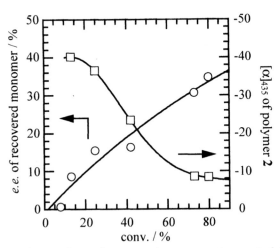

Figure 4. Dependence of e.e. of recovered monomer (open circle) and $[\alpha]_{435}$ of polymer 2(open square) on monomer conversion for polymerization of rac-1 using **3/Cu-1/l-S3**.

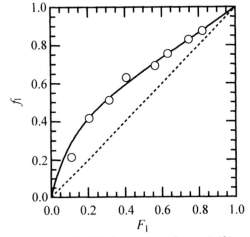

Figure 5. Mole fractions of the RR-1 units in polymer 2 (f_1) vs. mole fraction of RR-1 in the feed (F_1). The circles are experimental results, and the solid line is the fit of the Mayo-Lewis equation to the data with $r_S = 0.296$ and $r_R = 1.49$.

$$\text{~}S\text{–}S\bullet\ Cu^{II}Br_2/L_2\ +\ SS\text{-1}\ \xrightleftharpoons{\ k_{SS}\ }\ \text{~}S\text{–}S\text{–}S\text{–}S\bullet\ Cu^{II}Br_2/L_2 \qquad (3)$$

$$\text{~}S\text{–}S\bullet\ Cu^{II}Br_2/L_2\ +\ RR\text{-1}\ \xrightleftharpoons{\ k_{SR}\ }\ \text{~}S\text{–}S\text{–}R\text{–}R\bullet\ Cu^{II}Br_2/L_2 \qquad (4)$$

$$\text{~}R\text{–}R\bullet\ Cu^{II}Br_2/L_2\ +\ SS\text{-1}\ \xrightleftharpoons{\ k_{RS}\ }\ \text{~}R\text{–}R\text{–}S\text{–}S\bullet\ Cu^{II}Br_2/L_2 \qquad (5)$$

$$\text{~}R\text{–}R\bullet\ Cu^{II}Br_2/L_2\ +\ RR\text{-1}\ \xrightleftharpoons{\ k_{RR}\ }\ \text{~}R\text{–}R\text{–}R\text{–}R\bullet\ Cu^{II}Br_2/L_2 \qquad (6)$$

where the monomer reactivity ratios, r_S and r_R, are defined as k_{SS}/k_{SR} and k_{RR}/k_{RS}, respectively. Figure 5 shows the copolymerization composition curve for the copolymerization of RR-1 and SS-1 using 3/Cu-1/l-S3. The copolymerization reactivity of RR-1 was higher than that of SS-1, resulting in the mole fraction of the RR-1 unit in the copolymer being higher than that of the SS-1 unit for every monomer feed. The monomer reactivity ratios, which were calculated by the Kelen-Tüdős method (14), were $r_S = 0.296$ and $r_R = 1.49$.

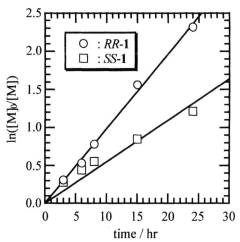

Figure 6. *Kinetic plots for homopolymerizations of RR-1 and SS-1 using 3/Cu-1/l-S3.*

Figure 6 shows the kinetic plots for the homopolymerizations of RR-1 and SS-1 using 3/Cu-1/l-S3. The consumption rate of RR-1 was grater than that of SS-1, i.e. $k_{RR} = 9.81 \times 10^{-2}$ hr^{-1} and $k_{SS} = 4.38 \times 10^{-2}$ hr^{-1}, and thus, the k_{RS} and k_{SR} values were calculated as 6.58×10^{-2} hr^{-1} and 1.48 1x 0^{-1} hr^{-1}, respectively. The rate constants of the four competing reactions, eq.s $3 \sim 6$, increased in the order of $k_{SR} \gg k_{RR} > k_{SS} > k_{RS}$, in particular, the k_{SR} value was highest among them, meaning that the growing end of the *SS*-1 unit predominantly reacted with *RR*-1 through eq. 4, resulting in the enantiomer-selective radical polymerization.

Ruthenium-catalyzed Cyclopolymerization of *rac*-1. Table II lists the results for the polymerization of *rac*-1 using the initiating system consisting of 3, **Ru-1**, and a chiral additive (**a-1** ~ **a-5**) in anisole at 60 °C. The polymerization of *rac*-1 proceeded through a cyclopolymerization mechanism to afford the polymer essentially consisting of cyclic constitutional repeating units, *i.e.*, the extent of the cyclization was ca. 100 %. The M_n values of the obtained polymers ranged from 15,600 to 35,600, and the M_w/M_n values were 1.19 ~ 2.89. For **a-1** as a chiral additive, the M_n value was greater than those for the other additives, but the SEC trace was bimodal ($M_w/M_n = 2.89$).

Table II. Enantiomer-selective polymerization of *rac*-1 using Ru-catalyzed chiral ATRP initiating system [a]

additive	time h	recovered monomer			polymer 2	
		conv. [b] %	e.e. [c] %	enriched isomer	M_n (M_w/M_n) [d]	$[\alpha]_{435}$ [e]
a-1	1	19.0	4.9	*SS*-1	35,600 (2.89)	-22.3°
a-2	2	24.3	3.9	*SS*-1	16,200 (1.52)	-7.6°
a-3	48	22.5	2.7	*RR*-1	15,600 (1.34)	+0.7°
a-4	20	22.6	16.9	*RR*-1	16,400 (1.19)	+40.3°
a-5	3	24.8	3.3	*SS*-1	19,800 (1.42)	-14.3°

[a] $[rac\text{-}1]_0 = 0.1$ mol L^{-1}; $[rac\text{-}1]_0/[3]_0/[\text{Ru-}1]_0/[\text{additive}]_0 = 200/2/1/4$; solvent, anisole; temp., 60 °C.
[b] Determined by HPLC equipped with CHIRALCEL OB-H column.
[c] Enantiomeric excess of recovered monomer.
[d] Determined by SEC in THF using polystyrene standards.
[e] $c = 0.3$ in chloroform at 28 °C

The *e.e.* value of the recovered monomers was 2.7 % ~ 16.9 %. Although *SS*-1 and *RR*-1 were in slight excess in the recovered monomer using **a-2** and **a-3**, respectively, *RR*-1 and *SS*-1 were predominantly consumed using **a-1** and **a-5** and **a-4**, respectively. The resulting polymers exhibited optical activity, and the specific rotations ($[\alpha]_{435}$) of the resulting polymers were –22.3° ~ +40.7°. The *r* values were 1.6 and 1.3 for **a-1** and **a-2** as chiral diamine additives, respectively. For **a-3** and **a-4** as chiral phenolic additives, **a-3**, an unidentate additive, slightly affected the enantiomer selectivity as *r* = ca. 1.0, while **a-4**, a bidentate additive, exhibited the highest *r* value of 3.4. The *r* value of **a-5** as a biphosphine additive was 1.3. These results indicated that the chiral bidentate phenolic compound

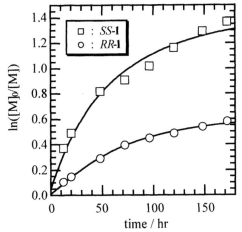

Figure 7. *Kinetic plots for polymerization of rac-1 using 3/Ru-1/a-4.*

was an effective additive for the ruthenium-catalyzed enantiomer-selective radical cyclopolymerization of the racemic monomer, *rac*-1.

Figure 7 shows the kinetic plots of the polymerization of *rac*-1 using **3/Ru-1/a-4** at 60 °C in anisole. Although both plots showed curvatures, the consumption rate of *SS*-1 was greater than that of *RR*-1, which was different from the result for the polymerization of *rac*-1 using **3/Cu-1//l-S3**, as shown in Figure 3. Figure 8 shows the relationships between the M_n and M_w/M_n of the resulting polymers *vs.* the monomer conversion. The M_n values linearly increased with the monomer conversion, though these values were greater than the calculated ones. The M_w/M_n values slightly increased with the monomer conversion.

Figure 9 shows the relationships between the *e.e.* values of the recovered monomers and the specific rotation ($[\alpha]_{435}$) of the resulting polymer *vs.* the monomer conversion. The *e.e.* values of the recovered monomers increased from 13.2 % to 37.5 % by increasing the monomer conversion. On the other hand, the specific rotations of the resulting polymers decreased from +40.7° to +26.3° with the increasing monomer conversion. These results were very similar to those for the Cu-catalyzed chiral ATRP, so that the chiral ruthenium complex also affected the addition of *rac*-1 to the growing chain end.

Ruthenium-catalyzed Copolymerization and Homopolymerization of RR-1 and SS-1. For the copolymerization of *RR*-1 and *SS*-1 using **3/Ru-1/a-2**, the mole fraction of the *RR*-1 units in the copolymer was higher than that of the *SS*-1 units for every monomer feed. The monomer reactivity ratios, r_R (= k_{RR}/k_{RS}) and r_S (= k_{SS}/k_{SR}), were determined as 1.83 and 0.80, respectively. In

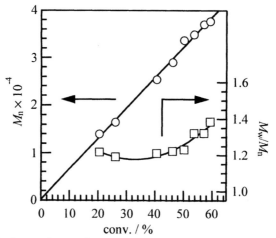

Figure 8. *Dependence of M_n (open circle) and M_w/M_n (open square) on monomer conversion for polymerization of rac-1 using 3/**Ru-1/a-4**.*

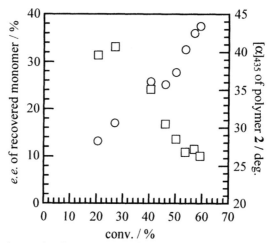

Figure 9. *Relationship between e.e. of the recovered monomer (open circle) and [α]$_{435}$ of the resulting polymer 2 (open square) vs. monomer conversion for polymerization of rac-1 using 3/**Ru-1/a-4**.*

$$\begin{array}{l}
\text{~}S\text{--}S\text{•RuX} \xrightarrow{\begin{array}{c}\overset{SS\text{-}1}{\underset{k_{SS}}{\longrightarrow}}\end{array}} \text{~}S\text{--}S\text{-}S\text{-}S\text{•RuX} \quad (7)\\[2em]
\phantom{\text{~}S\text{--}S\text{•RuX}}\xrightarrow{\begin{array}{c}\overset{RR\text{-}1}{\underset{k_{SR}}{\longrightarrow}}\end{array}} \text{~}S\text{--}S\text{-}R\text{-}R\text{•RuX} \quad (8)
\end{array}$$

$$\begin{array}{l}
\text{~}R\text{--}R\text{•RuX} \xrightarrow{\begin{array}{c}\overset{SS\text{-}1}{\underset{k_{RS}}{\longrightarrow}}\end{array}} \text{~}R\text{--}R\text{-}S\text{-}S\text{•RuX} \quad (9)\\[2em]
\phantom{\text{~}R\text{--}R\text{•RuX}}\xrightarrow{\begin{array}{c}\overset{RR\text{-}1}{\underset{k_{RR}}{\longrightarrow}}\end{array}} \text{~}R\text{--}R\text{-}R\text{-}R\text{•RuX} \quad (10)
\end{array}$$

addition, for the kinetic plots of the homopolymerizations of SS-1 and RR-1 using 3/**Ru-1/a-2**, the polymerization rates were first order with respect to the monomer concentration to yield $k_{RR} = 4.14 \times 10^{-2}$ hr^{-1} and $k_{SS} = 2.39 \times 10^{-2}$ hr^{-1}. Thus, k_{RS} and k_{SR} were calculated as 2.27×10^{-2} hr^{-1} and 2.97×10^{-2} hr^{-1}, respectively. Although these kinetic results differed from those for the Cu-catalyzed chiral ATRP, the growing end of the RR-1 unit predominantly reacted with RR-1 through eq. 10, resulting in the enantiomer-selective radical polymerization.

Conclusions

The cyclopolymerization of rac-2,4-pentanediyl dimethacrylate (rac-1) enantiomer-selectively proceeded using the chiral atom transfer radical polymerization (ATRP) initiating system. (S,S)-2,6-Bis(4-isopropyl-2-oxazolin-2-yl)pyridine was effective as a chiral ligand for the copper-catalyzed ATRP initiating system, and (S)-1,1'-bi-2-naphthol as a chiral ligand for the ruthenium-catalyzed ATRP initiating system. The enantiomeric excess ($e.e.$) of the recovered monomers increased with the increasing monomer conversion, and the specific rotation of the resulting polymers decreased. Although these kinetic results differed from those for the Cu-catalyzed chiral ATRP, there was an apparent difference among the addition reactions of the growing ends and the enantiomers of rac-1.

Literature Cited

(1) Okamoto, Y.; Ohta, K.; Yuki, H. *Macromolecules* **1978**, *11*, 724.
(2) Sepulchre, M.; Spassky, N. *Makromol. Chem.* **1981**, *182*, 2225.
(3) Coates, G. W.; Waymouth, R. M. *J. Am. Chem. Soc.* **1993**, *115*, 91.
(4) Wulff, G. *Angew. Chem. Int. Ed. Engl.* **1989**, *28*, 21.
(5) Kakuchi, T.; Obata, M. *Macromol. Rapid Commun.* **2002**, *23*, 395.

(6) Okamoto, Y.; Nishikawa, M.; Nakano, T.; Yashima, E.; Hatada, K. *Macromolecules* **1995**, *28*, 5135.
(7) Wang J. -S.; Matyjaszewski, K. *J. Am. Chem. Soc.* **1995**, *117*, 5614.
(8) Kato, M.; Kamigaito, M.; Sawamoto, M.; Higashimura T.; *Macromolecules* **1995**, *28*, 1721.
(9) Wang, J. -S.; Matyjaszewski, K. *Macromolecules* **1995**, *28*, 7901.
(10) Tsuji, M.; Sakai, R.; Satoh, T.; Kaga, H.; Kakuchi, T. *Macromolecules*, in press.
(11) Kwong, H.-L.; Lee, W.-S. *Tetrahedron: Asymmetry* **1999**, *10*, 3791.
(12) Kagan, H.B.; Fiaud, J.C. *Top. Stereochem.* **1988**, *18*, 249.
(13) Okamoto, Y.; Urakawa, K.; Yuki, H. *J. Polym. Sci. Polym. Chem. Ed.* **1981**, *19*, 1385.
(14) Tüdös, F.; Kelen, T.; Földes, -B., B.; Turcsanyi, B. *J. Macromol. Sci. Chem.* **1976**, *A10*, 1513.

Chapter 16

Organic Acids Used as New Ligands for Atom Transfer Radical Polymerization

Shenmin Zhu[1], Deyue Yan[1,2], and Marcel Van Beylen[2]

[1]College of Chemistry and Chemical Engineering, Shanghai Jiao Tong University, 800 Dongchuan Road, Shanghai 200240, China
[2]Department of Chemistry, Catholic University Leuven, Celestijnenlaan, 200F, B–3001 Heverlee, Leuven, Belgium

Organic acids have been successfully employed as new ligands in iron-mediated ATRP and reverse ATRP. Their applications in the preparation of block copolymers, graft polymers and graft block copolymers were also investigated. The new ligands, such as isophthalic acid, iminodiacetic acid, acetic acid and succinic acid, are much cheaper than conventional ligands used previously. Furthermore, non-toxic organic acids are safer for health. The controlled radical polymerization was carried out at 25-130 °C resulting in polymers with controlled molecular weight and narrow molecular weight distribution (M_w/M_n=1.2-1.5).

Introduction

Atom transfer radical polymerization (ATRP) is among the most promising approaches to controlled radical polymerization. Recent advances aimed at the design of new ligands and new metals that affect the activity and selectivity of the ATRP catalysts. The pioneering work in ATRP was reported by Matyjaszewski (*1-4*) who used copper as the transition-metal and bipyridine (bpy) or its derivatives as ligands, and by Sawamoto (*5*) who used ruthenium/organic phosphorus compounds as catalyst systems. Other catalyst systems involving different transition-metals, such as iron (*6,7*), nickel (*8*) and palladium (*8*) have also been reported. Iron-mediated ATRP has been successfully implemented with precise end functionality, predetermined molecular weights and low polydispersity (*9*). As far as Fe^{II} catalytic system is concerned, various ligands of nitrogen, phosphorus donors and mixed coordinating ligands have been successfully used in the ATRP of styrene (*10,11*). However organic amines and phosphorus are harmful to human beings and are rather expensive. Recently a new kind of ligands based on organic acids was developed in our laboratory.

Various acids such as acetic acid, iminodiacetic acid, succinic acid and isophthalic acid, have been successfully employed as new ligands in the iron-mediated atom transfer radical polymerization of vinyl monomers, such as styrene (St) and methyl methacrylate (MMA). The new ligands, i.e., organic acids, are much cheaper than conventional ligands used previously. Furthermore, non-toxic organic acids are safer for health than bpy, its derivatives and organic phosphorus compounds. The systems with different organic acids can react at 25 °C to 130 °C resulting in "living"/controlled radical polymerization with relatively narrow molecular weight distribution of the resulting polymers (M_w/M_n=1.2-1.5). The measured molecular weights are close to the calculated values for the polymerization of MMA and are somewhat lower than the theoretical ones for styrene.

As we know that ATRP has been performed in various solvents (*12*). Vairon reported that ATRP of styrene was implemented in the presence of a limited amount of DMF (~10 % v/v) (*13*). Even though block copolymers with polysulfones have been prepared in 1, 4-dimethoxybenzene (*14*), it is difficult to carry out ATRP of styrene in copper-mediated system using chloromethylated polysulfone as the macroinitiator which requires DMF as the solvent (>50 % v/v). Most recently, DMF was used as the solvent for ATRP of acrylonitrile (*15*). In our work, the catalyst of $FeCl_2$/isophthalic acid was used for the preparation of novel linear aromatic polyethersulfone (PSF)-based graft copolymers. GPC, DSC, IR, ^1H NMR were performed to characterize the graft polymers. Aromatic polyethersulfone is attractive for the rigid characteristics of the chains and PSF-based graft copolymers are promising to be used as a sort of molecular reinforced materials.

Results and Discussion

ATRP of Styrene Catalyzed by FeCl₂/Iminodiacetic Acid (IDA)

Iminodiacetic acid (IDA) is an effective ligand for ATRP of both St and MMA with Fe^{II} as the transition metal under heterogeneous conditions. Fig. 1 shows that the measured molecular weight linearly increases with increasing monomer conversion; however the measured molecular weight is lower than the theoretical value. As shown in Fig. 1, the increase of polydispersity index with conversion is observed especially at high monomer conversion.

In order to explain the reason, the polymerization of styrene with CCl_4 as the initiator was performed. The results are compared in Fig. 1. Although the system with a chlorine containing initiator gives polymers with higher polydispersity indices than in the system with a bromine containing initiator, M_w/M_n decreases slightly with the conversion and falls in with what is predicted for a living polymerization. It has been reported by Matyjaszewski (16) that most of chain ends of the polymer obtained are chlorinated (i.e., %R-Cl: 80~90 %). Decomposition may occur more easily in R-Br system than in R-Cl system due to the weaker C-X bond in the former during a long reaction time. The dehalogenation of the minority bromine on the active chain ends may result in the minor increase of M_w/M_n with conversion. A first–order kinetic plot of the polymerization is shown in Fig. 2.

The plot is almost linear, although a short induction period is observed. The short induction period may be caused by the limited solubility of the catalyst in the reaction medium. No induction period is observed in the chlorine-based system at the reaction temperature of 110 °C (Fig. 2). As it is known, usually high temperature is used to initiate the reaction employing CCl_4 as the initiator (17), which may sometimes result in side reactions. Therefore, MWD is broader at the beginning of the chloride-based system. The linear semilogarithmic plot of $Ln([M]_0/[M])$ versus time indicates that the polymerization of styrene in bulk is first order with respect to the monomer and the concentration of active centers remains constant throughout the polymerization. This infers that no termination reactions occurred during the polymerization process.

The end group of the resulting polystyrene with the number-average molecular weight equal to 4,100 was investigated using ^1H NMR. The signals of the triplets at 4.4 ppm are attributed to the methine proton geminal to the halide end group. From the ratio of the peak intensity of the end group at 4.4 ppm to that of the phenyl group at 6.4-7.2 ppm, the M_n is calculated to be 4,500 (Fig. 3).

The resulting PS with an α-halogen in the chain end can be used as a macroinitiator for block copolymerization. From the halogen terminated PS macroinitiator ($M_{n (GPC)} = 3,320$, $M_w/M_n = 1.26$), PS-b-PMMA copolymer, $M_n = 32,510$, $M_w/M_n = 1.51$, was prepared in DMF catalyzed by FeCl₂/IDA. In the FT-IR spectrum of the PS-b-PMMA block copolymer, the characteristic peaks at

224

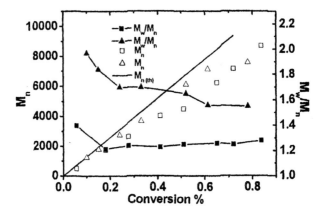

*Figure 1. Dependence of molecular weight and polydispersity on monomer conversion for ATRP of styrene in bulk, $[FeCl_2]_0 = [NH(CH_2COOH)_2]_0/2=0.067$ mole·L^{-1}, $[St]_0 = 8.7$ mole·L^{-1}, initiator $= \alpha$-bromoethylbenzene ($\square$,$\blacksquare$) at 70°C and CCl_4 ($\triangle$,$\blacktriangle$) at 110°C. (Reproduced with permission from J. Polym. Sci.-Chem. **2000**, 38, 4308-4314. Copyright 2000 Wiley.)*

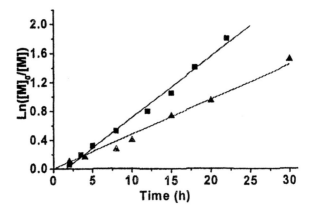

*Figure 2. Semilogarithmic kinetic plots for the bulk ATRP of styrene, $[FeCl_2]_0$: $[NH(CH_2COOH)_2]_0/2 = 0.067$ mole·L^{-1}, $[St]_0 = 8.7$ molL^{-1}, $[RBr]_0 = 0.067$ mole·L^{-1}, where initiator $= \alpha$-bromoethylbenzene ($\blacksquare$) at 70°C and CCl_4 ($\blacktriangle$) at 110°C. (Reproduced with permission from J. Polym. Sci.-Chem. **2000**, 38, 4308-4314. Copyright 2000 Wiley.)*

1731 cm^{-1}, 1193 cm^{-1} correspond to the ester group vibrations and 3060 cm^{-1}, 3026 cm^{-1}, 756 cm^{-1} and 700 cm^{-1} show the existence of the benzene ring in the block copolymer.

The GPC curve of the block copolymer shifts to the higher molecular weight side compared to the macroinitiator, which indicates that the PS-b-PMMA block copolymer has been successfully prepared. These experimental data demonstrate the living nature of the polymerization system.

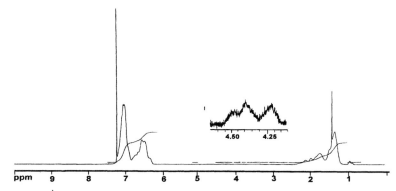

Figure 3. ^1H *NMR spectrum of polystyrene prepared using FeCl$_2$ coordinated by iminodiacetic acid as the catalyst, M$_n$ (GPC) = 4,100. (Reproduced with permission from J. Polym. Sci.-Chem. 2000, 38, 4308-4314. Copyright 2000 Wiley.)*

ATRP of MMA Catalyzed by FeCl$_2$/Iminodiacetic Acid (IDA)

Polymerization of MMA was carried out in bulk using FeCl$_2$/HN (CH$_2$COOH)$_2$ as the catalyst system under heterogeneous conditions with [MMA]:[RBr]:[FeCl$_2$]: [NH(CH$_2$COOH)$_2$] = 130:1:1:2, [MMA] = 9.4 mole·L^{-1}. A linear increase of the number average molecular weight vs monomer conversion up to 61 % was found (Fig. 4).

The molecular weight distribution of the resulting polymer is narrow (M$_w$/M$_n$ = 1.21) before the monomer conversion reaches about 50 %. However, the molecular weight distribution becomes broader (M$_w$/M$_n$~1.5) when the monomer conversion exceeds 50 %. The reaction system became very viscous and a small deviation of the molecular weight from linearity was found at high conversion. The M$_n$ (GPC) is close to the predetermined values. As can be seen from Fig. 4, PDI increases at relatively high conversion, it is owing to the side reactions in the system during a long reaction time (80 h.). The side reactions such as chain transfer and radical-radical coupling may exist. The non-symmetrical SEC curve of PMMA-X prepared with FeCl$_2$/HN(CH$_2$COOH)$_2$ as the catalyst shown in Fig. 6 further confirm the suggestion. The straight kinetic plot of Ln([M]$_0$/[M]) versus time in Fig. 5 reveals that no significant termination reactions occurred in the system.

Furthermore, the molecular weight estimated by GPC is close to that evaluated by ^1H NMR on the basis of the intensity ratio of methoxyl groups in the main chain to protons from residual initiator groups. The living nature can be further confirmed by synthesis of the block copolymer, PMMA-b-PMA. GPC curves of the block copolymer are represented in Fig. 6.

The number-average molecular weights of the block copolymer and the macroinitiator are 12,030 and 5,560, respectively. The block copolymer was characterized by differential scanning calorimetry (DSC). The DSC curve shows that there are two glass transition temperatures, one at 20 °C and one at 92 °C (Tgs of PMMA and PMA are 100 °C and 10 °C, respectively.). It indicates the formation of block copolymer of PMMA-b-PMA.

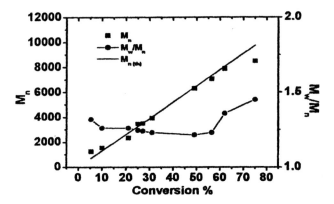

Figure 4. Molecular weight and molecular weight distribution dependence on monomer conversion for the ATRP of MMA, $[MMA]_0 : [RBr]_0 : [FeCl_2]_0 : [NH(CH_2COOH)_2]_0 = 130:1:1:2$, $[MMA]_0 = 9.4$ $mole \cdot L^{-1}$. (Reproduced with permission from J. Polym. Sci.-Chem. **2000**, 38, 4308-4314. Copyright 2000 Wiley.)

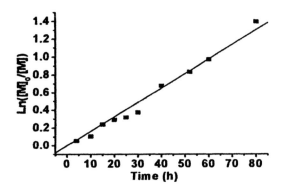

Figure 5. Semilogarithmic kinetic plot for the bulk ATRP of MMA at 90°C, $[MMA]_0 : [RBr]_0 : [FeCl_2]_0 : [NH(CH_2COOH)_2]_0 = 130:1:1:2$. (Reproduced with permission from J. Polym. Sci.-Chem. **2000**, 38, 4308-4314. Copyright 2000 Wiley.)

ATRP of Styrene Catalyzed by FeCl₂/ Succinic Acid

Two experiments were carried out to determine the effect of the new catalyst on ATRP, in the condition of [styrene]:[α-bromoethylbenzene]:[FeCl₂]: [HOOCCH₂CH₂COOH]=120:1:1:2, [St]=8.7mole·L⁻¹ in bulk at 70 and 40 °C, respectively. After reaction for 3 h, the conversion reached 80 % at 70 °C. A linear increase of the molecular weight with increasing monomer conversion throughout the reaction process was observed. However, the measured molecular weight is lower than the theoretical value when the conversion is larger than 40%. A similar phenomenon has been reported in ATRP of styrene using 1-PEBr/CuBr/bpy (18) and bis (1,10-phenanthroline)/copper bromide as catalyst

227

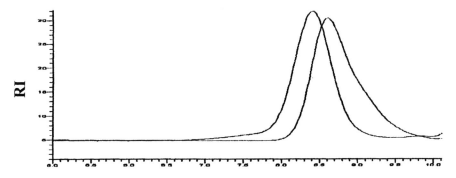

Elution Volume

Figure 6. GPC curves of PMMA-X and block copolymer PMMA-b-PMA.
(Reproduced with permission from J. Polym. Sci.-Chem. 2000, 38, 4308-4314.
Copyright 2000 Wiley.)

(19). Furthermore, similar deviations were also observed in $CCl_4/RuCl_2(PPh_3)_2/MeAl(ODBP)_2$ *(5)*, and $CCl_4/Ni(NCN')Br$ initiating systems *(8)*.

Matyjaszewski assumed that the deviation of the molecular weight was presumably due to chain transfer. The chain transfer constant of radicals to succinic acid, Cs, is 5.4×10^{-4} at 60 °C, and Cs=0.84×10^{-4} and 15.5×10^{-4} to p-xylene and ethyl acetate, respectively, at 60 °C. Both p-xylene and ethyl acetate are the conventional solvents for ATRP, so the effect of the chain transfer to succinic acid on the polymerization is negligible.

The polydispersity index is controlled rather well, i.e., $M_w/M_n \sim 1.30$. The calculated molecular weights are lower than theoretical ones under both reaction temperatures. It means chain transfer or termination may present.

The linearity of the semilogarithmic plots of $Ln([M]_0/[M])$ versus time indicates that the polymerization was first-order with respect to monomer, the concentration of growing radical remained constant before 65 % and 75 % monomer conversion at 70 °Cand 40 °C, respectively. At 40 °C the PDI of the product increases with monomer conversion. Generally, the rate constant of the side reaction, which results in an increase of the polydispersity index, is reduced at lower temperature. However, it needs a long time for the reaction to reach a higher conversion at 40 °C (Fig. 7, 8), probably more chains undergo the side reaction before the reaction system approaches the same conversion than the reaction at 70 °C. The fact that PDI increases with monomer conversion demonstrates chain transfer or other side reaction play an important role in the reaction system *(20)*.

ATRP of Styrene Catalyzed by FeCl₂/Acetic Acid

Polymerization of St with $[CCl_4]_0 = 0.044$ mole·L^{-1}, $[FeCl_2]_0 = 0.044$ mole·L^{-1}, $[acetic\ acid]_0 = 0.13$ mole·L^{-1} and $[St]_0 = 8.7$ mole·L^{-1} was carried out in bulk at 120 °C; the polymerization reached 75 % within 23 h at 120 °C to

228

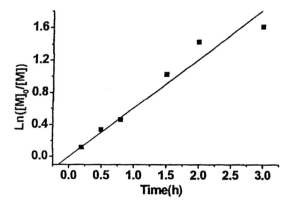

Figure 7. First-order time-conversion plots for bulk ATRP of styrene at 70°C catalyzed by FeCl₂/succinic acid. (Reproduced with permission from Macromol. Chem. Phys **2000**, 201, 2666-2699. *Copyright 2000.)*

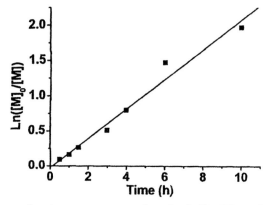

Figure 8. First-order time conversion plots for bulk ATRP of styrene at 40°C catalyzed by FeCl₂/succinic acid. (Reproduced with permission from Macromol. Chem. Phys **2000**, 201, 2666-2699. *Copyright 2000.)*

yield PS of $M_{n,exp}$ = 14,600 and M_w/M_n = 1.46. The measured molecular weight linearly increases with monomer conversion, and matches the theoretical value calculated.

In order to investigate whether the binary system of $FeCl_2/CH_3COOH$ can induce cationic polymerization or not, we carried out the polymerization in various conditions of $[CCl_4]_0:[CH_3COOH]_0:[St]_0$=1:3:200, $[CCl_4]_0:[FeCl_2]_0:[St]_0$ =1:1:200, and $[FeCl_2]_0:[CH_3COOH]_0:[St]_0$=1:3:200. It can be seen that the polymerization lacking one of the components of the initiating system results in a polymer with very high molecular weight ($>10^5$) and relatively wide molecular weight distribution (M_w/M_n > 2.0) from the beginning of the reaction. These are the characteristics of a conventional radical polymerization. On the contrary, controlled molecular weight and relatively low polydispersity (1.46) were obtained in the presence of CCl_4 with $FeCl_2/CH_3COOH$ as the catalyst. Furthermore, the inhibitor of cationic polymerization, triethylamine, was introduced into the polymerization system with $St/FeCl_2/CH_3COOH$ and without CCl_4 at 120 °C, after 5 h the monomer conversion reached 81 % and the number-average molecular weight was 2.6×10^5 with $M_w/M_n = 1.79$. It further verified that the polymerizations involved a conventional radical polymerization instead of a cationic one.

ATRP of MMA Catalyzed by $FeCl_2$/Isophthalic Acid

ATRP of MMA with [ethyl 2-bromopropionate]$_0$: [$FeCl_2$]$_0$: [isophthalic acid]$_0$: [MMA]$_0$ = 1:1:2:142 at 80 °C in DMF (3.23 % v/v) was fully controlled under heterogeneous conditions. During the reaction, the white color of the $FeCl_2$/isophthalic acid complex changed to yellowish-red as the reaction proceeded, which indicates that a Fe^{III} compound forms. The monomer conversion reached about 80 % within 10 h. The polymerization showed a linear increase in molecular weights with conversion and matched the theoretical values well (Fig. 9). The polydispersity (1.38) was relatively low at high monomer conversion of the polymerization.

Ethyl 2-bromopropionate is a bad initiator for MMA, unless halogen exchange happens (16,21). Herein, ethyl 2-bromopropionate was employed as the initiator in the ATRP with $FeCl_2$/isophthalic acid as the catalyst. The resultant PMMA was characterized by ^1H NMR spectroscopy. Characteristic resonances originated from the α-halocarbonyl moieties are visible at both 4.1 ppm and 3.8 ppm. The protons of $-OCH_3$ in the backbone are observed at 3.58 ppm. Furthermore, the molecular weight obtained by GPC equals 6,540, and is close to that estimated from ^1H NMR (5,880).

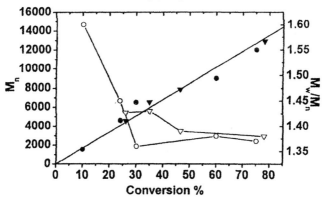

Figure 9. Evolution of molecular weight and polydispersity with conversion for ATRP of MMA catalyzed by $FeCl_2$/isophthalic acid. $[I]_0:[FeCl_2]_0:[ligand]_0:$ $[MMA]_0 = 1:1:2:142$ at 80°C, $[I]_0 = 6.4 \times 10^{-2}$ mole·L^{-1} in 3.23 % DMF (v/v). $\bigcirc/\triangledown$ and $\bullet/\blacktriangledown$ *represent the respective results from two experiments, and the reaction conditions of the two batches are identical with each other except the room temperatures (T=8°C and 20°C, respectively). (Reproduced from* Macromolecules **2000**, 33, 8233-8238. *Copyright 2000 American Chemical Society.)*

An additional method toward verifying the functionality of a polymer prepared by ATRP is its use as a macroinitiator for the same or another monomer. A chain extension of MMA with PMMA ($M_{n \, (GPC)}$ = 12,900, M_w/M_n = 1.40) as the macroinitiator was performed. The polymerization was carried out using $FeCl_2$/isophthlic acid as the catalyst in 50 % (v/v) DMF at 90 °C with $[MMA]_0$ = 4.7 mole·L^{-1}, $[FeCl_2]$ = [isophthlic acid]/2 = 0.02 mole·L^{-1}. A conversion of 65 % was achieved after polymerization for 12 h. The M_n of the chain-extended PMMA increased to 60,200; however, the polydispersity index (1.58) was a little higher than that of the macroinitiator (M_w/M_n = 1.40). A small part of the macroinitiator probably remains unreacted. The result shows the increase in the molecular weight and demonstrates the conclusion that the polymerization of MMA catalyzed by $FeCl_2$/isophthalic acid in the presence of DMF is controlled. To further characterize the elementary process leading to the formation of active species in DMF, the catalytic system was analyzed by UV-VIS spectroscopy. The corresponding $FeCl_2$/isophthlic acid complex does not show any characteristic UV-Visible adsorption between 260 nm to 600 nm. As the reaction proceeded, two absorption bands appeared; the main peak is centered at 360 nm, and the other one is located at 314 nm. The change in absorption spectra of the reaction system can be attributed to the formation of $FeCl_3$/isophthalic acid from $FeCl_2$/isophthalic acid through halogen transfer.

Influence of Different $[FeCl_2]$/[ligand] Ratios

An [isophthalic acid] to $[FeCl_2]$ ratio of 2 not only gave the best control of molecular weight and its distribution but also provides a rather rapid reaction rate in a controlled fashion. The experimental data reported in this work are different from those reported by Vairon (13) who used bpy and Matyjaszewski (22) who used monodentate amines as the ligands in copper mediated systems. They concluded that excess DMF affects the living nature of ATRP. In our ATRP system, the quantity of DMF has no significant effect on the living nature if there is sufficient ligand to complex the transition metal. It may be explained by the different properties of the complexes. In this system, we assumed that the two oxygen atoms of each carboxylate group chelate the iron atom in the equatorial plane, and two oxygen atoms from DMF molecules coordinate the iron atom in the axial direction, forming a six-coordinative geometry (Scheme 1).

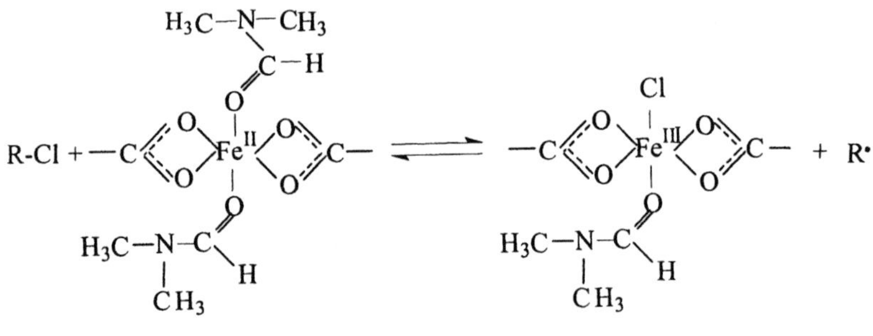

Scheme 1

The DMF molecules can easily exchange with other ligands such as chloride (Scheme 1). When there are halide species in the system, a reversible equilibrium between active species and dormant species is established via the ATRP mechanism. If there is enough stronger ligand to complex with the metal in this system, ligand exchange by DMF will be negligible. That means the coordination of DMF would affect the coordination of chloride ligands only. In other words, if DMF coordinates strongly, perhaps, there will be no empty coordination site at the iron center, and the activation step of the ATRP equilibrium would become very difficult; such a catalyst would be quite inactive and the process would become very slow. On the other hand, if the chloride once coordinated to iron (transforming it to Fe^{III}, immediately dissociates and is replaced by DMF, the deactivating species would not be able to react with the propagating radicals via the ATRP deactivation – this would lead to poorly controlled and fast reactions. In this case, active species cannot become sufficiently dormant through halogen transformation resulting in radical-radical termination. Therefore, the molecular weights are higher than the theoretical values.

Synthesis of Graft Copolymers Catalyzed by $FeCl_2$/Isophthalic Acid

The macroinitiator composed of side chloromethyl groups was derived from commercially available polysulfone (PSF). The chloromethylation reaction was carried out in the presence of chloromethyl methyl ether. The molecular weight of chloromethylated aromatic polyether sulfone (PSF-Cl) obtained was $M_n = 28,440$, and the polydispersity index was $M_w/M_n = 2.17$. Comparison of the ^1H NMR spectrum of PSF with that of the resultant polymer demonstrated that the functionalization did take place. The copolymerization was performed by means of ATRP using PSF-Cl as the macroinitiator and $FeCl_2$/ isophthalic acid as the catalyst system.

PSF-g-PMMA: The GPC traces of PSF-Cl and PSF-g-PMMA graft copolymers are demonstrated in Figure 10. When the monomer conversion increases from 23 % to 66 % during graft copolymerization, M_w/M_n almost remains constant (from 2.18 to 2.17), and the molecular weight increases from 64,900 to 96,600. The shapes of GPC traces can illustrate whether there is a side reaction or not besides the side chain growing.

One constraint condition of using controlled radical polymerization to grow the side chains from the backbone is that radical–radical coupling must be significantly suppressed, otherwise cross-linked polymers with multi-modal molecular weight distributions may result. To avoid this disadvantage, various conditions of catalyst systems, initiator concentration and polymerization temperature were optimized to prevent as much termination during polymerization as possible. In the graft polymerization of styrene and butyl acrylate from poly (2-(2-bromoisobutyryloxy)ethyl methacrylate) made by Matyjaszewski (*23*), $CuBr_2$ was added to avoid in situ radical termination. Conversion was usually 10-20 %, not more than 30 %, and the optimized

232

temperature was 70-80 °C. Higher conversion (>30 %) or temperature (>80 °C) led to significant termination.

GPC traces in Fig. 10 show that the molecular weight of the copolymer increased with increasing monomer conversion. The monomodal shape of the GPC traces of the products suggests the absence of the homopolymerization. The result indicated that side reactions especially radical–radical termination were negligible even at such high temperature (100 °C) and conversion (66 %). Graft copolymerization catalyzed by $FeCl_2$/isophthalic acid in DMF was successfully performed.

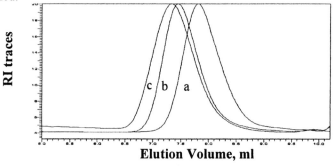

Elution Volume, ml

Figure 10. GPC traces of PSF-Cl and PSF-g-PMMA graft copolymers, a. PSF-Cl ($M_{n,GPC}$ = 28,440, M_w/M_n = 2.17); b. PSF-g-PMMA ($M_{n,GPC}$ = 64,900, M_w/M_n = 2.18, conversion = 23 %), c. PSF-g-PMMA ($M_{n,GPC}$ = 96,600, M_w/M_n = 2.17, conversion = 66 %).

Thermal analysis reveals that there is only one glass transition temperature T_g=122 °C for the copolymer of PSF-g-PMMA (the glass-transition temperatures of PMMA and PSF are 100 and 180 °C, respectively), indicating there is no macroscopic phase separation. The ^1H NMR spectra of the chloromethylated macroinitiator and typical graft copolymer are shown in Figure 11. In the analysis of ^1H NMR of the graft copolymer, the disappearance of the signals around 4.5 ppm corresponding to -CH_2Cl shows the complete initiation (Fig. 11b). Resonances at 0.8-1.1 ppm (-CH_3), 3.6 ppm (-OCH_3) represent the existence of MMA in the graft copolymer. So the successful formation of graft copolymers was supported by the results of ^1H NMR analysis.

The effect of the macroinitiator concentration on the graft copolymerization catalyzed by $FeCl_2$/isophthalic acid is compiled in Table I . The polymerization was implemented as mentioned above. After 16.5 h, the reaction was stopped.

When [PSF-Cl] = 2.25×10^{-6} mol·, the graft product with M_n = 96,650 and M_w/M_n = 1.78 is obtained (see entry no. 1, in table 1). When the concentration of PSF-Cl decreases from 1.13×10^{-6} mol to 3.52×10^{-7} mol, the molecular weight increases to 167,000 from 99,000. The various molecular weights of PSF-g-

PMMA samples resulting from different macroinitiator concentrations at 110 °C are prepared. The entire peaks are monomodal, and the polydispersity index (M_w/M_n) changes from 2.17 (macroinitiator) to 1.78, 1.83, 1.71, 1.85, respectively. It can be concluded that high temperature facilitates the graft copolymerization.

Table I. Graft Copolymerization of MMA from PSF-Cl Catalyzed by FeCl₂/Isophthalic Acid with Various Concentrations of Macroinitiator*

[Initiator]/mol (PSF-Cl)	Conversion (%)	M_n	M_w/M_n
2.25×10^{-6}	38.5	96,650	1.78
1.13×10^{-6}	34.8	99,000	1.83
5.65×10^{-7}	33.5	115,800	1.71
3.52×10^{-7}	33.5	167,000	1.85

*[MMA] = 2.69 mole·L⁻¹, [FeCl₂] = [isophthalic acid]/2 =0.0297 mole·L⁻¹, at 110 °C in 71% (v/v) DMF; reaction time = 16.5 h.

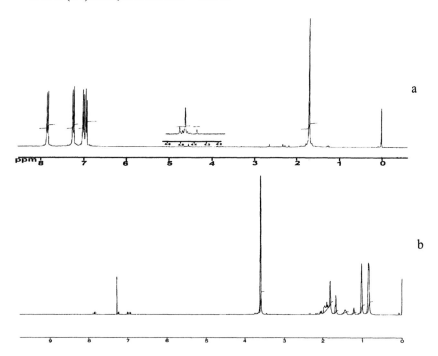

Figure 11. ¹H NMR spectra of chloromethylated aromatic polyethersulfone(a) and PSF-graft-PMMA (b) initiated by PSF-Cl/FeCl₂/isophthalic acid, MMA conversion = 66 %.

PSF-g-PBA and PSF-g-PMA Graft copolymers with elastic and glassy segments are expected to exhibit thermoplastic elastomer behavior. It is the goal of this work to synthesize graft copolymers with elastic and glassy segments such as PSF-g-PMA and PSF-g-PBA by ATRP of MA and BA using a rigid polymer (PSF-Cl) as the macroinitiator. The results of GPC, ^{1}H NMR, FT-IR, DSC tests illustrated the sucessful synthesis of PSF-g-PBA and PSF-g-PMA.

Experimental

Materials: Methyl methacrylate and styrene were vacuum distilled over CaH_2 just before polymerization. Ligands were recrystallized prior to use. $FeCl_2$ was washed with acetone and dried at 60 °C under vacuum before use. The initiators, 1-phenylethyl bromide and ethyl 2-bromopropionate, were used as received from Aldrich; CCl_4 was distilled before polymerization.

Polymerization Procedures: The general procedure of the polymerization was as follows: the catalyst, the ligand and the monomer were added to a flask with stirring; three cycles of vacuum-nitrogen are applied in order to remove oxygen; after the mixture was stirred at 25 °C for one hour, the initiator was added. Then the flask was immersed in an oil bath at the required temperature. After a given time, the flask was opened and a certain amount of tetrahydrofuran (THF) was added into the reaction system to dissolve the resulting polymer.

Measurements: The monomer conversion was determined by gravimetry. Molecular weight (M_n) and molecular weight distribution (MWD) were obtained by gel permeation chromatography (GPC) that was carried out with a PE200 instrument equipped with a mixed 5µ PS columns (refractive index detector). All samples were run in THF at 25 °C with a flow rate of 1.0ml/min and calibrated with polystyrene standards.^{1}HNMR spectrum was recorded on a BRUKER AVANCE500 500MHz NMR at room temperature in $CDCl_3$.

Conclusions

Organic acids, such as isophthalic acid, iminodiacetic acid, acetic acid and succinic acid, have been successfully used as a new kind of ligands in iron-mediated ATRP. They are much cheaper than conventional ligands used previously. Furthermore, organic acids of very low toxicity are safer for health. The controlled radical polymerization reactions were carried out at 25-130 °C resulting in polymers with controlled molecular weight and narrow molecular weight distribution.

Commercially available PSF was modified into an ATRP macroinitiator by chloromethylation. The $FeCl_2$/isophthalic acid catalyst system can be used to prepare the molecular reinforced materials such as PSF-g-PMMA, PSF-g-PMA and PSF-g-PBA from PSF-Cl in DMF.

References

1. Wang, J. S.; Matyjaszewski, K., *J. Am. Chem. Soc.* **1995**, *117*, 5614.
2. Wang, J. S.; Matyjaszewski, K., *Macromolecules* **1995**, *28*, 7901.
3. Patten, T. E.; Xia, J.; Abernathy, T.; Matyjaszewski, K., *Science* **1996**, *272*, 866.
4. Xia, J.; Gaynor, S. G.; Matyjaszewski, K., *Macromolecules* **1997**, *30*, 7697.
5. Kato, M.; Kamigaito, M.; Sawamoto, M.; Higashimura, T., *Macromolecules* **1995**, *28*, 1721.
6. Matyjaszewski, K.; Wei, M.; Xia, J.; McDermott, N. E., *Macromolecules* **1997**, *30*, 8161.
7. Teodorescu, M.; Gaynor, S. G.; Matyjaszewski, K., *Macromolecules* **2000**, *33*, 2335.
8. Granel, C.; Dubois, Ph.; Jérôme, R.; Teyssié, Ph., *Macromolecules* **1996**, *29*, 8576; Lecomte, P.; Drapier, I.; Dubois, Ph.; Jérôme, R.; Teyssié, Ph., *Macromolecules* **1997**, *30*, 7631.
9. Wei, M. L.; Xia, J. H.; Matyjaszewski, K., *Polym. Prep.* **1997**, *38(2)*, 233.
10. Ando, T.; Kamigaito, M.; Sawamoto, M., *Macromolecules* **1997**, *30*, 4507.
11. Moineau, G.; Granel, C.; Dubois, Ph.; Jérôme, R.; Teyssié, Ph., *Macromolecules* **1998**, *31*, 542.
12. Matyjaszewski, K.; Nakagawa, K.; Jasieczek, C. G., *Macromolecules* **1998**, *31*, 1535.
13. Pascual, S.; Coutin, B.; Tardi, M.; Polton, A.; Vairon, J.–P., *Macromolecules* **1999**, *32*, 1432.
14. Gaynor, S.; Matyjaszewski, K., *Macromolecules* **1997**, *30*, 4241.
15. Kowalewski, T.; Tsarevsky, N. V.; Matyjaszewski, K., *J. Am. Chem. Soc.* **2002**, *124*, 10632.
16. Matyjaszewski, K.; Shipp, D. A.; Wang, J.; Grimaud, T.; Patten, T. E., *Macromolecules* **1998**, *31*, 6836.
17. Matyjaszewski, K., *Macromolecules* **1998**, *31*, 4710.
18. Qiu, J.; Matyjaszewski, K., *Macromolecules* **1997**, *30*, 5643.
19. Cheng, G.; Hu, Ch.; Ying, Sh., *Macromol. Rapid Commun.* **1999**, *20*, 303.
20. Ahmad, N. M.; Heatley, F.; Lovell, P. A., *Macromolecules* **1998**, *31*, 2822.
21. Wang, J.-L.; Grimaud, T.; Shipp, D., *Macromolecules* **1998**, *31*, 1527.
22. Ziegler, M. J.; Paik, H.; Davis, K. A.; Gaynor, S. G.; Matyjaszewski, K., *Polym. Prep.* **1999**, *40*, 432.
23. Beers, K. L.; Gaynor, S. G.; Matyjaszewski, K., *Macromolecules* **1998**, *31*, 9413.

Chapter 17

Novel Fluorinated Polymer Materials Based on 2,3,5,6-Tetrafluoro-4-methoxystyrene

Søren Hvilsted[1], Sachin Borkar[1,2], Heinz W. Siesler[2], and Katja Jankova[1]

[1]Danish Polymer Centre, Department of Chemical Engineering, Technical University of Denmark, Building 423, DK–2800 Kgs. Lyngby, Denmark
[2]Department of Physical Chemistry, University of Essen, Schützenbahn 70, D–45117 Essen, Germany

2,3,5,6-Tetrafluoro-4-methoxystyrene (TFMS) has been poly-merized in bulk and in xylene solution by Atom Transfer Radical Polymerization (ATRP) in a conventional protocol at 110 °C. Relatively good control has been achieved with number-average molecular mass (M_n) up to 17,000 and corresponding polydispersity index (PDI) generally below 1.3. The ATRP of TFMS is the fastest observed for any substituted styrene in bulk or solution in organic solvents. The isolated poly(tetrafluoromethoxystyrene)s (PTFMS)s have been employed as macroinitiators for ATRP of 2,3,4,5,6-pentafluorostyrene (FS) and styrene (St) resulting in block copolymers with controlled characteristics. TFMS homo- and block copolymers with PS have better thermal stability than PS. The solubility of the PTFMS containing polymers is lower than that of PS. Furthermore, PTFMS has been demethylated and the resulting hydroxyl sites alkylated with different azobenzene side chains. The azobenzene derivatized polymer has additionally been copolymerized with St. Both homo- and block copolymers with azobenzene side chains form materials exhibiting liquid crystallinity.

Fluorinated polymers with varying fluorine content are materials that attract great attention due to a number of desirable properties such as high thermal, chemical, ageing and weather resistance and low surface energy. Moreover, thin film materials of fluorinated polymers with low permittivity or low dielectric constants (*1*), low flammability, excellent inertness, low refractive index and low-loss optical waveguiding potential (*2*) are desirable properties for the (opto)electronic industry. Recently, copolymers containing both fluorinated oligomer segments and polyethylene glycol have shown excellent performance as electrolyte materials for lithium ion conductivity in batteries (*3*). Thus, novel highly fluorinated monomers are potentially highly interesting.

We have recently demonstrated (*4, 5*) the versatility of 2,3,4,5,6-penta-fluorostyrene to be easily transformed by Atom Transfer Radical Polymerization (*6*) to polymers with controlled molecular weight characteristics. Both homopolymers (PFS) and block copolymers with PS could be prepared in a broad compositional range where the polydispersity index in all cases was kept below 1.3. Compared to PS all these novel PFS based materials achieved both much higher thermal stability and significantly increased chemical resistance as reflected in inferior solubility in a large number of chemically different solvents. Another heavily fluorinated styrene, 2,3,5,6-tetrafluoro-4-methoxystyrene, can be prepared from FS in a simple fashion. We here wish to report on the first polymerization of TFMS performed under ATRP conditions.

Furthermore, we show how PTFMS and block copolymers can be the basis of a number of novel materials by derivatization with azobenzene side chains. Such fluorinated materials can serve as models for novel optical storage materials. Materials in which the azobenzene content can be reduced (*7*) are potential candidates for optical storage by holographic multiplexing. Another possibility is the design of novel materials for optical waveguiding purposes (*2*). Materials based almost entirely on fluorine substituted carbons are transparent in the NIR range used for optical data transmission. Additionally, the azobenzene part can be addressed by laser light (*7*) causing material birefringence changes in such a manner that optical waveguiding patterns can be formed (*8*). Here also the azobenzene part can be highly fluorinated, since it is relatively easy to design the aromatic part fully fluorinated.

Results and Discussion

TFMS was prepared in high yield (86%) by a nucleophilic replacement reaction on FS with sodium methoxide in methanol as shown in Scheme 1 and purified by vacuum distillation as previously described (*2*). Neat TFMS can be polymerized in a conventional ATRP protocol at 110 °C by use of 1-phenylethyl bromide (PhEBr), CuBr and 2,2'-bipyridine (bipy). Under these conditions the

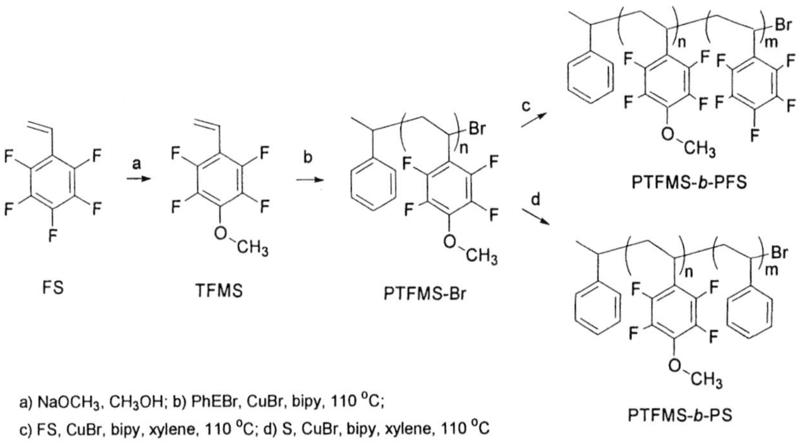

a) $NaOCH_3$, CH_3OH; b) PhEBr, CuBr, bipy, 110 °C;
c) FS, CuBr, bipy, xylene, 110 °C; d) S, CuBr, bipy, xylene, 110 °C

Scheme 1. Preparation and polymerization of TFMS and examples on block copolymerizations.

heterogeneous polymerization proceeds very fast, in fact, 97% conversion can be achieved within 23 min. In comparison, FS under similar conditions took almost 100 min. to reach that level of conversion (5). The results of a number of polymerizations performed in bulk and in xylene solution are listed in Table 1. In the bulk polymerizations 5 g of monomer was polymerized employing different monomer to initiator ratios necessary to reach the target molecular mass. In all experiments the initiator:CuBr:bipy was maintained as 1:1:3. The M_n was determined by ^1H NMR (Figure 1) using the equation $M_n = DP \cdot MW_{TFMS} + MW_{PhEBr}$, $DP = 5H_d/3H_a$, where MW_{TFMS} and MW_{PhEBr} are the corresponding

Table 1. Polymer Yield and Molecular Masses

Polymer	Time min.	Yield %	$M_{n,target}$[a]	NMR M_n	f_{Br}[b]	SEC M_n	PDI
PTFMS1-Br	10	93	2,000	2,900	0.88	3,400	1.36
PTFMS2-Br	3	60	1,500	3,300		4,400	1.21
PTFMS3-Br	15	67	10,000	9,700		8,800	1.26
PTFMS4-Br	25	55	20,000	12,200		13,700	1.30
PTFMS5-Br	20[c]	23	1,000	2,700	0.75	3,100	1.22
PTFMS6-Br	110[c]	42	25,000	11,100		10,500	1.13

[a] $M_{n,target} = [M]_0/[I]_0$ at 100 % conversion; [b] $f_{Br} = 3H_c/H_b$, where H_c & H_b are the areas of >$C\underline{H}Br$ and $-C\underline{H}_3$, respectively; [c] in 30% xylene solution.

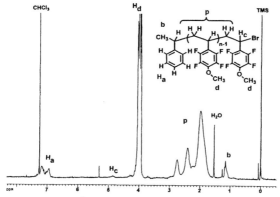

Figure 1. ¹H NMR spectrum of PTFMS1-Br.

molecular weights, and H_d and H_a are areas of methoxy and PhEBr aromatic protons, respectively. Similarily, the bromine end group functionality, f_{Br}, could be estimated, however, only in case of the polymers with the lowest M_ns and was between 0.75 and 0.88. Table 1 shows that M_ns up to 14,000 have been obtained both by bulk and solution polymerization. Furthermore, the PDI in all instances lie between 1.13 and 1.36, where in case of PFS (5) the PDI was generally lower, ≤1.2. The only explanation we can offer for the higher PDIs in case of PTFMS as well as the deviation of f_{Br} from unity is early termination.

The structure of the resulting PTFMS-Br as shown in Scheme 1 was elucidated by ¹H, ¹³C and ¹⁹F NMR spectroscopy. The particular ¹H spectrum shown in Figure 1 originates from a sample with a low degree of polymerization (DP = 13) thus allowing the initiator residue to be easily recognized.

In a series of experiments samples were withdrawn after increasing time and subjected to analyses of conversion by ¹H NMR as exemplified in Figure 2 and to molecular weight determination by size exclusion chromatography (SEC) as depicted in Figure 3. Based on the conversion data the first order plot in Figure 4 can be constructed. The plot implies the controlled ATRP of TFMS. Figure 4 additionally shows the same features for the ATRP of FS and St.

From the derived straight lines, the apparent rate coefficients $k_p{}^{app}$ = -d(ln[M])/dt, listed in Table 2 were calculated. With reference to these data it is evident that the replacement of the styrene aromatic hydrogens with 5 electron-withdrawing fluorines greatly enhances the rate of polymerization of FS as compared to that of St. However, substituting the 4-F in FS with an electron-donating methoxy group enhances the rate of polymerization of TFMS even further as compared to that of FS. Previously, the enhancing rate effect has

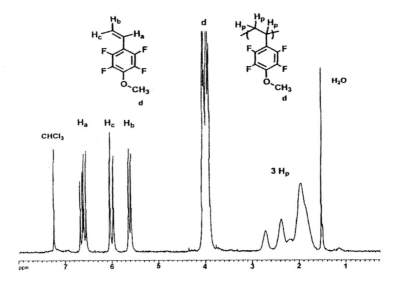

Figure 2. ^1H NMR of a crude sample for calculation of the monomer conversion.

been observed (9) in ATRP of substituted styrenes by introducing one electron-withdrawing substituent such as 3-CF$_3$ or 4-CF$_3$, whereas the electron-donating 4-OCH$_3$ had a slight retarding effect as also listed in Table 2; whereas a single F (4-F) only had a neglectable effect. Thus, in light of these earlier findings the apparent synergistic substitution of four F electron-withdrawing atoms and the one electron-donating 4-OCH$_3$ group in TFMS beneficial in terms of enhancing the rate of polymerization is surprising and unexpected. However, higher k_p or K_{eq} or lower solubility of Cu(II) can not be disregarded (6, 9) This unusual rate enhancing synergy will be the subject of a theoretical study of the reactivity.

Table 2. Apparent Rate Coefficients in ATRP of Substituted Styrenes

Substitution	$k_p^{app} \times 10^4$ (s^{-1})	Solvent	Reference
2,3,5,6-F$_4$-4-OCH$_3$	6.9	none	this work
2,3,4,5,6-F$_5$	3.0	none	5
none	1.41	none	this work
3-CF$_3$	1.44	DPE[a]	9
3-CF$_4$	1.25	DPE[a]	9
none	0.44	DPE[a]	9
4-F	0.39	DPE[a]	9
4-OCH$_3$	0.21	DPE[a]	9

[a] DPE: diphenyl ether, [M]$_0$ = 4.37 M,
[M]$_0$:[PhEBr]$_0$:[CuBr]$_0$:[bipy]$_0$ = 100:1:1:3 at 110°C.

241

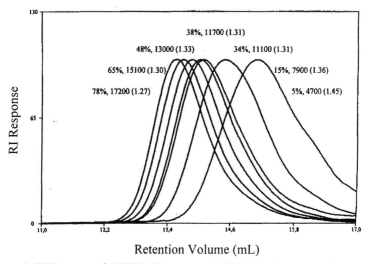

Retention Volume (mL)

Figure 3. SEC traces of PTFMS ($M_{n,target}$ 20,800) as a function of increasing conversion in %. Each trace has been added the calculated M_n and (PDI).

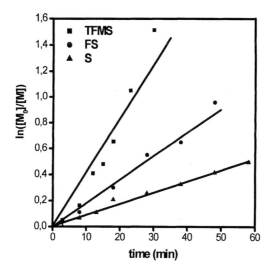

Figure 4. First order plot for ATRP of TFMS, FS, and St at 110 °C, $[M]_0:[PhEBr]_0:[CuBr]_0:[bipy]_0 = 100:1:1:3.$

The development of M_n and PDI as a function of conversion is shown in Figure 5. The facts that the molecular weights as determined by PS calibration are higher than the theoretical ones and a virtual line through the experimental points does not go through the origin suggest that a significant amount of radicals are terminating in the very beginning of the ATRP. Support for this behavior is the higher PDI of PTFMS polymers as compared to those of PFS ~1.2 (5) and of PS = 1.1 (9). Figure 3 also shows how the molecular weights develop with conversion. The initial M_w/M_n of PTFMS is 1.45 that decreases with increasing conversion to 1.27. PTFMS employed as macroinitiators and discussed in a subsequent section also tend to provide block copolymers with low molecular weight tailing indicating that termination reactions occur throughout the homopolymerization of TFMS.

Several of the recovered PTFMS-Br samples were used as initiators for ATRP in xylene solution of both FS and St, respectively. This approach afforded preparation of novel block copolymers as seen in Scheme 1. PFS-Br and PS-Br likewise prepared by ATRP were in the same manner employed as macroinitiators for the preparation of block copolymers with the complementary monomers as listed in Table 3. Nevertheless, the preferred synthetic pathway for block copolymers of TFMS and FS is the one starting from PTFMS-Br instead of PFS-Br due to the significantly better solubility in xylene of the former. No preference is needed due to solubility considerations in case of block copolymer synthesis involving PTFMS-Br or PS-Br.

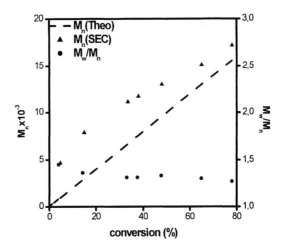

Figure 5. Molecular weights and polydispersities of PTFMS as a function of conversion for ATRP at 110 °C, $[M]_0:[PhEBr]_0:[CuBr]_0:[bipy]_0 = 100 :1:1:3$.

Table 3. Block Copolymers and the Seeding Macroinitiators

Polymer	PTFMS Wt-%	PFS Wt-%	PS Wt-%	$M_{n,SEC}$	PDI	$M_{n,NMR}$	T_g °C
PTFMS1-Br	100			3,400	1.36	2,900	41
PTFMS1-b-PS	10[a]		90[a]	34,000	1.32	34,000	90
PTFMS3-Br	100			8,800	1.26	9,700	90
PTFMS3-b-PFS	64[a]	36[a]		12,600	1.20	13,800	95
PFS-Br		100		11,400	1.21	13,000	95
PFS-b-PTFMS	71[a]	29[a]		35,000	1.40	39,300	90
PFS-b-PS		42[a]	58[a]	21,000	1.32	27,000	101
PS-Br			100	16,500	1.11	-	100
PS-b-PTFMS	57[a]		43[a]	35,000	1.54	38,400	99
PS-b-PFS		21[a]	79[a]	19,400	1.16	20,900	92

[a] By ^{1}H NMR.

The composition of the block copolymers as determined by ^{1}H NMR could be varied in a fairly large range while still maintaining the relatively narrow PDIs. Since the procedure is not sequential addition of monomers but in fact performed in two separate steps great design freedom is offered for both blocks. The content of Table 3 is by no means comprehensive and only in case of the PTFMS-Br the achievable molecular weight range is indicated (more PTFMS block copolymers are listed in Table 5). It should furthermore be stressed that $M_{n,SEC}$ and the corresponding PDI are only indicative. These values are obtained directly from the PS calibration and reported without any corrections. However, when the PS-Br was used as initiator the M_n of the block copolymer was determined by a combination of PS-Br molecular weight obtained from the PS calibration and the comonomer ratio derived from ^{1}H NMR. It can be seen from Table 3 that some discrepancy between the indicative $M_{n,SEC}$ and $M_{n,NMR}$ are found. For that reason, an effort to employ light scattering and MALDI-TOF-MS is presently in progress in order to assess absolute M_w determination.

Generally, the glass transition temperature, T_g, as listed in Table 3, is independent of whether the phenyl is substituted (PTFMS) or non-substituted (PS). On the other hand, the effect of molecular weight is important, such that short chain samples have lower T_gs. A similar behaviour was observed previously for PFS and PS (5). Finally, it is noted that the implied flexibility of block length design is considered important for future applications as further elaborated in a subsequent section.

The thermal stability of PTFMS was investigated by thermogravimetry (TGA) in N_2 atmosphere and compared to those of PFS (M_n 11,400) (5) and PS (M_n 14,800) as depicted in Figure 6. PTFMS (M_n 12,700) has a slightly higher

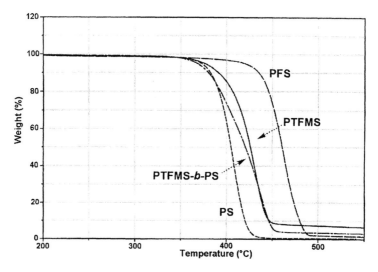

Figure 6. Thermal stability of PTFMS, PTFMS-b-PS, PFS and PS determined by TGA in N₂ by heating 5 °C/min.

thermal stability than PS as evident from the TGA traces. However, the thermal stability is inferior to that of PFS. A close inspection of the thermogramme also reveals that an approx. 8 wt-% residue is left at 550 °C suggesting that the methoxy group is thermo labile. Also the block copolymer, PTFMS-b-PS (M$_n$ 16,100, 50 wt-% PTFMS), leaves a small residue at the end of the analyses. In comparison, PFS and PS are seen to leave vertically no residues under the same experimental conditions. Most likely the loss of methoxy renders formation of tetrafluorophenyl radicals forming highly fluorinated, fused aromatic structures.

Solubility of polymers to a first approximation inversely reflects chemical resistance. The solubility of PTFMS was determined in a number of common chlorinated, oxygen containing or aromatic solvents as listed in Table 4 that also contains comparative information on PFS and PS. From Table 4 it is evident that the introduction of 4-OCH$_3$ enhances the solubility of PTFMS as compared to that of PFS in the investigated solvents with the exception of fluorobenzene. However, the general solvent resistance is still significantly higher than that of PS. Most likely the polarizability of fluorobenzene significantly better matches that of the pentafluorophenyl moiety of PFS than those of both phenyl residues in PTFMS and PS and thus accounts for this significant difference in solubility.

**Table 4. Solubility of PTFMS, PFS and PS in common solvents,
g polymer in 1 g solvent**

Solvent	PTFMS	PFS[a]	PS
Chloroform	0.206	0.027	>0.7[b]
Methylene chloride	0.148	0.009	>0.9[b]
Tetrahydrofuran	0.218	0.125	>0.9[b]
Methyl ethyl ketone	>0.9	0.143	>0.9[b]
Xylene	0.321	0.113	>1.3[a]
Fluorobenzene	0.194	0.413	0.106

[a] Ref. (5); [b] Ref. (10)

Conversion of the *p*-methoxy sites in PTFMS and PTFMS based block copolymers to the corresponding *p*-hydroxy analogues appeared very attractive. Basically, a significant change of the hydrophobic character of both homo- and block copolymers by a considerable introduction of hydrophilic sites is to be expected. Secondly, and not less important, a viable route to functional scaffold polymers also appears through potential derivatization by reactions on the hydroxy sites. The target was reached after several unsuccessful attempts with known deprotecting agents. Thus, both trimethylsilyl iodide (*11*) that works for the deprotection of poly(*p*-methoxystyrene) or concentrated hydrobromic acid (*12*) which similarly functions for poly(*p-tert*-butoxystyrene) failed in the case of PTFMS. On the other hand, the reported procedure (*2*) for the demethylation of TFMS to the corresponding hydroxy compound by use of boron tribromide (BBr$_3$) was adopted for the synthesized homo- and block copolymers. The degree of deprotection was subsequently improved with a molar ratio of -OCH$_3$: BBr$_3$ of 1:2. In the best case 97 % of the methoxy groups of PTFMS could be removed as demonstrated by ^{1}H NMR. Furthermore, in this case no influence on either M$_n$ or PDI was noticed (PTFH5-Br, Table 5). The consequences of demethylation of block copolymers with PS and one example of subsequent functionlization is elaborated below.

The strategy for preparation of azobenzene side-chain block copolymers that could have interesting applications for optical storage of information through polarization holography (*7*) is outlined in Scheme 2. Initially, route I was undertaken in which the plan was to prepare PTFHS-*b*-PS and then react suitable azobenzene side-chain precursors on the hydroxy sites. However, in this case some apparent degradation occurred as observed by both a significant reduction in M$_n$ associated with some broadening of PDI. On the other hand, supporting control experiments performed under the demethylation conditions with BBr$_3$ on narrow molecular weight PS standards prepared by anionic poly-merization and thus composed of only hydrocarbon fragments clearly

Table 5. Demethylation and Intermediates to the Azo-functionalized Block Copolymer, Routes I and II in Scheme 2

Route I	Route II	M_n	PDI	OH^a %	T_g °C	$T_{LC->I}^b$ °C	ΔH J/g
PTFMS1-Br		3,400	1.36		41		
PTFMS1-b-PS		34,000	1.32		96		
PTFHS1-b-PS		15,000	1.82	n.a.^{c}	n.a.^{c}		
	PTFMS5-Br	2,300	1.28		51		
	PTFHS5-Br	2,300	1.26	97	116		
PTF(azo-CN)S5-Br		6,200	1.37		67	126	6.3
PTF(azo-CN)S5-b-PS		11,400	1.65		93	n.i.^{d}	n.i.^{d}
	PTFMS3-Br	8,800	1.26		90		
	PTFHS3-Br	8,500	1.25	93	114		
PTF(azo-CN)S3		27,000	1.58		80	163	4.2
PTF(azo-F)S3		25,000	1.79		126	203	
PTF(azo-CF$_3$)S3		29,000	1.57		108	148	4.6

^{a} By ^{1}H NMR; ^{b} mesophase to isotropic transition; ^{c} not analyzed; ^{d} not possible to identify

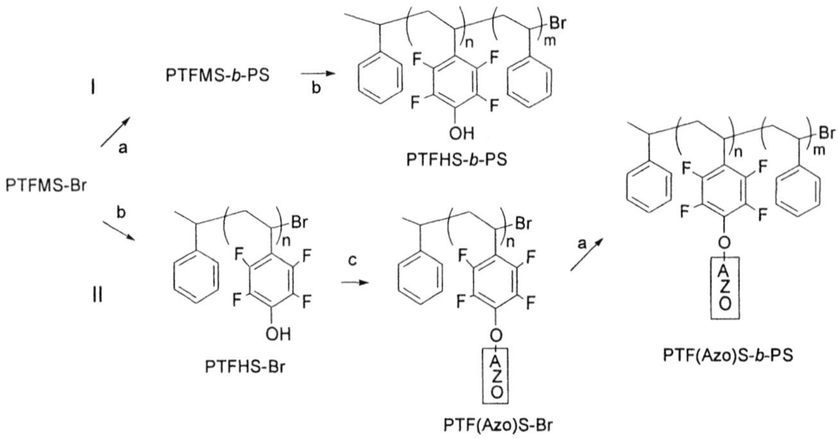

a) S, CuBr, bipy, xylene, 110 °C; b) BBr$_3$, CHCl$_3$,

c) Br(CH$_2$)$_6$OC$_6$H$_4$NNC$_6$H$_4$X (Br-Azo, X = CN, F, CF$_3$), toluene, NaOH, (n-C$_4$H$_9$)$_4$N$^+$Br$^-$.

Scheme 2. Demethylation and intermediates to azobenzene side-chain block copolymers.

demonstrated that PS partially degraded under these conditions. It was therefore concluded that the PS block of the copolymers was degraded but not the PTFHS block by the BBr$_3$ treatment. As a consequence the second route (II) was adopted. That no degradation of the PTFHS-Br homopolymers occurs is verified by the results reported in Table 5 where PTFH5-Br and PTFHS3-Br show no sign of changes of neither M$_n$ nor PDI. The next step, reacting the azobenzene side chains (AZO) onto the hydroxyl sites by alkylation under phase transfer conditions, resulted in a slight increase in PDI. Finally, the PS block was formed through ATRP in xylene solution (M$_{n,target}$ 84,000; PTF(azo-CN)S5-b-PS is the result of 14% conversion of St) with some increase in PDI. However, this increase is probably acceptable since the idea is to achieve microphase separation and form small liquid crystalline azobenzene domains in the amorphous PS phase. Already the difference between the fluorinated and the normal phenyl blocks of PS is expected to induce phase separation and the azobenzene derivatization likely enhances the expected microphase separation. The thermal transitions as detected by differential scanning calorimetry (DSC) indicate that all azobenzene derivatized polymers have developed a mesophase of varying broadness depending on both the block length and the substituent on azobenzene. The determined enthalpies of the formed mesophases could indicate the mesophases to be of a smectic type. Both the microphase separation and the mesophase nature are under further studies.

Conclusions

Neat TFMS polymerizes rapidly in a conventional ATRP protocol at 110 °C. Under identical conditions PTFMS is produced faster than FS. The polymerisation is relatively well controlled with M$_n$s up to 17,000 and corresponding PDIs are generally below 1.3. The recovered PTFMS can function as ATRP macroinitiator for both FS and St producing block copolymers while maintaining the controlled ATRP characteristics. In fact, block copolymers in all combinations of these three monomers have been prepared based on the macroinitiator concept. Homo- and block copolymers of PTFMS impart both better thermal stability and chemical resistance than that of PS, however, not to the same extent as that of PFS. On the other hand, the presence of the 4-methoxy group in PTFMS has been exploited for the possible demethylation that allows for post polymerization reactions. The resulting hydroxy functionality has been utilized for alkylations with different azobenzene side chains. Furthermore, the azobenzene derivatized polymers retain ATRP reactivity that has been employed in the preparation of an azobenzene PS block copolymer.

Experimental

Materials. Monomers. FS and St (Aldrich) were passed through a ready-to-use column AL-154 (Aldrich), stored over CaH_2 and vacuum distilled before polymerization. TFMS was synthesized according to Pitois *et al* (*2*); ^1H and ^{19}F NMR spectra confirmed (*2*) the structure of the pure *para* substituted isomer. Side-chain precursors. 6-Bromohexyloxy-1-azobenzenes differently substituted were prepared from the corresponding anilines through diazotation, phenol coupling and alkylation with 1,6-dibromohexane according to Crivello *et al* (*13*). Other reagents. CuBr, PhEBr and bipy (all from Aldrich) were used as received.

Synthesis. ATRP. Typically (PTFMS3-Br) a Schlenk tube was charged with 0.072 mL (0.509 mmol) of PhEBr, 0.075g (0.509 mmol) of CuBr and 0.238 g (1.526 mmol) of bipy. TFMS, 5.0g (24.272 mmol) was added, the system was degassed 3 times and then heated to 110 °C under N_2. The homopolymers were precipitated in CH_3OH and the yield was determined gravimetrically after vacuum drying. Block copolymers were obtained in the same manner starting from the macroinitiators in xylene.

Demethylation of PTFMS blocks was performed with BBr_3 in $CHCl_3$ under N_2 and reflux. The final molar ratio of $-OCH_3:BBr_3$ was 1:2. The degree of conversion of 4-methoxy groups was determined by ^1H NMR in DMSO-d_6.

Azobenzene functionalization. Alkylations were performed in a 1:1.1 molar ratio of hydroxy groups of PTFHS and azobenzene precursor in a 3:1 (v/v) mixture of toluene and 2 N aq. NaOH by phase transfer catalyst (0.1 eqv. (n-$Bu_4N^+Br^-$) under reflux for 26 h. After cooling the PTF(AZO)S's were extracted with CH_2Cl_2, concentrated and precipitated in CH_3OH, and purified by Soxhlet extraction with CH_3OH.

Analyses. NMR Spectroscopy. ^1H, ^{19}F and ^{13}C NMR measurements were carried out with a Bruker 250 MHz in $CDCl_3$ or DMSO-d_6. The three ^{19}F NMR resonances of PTFMS at -155, -141 and -140 ppm (ratio = 2:1:1) are shifted slightly upfield and broadened significantly as compared to those of TFMS (*2*) (C_6F_6: -163 ppm was used as a reference). Accordingly no coupling information can be extracted. The ^{13}C NMR of PTFMS is similarly dominated by small and broad featureless aromatic resonances due to the fluorines on the phenyl ring in addition to the normal aliphatic carbon resonances.

SEC in THF was performed on a Viscotek 200 equipped with 2 PLgel mixed D columns from Polymer Laboratories (PL) and by use of RI detection. Calibration was based on PS standards (M_n in the range $7 \cdot 10^2$ to $4 \cdot 10^5$) from PL and employing the TriSEC™ Software. TFMS purity was confirmed by use of an OligoPore column from PL.

DSC was performed with a Q1000 (TA Instruments) from -30 to 250 °C at a heating rate of 10 °C/min under N_2. The T_g and phase transition were calculated from the second heating trace.

TGA was performed on a Q500 (TA Instruments) measuring the total weight loss on approx. 8-9 mg samples from 30-550 °C at a rate of 5 °C/min in a N_2 flow of 90 mL/min.

Acknowledgements
Aage and Johanne Louis-Hansen's Foundation (DK); Materials Research, Danish Research Agency; and Ministry of Science and Education of North-Rhine Westfalia, Düsseldorf (SB) are gratefully acknowledged for financial support.

References
1. Han, L.M.; Timmons, R.B.; Lee, W.W.; Chen, Y.C.; Hu, Z. *J. Appl. Phys.* **1998**, *84*, 439-444.
2. Pitois C.; Wiesmann, D.; Lindgren, M.; Hult, A. *Adv. Mater.* **2001**, *13*, 1483-1487.
3. Gavelin, P.; Jannasch, P.; Wesslén, B. *J. Polym. Sci. Part A: Polym. Chem.* **2001**, *39*, 2223-2232; Gavelin, P., Ph.D. thesis, Lund Institute of Technology, Lund, SE, 2002.
4. Jankova, K.; Hvilsted, S. Book of Abstracts, IUPAC Intern. Symposium on Ionic Polymerization (IP'2001), Crete, Greece, October 2001, p. 145.
5. Jankova, K.; Hvilsted, S. *Macromolecules*, submitted.
6. Coessens, V.; Pintaur, T.; Matyjaszewski, K. *Prog. Polym. Sci.* **2001**, *26*, 337-377.
7. Hvilsted, S.; Andruzzi, F.; Kulinna, C.; Siesler, H.W.; Ramanujam P.S. *Macromolecules* **1995**, *28*, 2172-2183; Berg, R.H., Hvilsted, S.; Ramanujam, P.S. *Nature* **1996**, *383*, 505-508.
8. Sahlén, F.; Geisler, T.; Hvilsted, S; Holme, N.C.R.; Ramanujam, P.S.; Petersen, J.C. *Mat. Res. Soc. Symp. Proc.* **1999**, *561*, 57-62.
9. Qiu, J.; Matyjaszewski, K. *Macromolecules* **1997**, *30*, 5643-5648.
10. Suh, K.W.; Clarke, D.H. *J. Polym. Sci. A-1* **1967**, *5*, 1671-1681.
11. Xiang, M.; Jiang, M. *Macromol. Rapid Commun.* **1995**, *16*, 477-481.
12. Se, K.; Miyawaki, K.; Hirahara, K.; Takano, A.; Fujimoto, T. *J. Polym. Sci. Part A: Polym. Chem.* **1998**, *36*, 3021-3034.
13. Crivello, J.V.; Deptolla, M.; Ringsdorf, H. *Liq. Cryst.* **1988**, *3*, 235-247.

Chapter 18

Copper Removal in Atom Transfer Radical Polymerization

Mical E. Honigfort[1], Shingtza Liou[1], Jude Rademacher[1], Dennis Malaba[1], Todd Bosanac[2], Craig S. Wilcox[2], and William J. Brittain[1,*]

[1]Department of Polymer Science, The University of Akron, Akron, OH 44325–3909
[2]Department of Chemistry and The Combinatorial Chemistry Center, University of Pittsburgh, Pittsburgh, PA 15260

We have developed three methods for simple copper removal in atom transfer radical polymerization (ATRP). The first involved use of a polyethylene-ligand which is homogeneous under polymerization conditions, but precipitates upon cooling. Less than 1% of the original copper was left after simple decantation; however, long (24 h) polymerization times were observed. The second method involved the use of JandaJel™ ligands. Normal polymerization times were observed with this heterogeneous system, but copper removal was less efficient with 4-5% of the original copper remaining. The third method used precipiton ligands that are soluble in the cis-stilbene form, but precipitate when isomerized to the trans form. The precipitons were the most efficient at copper removal (<1% copper remaining), but cannot be recycled. For all three methods, ATRP was controlled as evidenced by monomer conversion and growth of molecular weight. We studied the polymerization of styrene, methyl methacrylate and N,N-dimethylaminoethyl methacrylate.

Introduction

Advances in atom transfer radical polymerization (ATRP) have made it an important method for the synthesis of a wide variety of polymers in a controlled fashion.(1) Despite many positive aspects, one of the major limitations of this technique is contamination of the polymer with a highly colored ligand/metal complex. In addition to requiring purification of the polymer solution after the reaction, the ligand/catalyst complex is not reusable. Purification of the polymer solution involves passing the solution through silica gel or alumina columns or precipitation into non-solvent. These procedures are practical on the laboratory scale but pose difficulties on a larger scale.

Surface-immobilized catalysts have shown promise in ATRP systems as a way to overcome this limitation.(2,3) General disadvantages of these systems, as compared to traditional ATRP conditions, are higher molecular weights than predicted, broader molecular weight distributions, and longer reaction times. A possible explanation for these sluggish reaction rates is the heterogeneous conditions that cause slow diffusion of the polymer chain to the silica or polystyrene bead surface.

Recently, Zhu and coworkers (4-6) reported using silica supported multidentate amine ligands in ATRP to afford low polydispersity and good molecular weight control for PMMA to about 80% conversion. Use of silica supported ligands with a poly(ethylene glycol) s pacer has also been investigated by Zhu et al.(5) demonstrating that the length of the spacer affected the rate of polymerization and the molecular weight control. They have also developed a continuous column reactor packed with silica gel for ATRP of MMA that enables simple catalyst removal.(7)

An immobilized/soluble hybrid catalyst system for ATRP, which overcomes the diffusion limitation of heterogeneous catalyst systems was developed by Matyjaszewski and coworkers.(8) Zhu and coworkers have also reported the use of soluble catalyst supports for ATRP. They have studied both polyethylene and polyethylene-b-poly(ethylene glycol) bound ligands for ATRP.(9,10)

This chapter details methods used by the Brittain research group to remove copper from ATRP reactions. First, a homogeneous system was investigated where a polyethylene (PE) supported ligand was synthesized for use as an ATRP ligand.(11) The solubility properties of low molecular weight linear PE (soluble in organic so lvents at the elevated reaction temperature and insoluble at room temperature) result in homogeneous reaction conditions coupled with simple removal of the insoluble metal/ligand complex after the reaction cools.

New swellable polystyrene resins, JandaJel™ resins, were also studied as ATRP catalyst supports. As compared to divinylbenzene (DVB) crosslinked resins, JandaJels™ have increased homogeneity, organic solvent compatibility, and site accessibility due to the flexible crosslinker.(12) For immobilized catalyst systems, a major factor controlling the polymerization is the deactivation step, which is affected by the mobility of the immobilized catalyst particles and the diffusion of the polymer chains in the reaction mixture. It was speculated that JandaJel™ resins functionalized with nitrogen ligands for ATRP would behave more like a homogeneous system than the DVB counterpart and would still allow for easy catalyst removal from ATRP reactions.

Compounds termed 'precipitons' were also used as supports for ATRP ligands.(13) These isomerizable compounds are attached to a reactant and after a reaction is complete they can be isomerized to cause precipitation of the attached product.(*14-17*) The *cis*-form of the stilbene is soluble whereas the *trans*-form is insoluble in common organic solvents. The precipitated product can be isolated by filtration or centrifugation. This strategy enables homogeneous reaction conditions to be combined with a facile method for removal of the copper catalyst.

Polyethylene Supported Ligands for ATRP

For synthesis of the PE-ligand (Scheme 1), we investigated poly(1,4-butadiene) (PBD) precursors because of the insolubility and characterization difficulties associated with a functional PE. PBD is soluble in common organic solvents at ambient temperature so functionalization reactions can be performed at lower temperatures and the polymeric intermediates can be easily characterized by GPC and NMR.

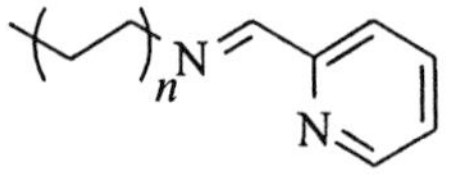

Scheme 1. PE-ligand for ATRP

The most desirable synthesis route to the PE-ligand precursor is based on the work of Grubbs and coworkers.(*18-20*) Ring-opening metathesis polymerization (ROMP) can be used to synthesize mono- or difunctional low molecular weight PBD with either –OH or –NH₂ functionality. This method results in 100% 1,4-PBD using standard dry box and Schlenk line techniques. Even though the polydispersity of this material is higher than other methods (PDI~1.4-1.5), it proved to be the simplest way to make functional PBD. The functional PBD is converted to the ATRP ligand as shown in Scheme 2.

ATRP Using PE-Ligand

Methyl methacrylate (MMA), styrene, and 2-(*N,N*-dimethylamino)ethyl methacrylate (DMAEMA) were polymerized via ATRP using the PE-imine ligand of M_n=2,000 g/mol. The results are presented in Table I. The ATRP reactions using these ligands were slow; reaction times of 18-26 h were required to r each high conversions. ATRP using *N*-(*n*-propyl)-2-pyridylmethanimine proceeds to high conversions in 4-6 h. The results with PE-ligand are comparable to ATRP using *N*-(*iso*-propyl)-2-pyridylmethanimine where 72% conversion was observed after 12 h.(*21*) In addition, t he PDIs for PMMA prepared using branched-alkyl pyridylmethanimine ligands was similar to the values we observed with the PE-ligand. There is clearly a steric effect involved in the ligation of Cu by the PE-ligand, similar to that reported by Haddleton and co-workers.(21)

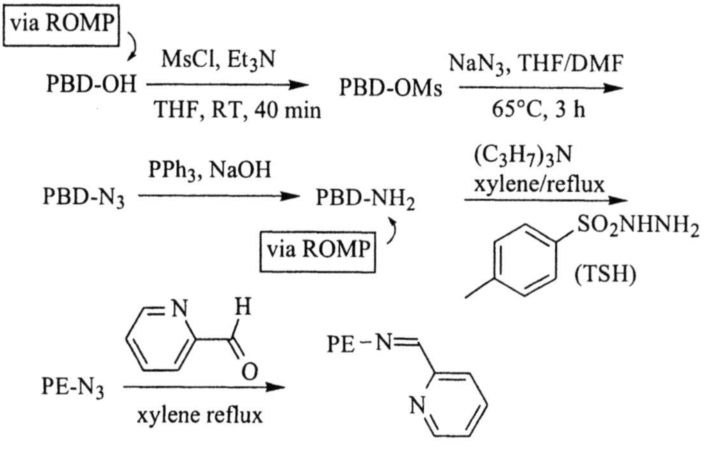

via ROMP

$$PBD\text{-}OH \xrightarrow[\text{THF, RT, 40 min}]{\text{MsCl, Et}_3\text{N}} PBD\text{-}OMs \xrightarrow[\text{65°C, 3 h}]{\text{NaN}_3,\ \text{THF/DMF}}$$

$$PBD\text{-}N_3 \xrightarrow[]{\text{PPh}_3,\ \text{NaOH}} PBD\text{-}NH_2 \xrightarrow[\text{xylene/reflux}]{\text{(C}_3\text{H}_7)_3\text{N}} \begin{array}{c} SO_2NHNH_2 \\ (TSH) \end{array}$$

via ROMP

$$PE\text{-}N_3 \xrightarrow[\text{xylene reflux}]{} PE\text{-}N\text{=}$$

Scheme 2. Synthesis of PE-ligand

The experimental molecular weights were close to the predicted values. The PDI of the resulting polymers increased over the course of the reaction and is broader than in a conventional system with a small ligand, such as N-propyl-(2-pyridyl)methanimine.(22,23) In conventional ATRP, the PDI narrows as the reaction progresses but for these reactions using the PE-ligands, the PDI broadened over the course of the reaction. The GPC traces of the final polymer from these reactions are monomodal and symmetrical.

U V-visible spectroscopy was used to analyze the polymer solution after the solid PE-ligand/CuBr complex was removed via filtration. Figure 1 shows the 500-700 nm range of the UV-visible spectrum for ATRP of MMA using the PE-ligand synthesized via ROMP compared to a copper standard solution of copper complexed with N-propyl-(2-pyridyl)methanimine ligand (the standard solution shown here contains 0.9 weight percent copper). The CuBr/imine ligand complex has a λ_{max} absorbance of 680 nm as shown in the UV-visible trace below. The UV-visible spectrum of the polymer solution shows that there is no absorbance at 680 nm where the CuBr/ligand would absorb. To quantify this, elemental analysis for residual copper was done on isolated polymer. For the MMA polymerizations, only 1% of the original copper remained in the unpurified polymer.

Table I. ATRP of MMA, Styrene, and DMAEMA Using the PE-Imine Ligand

Monomer	Time, h	% Conv.[c]	M_n(theo) (g/mol)	M_n(exp)[c] (g/mol)	PDI
MMA[a]	2	20	2,000	2,200	1.30
	5	32	3,200	3,600	1.38
	10	55	5,500	6,600	1.44
	14	70	7,000	7,800	1.45
	18	79	7,900	8,300	1.50
Styrene[a]	3	10	1,000	—	—
	6	25	2,500	2,700	1.29
	10	41	4,100	3,900	1.32
	12	50	5,000	5,400	1.42
	19	56	5,600	5,800	1.48
	23	62	6,200	6,700	1.50
DMAEMA[b]	2	33	5,200	6,200	1.42
	4	44	6,900	7,000	1.42
	6	56	8,800	8,800	1.53
	10	60	9,400	10,100	1.52
	16	75	11,800	11,300	1.54
	18	78	12,200	11,400	1.55

[a]Experimental conditions:[PE-ligand]:[ethyl 2-bromoisobutyrate initiator]: [CuBr]: [Styrene/MMA]=2:1:1:100; solvent = 2:1 v/v toluene/styrene; 100 °C. [b]Experimental conditions:[PE-ligand]:[ethyl 2-bromoisobutyrate initiator]: [CuBr] :[DMAEMA]=2:1:1:100; solvent = 2:1 v/v toluene/DMAEMA; 80 °C. [c]Conversion measured by ^{1}H-NMR relative to internal standard; M_n, M_w determined by GPC using universal calibration.

To determine if the catalyst is reusable, a second polymerization was attempted with used catalyst. After the first MMA polymerization was complete, the polymer solution was removed via syringe and while under argon atmosphere, the catalyst complex was rinsed with degassed solvent. The ATRP reaction for the second catalyst use was identical to the first catalyst use experiment with respect to concentration, time, and temperature and the results are in Table II. The catalyst had reduced activity as evidenced by the slower reaction time. This decrease in rate can be attributed to loss of catalyst activity which is likely due to excess $CuBr_2$ from the first polymerization. Given the long polymerization times, some termination is inevitable which will lead to excess deactivator ($CuBr_2$) in accord with "Persistent Radical Effect."(24) The reaction using the catalyst for the second time was slower but proceeded with the same control of molecular weight and polydispersity. To effectively recyle the catalyst without a loss in activity, the $CuBr_2$ should be reduced prior to the second polymerization.

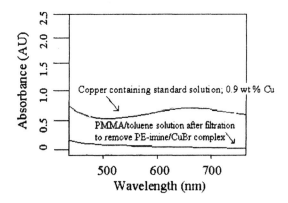

Figure 1. UV-visible Spectrum of PMMA/toluene Solution After Filtration to Remove the Catalyst Complex

Table II. Catalyst Reuse in ATRP of MMA Using PE-Imine[a]

	Time (h)	% Conv.	M_n(theo) (g/mol)	M_n(exp) (g/mol)	PDI
1st Catalyst Use	2	8	800	---	---
	4	31	3,100	2,500	1.32
	7	47	4,700	4,600	1.38
	10	63	6,300	7,500	1.38
	15	80	8,000	9,100	1.44
	21	87	8,700	9,400	1.52
2nd Catalyst Use	8	24	2,400	2,000	1.44
	16	57	5,700	6,300	1.50
	21	75	7,500	7,900	1.54

[a]Experimental Conditions:[PE-ligand]:[ethyl 2-bromoisobutyrate initiator]: [CuBr]:[MMA]=2:1:1:100; solvent = 2:1 v/v toluene/MMA; 100 °C; conversion measured by ^{1}H-NMR relative to internal standard; M_n, M_w determined by GPC relative to PMMA standards.

To synthesize a diblock copolymer (PMMA-*b*-PDMAEMA), solvent and residual monomer from the first polymerization were removed in vacuo after the first ATRP reaction had reached 70-80% conversion and a second aliquot of monomer and additional solvent were added to the reaction mixture. The PMMA-Br product from the first reaction acts as a macroinitiator. Figure 2 shows the GPC trace for the PMMA block and for the diblock. Each block had a theoretical molecular weight of 10,000 g/mol at 100% conversion. There is not good separation between the two peaks and the diblock peak has a small lower molecular weight tail. There is likely homopolymer still present in the mixture. The second block only reached 55% conversion and the molecular weight for the second block was 5,500 g/mol. This result is not surprising since the ATRP reactions using the PE-ligand were very long and thus, some termination is inevitable. The DMAEMA homopolymerization using the PE-ligand was also sluggish.

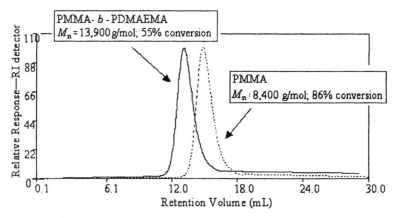

Figure 2. GPC Traces (RI Detector) for PMMA and PMMA-b-PD MAEMA Polymers Prepared via ATRP Using the PE-Imine Ligand

Understanding Long Reaction Times for PE-ligand System

The preliminary work using MMA as the monomer demonstrated that the PE-imine ligand successfully mediates a controlled ATRP reaction and effectively removes the copper catalyst. The reactions proceeded as well-controlled ATRP reactions with good molecular weight control and with relatively narrow polydispersities. However, the reaction times were over twice as long to reach high conversion compared to a conventional ATRP reaction. Although the PE-ligand system is homogeneous at reaction temperature like the control reaction, the difference for the PE-ligand system is the long PE chains that surround this catalyst system (depicted in Scheme 3). Similar to branched-alkyl pyridylmethanimine ligands, there is a steric effect that retards Cu complexation by the PE-ligand. This slows the polymerization and leads to increased termination. The increased termination produces excess deactivator in accord with the "Persistent Radical Effect." Another effect that may be operating in this system is the immiscibility of the growing PMMA chains with the non-polar PE-ligand/Cu complex.

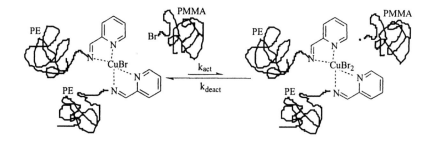

Scheme 3. General Schematic for ATRP Using the PE-Ligand

JandaJel™ Supported Ligands for ATRP

JandaJel™ resins were developed by Janda and coworkers (*12*) as an alternative to divinylbenzene (DVB) crosslinked polystyrene (PS) resins. These new resins are insoluble but swellable due to a more flexible crosslinker and are said to h ave increased homogeneity and site accessibility as compared with DVB crosslinked resins.

The synthesis and structures of the two JandaJel™ ligands that were used are shown in Scheme 4. The precursors to these ligands are commercially available resins functionalized with –OH or –NH₂ (1.0 mmol functionality per gram). The JandaJel™-imine ligand, as characterized by ¹H-NMR, had a functionality of 0.9 mmol/g. The JandaJel™-TEDETA ligand, characterized by elemental analysis for nitrogen had a functionality of 0.7 mmol/g.

ATRP Using JandaJel™ Ligands

The ATRP reactions for MMA, styrene, and DMAEMA were much faster for this system compared to the PE-ligand/CuBr catalyst system. For MMA, high conversions (up to 100% for some of the reactions) were reached in 6-8 h. One contributor to the fast rate could be the polar molecule u sed to crosslink the PS in the JandaJels™ since polar media are known to increase reaction rates in ATRP.(*25,26*) The polar nature of JandaJel™ could help increase solubility of the activating species (CuBr) and therefore increase reaction rate.

The polydispersity narrowed over the course of the ATRP reaction. Even though these PDIs are significantly higher than for a conventional ATRP of MMA, they are much narrower than those for ATRP of MMA using ligands supported by DVB crosslinked PS resins (PDI values up to 10 were reported).(*2*) With the JandaJel™ system, we would expect a higher PDI compared to a conventional reaction since there will be a diffusion effect on the deactivation reaction, which would not be present with a typical homogeneous

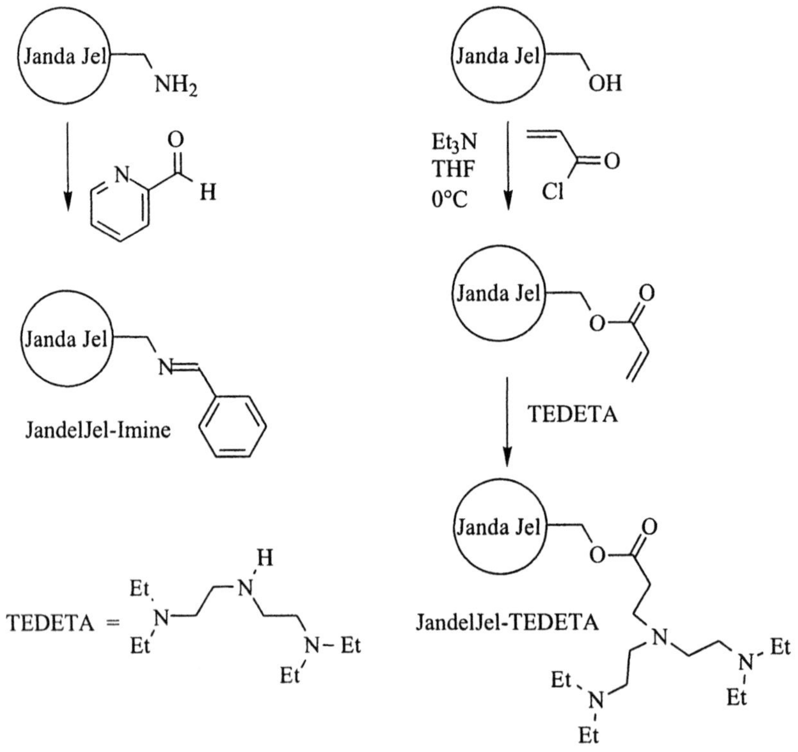

Scheme 4. Synthesis of JandaJel™ Ligands

ligand system. Also, some of the previously cited heterogeneous ATRP reactions have been reported as not very reproducible due to factors such as stirring rate (2) but these reactions with the JandaJel™ ligands were very reproducible.

Despite the broader polydispersity, these reactions proceeded with a linear increase of molecular weight with monomer conversion and good control of molecular weight for both catalyst uses (Table III, Figure 3). At high conversions the molecular weights were higher than predicted; this has been observed in the other reported heterogeneous ATRP systems.(2,3) There was ~20% decreased activity for the second catalyst usage, which can be attributed to excess CuBr₂ (deactivator) due to the "Persistent Radical Effect.". The first order kinetic plot was linear for both the first use and for the recycled catalyst (Figure 3). The JandaJel™-TEDETA ligand exhibited the same behavior as the JandaJel™-imine ligand in ATRP of MMA.

259

Table III. ATRP of MMA Using the JandaJel™-Imine Ligand[a]

	Time (h)	% Conv.	M_n(theo) (g/mol)	M_n(exp) (g/mol)	PDI
1st Catalyst Use	2	58	5,800	5,200	1.40
	3	72	7,200	7,500	1.36
	5	80	8,000	9,400	1.32
	8	96	9,600	10,800	1.29
2nd Catalyst Use	2	37	3,700	5,000	1.81
	3	64	6,400	7,500	1.76
	5	71	7,100	9,000	1.52
	6	85	8,500	9,900	1.50

[a] Experimental Conditions:[JandaJel™-ligand]:[ethyl 2-bromoisobutyrate initiator]:[CuBr]: [MMA] = 2:1:1:100; solvent = 3:1 v/v toluene/MMA; 100 °C. Conversion measured by ¹H-NMR relative to an internal standard; M_n, M_w determined by GPC using universal calibration.

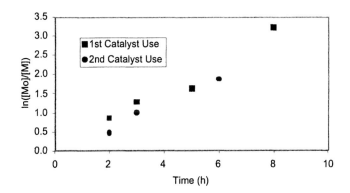

Figure 3. *1st Order Kinetic Plot for ATRP of MMA for 1st and 2nd Catalyst Uses for JandaJel™-Imine Ligand*

The ATRP of styrene and DMAEMA both showed controlled behavior in the polymerization. The experimental molecular weights were in agreement with the predicted values and the molecular weights increased linearly throughout the reaction (Table IV, Figure 4). The styrene reaction was sluggish and only reached 63% conversion while the DMAEMA polymerization was very fast, reaching 92% conversion in less than 3 h. The DMAEMA result is consistent with what is seen in conventional ATRP of this monomer.(*27*) Both reactions followed first order kinetics (Figure 5). The PDMAEMA had broader polydispersities than the other polymers (the PDI progressed from 1.8 at lower conversions to 1.6 at the end of the reaction) probably due to its polar nature, which helps to increase solubility of the activating species, shifting the atom transfer equilibrium and increasing the concentration of the active radicals in solution.

Table IV. ATRP of Styrene and DMAEMA Using the JandaJel^{TM}-Imine Ligand^a

	Time (h)	% Conv.	M_n(theo) (g/mol)	M_n(exp) (g/mol)	PDI
Styrene^b	2	15	1,500	1,500	1.70
	6	29	2,900	2,400	1.62
	10	63	6,300	5,800	1.54
DMAEMA^c	0.5	37	5,800	6,000	1.85
	2	69	10,800	9,000	1.77
	3	84	13,200	12,500	1.54
	6	92	14,400	14,100	1.50

^{a}Conversion measured by ^{1}H-NMR relative to internal standard; M_n, M_w determined by GPC using universal calibration or reference to PMMA or PS standards. ^{b}[JandaJel^{TM}-ligand]:[ethyl 2-bromoisobutyrate initiator]: [CuBr]:[styrene] = 2:1:1:100; solvent = 3:1 v/v toluene/styrene; 100 °C. ^{c}[JandaJel^{TM}-ligand]:[ethyl 2-bromo-isobutyrate initiator]:[CuBr]: [DMAEMA]=2:1:1:100; solvent = 3:1 v/v toluene/DMAEMA; 60 °C..

A diblock copolymer was synthesized using the JandaJel^{TM} ligand system. For the diblock experiment, solvent and residual monomer from the first polymerization were removed in vacuo after the first ATRP reaction has reached 70-80% conversion and a second aliquot of monomer and additional solvent are added to the reaction mixture. A PMMA-*b*-PDMAEMA diblock was synthesized; the GPC trace for the first block and the diblock is shown in Figure 6.

Copper removal from the JandaJel^{TM} ligand ATRP systems was determined in the same manner as described for the PE-ligand system. Elemental analysis for copper indicated 4-5% of the original copper remained in the unpurified polymers.

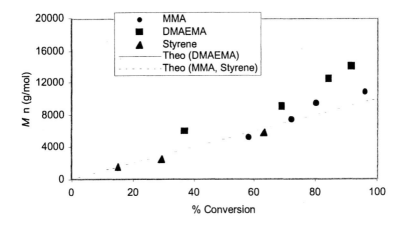

Figure 4. M_n vs. Conversion Comparing ATRP of MMA, DMAEMA, and Styrene Using the JandaJel^{TM}-Imine Ligand

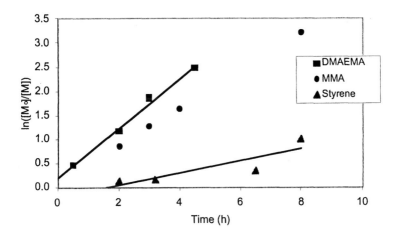

Figure 5. 1st Order Kinetic Plot Comparing ATRP of MMA, DMAEMA, and Styrene Using the JandaJel^{TM}-Imine Ligand

262

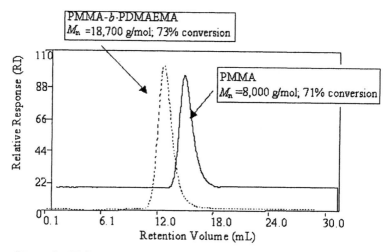

Figure 6. GPC traces (RI Detector) for PMMA and PMMA-b-PD MAEMA Polymers Prepared via ATRP Using JandaJel^{TM} Ligands

Precipitons for ATRP Catalyst Removal

Precipiton ligands-**1** and -**2**, shown in Scheme 5, were successfully used for ATRP of MMA mediated by CuBr, using toluene as the polymerization solvent and ethyl 2-bromoisobutyrate as the initiator (Table V).(*14*) Upon completion of the polymerization, the solution was cooled to room temperature and exposed to U V radiation for 2 h. The precipiton-ligand precipitated and remained complexed with the Cu catalyst. The precipitated product can be isolated by decantation, filtration, or centrifugation. Copper content of the polymer solution was determined by UV spectroscopy and ICP analysis. Less than 1% of the original copper was observed in the PMMA produced using either ligand-**1** or -**2**. The PMMA from this reaction required no purification other than simple decantation.

Compared to the various heterogeneous copper removal techniques, this precipiton-ligand system offered somewhat better control of molecular weight and molecular weight distribution in reasonable reaction times. The data in Table V indicate that M_n (exp) is in good agreement with M_n (theo) until high conversion. At high conversions, M_n(exp) is greater than expected, indicating some termination. The reactions achieved 90-93% conversion in 12 h while polydispersity narrowed slightly over the course of the reaction. Better control over polydispersity was obtained using monofunctional ligand-**1**.

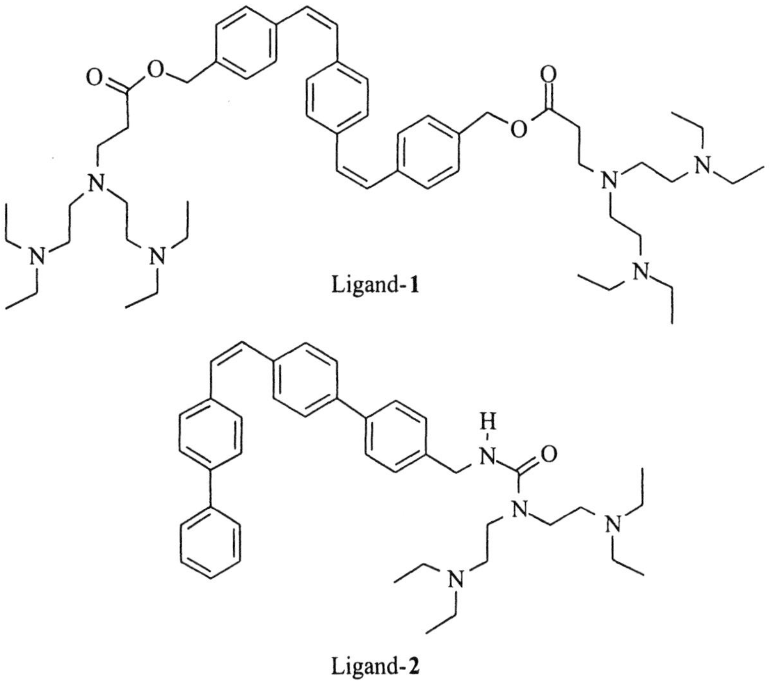

Ligand-1

Ligand-2

Scheme 5. Precipiton Ligands for ATRP

The precipiton-bound ligands successfully mediated the ATRP of MMA and allowed for easy and fast removal of the copper catalyst by exposure of the product solution to a UV light source. The present inability to reuse or recycle the ligand in this catalyst system is an undesirable feature. In other supported catalyst systems, the ligand can be recovered and reused, but the precipitons used in our experiments cannot be recycled. The results demonstrate that the method can be useful in specialty applications. If ongoing efforts to develop recyclable precipitons are successful, then this system will provide a general and economically attractive way to remove metals from ATRP systems.

Summary

Overall, the PE-ligands worked well in ATRP to produce homopolymers although reaction times were long and polydispersity increased over the course of the reaction. The system was able to produce a diblock copolymer although all the chains did not initiate polymerization of the second block. The catalyst complex retained its activity for a second catalyst reuse. The PE-ligands were very effective for copper removal, with only 1% of the original copper remaining in the unpurified polymer.

The JandaJel[TM]-ligands are applicable to various monomers and the ATRP reactions generally proceeded much faster as compared to the PE-ligands. Catalyst reuse is possible using these ligands. However, copper removal with the JandaJel[TM] system was less effective than for the PE-ligand systems or other reported ATRP catalyst removal systems.(2,3,8,28)

Table VI compares our results for copper removal with representative literature examples. Inspection of this table reveals that the precipiton ligand produced the lowest amount of residual copper. However, the inability to recycle these ligands mitigates this performance.

Table V. ATRP of MMA Using Precipiton Ligands 1 and 2[a]

Ligand	Time (min)	% Conv.	$M_n(theo)$ (g/mol)	$M_n(exp)$ (g/mol)	PDI
1	120	40	4,000	4,600	1.45
	300	54	5,400	5,800	1.45
	480	74	7,400	7,700	1.44
	600	78	7,800	8,100	1.42
	660	91	9,100	10,400	1.40
	800	93	9,300	13,700	1.40
2	240	27	2,700	3,200	1.22
	360	51	5,100	6,100	1.20
	510	68	6,800	7,500	1.20
	600	78	7,800	8,200	1.19
	720	90	9,000	10,100	1.19

[a] [Precipiton ligand]:[ethyl 2-bromoisobutyrate inititator]:[MMA] = 1.5:1:100; 50% (v/v) toluene; 90°C; conversion was measured by ^{1}H NMR relative to an internal standard; M_w, M_n determined by GPC by comparison to PMMA standards.

Table VI. Comparison of Copper Removal Techniques

Method	Percent Original Copper Remaining
PE-ligand (M_n = 2000 g/mol)	1%
Precipiton Ligand	<1%
JandaJel[TM] Imine	5%
JandaJel[TM] TEDTA	4%
PE-PEG-ligand (app. 1000 g/mol) (10)	2-3%
PS Supported Ligands (2)	3%

265

Acknowledgements

We would like to acknowledge ICI Paints for financial support of this work.

References

1. *Controlled Radical Polymerization*; Matyjaszewski, K., Ed.; ACS Symposium Series 685; American Chemical Society: Washington, DC, 1998.
2. Kickelbick, G.; Paik, H. J.; Matyjaszewski, K. *Macromolecules* **1999**, *32*, 2941.
3. Haddleton, D. M.; Kukulj, D.; Radigue, A. P. *Chem. Commun.* **1999**, 99.
4. Shen, Y.; Zhu, S.; Zeng, F.; Pelton, R. *J. Polym. Sci., Part A: Polym. Chem.* **2001**, *39*, 1051.
5. Shen, Y.; Zhu, S.; Pelton, R. *Macromolecules* **2001**, *34*, 5812.
6. Shen, Y.; Zhu, S.; Zeng, F.; Pelton, R. H. *Macromolecules* **2000**, *33*, 5427.
7. Shen, Y.; Zhu, S.; Pelton, R. *Macromol. Rapid Commun.* **2000**, *21*, 956.
8. Hong, S. C.; Matyjaszewski, K. *Macromolecules* **2002**, *35*, 7592.
9. Shen, Y.; Zhu, S.; Pelton, R. *Macromolecules* **2001**, *34*, 3182.
10. Shen, Y.; Zhu, S. *Macromolecules* **2001**, *34*, 8603.
11. Liou, S.; Rademacher, J. T.; Malaba, D.; Pallack, M. E.; Brittain, W. J. *Macromolecules* **2000**, *33*, 4295.
12. Toy, P. H.; Janda, K. D. *Tetrahedron Lett.* **1999**, *40*, 6329.
13. Honigfort, M. E.; Brittain, W. J.; Bosanac, T.; Wilcox, C. S. *Macromolecules* **2002**, *35*, 4849.
14. Bosanac, T.; Wilcox, C. S. *Tetrahedron Lett.* **2001**, 4309.
15. Bosanac, T.; Yang, J.; Wilcox, C. S. *Angew. Chem. Int. Ed.* **2001**, *40*, 1875.
16. Bosanac, T.; Wilcox, C. S. *Chem. Commun.* **2001**, 1618.
17. Bosanac, T.; Wilcox, C. S. *J. Am. Chem. Soc.* **2002**, *124*, 4194.
18. Morita, T.; Maughon, B. R.; Bielawski, C. W.; Grubbs, R. H. *Macromolecules* **2000**, *33*, 6621.
19. Hillmyer, M. A.; Grubbs, R. H. *Macromolecules* **1993**, *26*, 872.
20. Hillmyer, M. A.; Nguyen, S. T.; Grubbs, R. H. *Macromolecules* **1997**, *30*, 718.
21. Haddleton, D. M.; Crossman, M. C.; Dana, B. D.; Duncalf, D. J.; Heming, A. M.; Kukulj, D.; Shooter, A. J. *Macromolecules* **1999**, *32*, 2110.
22. Haddleton, D. M.; Duncalf, D. J.; Kukulj, D.; Crossman, M. C.; Jackson, S. G.; Bon, S. A. F.; Clark, A. J.; Shooter, A. J. *Eur. J. Inorg. Chem.* **1998**, 1799.

23. Haddleton, D. M.; Clark, A. J.; Crossman, M. C.; Duncalf, D. J.; Heming, A. M.; Morsley, S. R.; Shooter, A. J. *Chem. Commun.* **1997**, *13*, 1173.
24. Fischer, H. *Macromolecules* **1997**, *30*, 5666.
25. Matyjaszewski, K.; Nakagawa, Y.; Jasieczek, C. B. *Macromolecules* **1998**, *31*, 1535.
26. Haddleton, D. M.; Perrier, S.; Bon, S. A. F. *Macromolecules* **2000**, *33*, 8246.
27. Zhang, X.; Xia, J.; Matyjaszewski, K. *Macromolecules* **1998**, *31*, 5167.
28. Carmichael, A. J.; Haddleton, D. M.; Bon, S. A. F.; Seddon, K. R. *Chem. Commun.* **2000**, 1237.

Atom Transfer Radical Polymerization

Materials and Applications

Chapter 19

Synthesis and Properties of Copolymers with Tailored Sequence Distribution by Controlled/Living Radical Polymerization

Jean-François Lutz[1], Tadeusz Pakula[2], and Krzysztof Matyjaszewski[1]

[1]Center for Macromolecular Engineering, Department of Chemistry, Carnegie Mellon University, 4400 Fifth Avenue, Pittsburgh, PA 15213
[2]Max-Planck-Institute for Polymer Research, Postfach 3148, 55021 Mainz, Germany

The control of the overall composition and the sequence distribution in copolymers with both linear and branched architectures is described. The synthesis of alternating and gradient copolymers as well as graft copolymers with statistical, gradient and block architectures was accomplished by controlled/living radical polymerization (atom transfer radical polymerization and reversible addition-fragmentation transfer polymerization). The properties of these tailored copolymers indicate that adjustment of sequence distribution at the molecular level has important effects on the macroscopic properties of the obtained polymeric materials.

Introduction

Developing new materials with controlled properties is an important issue in polymer science. The macroscopic properties of polymeric materials depend on their nanoscale morphology and consequently on the macromolecular structure of the components. In that context, it is important to develop synthetic routes, which allow for precise control of the shape and composition of the macromolecules. Controlled/living radical polymerization (CRP) techniques *(1-3)* such as nitroxide mediated polymerization (NMP) *(4)*, atom transfer radical polymerization (ATRP) *(5-7)* or reversible addition-fragmentation transfer polymerization (RAFT) *(8)* are powerful methods for macromolecular engineering. Several examples of well-defined macromolecules prepared by CRP with controlled chain length, polydispersities, compositions, functionalities and architectures have been reported in the literature *(9,10)*. However, only a few studies describe the synthesis of copolymers possessing controlled comonomer sequences by CRP. In the case of both linear polymer chains and branched polymer chains, several types of sequence distribution are possible (Scheme 1). Among these various classes of polymers, some copolymers with special comonomer distributions such as gradient, periodic or alternating may exhibit unusual properties and consequently could provide a new avenue for the production of novel materials.

Recent advances in controlling comonomer sequence distribution will be discussed and examples of linear (alternating or gradient) and branched

Scheme 1. Different types of comonomer sequences distribution

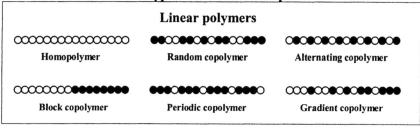

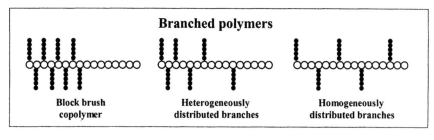

copolymers with controlled comonomer sequences prepared by CRP will be presented. The thermomechanical properties of some of these copolymers will be also discussed.

Linear polymers with controlled sequence distribution

Alternating copolymers: Alternating copolymers form a specific class of copolymers in which both comonomer units alternate in a regular fashion along the chain *(11)*. The ability of a comonomer pair to copolymerize spontaneously in an alternating fashion depends, mostly, on the polarity of the polymerizable double bonds. For example, acceptor monomers with a low electron density on the double bond preferentially react with a radical with an electron donor substituent rather than their own radical (assuming that the polarization of a radical is similar to a monomer). For example, maleic anhydride and styrene copolymerize spontaneously in an alternating fashion. However, for many comonomer pairs, the difference in polarity between acceptor and donor monomers is insufficient to result in alternation. In those cases, the tendency towards alternation may be enhanced by addition of a Lewis acid, including aluminum halides (EtAlCl$_2$, Et$_2$AlCl, Et$_3$Al$_2$Cl$_3$), and ZnCl$_2$, TiCl$_4$, BCl$_3$ or SnCl$_4$ *(12,13)*. The Lewis acid complexes with the acceptor monomers reduce electron density at the conjugated double bonds and favor alternating copolymerization.

Only a few examples of well-defined alternating copolymers synthesized by CRP were reported in literature. Most of those studies reported copolymers from a combination of comonomers with a spontaneous tendency for alternation: a strong electron accepting monomer (maleic anhydride *(14,15)*, *N*-butylmaleimide *(16)*, or *N*-phenylmaleimide *(17-19)*) and an electron donating monomer (styrene). Therefore, it was considered to be important to investigate the CRP synthesis of alternating copolymers from comonomer pairs without this inherent tendency. Two systems were investigated: ATRP of acrylates with isobutene, a less reactive electron rich monomer, and RAFT copolymerization of methyl methacrylate with styrene in the presence of Lewis acids.

ATRP Synthesis of well-defined alternating copolymers of acrylates and isobutene: Alternating copolymers based on acrylates (methyl acrylate (MA) or *n*-butyl acrylate (BA)) and isobutene (IB) are thermoplastic elastomers with interesting properties such as high tensile strength, high thermal decomposition temperature and good resistance to hydrolysis *(20)*. When polymerized via a conventional radical process, copolymers from IB and acrylic esters exhibit a relatively low content of IB (~20%) and low molecular weights because of the degradative chain transfer of isobutene. However, alternating copolymers p(acrylate-*alt*-IB) containing ~50% of IB were synthesized in the presence of Lewis acids such as aluminum or boron halides *(21-23)*. No control of molecular weight and polydispersity was possible in these systems. In an earlier communication, we reported that well defined copolymers p(MA-*alt*-IB) could be prepared by ATRP, using a large excess of IB with respect to MA in the feed *(24)*. These copolymers exhibited controlled molecular weight (up to M_n=50000 g.mol^{-1}), relatively low polydispersities (M_w/M_n < 1.6) and high content of IB (~45%). Sen *et al.* also reported the synthesis of well-defined copolymers based on MA and non-polar alkenes *(25)*. The copolymerization of BA with IB was also investigated in the presence of several ATRP catalyst systems (Table I) *(26)*. When the reaction was conducted in homogeneous systems; CuBr/4,4'-di-(*tert*-butyl)-2,2'-bipyridine (dtBbpy)$_2$ in 1,2-dimethoxybenzene solution, or CuBr/4,4'-di-(5-nonyl)-2,2'-bipyridine (dNbpy)$_2$ in bulk, the polymerization was well controlled yielding polymers with molecular weights close to theory and low polydispersities (Table I, entry 1 and 2).

Table I: Copolymerization of BA and IB with various ATRP catalysts

	Catalyst	Solvent	Time (h)	Conv. BA (%)	F_{IB}	$M_{n\ th}$ (g.mol^{-1})	$M_{n\ exp}$ (g.mol^{-1})	M_w/M_n
1	CuBr/dtBbpy$_2$	DMB^a	107	24.1	0.21^c	1530	1810	1.35
2	CuBr/dNbpy$_2$	-	89	29.4	0.27^c	1870	1690	1.47
3	CuBr/dtBbpy$_2$	-	112	68.9	0.36^c	3380	9480	1.93
4	CuBr/bpy$_2$	-	17.5	94	0.40^b		19730	2.42
5	CuBr/PMDETA	-	7	100	0.41^b	4520	4820	1.40

Experimental conditions: 50°C, catalyst/methyl 2-bromopropionate/BA/IB = 1/1/25/75
^{a} 1,2 dimethoxybenzene ; ^{b} obtained gravimetrically ; ^{c} obtained by ^{1}H NMR

Limited BA conversions and a low fraction of isobutene in the final copolymer F_{IB} were observed with these catalysts. This could be due to differences in the reactivities of the two dormant polymeric halides present in the copolymerization. Model studies confirmed this hypothesis *(26)*, and indicated that the absence of resonance stabilizing group on the terminal carbon of pIB-Br chain end, resulted in a much slower activation by the copper catalyst of this moiety than of the acrylate analogue. Therefore, during the copolymerization, the chains deactivated after addition of an IB unit could be considered as dead chains (at least comparatively to pMA-Br). In the presence of heterogeneous catalyst systems such as CuBr/bpy$_2$ or CuBr/dtBbpy$_2$ in bulk, due to the low solubility of the catalyst in the reaction mixture, polymerizations behave partially as a conventional radical process and much higher conversions and F_{IB} were observed (Table I, entry 3 and 4). However, broad molecular weight distributions and significant discrepancies between experimental and theoretical molecular weights were observed. A compromise was found using a slightly heterogeneous and more active catalyst system CuBr/PMDETA (Table I, entry 5). With this catalyst, well defined copolymers p(BA-*co*-IB) with controlled molecular weight, relatively low polydispersities and high content of IB (41%)

were obtained. This value of F_{IB} suggests that the obtained copolymers are not pure alternating copolymers, but could be represented by the sequences of alternating p(BA-*alt*-IB) sections randomly interrupted by short BA sequences.

Synthesis of well-defined alternating copolymers p(MMA-alt-S) by RAFT polymerization in the presence of Lewis acids. RAFT copolymerization of methyl methacrylate (MMA) and styrene (S) in the presence of a range of Lewis acids: tin (IV) chloride ($SnCl_4$), zinc chloride ($ZnCl_2$), ethylaluminum sesquichloride (EASC) and diethylaluminum chloride, was studied at 60 °C and at 40 °C (Table II).

Table II: RAFT copolymerization of MMA and S with Lewis acids

	Lewis acid	Temperature (°C)	Time (mn)	Overall Conversion [a]	$M_{n\ exp}$ (g.mol^{-1})	$M_{n\ th}$ [b] (g.mol^{-1})	M_w/M_n
1	$SnCl_4$	40	120	-	2800	-	2.50
2	$ZnCl_2$	40	125	-	10800	-	1.80
3	$ZnCl_2$	60	15	-	7600	-	3.74
4	EASC	40	80	73%	24500	23500	1.22
5	EASC	60	9	-	31500	-	1.45
6	Et_2AlCl	60	100	62.5%	20200	20000	1.38

Experimental conditions: Bulk, $[S]_0/[CDB]_0 = [MMA]_0/[CDB]_0 = 190$; $[Lewis\ acid]_0/[M]_0 = 0.4$; $[CDB]_0/[azo\text{-}bis\ isobutyronitrile]_0 = 10$
[a] Measured by gravimetry
[b] M_n theoretical $= ([M]_0+[S]_0)(conv.)(104.15+100)/(2(2[AIBN]_0+[CDB]_0))$

RAFT was selected as the preferred CRP method, since Lewis acids interact less with the dithioester control agent than with ATRP ligands or nitroxide radicals. In the presence of strong Lewis acids ($SnCl_4$ and $ZnCl_2$), poor control of the polymerization was observed (Table II, entry 1-3). These strong Lewis acids may form a complex with the cumyl dithiobenzoate (CDB) and may also lead to decomposition of CDB. The alkylaluminum halide Lewis acids (EASC and Et_2AlCl) could also form complexes with CDB, as evidenced by the formation of an intense orange color (instead of the pink uncomplexed CDB). However, in the presence of EASC at 40°C or Et_2AlCl at 60°C, the copolymerization of MMA with S exhibited a controlled/living behavior with experimental molecular weights close to the theoretical values and narrow molecular weight distribution (Table II, entry 4-6).

In the presence of both aluminum halides, the kinetics of the copolymerization were much faster (about 40 times) that in the case of RAFT copolymerization without Lewis acids. This results from higher values of the cross-propagation rate constants in the presence of Lewis acid than in their absence. Lewis acid could also accelerate initiation *(13,27)*. The comonomer sequence distribution of the synthesized copolymers was investigated by 600

MHz ^{1}H NMR. The percentage of various methyl methacrylate centered triads was calculated from the integrations of the regions of absorption of the methoxy groups *(28-30)*. The copolymers synthesized with EASC or Et$_2$AlCl contain ~90% alternating triads S-MMA-S, whereas the copolymers synthesized in the absence of Lewis acid <50% alternating triads. RAFT in the presence of aluminum halides allows synthesis of well-defined alternating copolymers with controlled molecular weight, low polydispersities and controlled comonomer sequence distribution. Alternating copolymerization of MMA with S with targeted M_n= 10,000 g.mol^{-1} to 100,000 g.mol^{-1} with Et$_2$AlCl at 60°C and at 40°C with EASC was studied. With Et$_2$AlCl, for M_n>50000 g.mol^{-1}, the copolymers had higher polydispersities (M_w/M_n~1.5) and the experimental M_n were lower than theoretical values. In the presence of EASC, copolymers with lower polydispersities (M_w/M_n<1.3) and good agreement between experimental and theoretical M_n (up to 70000 g.mol^{-1}) were obtained (Figure 1). In all cases, good control over comonomer sequence distribution was observed by ^{1}H NMR.

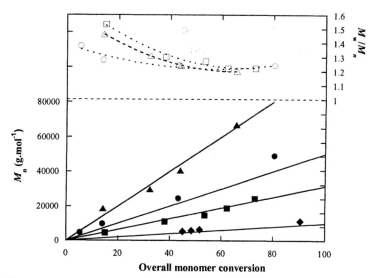

Figure 1: Number-average molecular weight M$_n$ *(full symbols) and polydispersity index* M$_w$/M$_n$ *(open symbols) as a function of overall monomer conversion (conv.) for the copolymerization of methyl methacrylate and styrene at 40°C in bulk: [S]$_0$ = 3.2 mol.L^{-1} ; [M]$_0$ = 3.2 mol.L^{-1} ; [EASC]$_0$ = 1.28 mol.L$^-$*
^{1} ; [CDB]$_0$ = 5.32×10^{-2} mol.L^{-1} (◆, ◇), 1.69×10^{-2} mol.L^{-1} (■, □), 1.06×10^{-2}
mol.L^{-1} (●, O), 5.3×10^{-3} mol.L^{-1} (▲, △), [CDB]$_0$/[AIBN]$_0$ = 10. Theoretical
M$_n$=([M]$_0$+[S]$_0$)(conv) (104.15+100)/(2(2[AIBN]$_0$+[CDB]$_0$)).

The alternating copolymers formed in the presence of EASC retained chain end functionality and were used as macroinitiators for the synthesis of well-defined diblock copolymers P(MMA-*alt*-S)-*b*-PS.

Gradient copolymers: Gradient copolymers exhibit a continuous change in composition along the chains *(31)*. Since in CRP systems all chains are initiated simultaneously and survive until the end of polymerization a gradient distribution can be produced spontaneously along the chains due to feed composition drift that can occur during the reaction. In a batch process the gradient distribution is fixed by the reactivity ratios and by the initial monomer feed. Using a semi-batch process (i.e. continuous addition of one comonomer) the gradient distribution of the comonomers in the copolymer may be precisely adjusted. Several types of gradient copolymers were prepared by ATRP: p(S-*grad*-acrylate) *(32)*, p(S-*grad*-acrylonitrile) *(33)* or p(MMA-*grad*-acrylate) *(34)*. The resulting gradient copolymers possess thermal and mechanical properties different from those of block or random copolymers *(31,35,36)*. These unique properties can be exploited for the synthesis of improved polymeric materials. For example, it was shown that triblock p(MMA-*grad*-BA)-b-pBA-b-p(MMA-*grad*-BA) copolymers are thermoplastic elastomers with much higher tensile elongation than the pure pMMA-b-pBA-b-pMMA triblock copolymers *(37)*.

The properties of diblock copolymers PBA-*b*-(PMMA-*grad*-PBA) prepared by ATRP (Table III) were investigated. In this series, the PBA content of the second block increases. Consequently, this series exhibits a gradual crossover from the properties of a diblock to those of a triblock copolymer.

Table III: Diblock copolymers pBA-*b*-p(MMA-*grad*-BA)

Sample	F_{BA} block 1	M_n (g.mol^{-1}) block 1	M_w/M_n block 1	F_{BA} block 2	M_n (g.mol^{-1}) block 2	M_w/M_n block 2
1	100%	14200	1.20	0	29000	1.23
2	100%	23300	1.22	10	52000	1.22
3	100%	23300	1.22	20	48500	1.25
4	100%	23300	1.22	50	48500	1.30

This behavior is illustrated in Figure 2, which shows the results of dynamic mechanical testing and small angle X-ray scattering (SAXS) for these samples. The evolution of the storage and loss moduli versus temperature shows two distinct segmental relaxations. The first one which appears in the subambient temperature range and is representative of the soft pBA first block. The second one, representative of the p(MMA-*grad*-BA) second block, appears at higher temperature depending on the BA content in this block. For high BA content, a very broad segmental relaxation (typical of gradient copolymer) is seen as a result of a continuous composition change across the interfaces in the microphase separated state of the copolymer. This behavior is also confirmed by the SAXS measurements. The scattered intensity decreases with increasing BA content in the second block.

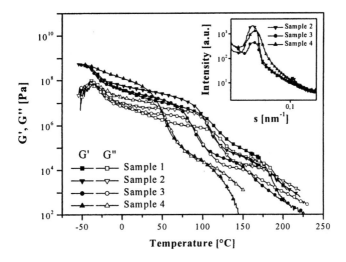

Figure 2: Temperature dependencies of the storage (G') and loss moduli (G'') for diblock copolymers pBA-b-p(MMA-grad-BA). The inset shows the small angle x-ray scattering intensities for these samples.

Branched polymers with controlled sequences distribution

Graft copolymers are segmented copolymers which were used as impact resistant materials, thermoplastic elastomers, compatibilizers and emulsifiers. They were prepared using several synthetic routes *(38,39)*. The "grafting through" method employs copolymerization of a low molar mass monomer with a macromonomer.*(40)* This method permits incorporation of various macromonomers such as polyethylene *(41)*, poly(ethylene oxide) *(42,43)*, poly(dimethyl siloxane) (PDMS) *(44,45)*, poly(lactic acid) *(46)* into thermodynamically incompatible polystyrene or poly (meth)acrylate backbones. The combination of the "grafting through" method with CRP allows control of polydispersity, functionality, copolymer composition, backbone length, branch length and branch spacing *(44-47)*. Such control over macromolecular structure may impact the macroscopic properties of the polymer material. This is illustrated by the mechanical properties of a series of graft copolymers (PMMA-*graft*-PDMS) prepared using various techniques. Table IV shows the experimental conditions and the apparent reactivity ratios r_{MMA} determined by the Jaacks method for these different processes *(48)*.

The first sample was prepared using a conventional free radical process (FRP); the measured reactivity ratio r_{MMA} suggests a low reactivity of the PDMS macromonomer. Since initiation continues throughout FRP, the sample contains chains with different compositions. The chains initiated early in the process are rich in PMMA, while the chains initiated at the end of the polymerization possess a high content of PDMS. The second sample was prepared by RAFT and the reactivity ratio also indicates low reactivity of the macromonomer. However, since all the chains were initiated simultaneously, the same heterogeneous (or gradient) distribution of segments was formed. Sample 3 was synthesized by ATRP with a fast initiation. In this case, the reactivity ratio was close to 1, suggesting that all chains have homogeneous distribution of the segments.

Table IV: Graft copolymers PMMA-*graft*-PDMS

Sample	technique	Temperature	solvent	r_{MMA}
1	FRP [a]	60°C	Xylene 31weight %	3.07
2	RAFT [a, b]	60°C	Xylene 31weight %	1.75
3	ATRP [a, c]	90°C	Xylene 31weight %	1.24

Macromonomer PDMS-M: M_n=2300 g.mol^{-1} ; M_w/M_n =1.25 ; functionality~1

[a] MMA/PDMS-M = 95/5 mol % = 50/50 weight % in the feed.

[b] MMA/PDMS-M/CDB/AIBN=285/15/1/0.5

[c] MMA/PDMS-M/Ethyl 2-bromoisobutyrate/CuCl/d-n-bipy=285/15/1/1/2

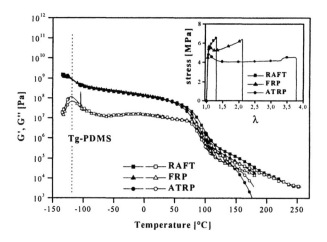

Figure 3: Temperature dependence of the storage (G') and loss moduli (G'') for graft copolymers (PMMA-graft-PDMS). The inset shows the stress-strain curves for the same copolymer samples recorded at 25 °C during drawing of films with a constant rate of 1 mm/mn.

Figure 3 shows that in all samples, two different segmental relaxations were evident in the dynamic mechanical analysis. This indicates, that in all cases, the graft copolymer exists in a strongly microphase segregated state. Moreover, significant differences in elongation at break were observed. The sample synthesized by RAFT breaks at 30% elongation whereas the one prepared by FRP and ATRP break respectively at 115% and 280%. These differences indicate that the mechanical properties strongly depend on the molecular structure, since all copolymers had similar molecular weight (M_n~90,000 g.mol^{-1}) and similar overall composition (~50% PMMA)

A new terpolymer, pMMA-*graft*-pLA/pDMS was prepared by ATRP. This copolymer is an organic/inorganic hybrid *(49)*, but also has a potential for biodegradability, due to incorporation of the poly(D-lactic acid) moiety. *(50)*. Two approaches to graft terpolymers were investigated (Scheme 2). The first one was a "one step approach" in which the low molecular weight methacrylate monomer (MMA or BMA), the methacrylate functionalized poly(D-lactic acid) macromonomer (PLA-M) and the methacrylate functionalized poly(dimethyl siloxane) macromonomer (PDMS-M) were simultaneously copolymerized. The second strategy was a "two step approach" where a graft copolymer containing one macromonomer was chain extended by a copolymerization of the second macromonomer with a methacrylate. In both cases, the targeted molecular weight of the final copolymers was M_n=60,000 g.mol^{-1}: DP_{target}=300 for methyl methacrylate or butyl methacrylate (BMA), DP_{target}=5 for PLA-M (M_n=3,000 g.mol^{-1}), DP_{target}=5 for PDMS-M (M_n=3,000 g.mol^{-1}).

Table V shows the values of final polydispersities and molecular weights for the synthesized terpolymers. All the copolymers possess a narrow molecular weight distribution and experimental molecular weights close to the calculated values. However, depending on the synthesis, the distribution of the PLA and PDMS segments along the polymer chain are rather different.

When a one step approach was used, the synthesized terpolymer possessed a gradient distribution of segments (i.e. branch spacing of each macromonomer depends on the reactivity ratio of the three monomers). The overall reactivity ratios calculated for this terpolymerization system (Table V, Entry 1) using the Jaacks method *(48)* show that PLA-M has a higher reactivity than PDMS-M vs. MMA. This behavior may be explained by the thermodynamic incompatibility between the inorganic macromonomer and organic backbone. In order to synthesize other types of distribution, a PDMS macroinitiator (M_n=8400 g.mol^{-1}, M_w/M_n = 1.2) was additionally employed in a one step synthesis. It was reported that the reactivity ratio of PDMS-M vs. MMA was increased by using a PDMS macroinitiator by decreasing the thermodynamic repulsion between the growing organic backbone and PDMS-M *(44)*. Nevertheless, in the presence of the PDMS macroinitiator (Table V, Entry 2) the measured reactivity ratio r$_{MMA/PDMS}$ was nearly the same as when a low molar mass initiator was used (Table V, Entry 1). This could be explained by the presence of PLA-M which is preferentially incorporated into the backbone and plausibly decreases the compatibilizing effect of the PDMS macroinitiator. The one step

278

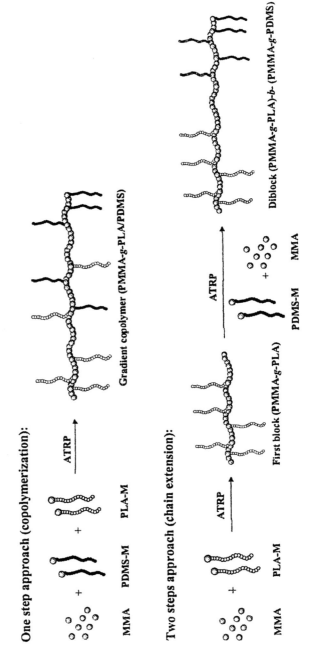

Scheme 2: Graft terpolymers by ATRP

One step approach (copolymerization):

MMA   PDMS-M   +   PLA-M $\xrightarrow{\text{ATRP}}$ Gradient copolymer (PMMA-g-PLA/PDMS)

Two steps approach (chain extension):

MMA   PLA-M $\xrightarrow{\text{ATRP}}$ First block (PMMA-g-PLA)   PDMS-M   +   MMA $\xrightarrow{\text{ATRP}}$ Diblock (PMMA-g-PLA)-b- (PMMA-g-PDMS)

Table V: Graft terpolymers poly(alkyl methacrylate)-*graft*-PLA/PDMS prepared by ATRP

	Terpolymer	$M_{n\,exp}$	$M_{n\,th}$	M_w/M_n	$r_{MMA/PL}$	$r_{MMA/PDMS}$
					A	
1	(PMMA-*graft*-PLA/PDMS)[a]	52200	51600	1.20	0.8	1.4
2	PDMS-*b*-(PMMA-*graft*-PLA/PDMS)[a]	68400	58050	1.30	0.9	1.7
3	(PBMA-*graft*-PLA/PDMS)[a]	60000	61260	1.25	0.5	0.5
4	(PMMA-*g*-PLA)-*b*-(PMMA-*g*-PDMS)[b]	61900	60950	1.35	0.6	2.7
5	(PMMA-*g*-PDMS)-*b*-(PMMA-*g*-PLA)[b]	66227	59200	1.25	0.5	1.3

[a] one step approach ; [b] two steps approach

copolymerization of PLA-M, PDMS-M and MMA (Table V, Entry 1) was also compared to the one of PLA-M, PDMS-M and BMA (Table V, Entry 3). Both copolymerizations were controlled. Both PDMS-M and PLA-M showed a higher reactivity vs. BMA.

When the two step approach was used, both macromonomers PLA-M and PDMS-M were grafted at different regions of the backbone. The spacing between PDMS branches varied, depending on the synthesis conditions. When the PMMA-*graft*-PLA macroinitiator was used to initiate the copolymerization of MMA and PDMS-M (Table V, Entry 4), the PDMS-M exhibited a low reactivity vs. MMA, suggesting an irregular branch distribution. This could be due to a strong incompatibility between the organic grafted macroinitiator and the inorganic macromonomer. When the PMMA-*graft*-PDMS block was synthesized first (Table V, Entry 5), a relatively higher reactivity was observed for PDMS-M vs. MMA, suggesting a more regular branch spacing. The reactivity of PLA-M with MMA was found to be the same in every case. As a conclusion, depending on the synthesis conditions (one step or two step approach) it was possible to adjust the branch spacing of the copolymer in different ways. The effect of this tailored branch spacing on the morphologies and properties of the terpolymers is currently under investigation.

Conclusions

Various synthetic pathways for controlling the sequence distribution of copolymers with either linear or branched topology were explored. Due to their living character, CRP techniques allow relatively simple synthesis of copolymers with controlled sequence distributions such as alternating, gradient or block copolymers as well as graft copolymers with adjustable branch spacing. The control over sequence distribution could be enhanced by extending CRP to a semi-batch process or combining with complexation techniques by Lewis acids. For example, well defined alternating copolymers p(MMA-*alt*-S) were formed by RAFT polymerization in the presence of Lewis acids. The results reported in this chapter also illustrate the potential impact of controlled comonomer distribution on copolymer macroscopic properties. Depending on the size range at which the control extends, the molecular self-organization can be influenced at various structural levels. For the alternating copolymers, the material remains homogeneous and the main effect is related to the local segmental mobility in the system and consequently to the location of the glass transition on the temperature scale. In the other cases, when control of composition extends to the entire macromolecules, the microphase separation process may occur. Additionally, combining controlled variation of intramolecular composition with the control of topology, as illustrated in the last example, one can obtain specific architectures, which can give further possibilities for material modification. Consequently, in the future, copolymers with tailored sequence distribution could play an increasingly important role in material science.

Acknowledgments: The authors thank the National Science Foundation (DMR-00-09409) and the members of the CRP Consortium at Carnegie Mellon University for financial support. Phil Costanzo, Mathias Destarac, Betül Kırcı and Hosei Shinoda are also gratefully acknowledged for their contributions to this study.

References

(1) Matyjaszewski, K.; Davis, T. P., Eds. *Handbook of Radical Polymerization*; Wiley Interscience: Hoboken, 2002.
(2) Matyjaszewski, K., Ed. *Controlled Radical Polymerization. ACS Symp. Ser.*; ACS: Washington D.C., 1998; Vol. 685.
(3) Matyjaszewski, K., Ed. *Controlled/Living Radical Polymerization. Progress in ATRP, NMP, and RAFT. ACS Symp. Ser.*; ACS: Washington D.C., 2000; Vol. 768.
(4) Hawker, C. J.; Bosman, A. W.; Harth, E. *Chem. Rev.* **2001**, *101*, 3661-3688.
(5) Wang, J.-S.; Matyjaszewski, K. *J. Am. Chem. Soc.* **1995**, *117*, 5614-5615.
(6) Matyjaszewski, K.; Xia, J. *Chem. Rev.* **2001**, *101*, 2921-2990.
(7) Kamigaito, M.; Ando, T.; Sawamoto, M. *Chem. Rev.* **2001**, *101*, 3689-3745.
(8) Chiefari, J.; Rizzardo, E. In *Handbook of Radical Polymerization*; Matyjaszewski, K., Davis, T. P., Eds.; Wiley Interscience: Hoboken, 2002; pp 621-690.
(9) Patten, T. E.; Matyjaszewski, K. *Adv. Mater.* **1998**, *10*, 901-915.
(10) Davis, K. A.; Matyjaszewski, K. *Adv. Polym. Sci.* **2002**, *159*, 2-166.
(11) Cowie, J. M. G., Ed. *Alternating Copolymers*; Plenum Press: New York, 1985.
(12) Hirai, H. *J. Polym. Sci., Macromol. Rev.* **1976**, *11*, 47-91.
(13) Bamford, C. H. In *Alternating Copolymers*; Cowie, J. M. G., Ed.; Plenum Press: New York, 1985; pp 75-152.
(14) Benoit, D.; Hawker, C. J.; Huang, E. E.; Lin, Z.; Russell, T. P. *Macromolecules* **2000**, *33*, 1505-1507.

(15) De Brouwer, H.; Schellekens, M. A. J.; Klumperman, B.; Monteiro, M. J.; German, A. L. *J. Polym. Sci., Part A: Polym. Chem.* **2000**, *38*, 3596-3603.

(16) Lokaj, J.; Vlcek, P.; Kriz, J. *J. Appl. Polym. Sci.* **1999**, *74*, 2378-2385.

(17) Chen, G.-Q.; Wu, Z.-Q.; Wu, J.-R.; Li, Z.-C.; Li, F.-M. *Macromolecules* **2000**, *33*, 232-234.

(18) Li, F.-M.; Chen, G.-Q.; Zhu, M.-Q.; Zhou, P.; Du, F.-S.; Li, Z.-C. *ACS Symp. Ser.* **2000**, *768*, 384-393.

(19) Lokaj, J.; Holler, P.; Kriz, J. *J. Appl. Polym. Sci.* **2000**, *76*, 1093-1099.

(20) Mashita, K.; Hirooka, M. *Polymer* **1995**, *36*, 2983-2988.

(21) Hirooka, M.; Mashita, K.; Imai, S.; Kato, T. *Rubber Chem. Technol.* **1973**, *46*, 1068-1076.

(22) Kuntz, I.; Chamberlain, N. F.; Stehling, F. J. *J. Polym. Sci., Polym. Chem. Ed.* **1978**, *16*, 1747-1753.

(23) Mashita, K.; Yasui, S.; Hirooka, M. *Polymer* **1995**, *36*, 2973-2982.

(24) Coca, S.; Matyjaszewski, K. *Polym. Prepr. (Am. Chem. Soc., Div. Polym. Chem.)* **1996**, *37*, 573-574.

(25) Liu, S.; Elyashiv, S.; Sen, A. *J. Am. Chem. Soc.* **2001**, *123*, 12738-12739.

(26) Destarac, M.; Matyjaszewski, K. *to be published.*

(27) Pakravani, M. M.; Tardi, M.; Polton, A.; Sigwalt, P. *C. R. Acad. Sci., Ser. 2* **1984**, *299*, 789-794.

(28) Ito, K.; Yamashita, Y. *J. Polym. Sci., Polym. Lett. Ed.* **1965**, *3*, 625-630.

(29) Ito, K.; Iwase, S.; Umehara, K.; Yamashita, Y. *J. Macromol. Sci., Part A* **1967**, *1*, 891-908.

(30) Kirci, B.; Lutz, J.-F.; Matyjaszewski, K. *Macromolecules* **2002**, *35*, 2448-2451.

(31) Matyjaszewski, K.; Ziegler, M. J.; Arehart, S. V.; Greszta, D.; Pakula, T. *J. Phys. Org. Chem.* **2000**, *13*, 775-786.

(32) Arehart, S. V.; Matyjaszewski, K. *Macromolecules* **1999**, *32*, 2221-2231.

(33) Greszta, D.; Matyjaszewski, K.; Pakula, T. *Polym. Prepr. (Am. Chem. Soc., Div. Polym. Chem.)* **1997**, *38*, 709-710.

(34) Arehart, S. V.; Matyjaszewski, K. *Polym. Prepr. (Am. Chem. Soc., Div. Polym. Chem.)* **1999**, *40*, 458-459.

(35) Pakula, T. *Macromol. Theory Simul.* **1996**, *5*, 987-1006.

(36) Kryszewski, M. *Polym. Adv. Technol.* **1998**, *9*, 244-259.

(37) Matyjaszewski, K.; Shipp, D. A.; McMurtry, G. P.; Gaynor, S. G.; Pakula, T. *J. Polym. Sci., Part A: Polym. Chem.* **2000**, *38*, 2023-2031.

(38) Pitsikalis, M.; Pispas, S.; Mays, J. W.; Hadjichristidis, N. *Adv. Polym. Sci.* **1998**, *135*, 1-137.

(39) Borner, H. G.; Matyjaszewski, K. *Macromol. Symp.* **2002**, *177*, 1-15.

(40) Meijs, G. F.; Rizzardo, E. *J. Macromol.r Sci., Rev. Macromol. Chem. Phys.* **1990**, *C30*, 305-377.

(41) Hong, S. C.; Jia, S.; Teodorescu, M.; Kowalewski, T.; Matyjaszewski, K.; Gottfried, A. C.; Brookhart, M. *J. Polym. Sci., Part A: Polym. Chem.* **2002**, *40*, 2736-2749.

(42) Wang, Y.; Huang, J. *Macromolecules* **1998**, *31*, 4057-4060.

(43) Neugebauer, D.; Matyjaszewski, K. *Polymer Preprints* **2002**, *43*, 241-242.

(44) Shinoda, H.; Miller, P. J.; Matyjaszewski, K. *Macromolecules* **2001**, *34*, 3186-3194.

(45) Shinoda, H.; Matyjaszewski, K. *Macromol. Rapid Commun.* **2001**, *22*, 1176-1181.

(46) Shinoda, H.; Matyjaszewski, K. *Macromolecules* **2001**, *34*, 6243-6248.

(47) Roos, S. G.; Muller, A. H. E.; Matyjaszewski, K. *ACS Symp. Ser.* **2000**, *768*, 361-371.

(48) Jaacks, V. *Makromol. Chem.* **1972**, *161*, 161-172.

(49) Pyun, J.; Matyjaszewski, K. *Chem. Mater.* **2001**, *13*, 3436-3448.

(50) Ikada, Y.; Tsuji, H. *Macromol. Rapid Commun.* **2000**, *21*, 117-132.

Chapter 20

Controlled Synthesis of Amphiphilic Poly(methacrylate)-*g*-[poly(ester)/poly(ether)] Graft Terpolymers

Isabelle Ydens[1], Philippe Degée[1], Jan Libiszowski[2], Andrzej Duda[2], Stanislaw Penczek[2], and Philippe Dubois[1,*]

[1]Laboratory of Polymeric and Composite Materials (LPCM), University of Mons-Hainaut, Place du Parc 20, B–7000 Mons, Belgium
[2]Department of Polymer Chemistry, Center of Molecular and Macromolecular Studies (CMMS), Polish Academy of Sciences, Sienkiewicza 112, PL–90–363 Lodz, Poland

Coupling atom transfer radical polymerization (ATRP) and coordination-insertion ring-opening polymerization (ROP) allows access to well-defined poly(methacrylate)-g-[poly(ester)/poly(ether)] graft terpolymers according to a two-step procedure. In the first step, the controlled copolymerization of methyl methacrylate (MMA), 2-hydroxyethyl methacrylate (HEMA) and poly(ethylene glycol) methyl ether methacrylate (PEGMA) was carried out by using ethyl 2-bromoisobutyrate and $NiBr_2(PPh_3)_2$ as initiator and catalyst, respectively. The second step consisted of the ring-opening polymerization (ROP) of ε-caprolactone (CL) or L,L-dilactide (LA) initiated by poly(MMA-co-HEMA-co-PEGMA) in the presence of either tin (II) bis(2-ethylhexanoate) $(Sn(Oct)_2)$ or triethylaluminum $(AlEt_3)$. The resulting graft copolymers as well as the intermediate poly(MMA-co-HEMA-co-PEGMA) terpolymers proved to be efficient surfactants as evidenced by dynamic interfacial tension measurements.

283

Introduction

In analogy to low molecular weight surfactants, amphiphilic block and graft copolymers consisting of hydrophilic and hydrophobic segments can be used for stabilizing aqueous dispersions and emulsions (*1,2*), for compatibilizing the interface in polymer blends and composites (*3*), and for tuning the wettability of polymeric materials by surface modification, especially in the biomedical field (*4,5*). The factors that influence the tensioactive properties of amphiphilic copolymers include composition, structure, molar mass, copolymer-solvent interactions, concentration, temperature and preparation methods. In that respect graft amphiphilic coplymers represent valuable materials with the unique ability to vary a large variety of molecular parameters provided that they are obtained by controlled polymerization techniques, for example controlled anionic (*6*), cationic (*7*), radical (*8-11*) or ring-opening (metathesis) polymerization (*12*).

In a recent paper (*13*), we reported on the synthesis of poly(methyl methacrylate-co-2-hydroxyethyl methacrylate) copolymers prepared by atom transfer polymerization (ATRP) and used as precursors for the ring-opening polymerization (ROP) of (di)lactones. The combination of these two consecutive polymerization processes allowed to access to well-defined brush like poly(methacrylate)-g-poly(aliphatic ester) copolymers of a wide range of molar mass and composition (*14,15*). However, both the polymethacrylate backbone and the polyester branches were hydrophobic which prompted us to incorporate a third functional (hydrophilic) comonomer, i.e. poly(ethylene glycol) methyl ether methacrylate (PEGMA). This paper thus focuses on the synthesis and characterization of poly(methacrylate)-g-[poly(ester)/poly(ether)] graft terpolymers (Figure 1) by controlled radical terpolymerization of methyl methacrylate (MMA), 2-hydroxyethyl methacrylate (HEMA), and poly(ethylene glycol) methyl ether methacrylate (PEGMA), followed by the graft copolymerization of ε-caprolactone (CL) or L,L-dilactide (LA) initiated from the free hydroxyl groups of the HEMA units in the presence of tin (II) bis(2-ethylhexanoate) $(Sn(Oct)_2)$ or triethylaluminum $(Al(Et)_3)$. Preliminary dynamic interfacial tension experiments have been carried out on the copolymers.

Experimental Section

Synthesis of Poly(MMA-co-HEMA-co-PEGMA)

All experiments were conducted according to the Schlenck method. $NiBr_2(PPh_3)_2$ and a magnetic bar were introduced in open air into a pre-weighed glass tube, which was then closed by a three-way stopcock and purged by three repeated vacuum/nitrogen cycles. In a 50ml flask, MMA, dried HEMA, PEGMA and the desired volume of toluene were introduced and bubbled with nitrogen for

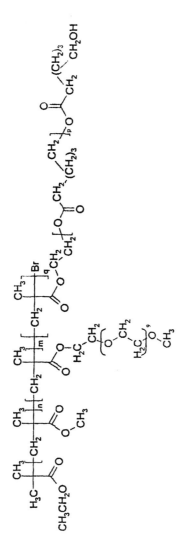

Figure 1. Poly(MMA-co-HEMA-co-PEGMA)-g-poly(ε-caprolactone)

10 minutes. before transferring the mixture into the glass tube. Then, the initiator (ethyl 2-bromoisobutyrate, (Eib)Br) (0.27 moll^{-1} in dried toluene) was added under nitrogen with a syringe. The glass tube was then immerged into an oil bath maintained at a defined temperature. After a predetermined polymerization time, the glass tube was rapidly cooled down to room temperature and its content was dissolved in THF. The terpolymer was then selectively recovered by precipitation from heptane. The conversion in terpolymer was determined by weighing after drying at 80°C for 24h under reduced pressure. In order to remove out the nickel catalyst, the terpolymer was dissolved in tetrahydrofuran and passed through a column of basic alumina. The purified terpolymer was recovered by precipitation from heptane, filtration and drying until constant weight.

Synthesis of Poly(methacrylate)-g-[poly(ester)/poly(ether)] Graft Terpolymers

Polymerization was carried out with purified and dried reactants, using standard high vacuum technique (*17*).

Typically, separate ampoules equipped with break-seals containing respectively, the desired cyclic ester (LA or CL), the poly(MMA-co-HEMA-co-PEGMA) terpolymer and Sn(Oct)$_2$ were sealed to a reacting glass ampoule (30 mL) together with a polarimetric (5 mL) or a dilatometric (15 mL) cell. A known volume of THF was distilled into the reacting glass ampoule which was then sealed off. The break-seals were broken, and when all components were dissolved, the resulting solution was transferred into either the polarimetric (LA case) or dilatometric cell (CL case), and into the reacting ampoule. The cell and ampoule containing the reacting mixture were separately sealed off. The cell was first placed into a bath kept at 80°C so as kinetic data could be collected. Then, the ampoule was allowed to polymerize at 80°C for the time required to reach maximum monomer conversion. The ampoule was opened and a drop of the crude reacting mixture was injected in a SEC apparatus to determine the conversion and molecular weight parameters. The rest of the reacting medium was precipitated into cold methanol, separated by filtration and dried under vacuum.

In an alternative procedure, the poly(MMA-co-HEMA-co-PEGMA) terpolymers were transferred in open air into a previously dried two-necked round bottom flask equipped with a stopcock and a rubber septum, dried by three successive azeotropic distillations of toluene, then dissolved in dry toluene and added with a defined volume of the activator : Sn(Oct)$_2$ or Al(Et)$_3$. As far as Sn(Oct)$_2$ is concerned, the activator and the monomer were added successively to the terpolymer solution in toluene and the temperature was raised up to 110 °C. When triethylaluminum (Al(Et)$_3$) was used as the activator, the reaction flask was equipped with an oil valve for ethane evolution and the reaction was stirred for 4 hours at r.t. before adding CL and heating up to desired polymerization temperature. Whatever the activator, the reaction product was recovered by precipitation from heptane, filtration and drying under vacuum until constant weight.

Instrumentation

^{1}H NMR spectra were recorded using a Bruker AMX300 or a Bruker DRX 500 spectrometer at room temperature in CDCl$_3$. Size exclusion chromatography (SEC) was performed either in THF at 35°C or in CH$_2$Cl$_2$. The actual number-average molar masses (M$_n$) of the copolymers were determined in chloroform with a Knauer vapor phase pressure osmometer and a Knauer membrane osmometer for M$_n$ $\leq$ 3.5 x 10^4 and $\geq$ 3.5 x 10^4, respectively. Interfacial tensions were obtained using a DSA 10-MK2 tensiometer (Wilten Fysika) at 20°C according to the pendant drop method.

Interfacial tension values were determined from the shape of the pendant drop by fitting the Gauss-Laplace equation to the experimental drop shape coordinates (16). All reported data points are average values of triplicate measurements (with a max. variation < 2%).

Results and Discussion

Poly(MMA-co-HEMA-co-PEGMA)

Poly(MMA-co-HEMA-co-PEGMA) terpolymers have been synthesized by atom transfer radical copolymerization of MMA, HEMA and PEGMA in toluene at 80 or 85 °C using NiBr$_2$(PPh$_3$)$_2$ as catalyst and ethyl 2-bromoisobutyrate ((EiB)Br) as initiator.

Tables I and II show the experimental conditions for the synthesis of the terpolymers, which will be used as precursors for the grafting ROP reaction, and their molecular characteristics, respectively. The initial content of PEGMA (M$_n$ = 455) ranged from 3 to 11 mol % while the HEMA molar fraction was deliberately maintained below 11 mol % in order to prepare graft copolymers with a low branching density. Compared to the initial molar fractions in comonomers, the final composition of poly(MMA-co-HEMA-co-PEGMA) terpolymers as determined by ^{1}H NMR shows that the functional monomers, i.e. HEMA and PEGMA, are preferably incorporated into the growing chains. For instance, molar fractions in T3 reach 0.75 for MMA, 0.15 for HEMA and 0.10 for PEGMA starting from 0.82, 0.10, and 0.08, respectively.

Table I. Experimental conditions for the synthesis of methacrylic terpolymers

Sample	$[MMA]_0$ $(molL^{-1})$	$[HEMA]_0$ $(molL^{-1})$	$[PEGMA]_0$ $(molL^{-1})$	Conv. (%)
T1	2.56	0.12	0.09	68
T2	2.61	0.17	0.29	76
T3	2.62	0.33	0.25	74
T4*	2.59	0.17	0.33	83
T5**	2.59	0.17	0.33	64

Toluene, 16h, 85°C, $[Mtotal]_0/[(EiB)Br]_0/[NiBr_2(PPh_3)_2]_0 = 100/1/0.5$
*T = 80°C
**T = 80°C, $[Mtotal]_0/[(EiB)Br]_0/[NiBr_2(PPh_3)_2]_0 = 200/1/0.5$

Table II. Molecular characteristics of methacrylic terpolymers

Sample	$M_n(VPO)^a$	M_n^b	M_w/M_n^b	n_{MMA}^c	n_{OH}^c	n_{PEGMA}^c
T1	11,100	10,000	1.20	80.9	6.7	3.8
T2	16,100	13,900	1.18	77.5	8.7	13.6
T3	14,600	14,600	1.19	72.9	14.2	10.2
T4	13,000	14,600	1.27	73.1	7.1	9.9
T5	16,000	20,600	1.32	88.6	9.3	12.4

[a]As determined by vapour phase pressure osmometry in methylene chloride.
[b]As determined by SEC in THF with reference to PMMA standards.
[c]n_i denotes the average number of repetitive units per terpolymer chain : $n_{OH} = (M_n(VPO)-M_{(EiB)Br})/[(2I_d/3I_f) \times M_{MMA} + (2I_f/3I_f) \times M_{PEGMA} + M_{HEMA}]$; $n_{MMA} = n_{OH} \times (2I_d/3I_f)$; $n_{PEGMA} = n_{OH} \times (2I_f/3I_f)$ (see Figure 2).

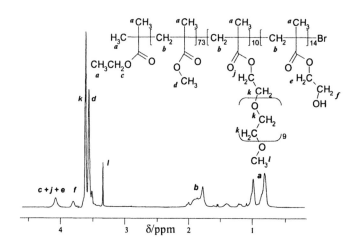

Figure 2. ^1H NMR (500MHz) spectrum of a poly(MMA-co-HEMA-co-PEGMA) terpolymer (sample T3 in Tables 1 and 2).

As recently reported (*13*) and in perfect agreement with previous publications (*18,19*), the copolymerization of MMA and HEMA under very similar experimental conditions has proved to be totally controlled as evidenced by kinetic measurements (linear time dependencies of $\ln([M]_0/[M])$ where $[M]_0$ and $[M]$ represent the comonomer concentration at start and at a given polymerization time, respectively) and a linear increase of M_n as a function of conversion.

The molar mass can be predicted from the initial comonomer-to-initiator molar ratio corrected by comonomers consumption degree (α) and the molar mass distribution is remarkably narrow ($M_w/M_n \leq 1.25$), at least for $\alpha \leq 0.85$.

Poly(MMA-co-HEMA-co-PEGMA)-g-aliphatic polyesters

Poly(methacrylate)-g-[poly(ester)/poly(ether)] graft copolymers have been obtained by initiating the ring-opening polymerization (ROP) of either ε-caprolactone (CL) or L,L-dilactide (LA) from the hydroxyl functions pending along the poly(MMA-co-HEMA-co-PEGMA) terpolymers, previously activated by tin(II) bis(2-ethyl hexanoate), $Sn(Oct)_2$, in THF at 80 °C for an initial concentration in $Sn(Oct)_2$ of $5 \cdot 10^{-3}$ molL^{-1}. Under such conditions, $Sn(Oct)_2$ has proved to react with hydroxyl functions providing initiating species such as tin(II) alkoxides (*20-22*). Table III and IV list the experimental conditions and molecular characteristics of the synthesized graft copolymers. Cyclic ester conversion was determined by SEC analysis as reported elsewhere (*20,21*). The molar mass of both the graft copolymer and the polyester branches, as determined by osmometry and ^{1}H NMR spectroscopy, respectively, are in quite good agreement with the values expected from the initial monomer-to-macroinitiator molar ratio.

Table III. Conditions for the synthesis of poly(MMA-co-HEMA-co-PEGMA)-g-aliphatic polyester in THF at 80 °C under high vacuum technique

Sample	M^a	MI^a	$[M]_0$ (molL^{-1})	$[OH]_0^b$ (molL^{-1})	Time (min.)	Conv. (%)
P1	LA	T1	0.82	0.0751	256	92.3
P2	LA	T1	1.09	0.0104	776	89.0
P3	CL	T1	1.71	0.0128	1,651	90.2
P4	LA	T2	1.00	0.0082	497	88.3
P5	LA	T3	1.01	0.0156	264	75.0

[a] M and MI denote the monomer and the terpolymer macroinitiator, respectively.
[b] $[OH]_0 = (1000 \times m_{Terpolymer} \times n_{OH})/(M_n(VPO)_{Terpolymer} \times$ total solution volume (ml)).

Table IV. Molecular characteristics of poly(MMA-co-HEMA-co-PEGMA)-g-aliphatic polyester

Sample	M_n (calc)[a]	M_n (osmo)	M_n (SEC)[b]	M_w/M_n	M_n (calc, branch)[c]	M_n (NMR, branch)[d]
P1	20,000	21,300	18,700	1.22	1,350	1,150
P2	99,600	73,300	55,100	1.67	13,200	13,600
P3	109,500	109,600	80,400	1.99	6,700	5,800
P4	148,700	97,800	89,400	1.75	15,250	15,100
P5	101,900	86,650	69,600	1.59	13,550	13,350

[a] $M_n(calc) = ((m_{Monomer}/m_{Terpolymer})+1) \times M_n(VPO)_{Terpolymer} \times conv.$

[b] As determined by SEC in THF with reference to PMMA standards.

[c] M_n calc, branch $= (M_n \, calc - M_n(VPO)_{Terpolymer})/n_{OH}.$

[d] M_n NMR, PCL branch $= ((I_{m+j+c} - 2/3 \, I_l) / (I_{n+f} - I_e)) \times M_{CL}$ (see Figure 3) and M_n NMR, PLA branch $= (I_h / (I_{f+m} - I_{j+c+e} + 2/3 \, I_l)) \times M_{LA}/2$ (see Figure 4).

Practically, the number-average molecular weight of the poly(ε-caprolactone) (PCL) and poly(L,L-lactide) (PLA) branches were determined from the relative intensities of the protons of the repetitive units (-C$\underline{H}_2$-O-CO- at 4.06 ppm for PCL and –C$\underline{H}$-O-CO- at 5.15 ppm for PLA) and the protons of the polyester hydroxyl end-groups (-C$\underline{H}_2$-OH at 3.63 ppm for PCL and –C$\underline{H}$-OH at 4.35 ppm for PLA) (Figures 3 and 4). It is worth mentioning that for poly(MMA-co-HEMA-co-PEGMA)-g-PCL, the assignment has been facilitated by the addition of a few drops of trichloroacetylisocyanate (TCAI) directly in the NMR tube. TCAI is known to quantitatively react with hydroxyl functions to form trichloroacetylcarbamate derivatives, thereby provoking a down-field shift of approximately 1 ppm for the α-hydroxymethylene protons. The corresponding α-trichloroacetylcarbamate methylene protons are thus observed at 4.30 ppm. It must also be emphasized that the chemical shift of terpolymer α-hydroxymethylene protons H_f from 3.85 (see Figure 2) up to 4.28 ppm (Figure 3), attests for the grafting efficiency, meaning that every pendant hydroxyl group of the poly(MMA-co-HEMA-co-PEGMA) precursors has actually initiated the polymerization of both CL and LA.

The kinetics of cyclic esters polymerization initiated with $Sn(Oct)_2$/macroalcohol system have been studied by measuring the monomer conversion by either dilatometry (CL) (20) or polarometry (LA) (21) as previously reported.

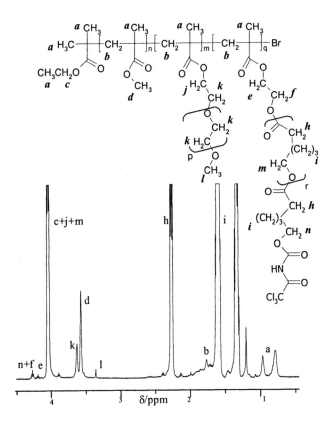

Figure 3. ^1H *NMR (500MHz) spectrum of poly(MMA-co-HEMA-co-PEGMA)-g-poly(ε-caprolactone) with the polyester ω-hydroxyl end-groups transformed into trichloroacetylcarbamate (sample P3 in Tables III and IV).*

Figure 5 shows semilogarithmic plots for various polymerizations carried out in THF at 80 °C with poly(MMA-co-HEMA-co-PEGMA) as multifunctional macroinitiator activated by Sn(Oct)$_2$. For the sake of comparison, the kinetics of LA polymerization initiated by n-butyl alcohol ([BuOH]$_0$ = 1.5 10^{-2}molL^{-1}) in the presence of Sn(Oct)$_2$ ([Sn(Oct)$_2$]$_0$ = 5 10^{-3} molL^{-1}) is also presented. After an induction period, linear time dependencies of ln([M]$_0$/[M]) are observed whatever the (macro)initiator and monomer which demonstrate that both CL and LA polymerizations proceed without significant change in the number of active species.

292

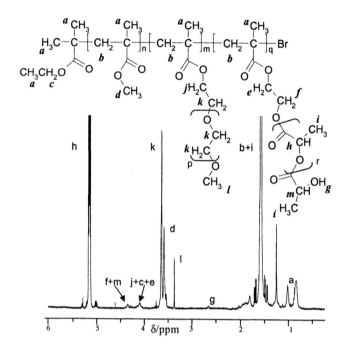

Figure 4. ^{1}H *NMR (500 MHz) spectrum of poly(MMA-co-HEMA-co-PEGMA)-g-poly(L,L-lactide) (sample P4 in Tables III and IV).*

In the BuOH and T3 coinitiated LA polymerizations ((P5 in Tables III and IV), starting concentrations of LA ($[LA]_0 = 1$ moll^{-1}), Sn(Oct)$_2$ ($[Sn(Oct)_2]_0 = 5$ 10^{-3} moll^{-1}) and hydroxyl groups ($[OH]_0 = 1.5$ 10^{-2} moll^{-1}) are identical but the polymerization rate for T3 is significantly depressed. As all OH groups pendant along the polymethacrylic backbone have proved to initiated the lactone polymerization, the decreased polymerization rate can be attributed to a lower accessibility of the OH groups in the macroinitiator, and then in the growing polyester pendant chains, in perfect agreement with previously reported data involving poly(MMA-co-HEMA) macroinitiator (*13*). It has to be stressed that the specific effects of the hydroxyl groups concentration, the macroinitiator composition and the molar mass on the polymerization rate have not been investigated in the present contribution, the main issues being the controlled synthesis of poly(methacrylate)-g-[poly(ester)/poly(ether)] graft copolymers and their use as potential tensioactive agents.

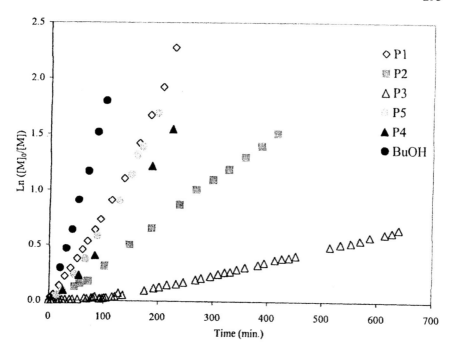

Figure 5. Kinetics of LA and CL polymerizations initiated by Sn(Oct)$_2$/ROH system. Experimental conditions are listed in Table III except for Sn(Oct)$_2$/BuOH initiated LA polymerization ([LA]$_0$ = 1 molL^{-1}, [BuOH]$_0$ = 1.5 10^{-2} molL^{-1} and [Sn(Oct)$_2$]$_0$ = 5 10^{-3} molL^{-1}).

However, under the aforementioned conditions, it must be noticed that molar masses distributions can be relatively broad with polydispersity indices which can be as high as 2 (sample P3 in Table IV). Smaller values were obtained when CL was substituted for LA (P2 and P3 in Tables III and IV) and the initial hydroxyl/Sn(Oct)$_2$ molar ratio increased up to 15 (P1 in Tables III and IV). The broadening of molecular weight distribution is thus likely consistent with the occurrence of transesterification reactions that can be prevented, or at least limited, by using lower (catalytic) amounts of activators as exemplified hereafter.

In this respect, the ROP of CL starting from poly(MMA-co-HEMA-co-PEGMA) terpolymers has been carried out in toluene at 60 or 110 °C by varying the nature and the relative content of the activator, i.e. Al(Et)$_3$ or Sn(Oct)$_2$. Tables V and VI show the experimental conditions and the molecular characteristics of the synthesized graft terpolymers. As well-known, hydroxyl groups readily react with Al(Et)$_3$ to form mono-, di-, and/or trialkoxides depending on the reaction stoechiometry. In the current study, when Al(Et)$_3$ was

Table V. Conditions for the synthesis of poly(MMA-co-HEMA-co-PEGMA)-g-poly(ε-caprolactone) in toluene using either $Sn(Oct)_2$ or $AlEt_3$ as activators (X).

Sample	MI^a	$[M]_0$ $(mol.L^{-1})$	$[OH]_0^b$ $(mol.L^{-1})$	X	$[X]_0/$ $[OH]_0$	T $(°C)$	Time (min)	Conv.^{c} $(\%)$
P6	T4	2.03	0.0140	$Al(Et)_3$	1.1	60	40	95
P7	T4	2.01	0.0139	$Al(Et)_3$	1/21	60	930	97
P8	T4	3.53	0.0832	$Sn(Oct)_2$	1/105	110	861	65
P9	T4	7.78	0.0901	$Sn(Oct)_2$	1/231	110	946	64
P10	T5	7.76	0.0961	$Sn(Oct)_2$	1/232	110	946	85

^{a}MI = macroinitiator ^{b} $[OH]_0 = (1000$ x $m_{Terpolymer}$ x $n_{OH})/(M_n(VPO)_{Terpolymer}$ x total solution volume (ml)). ^{c} As determined by gravimetry

Table VI. Molecular characteristics of poly(MMA-co-PEGMA-co-HEMA)-g-poly(ε-caprolactone)

Sample	$M_n (calc)^a$	$M_n SEC^b$	M_w/M_n^b	M_n $(calc,$ $branch)^c$	$M_n (NMR$ $branch)^d$
P6	124,500	105,800	2.25	15,700	ND
P7	126,500	175,700	1.37	16,000	11,500
P8	21,200	30,300	1.32	1,200	1,400
P9	58,500	95,300	1.32	6,400	7,200
P10	89,500	181,900	1.29	7,900	ND

^{a} M_n (calc) = $M_n(VPO)_{terpolymer}$ + (n_{OH} x M_n calc branch^c). ^{b} As determined by SEC in THF with reference to PMMA standards. ^{c} M_n (calc, branch) = ([CL]$_0$/[OH]$_0$) x conv x M_{CL}. ^{d} M_n (NMR, branch) = [($I_{PCL\ 2.3ppm}$ x 3 x n_{PEGMA})/($I_{PEGMA\ 3.35ppm}$ x 2 x n_{OH})] x M_{CL}.

used in slight excess with respect to the number of hydroxyl groups along the polymethacrylic backbone ([AlEt$_3$]/[OH] = 1.1), the size exclusion chromatogram displayed a rather broad molecular weight distribution (M_w/M_n = 2.25) (P6 in Tables V and VI).

As previously reported by some of us (23, 24), such a behavior can be explained by the formation of ethylaluminum dialkoxide together with the expected diethylaluminum monoalkoxide, particularly when the hydroxyl functions are localized close to each other. As a result, two different active species coexist and can promote the growth of polyester grafts with different kinetics, which could lead to very broad molecular weight distribution. In contrast, when a catalytic amount of Al(Et)$_3$ ([AlEt$_3$]/[OH] =4.8 10^{-2}) is used to initiate the ROP (P7 in Tables V and VI), the polymerization of CL is slower but the molar masses distribution is narrower (M_w/M_n = 1.37).

Such a behavior can be explained by a faster alcohol-alkoxide interchange reaction than the propagation so that one type of active aluminium alkoxide actually initiates and propagates the ROP.

Similarly, the polymerization of CL has been carried out in toluene at 110 °C in the presence of a catalytic amount of $Sn(Oct)_2$ ($4.3 \ 10^{-3} \leq [Sn(Oct)_2]_0/[OH]_0 \leq 9.5 \ 10^{-3}$). Under such experimental conditions, the resulting graft copolymers show narrow molecular weight distributions and molar masses in good agreement with the initial monomer-to-macroinitiator molar ratios (P8-10 in Tables V and VI). Note that similar observations have already been reported by Hedrick and al. in the frame of hyperbranched polyester synthesis (25).

Interfacial Tension Measurements

In order to get a better insight into the potential of poly(MMA-co-HEMA-co-PEGMA)-g-aliphatic polyester graft terpolymers as surfactants, preliminary dynamic interfacial tension experiments have been performed using the pendant-drop method. Typically, a drop of water hanging on the tip of a capillary has been immersed into a quartz cell containing copolymer solutions in toluene. Interfacial tension (IFT) has been determined from the shape of the pendant drop by using the Gauss-Laplace equation. Figure 6 shows the concentration dependence of the IFT for two copolymers : a poly(MMA-co-HEMA-co-PEGMA) terpolymer with HEMA and PEGMA molar fractions of 7.9 and 11.0 %, respectively (T4 in Tables III and IV) and a poly(MMA-co-HEMA-co-PEGMA)-g-PCL graft copolymer characterized by a weight fraction in polyester chains of 39 % (P8 in Tables V and VI). Interestingly, both copolymers show a critical micelle concentration (CMC) and substantially decrease the initial IFT of water in toluene. The dependence of the CMC on the copolymers composition ($6.4 \ 10^{-4} \ gL^{-1}$ for T4 and $3.2 \ 10^{-3} \ gL^{-1}$ for P8) is consistent with the formation of hydrophilic domains as the driving force for reverse micellization in apolar organic solution. When the initial copolymer concentration increases and approaches $0.1 \ gL^{-1}$, the IFT drops from 37.0 to 1.3 mNm^{-1} for T4 while it reaches 10.0 mNm^{-1} for P8. Such a difference can be explained by the higher steric hindrance and lower flexibility of the graft copolymer compared to the amphiphilic poly(MMA-co-HEMA-co-PEGMA) terpolymer. A thorough study of the dynamic interfacial properties as a function of the copolymer nature and composition is under current investigation.

Conclusions

A two-step procedure has been developed which allows the controlled synthesis of amphiphilic poly(methacrylate)-g-[(polyester)/poly(eher)] graft copolymers. It consists of the atom transfer radical polymerization (ATRP) of methyl methacrylate (MMA), 2-hydroxyethyl methacrylate (HEMA) and

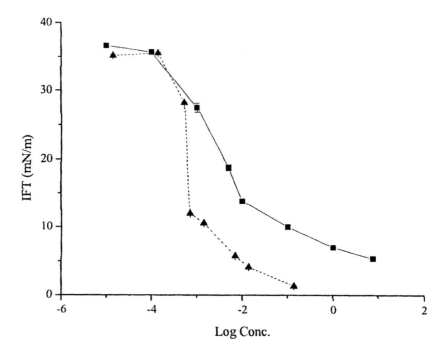

Figure 6. Semi-logarithmic concentration (gL⁻¹) dependence of the interfacial tension (IFT) for T4 terpolymer (▲) (Tables III and IV) and P8 graft copolymer (■) (Tables V and VI)

poly(ethylene glycol) methacrylate (PEGMA) to form poly(MMA-co-HEMA-co-PEGMA) terpolymers, followed by the ring-opening polymerization (ROP) of ε-caprolactone (CL) or L,L-dilactide (LA) initiated by the hydroxyl groups of the HEMA units, previously activated by $Al(Et)_3$ or $Sn(Oct)_2$.

In the first step, a preferential incorporation of the functional monomers, i.e. HEMA and PEGMA, within the terpolymer have been evidenced which would privilege the formation of a bottle-brush architecture with hydrophilic bristles rather than a comb-like structure. The second step of the synthesis has highlighted that the ROP of LA and CL can be advantageously performed by activating the hydroxyl groups along the poly(MMA-co-HEMA-co-PEGMA) terpolymers with a catalytic amount of either $Sn(Oct)_2$ or $Al(Et)_3$.

Indeed, graft terpolymers with controlled molecular weight and narrow molecular weight distribution can be prepared under such prevailing conditions. Preliminary interfacial tension experiments have been undertaken and give evidence for the amphiphilic character of the investigated (graft) terpolymers.

Acknowledgements

This work was realized partly in the frame of the PAN (Poland) and FNRS (Belgium) cooperation agreement. The financial supports of European Commission within the 5-th Framework Program -Contract ICA1-CT-2000-70021-, and the *Région Wallonne and Fonds Social Européen* in the frame of *Objectif 1-Hainaut : Materia Nova* program, are also gratefully acknowledged. LCPM thanks the *Services Fédéraux des Affaires Scientifiques, Techniques et Culturelles* for general support in the frame of the PAI-5/03. I. Ydens is grateful to F.R.I.A. for her PhD grant.

References

1. Xie, H.; Zhou, P. *Adv. Chem. Ser.* **1986**, *211*, 139-150.
2. Baines, F.L.; Dionisio, S.; Billingham, N.C.; Armes, S.P. *Macromolecules* **1996**, *29*, 3096-3102.
3. Rutot, D.; Duquesne, E.; Ydens, I.; Degée, Ph.; Dubois, Ph. *Polymer Degrad. Stab.* **2001**, *73*, 561-566.
4. Vulic, I.; Okano, T.; Kim, S.W.; Feijen, J. *J. Polym. Sc. [A] Polym. Chem.* **1988**, *26*, 381-391.
5. Wesslen, B.; Kober, M.; Freij-Larsson, C.; Ljungh, A.; Paulsson, M. *Biomaterials* **1994**, *15*, 278-284.
6. Dubois, Ph.; Jérôme, R.; Teyssié, Ph. In *Synthesis of Polymers : Materials Science and Technologies – a Comprehensive Treatment*; Schlüter, A.D., Ed.; Wiley **1999**; pp 196-229.
7. Puskas, J.E.; Kaszas, G., *Prog. Polym. Sci.* **2000**, *25*, 403-452.
8. Kamagaito, M.; Ando, T.; Sawamoto, M. *Chem Rev.* **2001**, *101*, 3689-3745.
9. Matyjaszewski, K.; Xia, J. *Chem Rev.* **2001**, *101*, 2921-2990.
10. Hawker, C. J.; Bosman, A.W.; Harth, E. *Chem. Rev.* **2001**, *101*, 3661-3688.
11. Yagci, Y., Reetz, I., *React. Funct. Polym.* **1999**, *42*, 255-264.
12. Mecerreyes, D.; Jérôme, R.; Dubois, Ph. *Adv Polym. Sci.* **1999**, *147*, 1-59.
13. Ydens, I.; Degée, Ph.; Dubois, Ph.; Libiszowski, J.; Duda, A.; Penczek, S. *Macromol. Chem. Phys, in press*
14. Hawker, C.J.; Hedrick, J.L.; Malmström, E.E.; Trollsås, M.; Mecerreyes, D.; Dubois, Ph.; Jérôme, R. *Macromolecules* **1998**, *31*, 213-219.
15. Mecerreyes, D.; Moineau, G.; Dubois, Ph.; Jérôme, R.; Hedrick, J.L.; Hawker, C.J.; Malmström, E.E.; Trollsås, M. *Angew. Chem., Int. Ed. Engl.* **1998**, *37*, 1274-1276.
16. Arashiro, E.Y.; Demarquette, N.R. *Materials Research* **1999**, *2*, 23-32.

298

17. Ydens, I.; Degée, Ph.; Dubois, Ph.; Libiszowski, J.; Duda, A.; Penczek, S. *Polym. Prepr.(Am. Chem. Soc., Div. Polym. Chem.)* **2002**, *43 (2)*, 38-39.
18. Wang, J.-S.; Matyjaszewski, K. *Macromolecules* **1995**, *28*, 7572-7573.
19. Beers, K.L.; Boo, S.; Gaynor, S.G.; Matyjaszewski, K. *Macromolecules* **1999**, *32*, 5772-5776.
20. Kowalski, A.; Duda, A.; Penczek, S. *Macromol. Rapid Commun.* **1998**, *19*, 567-572.
21. Kowalski, A.; Duda, A.; Penczek, S. *Macromolecules* **2000**, *33*, 7359-7370.
22. Kowalski, A.; Duda, A.; Penczek, S. *Macromolecules* **2000**, *33*, 689-695.
23. Dubois, Ph.; Degée, Ph.; Jérôme, R.; Teyssié, Ph. *Macromolecules* **1993**, *26*, 2730-2735.
24. Ydens, I.; Rutot, D.; Degée, Ph.; Six, J-L.; Dellacherie, E.; Dubois, Ph. *Macromolecules* **2000**, *33*, 6713-6721.
25. Trollsås, M.; Hedrick, J.L.; Mecerreyes, D.; Dubois, Ph.; Jérôme, R.; Ihre, H.; Hult, A. *Macromolecules* **1998**, *31*, 2756-2763.

Chapter 21

New Materials Using Atom Transfer Radical Polymerization: Microcapsules Containing Polar Core Oil

Mir Mukkaram Ali and Harald D. H. Stöver

Department of Chemistry, McMaster University, 1280 Main Street West, Hamilton, Ontario L8S 4M1, Canada

A technique for encapsulation of polar organic solvents using Atom Transfer Radical Polymerization (ATRP) by suspension polymerization is presented. ATRP was used to synthesize crosslinked amphiphilic copolymers based on oil soluble and water soluble comonomers by suspension polymerization. We demonstrate that the slow rate of ATRP crosslinking reactions allows enough time for polymer to migrate in the oil droplets from the interior to the oil-water interface, yielding the thermodynamically favored capsular morphology. This finding is relevant for the encapsulation of polar core oils where the driving force for polymer migration to the oil-water interface is weak. Using less polar xylene as core oil we have shown that polymer architecture can strongly influence its solubility in the oil phase and therefore the morphology of the suspension polymer particles.

Polymeric capsules and hollow particles can be prepared both from monomeric starting materials as well as from oligomers and pre-formed polymers (*1*). In most cases, the process involves a disperse oil phase in an aqueous continuous phase, and the precipitation of polymeric material at the oil – water interface causing each oil droplet to be enclosed within a polymeric shell. Interfacial polycondensation is used to prepare poly(urea) (*2*), poly(amide), or poly(ester) capsules (*3*), for instance, by reaction between an oil soluble monomer and a water soluble monomer at the oil – water interface. On the other hand, vinyl polymers such as poly(styrene), acrylates and methacrylates prepared by free radical polymerization under suspension (*4*) or emulsion (*5,6*) conditions have been used to prepare hollow or capsular polymer particles. In this approach, the dispersed oil phase usually serves as the polymerization solvent. The oil phase is chosen so as to be a good solvent for the monomeric starting materials but a non-solvent for the product polymer. Therefore, upon polymerization the system is comprised of three mutually immiscible phases. Sundberg *et. al.* (*7*) published a theoretical model based on the Gibbs free energy change of the process of morphology development when three immiscible phases are brought together. Using oil, polymer and water as three immiscible phases they showed that the Gibbs free energy change per unit area for the process leading to a core shell morphology (with oil encapsulated within the polymer phase), is given by $\Delta G = \gamma_{op} + \gamma_{pw} (1 - \phi_p)^{-2/3} - \gamma_{ow}$. Where γ_{op}, γ_{pw}, and γ_{ow} are the oil-polymer, polymer-water and oil-water interfacial tensions and ϕ_p is the volume fraction of the polymer (in polymer plus oil "combined phase"). In the limit as ϕ_p tends to zero, $\Delta G = (\gamma_{op} + \gamma_{pw})- \gamma_{ow}$. Thus, when $\gamma_{ow} > (\gamma_{op} + \gamma_{pw})$, the core shell morphology with the core oil being engulfed by the polymer is the thermodynamically stable morphology. Using these expressions the authors were able to predict the expected morphologies for a given set of interfacial conditions. The predictions were checked and confirmed by experiment.

The work of the four major research groups in the area, i.e., Kasai *et. al.* (*8,9*), Okubo *et. al.* (*10–16*), McDonald *et. al.* (*5,6*), and Sundberg *et. al.* (*7*), shows that present techniques allow only the encapsulation of relatively hydrophobic solvents. McDonald *et. al.* and Sundberg *et. al.* have encapsulated highly non-polar core oils such as decane and octane. Kasai *et. al.* and Okubo *et. al.* have encapsulated slightly more polar materials such as benzene, toluene and xylene. Since these groups set out to synthesize hollow polymer particles, the nature of the core oil has not been of relevance. However, if core-shell particles are intended for encapsulation of the core material, then it becomes desirable that the technique allows encapsulation of both hydrophobic and hydrophilic core materials.

Core-shell particles, with polymer engulfing an oil core, only form if the sum of the oil/polymer and polymer/water interfacial tensions is less than the oil/water interfacial tension. Consequently, encapsulation of polar oil requires sufficiently amphiphilic polymers to satisfy this interfacial requirement. Furthermore, the driving force for polymer migration to the oil-water interface is weak when the core oil is polar compared to non-polar core oils. Thus, polymer migration to the oil water interface may require more time, introducing a kinetic element to the encapsulation of polar oil. In this work we have demonstrated the advantages of using a living polymerization for the encapsulation of polar oils. The use of a living polymerization method is expected to have two interesting consequences. First, using a living polymerization it should be possible to prepare amphiphilic copolymers based on oil soluble and a water soluble comonomers by suspension polymerization. This hypothesis is based on the reasoning that despite the partitioning of the polar comonomer between the two phases, its consumption in the organic phase should continuously drive this monomer back into the oil phase. Since all polymer chains remain alive throughout the polymerization, the polar monomer would be incorporated statistically into all polymer chains imparting the required polarity to the copolymer. Second, the relatively slow rate of ATRP crosslinking reactions (*17,18*) should allow more time for polymer migration to the oil water interface, thereby helping to meet the above mentioned kinetic requirement in the encapsulation of polar solvents. Our choice of ATRP for living polymerization (*19*) is based on its versatility and success in both homogeneous and heterogeneous aqueous systems (*20*).

We describe the *in situ* synthesis of an amphiphilic co-polymer, poly(methyl methacrylate-*co*-poly(ethylene glycol monomethyl ether) methacrylate)) P(MMA-co-PegMA) by atom transfer radical polymerization (ATRP) in suspension polymerization conditions using diphenyl ether as a polar oil phase. The crosslinking of this copolymer, with diethylene glycol dimethacrylate (DegDMA) yields capsular polymer particles at a PegMA content of 31 mol %. The same crosslinked terpolymer when prepared using conventional free radical polymerization yielded matrix type particles. We attribute this difference to the slow crosslinking reaction in ATRP relative to conventional free radical polymerization, which permits sufficient time for polymer migration to oil-water interface. When a less polar solvent, xylene, is used as the oil phase, the driving force for polymer migration to the oil water interface is strong and PMMA crosslinked with DegDMA is sufficiently polar to yield capsular particles. Hence, the thermodynamically favored morphology in this case is capsular even in the absence of the polar comonomer. Furthermore, we have found that the solubility of this crosslinked polymer formed in xylene depends on the degree of polymerization of the PMMA at the time of crosslinker addition. If PMMA prepolymers with a low degree of polymerization are crosslinked, they precipitate early yielding the thermodynamically favorable capsular morphology. However, crosslinking of PMMA prepolymers of a higher degree of polymerization yields star like microgels that are sterically stabilized by the linear preformed polymer chains. Due to the high solubility of star like microgels in the oil phase, there is no tendency for polymer migration to the oil-water interface. This work therefore illustrates that the slow crosslinking in ATRP favors encapsulation of polar core oils where the driving force for polymer migration to the oil water interface is weak. Also, we have shown that the crosslinked polymer architecture affects microgel colloidal stability in the oil phase and the resulting suspension particle morphology.

Results (21) and Disussion

Copolymer synthesis by solution polymerization. The ATRP synthesis of poly(methyl methacrylate-*co*-poly(ethylene glycol monomethyl ether) methacrylate)) (P(MMA-co-PegMA)) was first developed in solution polymerization conditions. A series of copolymers comprised of methyl methacrylate (MMA) and 9.5, 18, and 39 mol% poly(ethylene glycol monomethyl ether) methacrylate (PegMA) were prepared in 25% solutions of diphenyl ether (DPE, a polar water immiscible solvent, solubility parameter δ = 20.9 $MPa^{1/2}$). Also, PMMA and the copolymer containing 18 mol% PegMA were prepared by ATRP in xylene (a model non-polar water immiscible solvent, solubility parameter δ = 18.0 $MPa^{1/2}$). DPE is a good ATRP solvent owing to its low chain transfer constant and its use for the ATRP of methacrylates has been reported previously (22). Toluene sulfonyl chloride (TSC) was used as initiator together with a catalyst based on Cu(I)Br and 4,4'-dinonyl-2,2'-bipyridine (dNBpy), and the polymerizations were run at 70 °C. The modified bipyridine ligand ensures catalyst solubility in diphenyl ether and xylene as well as favorable partitioning of the catalyst into the oil phase during suspension polymerizations (23).

Results of the ATRP synthesis of PMMA and the copolymers containing 9.5, 18, and 39 mol% PegMA in DPE are shown in Figures 1, 2 and 3. Polymerizations proceeded to high conversions in each case and polydispersities remained low confirming the living character of the polymerization. We attribute the difference between the experimental molecular weight (M_n(SEC)) and the theoretical molecular weight (M_n(Theo)) of the copolymer containing 39 mol % PegMA to structural differences between the SEC calibration standard i.e. poly(styrene), and the copolymer. Figures 4 and 5 show the results of ATRP synthesis of PMMA and the copolymer containing 18 mol% PegMA in xylene. The results point to the living nature of the polymerization.

Copolymer synthesis by suspension polymerization. Next, the ATRP synthesis of P(MMA-co-PegMA) was performed, in part, in suspension conditions. The purpose of this experiment was two-fold. First, to show that P(MMA-co-PegMA) can be prepared by ATRP in suspension polymerization conditions. Secondly, in our experiments, 20 wt % PegMA partitions into the water phase (21(a)), thus it was interesting to check if its incorporation in the copolymer occurs to the same extent as in the solution copolymerization. The copolymer syntheses were initiated in solution and allowed to proceed to 25 and

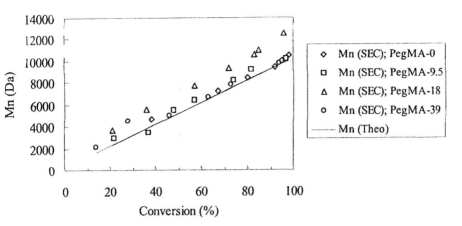

Figure 1. Plots of experimental molecular weights (M_n(SEC)) vs conversion for homo-polymerization of MMA and copolymerization of MMA and PegMA in DPE at 25 wt % total monomer loading and 70 °C. Homo-polymerization composition: [TSC]_o: [Cu(dNBpy)_2Br]_o : [MMA]_o = 1:1:100. Copolymerization compositions: [TSC]_o: [Cu(dNBpy)_2Br]_o : [MMA]_o: [PegMA]_o = 1:1:80:8 (9.5 mol% PegMA); 1:1:60:13 (18 mol % PegMA); 1:1:34:22 (39 mol % PegMA). (Reproduced from Macromolecules 2003. Copyright 2003 American Chemical Society.)

50% conversion in xylene and diphenyl ether, respectively. At these points, the solution polymerization mixtures were each transferred to a four-fold excess (by volume) of 1% aqueous PVA at 70 °C and mechanically stirred at 1000 rpm for 30 minutes and at 500 rpm thereafter to yield oil in water suspensions. Results of the suspension reaction for 25% monomer loading in oil phase are presented in Figure 6 and 7. Figure 6 shows time-conversion curves. Figure 7 shows that the molecular weight increased linearly with conversion and that the polydispersity remained below 1.2, suggesting good living character. Hence, the ATRP synthesis of the amphiphilic copolymer was achieved (with over 80% monomer conversion) in suspension polymerization conditions. The incorporation of PegMA in samples prepared by solution polymerization as well as suspension polymerization with DPE as oil phase containing 18 mol% PegMA in the monomer feed, was shown to be 17 mol % using Nuclear Magnetic Resonance Spectroscopy. This confirmed that PegMA was incorporated into the copolymer to the same extent in the suspension polymerization as it was in the corresponding solution polymerization.

Encapsulation of xylene and diphenyl ether. The encapsulation process consisted of three steps: synthesis of low molecular weight amphiphilic copolymers by solution ATRP; addition of the cross-linking monomer to the reaction solution followed by a 10 minute mixing period to ensure homogeneous distribution of the crosslinking monomer; and transfer of this oil phase to a four

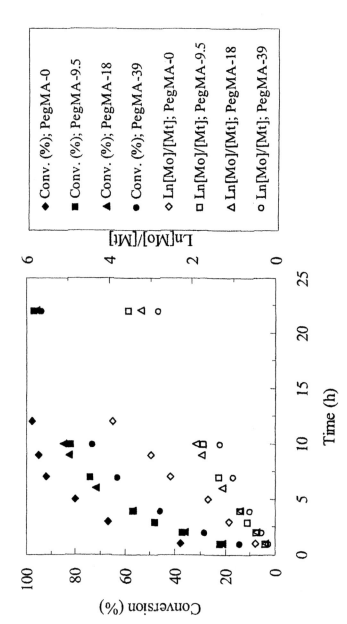

Figure 2. Kinetic profile for homo-polymerization of MMA and copolymerization of MMA and PegMA in DPE;conditions as in Figure 1. (Reproduced from Macromolecules **2003**. Copyright 2003 American Chemical Society.)

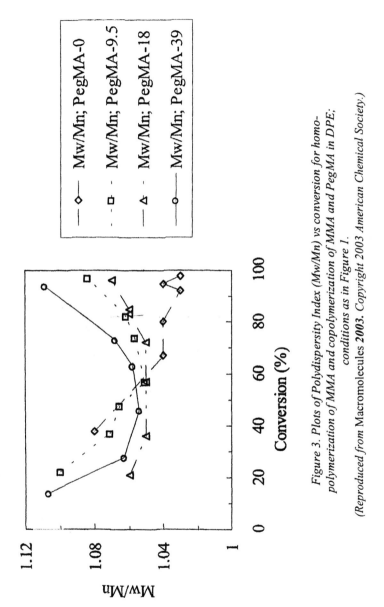

Figure 3. Plots of Polydispersity Index (Mw/Mn) vs conversion for homopolymerization of MMA and copolymerization of MMA and PegMA in DPE;
conditions as in Figure 1.
(Reproduced from Macromolecules 2003. Copyright 2003 American Chemical Society.)

306

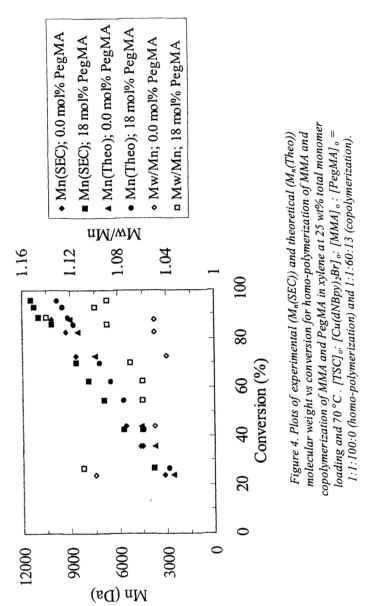

Figure 4. Plots of experimental (M_n(SEC)) and theoretical (M_n(Theo)) molecular weight vs conversion for homo-polymerization of MMA and copolymerization of MMA and PegMA in xylene at 25 wt% total monomer loading and 70 °C. [TSC]$_o$: [Cu(dNBpy)$_2$Br]$_o$: [MMA]$_o$: [PegMA]$_o$ = 1:1:100:0 (homo-polymerization) and 1:1:60:13 (copolymerization).

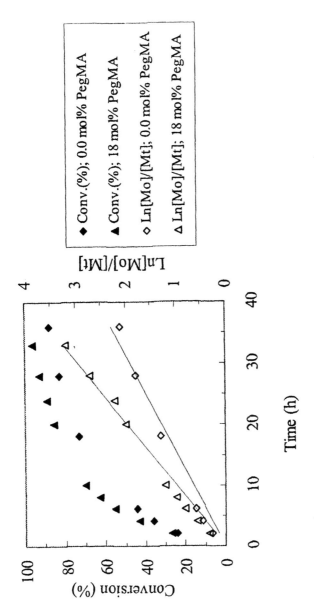

Figure 5. Kinetic profiles for homo-polymerization and copolymerization of MMA and PegMA in xylene; conditions as in Figure 4.

fold excess of 1-3 % (w/w) aqueous poly(vinyl alcohol) (PVA) in a baffled reactor. The resulting suspension was mechanically stirred with a propeller type mixer at 1000 rpm for 30 minutes to 2 hours and subsequently at 500 rpm to the end of the suspension polymerization. Table 1 gives details of the experimental formulations for suspension conventional free radical polymerizations (CFRP) and suspension ATRP for both xylene (entries 1-5) and DPE (entries 6-11). In entries 1 and 2, suspension polymer particles were prepared by cross-linking MMA with 20 mol% DegDMA in xylene oil phase using CFRP and ATRP, respectively. Both experiments yielded capsular particles showing that MMA crosslinked with DegDMA phase separates from xylene and migrates to the oil water interface due to the strong driving force for polymer migration in non-polar xylene. In experiments 2 and 3, the terpolymer composition and core solvent are identical, but the PMMA prepolymers were transferred from solution to suspension at 25% (X_n~25 (PMMA-25)) and 75% (Xn~75 (PMMA-75)), respectively. Hence, upon crosslinker addition and subsequent transfer to water phase, the forming cross-linked terpolymers differ in their architecture, the latter being sterically stabilized star like microgels. The resulting suspension polymer particles have starkly different morphologies. Exp. 2 gave capsular particles, while matrix particles formed in exp. 3. The Transmission Electron Micrographs (TEM) and Environmental Scanning Electron Micrographs (ESEM) of particles prepared with xylene as oil phase are given in Figure 8.

Table 1. Suspension polymerizations for encapsulation of xylene and diphenyl ether

Exp	Polym. method	Core sol.	Mole % PegMA	Mole % DegDMA	Conv. at trans.[5] (% w/w)	Final conv.[6] (% w/w)	Mrph.[7]
1	CFRP[1]	Xyl[3]	0	20	0	100	Cap[8]
2	ATRP[2]	Xyl	0	20	24	100	Cap
3	ATRP	Xyl	0	20	73	93	Mtx[9]
4	ATRP	Xyl	15	20	25	96	Cap
5	ATRP	Xyl	15	20	45	96	Mh[10]
6	CFRP	Dpe[4]	0	20	0	93	Mtx
7	CFRP	Dpe	31	20	0	100	Mtx
8	ATRP	Dpe	0	20	50	100	Mtx
9	ATRP	Dpe	8	20	35	98	Mtx
10	ATRP	Dpe	15	20	57	100	Mtx
11	ATRP	Dpe	31	20	46	99	Cap

[1]CFRP = Conv. free radical polym. ; [2]ATRP = Atom transfer radical polym.; [3]Xylene; [4]Diphenyl ether; [5] Conversion at time of transfer to suspension; [6] Final conversion to polymer; [7] Particle morphology as determined by ESEM and TEM; [8] Capsule; [9] Matrix; [10] Multi-hollow.

Source: Adapted from Macromolecules **2003**. Copyright 2003 American Chemical Society.

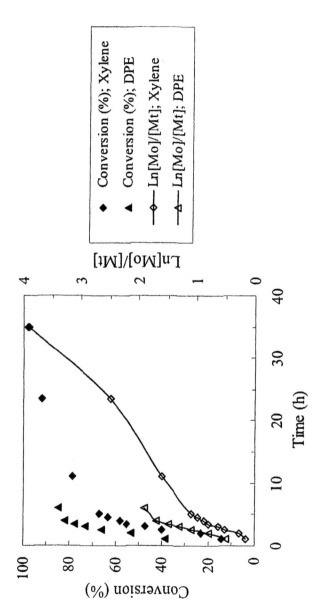

Figure 6. Kinetics of suspension copolymerization of P(MMA-co-PegMA).
$[MMA]_o : [PegMA]_o : [TSC]_o : [Cu(dNBpy)_2Br]_o = 30:7:1:1; 25\%$ *total monomer in xylene and diphenyl ether (DPE) oil phase; Water(1% PVA) : Oil = 4:1 at 70 °C. The solid lines are only meant to aid the eye.*

(Adapted from Macromolecules 2003. Copyright 2003 American Chemical Society.)

310

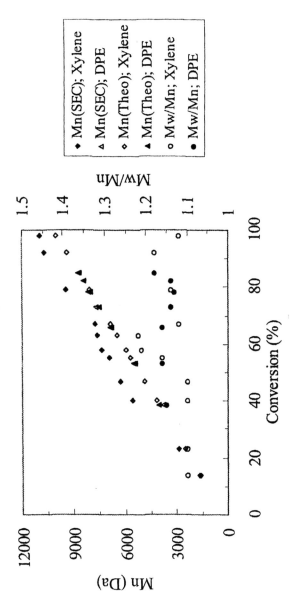

Figure 7. Plots of molecular weight and polydispersity versus conversion for suspension copolymerization of MMA and PegMA; conditions as in Figure 6. (Adapted from Macromolecules 2003. Copyright 2003 American Chemical Society.)

To rationalize this result we studied the crosslinking reaction in solution conditions and determined the conversion at gel point and the time elapsed before gel formation. The results are shown in Figure 9. An early sol-gel transition in the PMMA-25 case is suggestive of early formation of a macroscopic network, and since in the encapsulation experiments polymer phase separation in the oil phase is driven by crosslinking, it is fair to assume that in this case polymer precipitates early, at a low conversion, and subsequently migrates to the oil water interface forming the capsular wall. In the PMMA-75 case, macroscopic gel forms after 100% conversion, plausibly due to the high colloidal stability of the star like microgels. Thus, polymer phase separation occurs after a three dimensional crosslinked network has formed, physically preventing polymer migration to the oil water interface. That is, the thermodynamically stable morphology in either case is the capsular morphology but in the PMMA-75 case it is not achieved because the forming polymer is kinetically trapped in the entire volume of the oil droplet at the time of phase separation.

Mixtures of 85 mol% MMA and 15 mol% PEGMA were polymerized in xylene to 25 % (exp. 4) and 45% (exp. 5) conversion, prior to addition of DEGDMA and transfer to suspension conditions. A similar trend in the morphology was seen in this series (Figure 10). Despite the higher polarity of the crosslinked terpolymer in this series, multi-hollow particles were observed when Copoly-45 was used. Hence, these results illustrate the kinetic requirement for capsule formation that polymer precipitation must occur prior to formation of a crosslinked network that spans the entire volume of the oil droplet thereby allowing the precipitating microgels to migrate to the oil water interface and form the capsule wall. The TEM of the suspension polymer particles prepared with a DPE oil phase (experiments 6-11, Table 1) are shown in Figure 11. In this case, suspension particles prepared by either CFRP or ATRP and containing no PegMA, gave matrix type particles (Figure 11A and B). This illustrates that with DPE as core oil, PMMA crosslinked with DegDMA is not polar enough to yield the capsular morphology since upon phase separation in the oil phase the copolymer does not migrate compleately to the oil water interface. While crosslinked terpolymer containing 8 and 15 mol% water soluble monomer (Figure 11C and D) also yields matrix type suspension particles, such particles containing 31 mol% PegMA exhibit capsule morphology with either a single or several small hollow domains in the interior (Figure 11E), showing that at this composition, the crosslinked terpolymer is sufficiently polar to largely migrate to the oil-water interface upon phase separation in the oil droplets. The analogous CFRP suspension particles containing 31 mol% PegMA (Figure 11F) yielded matrix particles. We propose that the high crosslinking reaction rate in CFRP relative to ATRP allows no time for polymer migration to the oil water interface prior to the formation of a three dimensional network that spans the entire

312

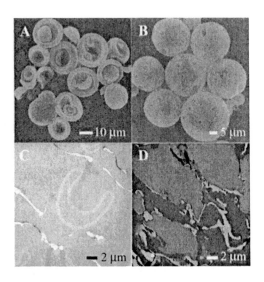

Figure 8. ESEM/TEM micro graphs showing surface and internal morphology of suspension polymerization particles prepared from P(MMA-co-DegMA) using xylene as oil phase and, PMMA-25 prepolymers (A and C) and PMMA-75 prepolymers (B and D).

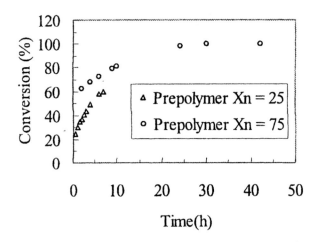

Figure 9. Conversion at gel point; and gel-time; for PMMA prep olymers crosslinked with Deg DMA in xylene solution.

313

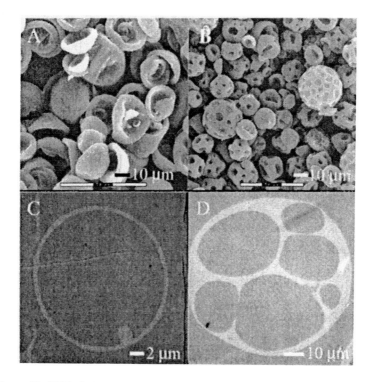

Figure 10. ESEM/TEM micrographs showing surface and internal morphology of suspension polymerization particles prepared from PMMA-co-PegMA-co-DegMA (65:15:20 mol %) using xylene as oil phase and PMMA-25 prepolymers (A and C) and PMMA-45 prepolymers (B and D).

volume of the oil droplet. Thus, the forming polymer is kinetically trapped in the oil droplet yielding the matrix morphology. The above results illustrate the thermodynamic requirement for encapsulation of the oil phase in the polymer phase i.e., for a relatively non-polar solvent (xylene) a less polar copolymer composition is sufficient to give capsular particles, while for a relatively polar oil phase, DPE, a more polar terpolymer composition is necessary to achieve the capsular morphology.

Conclusions

Encapsulation by suspension polymerization is based on the *in situ* synthesis of amphiphilic copolymers that are more hydrophilic than the oil being encapsulated. Thus as the polymer forms and precipitates in the oil droplets it migrates to the oil-water interface replacing the oil-water interface with two new interfaces, i.e., the oil-polymer and polymer-water interface thereby minimizing the interfacial energy. Our results with polar diphenyl ether core oil show that encapsulation of polar core-oil requires the synthesis of sufficiently amphiphilic

314

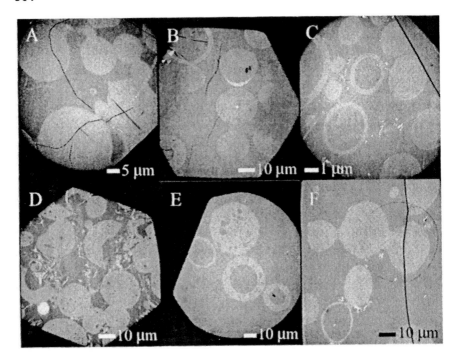

*Figure 11. TEM micrographs showing internal morphology of suspension
particles using DPE as oil phase: A. PegMA-0 particles (conv.= 90%) prepared
by CFRP. B. PegMA-0 particles (conv.= 99%) prepared by ATRP. C. PegMA-8
particles (conv.= 98%) prepared by ATRP. D. PegMA- 15 particles (conv.=
98%) prepared by ATRP. E. PegMA- 31 particles (conv.= 92%) prepared by
ATRP. F. PegMA- 31 particles (conv.= 100%) prepared by CFRP.*
(Reproduced from Macromolecules *2003.* Copyright 2003 American Chemical Society.)

polymers (31% PegMA content for DPE). Also, if polymer precipitation in the
oil phase is crosslinking induced, then the rate of the crosslinking reaction (and
consequently the rate of precipitation) must be lower than the rate of polymer
migration to the oil water interface. This condition is particularly important for
polar core oils where the driving force for polymer migration to the oil water
interface is weak and the polymer migration rate is low. Hence, for polar core
oils the use of ATRP as the polymerization method favors capsule formation,
owing to the slow rate of ATRP crosslinking reactions. Furthermore, our results
using non-polar xylene core oil, show that the architecture of the crosslinked
polymer stongly affects its solubility. Higher solubility of star like microgels in
the oil phase delays polymer precipitation. When sufficient inter-microgel
crosslinks to induce polymer precitation have formed, a three dimensional
polymer network spanning the entire volume of the oil droplet is in place,
thereby causing the formation of matrix particles.

References

1. Arshady, R. *Microspheres, Microcapsules Liposomes* **1999**, 1, 1461-1732.
2. Beestman, G. B.; Deming, J. M. U.S. Patent No. 4, 417, 916, Nov. 29, 1983.
3. Arshady, R. *J. Microencapsulation* **1989**, Vol. 6, No. 1, 13-28.
4. Kasai, K.; Hattori, M.; Takeuchi, H.; Sakurai, N. U.S. Patent No. 4, 908, 271, Mar.13, 1990.
5. McDonald, C; Chonde, Y.; Cohrs, W. E.; MacWilliams, D. C. U.S. patent No. 4, 973, 670, Nov. 27, 1990.
6. McDonald, C. J.; Bouck, K. J.;Chaput, A. B.; Stevens, C. J.; *Macromolecules* **2000**, 33, 1593-1605.
7. Sundberg, D. C.; Casassa, A. P.; Pantazopoulos, J.; Muscato, M.R. *J. Appl. Polym. Sci.* **1990**, 41, 1425.
8. Kasai, K.; Hattori, M.; Takeuchi, H.; Sakurai, N. US Patent: 4,798,691 (Jan. 17, 1989).
9. Itou, N.; Masukawa, T.; Ozaki, I.; Hattori, M., Kasai, K.; *Colloids Surf. A: Physiochem. and Eng. Aspects* **1999**, 153, 311-316.
10. Okubo, M.; Minami, H.; Yamashita, T. *Macromol. Symp.* **1996**, 101, 509.
11. Okubo, M.; Minami, H.; *Colloid Polym. Sci.* **1996**, 274, 433-438.
12. Okubo, M.; Minami, H. *Colloid Polym. Sci.* **1997**, 275, 992-997.
13. Okubo, M.; Konishi, Y.; Minami, H. *Colloid Polym. Sci.* **1998**, 276, 638.
14. Okubo, M.; Konishi, Y.; Minami, H. *Colloid Polym. Sci.* **2000**, 278, 659-664.
15. Okubo, M.; Konishi, Y.; Minami, H. *Colloid Polym. Sci.* **2001**, 279, 519-523.
16. Okubo, M.; Konishi, Y.; Inohara, T; Minami, H. *Macromol. Symp.* **2001**, 175, 321-328.
17. Yu, Q.; Zeng, F.; Zhu, S.; *Macromolecules* **2001**, 34, 1612-1618.
18. Jiang, C.; Shen, Y.; Zhu, S.; Hunkeler, D.; *J. Polym. Sci., Part A: Polym. Chem.* **2001**, 39, 3780-3788.
19. Wang, J.S.; Matyjaszewski, K. *J. Am. Chem. Soc.* **1995**, 117, 5614-5615.
20. Qiu, J.; Charleux, B.; Matyjaszewski, K. *Prog. Polym. Sci.* **2001**, 26, 2083-2134.
21. Experimental details of the results presented here may be found in: (a) Ali, M.M.; Stöver, H. D. H.; Accepted for publication in *Macromolecules*, and (b) Ali, M.M.; Stöver H.D.H.; *Polymer Preprints* 2002, 43 (2), 59 – 60.
22. Wang, J.; Grimaud, T.; Matyjaszewski, K.; *Macromolecules* **1997**, 30, 6507-6512.
23. Gaynor, S. G.; Qiu, J.; Matyjaszewski, K.; *Macromolecules* **1998**, 31, 5951-5954.

Chapter 22

Synthesis of Poly(*N,N*-Dimethylacrylamide) Brushes from Functionalized Latex Surfaces by Aqueous Atom Transfer Radical Polymerization

Jayachandran N. Kizhakkedathu[1], Diane Goodman[2], and Donald E. Brooks[1,2,*]

Departments of [1]Pathology and Laboratory Medicine and [2]Chemistry, University of British Columbia, Vancouver, British Columbia V6T 2B5, Canada

Negatively charged, ATRP initiator functionalized polystyrene latexes were synthesized and N,N-dimethylacrylamide was polymerized from their surfaces by aqueous ATRP using several catalyst combinations. Very high grafting densities and molecular weights were achieved with good polydispersities. The brushes exhibited a strong repulsive force that increased with molecular weight in atomic force microscopy measurements. Hydrodynamic thicknesses were consistent with theory. Polymerization from these surfaces is very different from ATRP in solution and depends on monomer concentration in solution, type of catalyst, and surface initiator concentration. The observations were explained with an electrostatic model of the surface region that predicts an increase in the local concentration of Cu(I) and Cu(II) catalyst complexes.

Introduction

Surfaces bearing high concentrations of tethered polymer, known as polymer brushes, have relevance in many fields including reduction of protein adsorption, polymer electronics, and wetting control and colloidal stability (*1,2*). The properties of the polymer brush system can be changed by varying the molecular weight and grafting density of the chains on the surface. These two parameters, which determine the crowding of the polymer chains, are very important for a polymer brush system (*3*). Several methods such as cationic (*4a*), anionic (*4b*), free radical (*4c*) and living free-radical polymerization (*2,5,6*) have been used for the synthesis of polymer brushes from different materials. Although there are a number of publications in this area, there is no clear understanding of the mechanism and nature of the chain growth from different surfaces. Among the methods used for surface grafting, ATRP is of particular interest because of its tolerance to a wide variety of reaction conditions, monomers and substrates (*5c,6*).

The polystyrene latex (PSL) surface is a good model system (*6d,7*) for studying surface initiated polymerization due to the large surface area, narrow size distribution and the synthetic control over surface properties offered by surfactant-free latex synthesis. In this contribution we report the surface initiated aqueous ATRP from these latexes with different initiator densities to synthesize polymer brushes with varying molecular weight and grafting density. We also studied the effect of various reaction parameters on $\overline{M}_n$ and $\overline{M}_w/\overline{M}_n$ and on grafting density. A model for the surface is also outlined to explain the unusual polymerization behavior as being due to charges on the surface and catalyst and restricted diffusion of monomer to the reaction site due to the hydrophobic nature and high concentration of copolymer chains near the surface.

We also studied the compression of these grafted layers by atomic force microscopy (AFM) in aqueous environment. The technique, which was initially developed to image surface topography (*8*), has become an increasingly common method of probing interaction forces exerted by grafted polymer layers (*9-12*). In this work we have examined the hydrated structure of brushes of varying molecular weight and grafting density.

Experimental

All commercial reagents were purchased from Aldrich and used without further purification unless noted. N,N,N′,N′,N″-pentamethyldiethylene-triamine (PMDETA) (Aldrich 99%) and 1,1,4,7,10,10- hexamethyltriethylene-tetramine (Aldrich 97%) (HMTETA) were used as such. Tris[2-(dimethylamino)ethyl] amine (Me$_6$TREN) was prepared by one step synthesis (*13*) from tris-(2-amino- ethyl) amine (Aldrich 96%). Styrene (Aldrich 99%),

N,N-dimethylacrylamide (DMA) (Aldrich 99%) were purified by vacuum distillation and stored at -80°C until use. 2-(methyl 2'- chloropropionato) ethylacrylate (HEA-Cl) was synthesized and purified by our reported procedure (*6d*). All water used was purified using a Milli-Q Plus system (Millipore Corp., Bedford, MA). Two sets of narrowly dispersed polystyrene seed latexes were synthesized and characterized as in our earlier report (11g) and were used for shell-growth polymerization. The hydrodynamic sizes of the seed latexes were 509 and 551 nm and are given in Table I. ^{1}H NMR measurements were done on a Bruker 300 machine. Conductometric titrations were done with a YSI model 35 conductance meter and 3403 cell with platinum electrodes at 25°C.

Table I. **Reaction Conditions for the Functionalization of Latex Particles by Shell-Growth Polymerization**

Latex code	Seed latex (g) (3.33 wt%)	Styrene (mol)	HEA-Cl (mol) $\times 10^5$	KPS^a (mmol)	HD^b (nm)
L1	166	0.019	2.42	23	509
L2	166	0.016	9.71	23	509
L3	166	0.016	48.5	23	509
L4	166	0.016	97.5	23	509
L5	265	0.025	784	37	551

1. Potassium persulfate,
2. Hydrodynamic diameter of seed latex determined by particle size analyzer

Measurements of the hydrodynamic diameter distribution of particle suspensions utilized a Beckman Coulter N4 Plus multi angle particle size analyzer working on the PCS principle; conditions were described earlier (*6d*). Molecular weights were determined by gel permeation chromatography (GPC) on a Waters 2690 separation module fitted with a DAWN EOS multi-angle laser light scattering (MALLS) detector from Wyatt Technology Corp. (laser wavelength λ=690 nm) and a refractive index detector from Viscotek Corp operated at λ=620 nm. Instrument configuration and conditions are given in our earlier report (*6d*).

Shell Growth Polymerization: Synthesis of the ATRP Initiator Layer

A suspension of PSL seed (3.33 wt%) stirring at 350 rpm was heated to 70 °C , degassed and purged with argon (99.99%). Styrene and HEA-Cl were added successively to the suspension at 10 min intervals and shell polymerization initiated with potassium persulfate (KPS) 5 min later. The concentrations and amounts of reagents used in the shell growth polymerizations are given in Table I. The reaction was continued for 6 h and the latex was cleaned by dialysis against water (30 L dialysis tank), with daily changes in water for one week followed by 5 cycles of centrifugation and resuspension. The solid content was determined by freeze drying. The total negative charge on the surface of saponified latex was calculated from a conductometric titration as reported earlier (6d,14). A control titration of latex without saponification gives the inherent negative charge on the surface. The difference between these gives the surface initiator concentration. The total concentration of HEA-Cl incorporated into the latex was determined from ^{1}H NMR spectra of freeze dried samples dissolved in $CDCl_3$ utilizing the peaks at 4.1-4.3 ppm (HEA-Cl) and aromatic peaks of styrene (6d). Hydrodynamic sizes of the particles were determined with the N4 Plus and found to fit a unimodal distribution function. When compared with SEM images of preparations made by the above method the N4 Plus gave comparable means and somewhat larger standard deviations. Characteristics of the latexes are presented in Table II.

Table II. Characteristics of the Functionalized Latexes

La-tex	Feed conc. HEA-Cl[a] mol/g × 10[5]	Surface conc. of HEA-Cl[b]		Surface charge conc.[b]		HD (SD)[c] nm	ST[d] nm	AT[e] nm
		mol/ m^2 × 10[7]	Area/ Initiator nm^2	mol/ m^2 × 10[7]	Area/ charge nm^2			
L1	0.32	0.50	32.80	5.78	2.85	576 (115)	34	5.2
L2	1.35	2.11	8.13	4.99	3.29	578 (125)	35	5.2
L3	6.67	2.81	5.92	2.17	7.58	578 (112)	35	1.4
L4	13.2	4.47	3.72	2.17	7.47	586 (137)	39	1.3
L5	59.9	20.0	0.83	2.75	5.19	669 (152)	59	1.6

a. Calculated from the total amount of solid PSL, styrene and HEA-Cl added.

b. By conductometric titration

c. HD = hydrodynamic diameter, SD = standard deviation from the N4 Plus software

d. ST = shell thickness = half the increase in HD when shell added

e. Estimate of depth of region accessible to aqueous reagents

General Method for Grafting of Poly(N, N-dimethylacrylamide) (PDMA) by Aqueous ATRP from Latexes

All reactions were performed in a glove box under argon, following the general methods of polymerization reported in our earlier paper (*6d*). In a typical reaction, DMA (3.5 mM) was stirred with HMTETA (20 μM), CuCl (16 μM), CuCl$_2$ (2.4 μM), and Cu powder (19μM) for 3 minutes to form a solution; 3.5 g of degassed, 3 % w/w Brij-35 stabilized shell latex L5 was added to this with stirring at room temperature (22 °C) for 12 h. Supernatants were analyzed for monomer conversion by reverse phase HPLC. Repeated sequential centrifugation and resuspension in water, then NaHSO$_3$ (50mM) then water was done to clean the grafted latex. The same reaction conditions and reagents were used for reactions using latex L1 to L4 with a 24 hr reaction time.

Analysis of the Grafted Latex

The polymer chains grafted onto the latex surface were cleaved by quantitative hydrolysis with NaOH; 2 g of the grafted latex was stirred with 1 ml 2N NaOH for two weeks. Since the monomer, N,N-dimethylacrylamide is an N-substituted amide, the polymer PDMA is hydrolytically very stable in alkali solution at room temperature (*15*). Thus the ester groups between the polymer chains and surface can be selectively hydrolysed without afftecting the main chain. The released grafts were collected and analyzed by GPC to determine the molecular weight, molecular weight distribution, mass of the grafted chains per unit area of latex and radius of gyration, s.

The brush layer was characterized using a particle size analyzer to determine the hydrodynamic thickness. The particle diffusion constant of the grafted latex is less than that of the bare latex, which is interpreted as being caused by a uniform increase in particle diameter. We refer to the difference between the grafted and bare latex radii as the hydrodynamic thickness of the grafted layer.

Force measurements were performed on a Nanoscope IIIa MultiMode (Digital Instruments) using V-shaped silicon nitride tips (Digital Instruments) of nominal spring constant ~0.58 N/m and radius of curvature ~20-60 nm, as quoted by the manufacturer. Tips were cleaned in chromic acid and rinsed thoroughly with water prior to use. Force curves taken on a clean glass substrate before and after sample measurements exhibited no detectable adhesion, confirming the tip was free of contaminant. Sample films were prepared by drying grafted latexes onto clean glass substrates. Films were sonicated and rinsed thoroughly to remove any latex which had not been physisorbed. The films were then rehydrated and allowed to equilibrate. Measurements were taken in aqueous 150 mM NaCl solution in a fluid cell to minimize electrostatic interactions due to the negatively charged latex. Typically, a scan size of 2 µm at a rate of 0.5 Hz was used. Silicon calibration standards having 200 ± 3% nm vertical features were scanned to calibrate the z-piezoelectric crystal, using the calibration software provided by the manufacturer.

Forces were calculated from the measured tip deflection according to Hooke's law, using the spring constant quoted by the manufacturer. Tip-sample separations were obtained according to the method of Ducker et. al. (15) by subtracting the change in tip deflection (Δd) from the measured relative sample position (Δz). Zero separation corresponded to the position at which the force exerted by the sample upon further compression was indistinguishable from a hard surface, commonly known as the constant compliance regime. The equilibrium thickness was taken as the separation at which the repulsive force consistently deviated from the baseline by ~0.02 nN.

Results and Discussion

Model of the Surface Region

We have presented a physical model for the surface region of shell L5 (6d) in which chain initiation occurs in a region of small but non-zero depth, rather than from a strictly solid surface. The reaction occurs in a region of high copolymer concentration and low water content; the non-uniform local distribution of positively charged catalyst complex is determined by the negative surface charge density of the PSL, which is significantly lower than the concentration of initiator. These factors make the surface-initiated ATRP reaction complex.

We have shown earlier that styrene and HEA-Cl form a random copolymer under the conditions used (6d). The aqueous accessible depth (or degree of penetration of water soluble reagents) for latex L5 is calculated from the titratable initiator concentration, total initiator incorporated in the shell calculated from the ^{1}H NMR analysis and shell thickness determined by the hydrodynamic thickness measurement and, assuming a uniform initiator distribution throughout the shell, is found to be 1.51 nm. The ATRP grafting reaction must be occurring in this region as all the reagents used in the reaction are water-soluble. The calculated volume concentration of initiator and negative charges is 1.33 mol/L and 0.213 mol/L respectively for latex L5.

Since the catalyst concentration is much lower than the initiator concentration of 1.33 mol/L within the shell, the reactions are carried out under conditions of limiting ligand/Cu (I) rather than the 1:1 mole ratio typical of solution reactions (17). Although the complex is at a lower mole ratio with respect to initiator than in solution ATRP, its surface concentration is higher than the bulk solution.

In the case of latexes L1 to L4, the situation is slightly different. Some of them have more charges than initiator per unit area, creating conditions that are more favorable for ATRP reactions. In these cases, it is the concentration of catalyst at the surface and finite diffusion of monomer in the surface region that predominantly govern the reaction. Calculations show that latex L1 and L2 have much more water accessibility than the other three latexes, due to the high concentration of heavily hydrated sulfonic acid residues (*18*). This may have an influence on the nature of the polymerization. The concentration of negative charges on the surface could have a significant effect on initiation. A small charge concentration on the surface would enable the catalyst to concentrate near the surface and facilitate the initiation process. A highly charged surface could inhibit the reaction by somewhat immobilizing catalyst electrostatically, making it less available for initiation. An optimum charge on the surface therefore is required to achieve controlled reactions.

Effect of Monomer Concentration and Ligand Type

We have carried out a set of reactions with monomer concentrations from 1.5 to 10.3 wt% with latex L5 under otherwise similar experimental conditions. Three ligands HMTETA, PMDETA, Me$_6$TREN with CuCl were used as catalysts (6d). The M_n increases with monomer concentration (Fig. 1a, Table III) at constant initiator concentration for catalyst systems HMTETA/CuCl and PMDETA/CuCl, although the molecular weights differ significantly. This trend is not observed in the Me$_6$TREN/CuCl system. Reasonably good polydispersities are obtained using the first two catalyst combinations but the third one produces highly polydispersed chains (Fig.1b).

The number of chains per unit area (N_c) is calculated from the total mass of the graft released per unit area of the latex and $\overline{M}_n$. Assuming the chains are grafted in a square lattice, the average separations (D) are given by $(N_c)^{-1/2}$ and presented in Table III for the HMTETA/CuCl and PMDETA/CuCl systems. Due to the high polydispersity of the polymer formed using Me$_6$TREN/CuCl, we have not calculated the grafting density for this system. Unexpectedly, the grafting density is found to depend on monomer concentration and the type of catalyst used (Table III). The PMDETA/CuCl produces a higher grafting density than HMTETA/CuCl. PMDETA/CuCl at a 10.3 wt% monomer concentration gives one of the highest graft density brushes reported in the literature.

The number of chains initiated from the surface was found to increase with monomer concentration (Table III). We also found some solution polymerization along with graft polymerization, which depends on the monomer concentration. With increase in monomer concentration, the density of chains

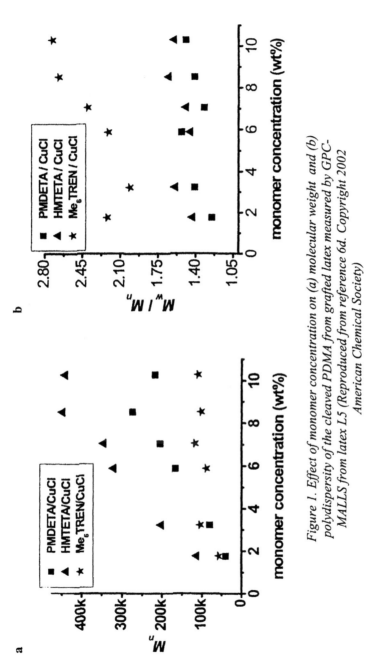

Figure 1. Effect of monomer concentration on (a) molecular weight and (b) polydispersity of the cleaved PDMA from grafted latex measured by GPC-MALLS from latex L5 (Reproduced from reference 6d. Copyright 2002 American Chemical Society)

initiated from the surface increased and the amount of solution polymerization decreased. This unusual observation can be discussed in terms of our surface model. Surface initiation and growth of chains is highly restricted by diffusion of monomer and reagents to the reaction sites due to the high volume fraction of copolymer chains in the surface region. For a given experimental condition, the chain growth near the surface is governed by the monomer accessibility to the surface region. At high monomer concentration, there will be sufficient monomer in the surface region to facilitate the chain growth with minimal chain transfer to solvent or catalyst, as discussed in detail in our earlier report (6d). As the monomer concentration decreases in the bulk, more and more free radicals will transfer to solvent or catalyst from the surface in the absence of monomer in the region and start polymerizing in solution. Thus, with decrease in monomer concentration, there is an increase in solution polymerization. The similar molecular weight of PDMA chains grown from the surface and in solution supports this idea.

In these surface polymerization reactions, catalyst type is also very important. This is evident from the differences in molecular weight and grafting density of the chains grown from the surfaces. The difference in the surface concentration of these catalyst complexes and their inherent reactivity under the conditions of low solvent volume fraction has influence on this observation. The higher propagation rate of HMTETA/CuCl catalyst may also contribute to the high molecular weight of PDMA grafts obtained using this catalyst. The results suggests that surface polymerization depends on many more parameters than does regular solution polymerization. The grafted PDMA layers have large hydrodynamic thickness as evidenced by the increase in the size of the grafted latex. By increasing monomer concentration, we were able to change graft thickness from ~67 to 800 nm using latex L5. The hydrodynamic thickness increases linearly with monomer concentration.

Effect of Surface Initiator Concentration

It has been reported (19,20) that the grafting density of polymer brushes synthesized from silica surfaces can be changed by varying surface initiator density. We have successfully synthesized negative surfaces with differences in the initiator density by varying the initial monomer feed ratio (18) of styrene and HEA-Cl during shell-growth polymerization. The values for the surface initiator concentration and surface negative charges are given in Table II. We were unsuccessful in determining the total initiator concentration by ^{1}H NMR for L1 to L3 as they have a very low mol percent of HEA-Cl compared to polystyrene. A series of reactions was done where the monomer concentration for different latexes was changed under identical catalyst concentrations. The results are shown in Figs 2 and 3.

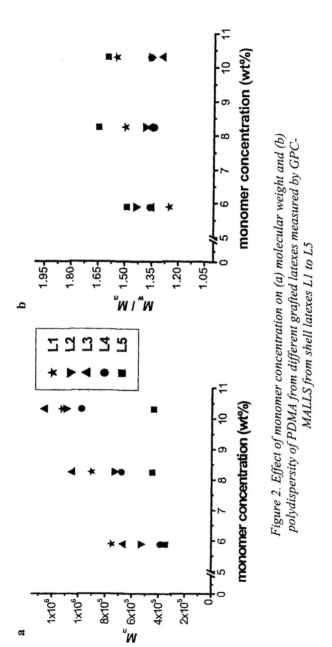

Figure 2. Effect of monomer concentration on (a) molecular weight and (b) polydispersity of PDMA from different grafted latexes measured by GPC-MALLS from shell latexes L1 to L5

Table III: Properties of the Grafted Latex Synthesized from Shell Latex L5

	[mon.] wt%	conversion (wt%)	$M_n \times 10^{-3}$	M_w/M_n	D (Å)	D/s	Thickness[a] (nm) AFM	HDT[b]	HD/l_c[c]
PMDETA/CuCl	1.5	42	39.6	1.25	39	0.561	95	67	0.673
	2.9	51	80.4	1.41	22	0.183	132	148	0.730
	4.4	48	166.0	1.53	22	0.118	200	296	0.705
	5.9	74	204.6	1.32	21	0.107	207	377	0.731
	8.2	ND	273.0	1.41	15	0.064	ND	607	0.883
	10.3	ND	217.0	1.49	11	0.050	ND	680	1.245
HMTETA/CuCl	1.5	68	112.9	1.43	29	0.214	ND	192	0.674
	2.9	74	202.0	1.59	27	0.147	ND	345	0.677
	4.4	ND	320.7	1.45	25	0.099	ND	488	0.603
	5.9	ND	344.7	1.49	25	0.097	ND	619	0.712
	8.2	ND	447.2	1.65	21	0.063	ND	800	0.710
	10.3	ND	440.6	1.60	19	0.058	ND	817	0.735

a. Thickness of the polymer film in the aqueous medium

b. HD = hydrodynamic thickness calculated from particle size analyzer data.

c. l_c = contour length calculated from M_n using an effective bond length of 0.25 nm per repeating unit.

ND: not determined

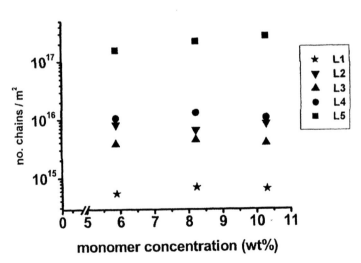

Figure 3. Effect of monomer concentration on the grafting density of PDMA for shell latexes L1 to L5 carrying different surface initiator concentrations.

The molecular weight of the PDMA grafts (Fig. 2a) increases with decrease in the surface initiator concentration at constant monomer concentration.molecular weight also increases with monomer concentration. The surface with lowest initiator concentration gives molecular weight of ~1.3 × 10^6 at the highest monomer concentration studied which is the highest value reported for a surface initiator polymerization. The PDI values decreased with decrease in the surface initiator concentration and are in the range 1.25 to 1.6.

The grafting density of the PDMA layers is shown in Fig. 3. We could achieve inter-chain separations from 1.9 to 40 nm using different surface concentrations. Chain initiation for the surface with low initiator concentration is independent of monomer concentration. The grafting density remains constant with monomer concentration for latexes L1 to L4. In the case of latex L5 which has a large initiator concentration on the surface a different behavior is observed. At lower surface initiator concentrations, monomer depletion near the surface region due to polymerization is not large enough to produce a variation in the grafting density, as explained earlier. At high surface initiator concentration, this effect is more evident and the grafting density shows a change with monomer concentration.

Generally, initiator efficiency for successful grafts is much lower for surface polymerization than solution polymerization (18). In our case initiator efficiency decreases with decreasing surface initiator concentration. For latexes L1 to L4 only a few percent of the initiators form successful grafts (~2-7 %). But in the case of latex L5, initiator efficiency reaches a value of ~65% at 10.3wt % monomer concentration, which is the highest value reported for a surface polymerization. For a given monomer concentration, solution polymerization increases with decreasing surface initiator concentration. At low initiator concentrations with equal or higher charge concentrations, the concentration of catalyst in the surface region will be greater than that of the free radicals produced. This might facilitate more chain transfer under the conditions of restricted monomer diffusion near the surface. Thus increasing charge density and decreasing surface initiator concentration produces more solution polymerization.

Brush Properties

The very tight configurations of these grafted layers are evident from the large hydrodynamic thickness and the low (D/s) values in Table III, where s is the radius of gyration of the chain in solution. In some cases, the inter-chain separation reaches as low as 5 % of s and the hydrodynamic thickness approaches the value of the contour length calculated from the molecular weight of the polymer. The hydrodynamic thickness of the grafted layer increases linearly with monomer concentration in solution.

328

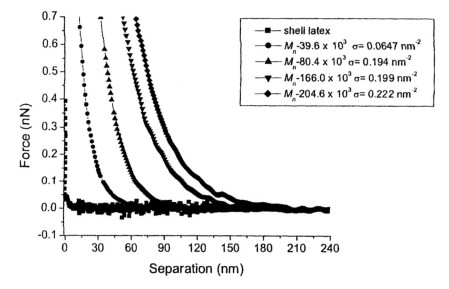

Figure 4. Force-distance curves for the shell latex L5 and grafted latex with different molecular weight PDMA using PMDETA/CuCl catalyst measured by AFM; σ = chain density on surface

The force profiles corresponding to a series of brushes of varying molecular weight and grafting densities are presented in Figure 4. The characteristic force curve of the polymer brushes is a monotonically increasing repulsive force on approach which, at long range, can be attributed mainly to excluded volume effects. Upon further approach, there is a constant compliance region where the sample is no longer compressible. At large tip-sample separations, the force curve consists of a region of zero deflection where no interaction is observed. Experimentally, only the relative sample position is determined, and a zero position must be defined to obtain an absolute measure of tip-sample separation. This position is taken as the onset of constant compliance, at which point the sample behaves as a hard surface upon further approach. The equilibrium thickness of the grafted layer is defined as the separation where an interaction between the tip and polymer is first observed, at which point a repulsive force can be distinguished from the noise in the baseline.

A definite trend toward larger repulsive forces at greater separations with increasing molecular weight and grafting density is observed, as a result of the increasingly extended equilibrium structure. The equilibrium thicknesses given in Table III represent the thickness of the grafted layer between the tip and the polystyrene latex. While it is possible that compression of the grafted layer between the polystyrene and glass substrate may result in an overestimated thickness, the motion of this layer is highly restricted and we therefore believe its contribution is insignificant. The latex dries on the glass surface in local hexagonally packed arrays. These structures are resistant to the large shear forces present during sonication, which suggests that the underlying grafted layer behaves as a rigid body.

The thicknesses measured by AFM are typically smaller than the values obtained by hydrodynamic thickness measurements. The hydrodynamic resistance associated with particle diffusion arises from the integrated effects of the drag produced by all of the chains on the surface; it is this resistance that is attributed to that of an equivalent hard sphere whose diameter is provided by the measurement. The resistance is particularly sensitive the longest chains on the surface (21). In contrast, the AFM tip samples only a small fraction of the total surface area, with a low probability of sampling the very longest chains. While there is this discrepancy in the two sets of measurements, we stress that the general trend toward increasing layer thicknesses with molecular weight is the same in both methods.

For all AFM results presented, it was possible to compress the chains to a point where the repulsion was like that of a hard surface, at which point we assume the tip is at the latex surface. A constant compliance region for the higher molecular weight brushes in the series ($\overline{M}_n \sim$ 217,000 and 273,000) could not be obtained using the same AFM tips (k $\sim$ 0.58 N/m) as those used on the lower molecular weight brushes. It is important to note that the equilibrium thickness measured by AFM is dependent on the spring constant of the probe. The ability to fully compress the higher molecular weight brushes may be achieved using a stiffer tip, but sensitivity to small repulsive forces will be reduced and the equilibrium thickness underestimated.

Summary

We have successfully synthesized PSL surfaces with different initiator concentrations by changing the feed ratio of styrene to HEA-Cl in a series of shell-growth polymerizations. We have investigated the role of monomer concentration, type of ligands and surface initiator concentration on the polymerization of N,N-dimethylacrylamide from latex surfaces by aqueous ATRP at room temperature. Results indicate that surface polymerization is very different from solution ATRP reactions. True brushes are obtained with highly extended configurations, molecular weights up to $\sim 1.3 \times 10^6$ and hydrodynamic thicknesses from a few nanometers to 800 nm. The grafted polymer layers are highly repulsive to an AFM tip.

Acknowledgments: We thank CIHR, NSERC of Canada, the Canadian Blood Services and the CFI for financial support.

330

References

1. Milner, S.T. *Science* **1991**, 251, 905.
2. Zhao, B. Brittian,W.J. *Progress in Polym.Sci.* **2000**, 25, 677.
3. a)de Gennes, P.G. *J. Phys. (Paris)* **1976**, 37, 1445. *Macromolecules* **1980**, 13, 1069. b) Alexander, S.J. *J. Phys. (Paris)* **1977**, 38, 983.
4. a) Ingall, M.D.K.; Honeyman, C.H.; Mercure, J.V.; Bianconi, P.A.; Kunz, R.R. *J. Am. Chem. Soc.* **1999**, 121, 3607. b) Jordan, R.; Ulman, A.; Kang, J.F.; Rafailovich, M.H.; Sokolov, J. *J. Am. Chem. Soc.* **1999**, 121, 1016. c)Prucker, O.; Ruhe, J. *Macromolecules* **1998**, 31, 592.
5. a)Husseman, M.; Malmstrom, E.E.; McNamara,M.; Mate,M.; Mecerreyes, D.; Benoit, D.G.; Herdrick, J.L.; Mansky, P.; Huang,E.; Russell, T.P.; Hawker, C.J. *Macromolecules* **1999**,32, 1424. b) Matyjaszewski, K.; Miller, P.J.; Shukla, N.; Immaraporn, B.; Gelman , A.; Luokala, B.B.; Siclovan, T.M.; Kickelbick, G.; Vallant, T.; Hoffmann, H.; Pakula, T. *Macromolecules*, **1999**, 32, 8716.
6. a)Kim, J.; Bruening, M.L.; Baker G.L. *J. Am. Chem. Soc.* **2000**, 122, 7616. b) Jones, D.M.; Huck, W.T.S. *Adv. Mater.* **2001**, 13, 1256. c) Pyun J, Matyjaszewski K, Kowalewski T, Savin D, Patterson G, Kickelbick G, Huesing N. *J. Am. Chem. Soc.* **2001**, 123, 9445. d) Jayachandran, K.N.; Takacs-Cox, A.; Brooks, D.E. *Macromolecules* **2002**, 35, 4247. *Macromolecules* **2002**, 35, 6070.
7. Guerrrinni, M.M.; Charleux, B.; Vairon, J.P. *Macromol. Rapid. Comun.* **2000**, 21, 669.
8. Binnig, G.; Quate, C.; Gerber, G. *Phys. Rev.Lett.* **1986**, 56, 930.
9. Prescott, S. W.; Fellows, C.M.; Considine, R.F.; Drummond, C.J.; Gilbert, R.G. *Polymer* **2002**, 43, 3191.
10. Yamamoto, S.; Ejaz, M.; Tsujii, Y.; Matsumoto, M.; Fukuda, T. *Macromolecules* **2000**, 33, 5602.
11. Kidoaki, S.;Ohya, S.;Nakayama, Y.; Matsuda, T. *Langmuir*, **2001**, 17, 2402.
12. Kelley, T.W.; Schorr, P.A.; Johnson, K.D.; Tirell, M.; Frisbie, C.D. *Macromolecules***1998**, 31, 4297.
13. Ciampolini,M.; Nardi, N. *Inorg. Chem.* **1966**, 5, 41.
14. Hritcu, D.; Muller, W.; Brooks, D.E. *Macromolecules* **1999**, 32, 565.
15. Barton,D.; Ollis, W.D. in *Comprehensive Organic Chemistry*; Pergamon Press, Oxford, UK,1979; p 1004.
16. Ducker, W.; Sendon, T.; Pashley, R. *Nature* **1991**, 353, 239.
17. Wang, J.S.; Matyjaszewski, K. *J. Am. Chem. Soc.* **1995**, 117,5614.
18. Kizhakkedathu, J.N.; Brooks, D.E. *Macromolecules* (in press) **2003**
19. Jones, D.M.; Brown, A.A.; Huck, W.T.S. *Langmuir* **2002**, 18, 1265.
20. Yamamoto,S.;Ejaz, M.;Tsujii,Y.;Fukuda,T.*Macromolecules* **2000**, 33, 5608.
21. Fleer, G.J.;Cohen Sutart, M.A.; Scheutjens, J.M.H.M.; Cosgrov, T.; Vincent, B. *Polymers at Intefaces* Chapman & Hall; London, 1993; p255

Chapter 23

End-Functional Polystyrenes via Quasiliving Atom Transfer Radical Polymerization and New Polymer Structures Therefrom

Béla Iván[1], Tamás Fónagy[1], Tibor Erdey-Grúz[1],
György Holló-Szabó[1], Márta Szesztay[1], Ulrich Schulze[2],
and Jürgen Pionteck[2]

[1]Department of Polymer Chemistry and Material Science, Chemical
Research Center, Hungarian Academy of Sciences, H–1525 Budapest,
Pusztaszeri u. 59–67, P.O. Box 17, Hungary
[2]Institute of Polymer Research, Dresden, Hohe Strasse 6, D–01069 Dresden,
Germany

Polystyrenes with 2-chloro-2-phenylethyl chain end (PSt-Cl)
prepared by quasiliving atom transfer radical polymerization
(ATRP) are useful intermediates for the synthesis of a variety
of well-defined macromolecular architectures. The preparation
of star-shaped polystyrenes by quasiliving ATRP of
divinylbenzene initiated by PSt-Cl and hyperbranched
polystyrene by self-grafting of PSt-Cl mediated with $TiCl_4$ is
carried out in the course of our studies. Allyl-terminated
macromonomers were also prepared by reacting PSt-Cl with
allyltrimethylsilane in the presence of $TiCl_4$. This PSt-allyl
was copolymerized with propylene by metallocene catalysts
yielding novel poly(propylene-g-styrene) (PP-g-PSt) graft
copolymers.

331

Introduction

There have been significant developments in the field of quasiliving polymerizations (*1*) during the last decade. As a consequence of this evolution the possibilities for the synthesis of well-defined macromolecules have considerably broadened. Quasiliving radical polymerization processes, such as ATRP (*2,3*), SFRP (*4*) and RAFT (*5*), have led to a variety of polymers with predetermined molecular weight, relatively low polydispersity and high chain end functionality unavailable by other techniques.

Quasiliving atom transfer radical polymerization (ATRP) is a transition metal catalyzed process yielding polymers with halogen at chain ends. These terminal halogen atoms are very useful as intermediates in the synthesis of complex macromolecular architectures. For example, the reactive endgroup can initiate the ATRP of a second monomer leading to block copolymers (*6,7*), or it can be transformed by nucleophilic substitution, elimination or radical chemistry to other advantageous groups. Several functional groups, such as allyl (*8*), hydroxyl (*9*) and amino (*10*) groups, were synthesized by these ways resulting in potential macromonomers or macroinitiators both in step- and chain-growth polymerizations.

This study deals with our recent results on the synthesis and applications of endfunctional polystyrenes obtained via quasiliving ATRP.

Experimental Section

Polystyrenes with 2-chloro-2-phenylethyl chain end (PSt-Cl) were synthesized by quasiliving ATRP of styrene with (1-chloroethyl)benzene initiator and copper(I) chloride catalyst complexed with 2,2'-bipyridine (bpy) in bulk at 130 °C. The molar ratio of initiator/CuCl/bpy was 1:1:2.5. The initiator/styrene ratio was calculated from the desired molecular weight of the polymer. The resulting polymers were purified by chromatography and precipitation.

PSt-Cls with average molecular weight of 1,800 was used as macroinitiator in ATRP of divinylbenzene (DVB) for the synthesis of star polymers. The polymerizations were carried out in xylene with CuCl/3bpy catalyst system, the ratio of CuCl to PSt-Cl was 1:1. The resulting polymers were purified by chromatography and precipitation. The unreacted linear polystyrene was removed by passing the polymer in carbon tetrachloride solution through a chromatography column filled with neutral aluminum oxide followed by elution of adsorbed PSt-Cl with tetrahydrofuran.

Hyperbranched polymers were obtained by self-grafting of PSt-Cl. PSt-Cl with MW of 2000 was dissolved in dichloromethane or in mixture of dichloromethane and hexane (40/60) in concentration of 0.1 g/ml, then 8 eq. titanium tetrachloride was added under argon atmosphere. At predetermined reaction times some solution was withdrawn. The samples were quenched with 2 ml methanol saturated with ammonia, then THF was added, left for 24 hours at ambient temperature and filtered. Finally the polystyrene samples solved in THF were precipitated into 10-fold amount of methanol, let to settle, filtered and dried in vacuum.

The chain end of PSt-Cl was transformed to allyl terminus by reacting it with allyltrimethylsilane (ATMS) in the presence of $TiCl_4$ in 0.1 mol/dm^3 dichloromethane solution under argon atmosphere at 0 °C.

The resulting allyl-terminated polystyrene (PSt-allyl) was copolymerized with propylene in the presence of metallocenes. Polymerizations were carried out in 200 ml toluene in a 1 liter glass autoclave (Büchi) equipped with a stirrer, manometer, thermocouple, heating and cooling units. The polymerization temperature was varied between 30 °C and 70 °C, and the total pressure was set in the range of 1 bar to 3 bar. The catalysts were either $Me_2Si(2-Me-4,5-Benzind)_2ZrCl_2$ (MBI), $Et[Ind]_2ZrCl_2$ (EtInd) or $Me_2Si[(t-Bu-N)(Me)_4Cp]TiCl_2$ (TiN). For the copolymerizations, the Al/Zr molar ratio was 4,000 and catalyst concentration was $8×10^{-6}$ mol/l. After consumption of 0.5 mol propylene the polymerization was terminated by injecting a small amount of ethanol. The reaction solution of the copolymer was precipitated into a mixture of ethanol, water and some hydrochloric acid. The resulting copolymer was filtered, washed with water and ethanol, and dried overnight in a vacuum oven at 70 °C. The unreacted polystyrene was removed by precipitation from hot xylene into cold acetone, followed by drying in vacuum oven at 70 °C.

Characterization. ^1H NMR and ^{13}C NMR spectra were recorded on a Bruker DRX 500 and a Varian 400 MHz spectrometers. Polystyrenes and polypropylene containing samples were measured at room temperature in $CDCl_3$ and at 120 °C in $C_2D_2Cl_4$, respectively.

Molecular weight distributions of polystyrenes were determined by gel permeation chromatography (GPC) with a Waters/Millipore liquid chromatograph equipped with a Waters 510 pump, Ultrastyragel columns of pore sizes $1×10^5$, $1×10^4$, $1×10^3$ and 500 Å, a Viscotek parallel differential refractometer/viscometer and a laser light scattering (miniDawn®, Wyatt Technology Co.) detectors. Tetrahydrofuran was used as the mobile phase with a flow rate of 1.5 ml/min. Calibration was made with narrow MWD polystyrene standards. Molecular weights of PP-g-PSts were determined using high temperature size exclusion chromatography (Polymer Laboratories 210 GPC) operated at 135 °C. The GPC apparatus had a column set with four columns (PL gel 20 μm Mixed-A). The mobile phase was 1,2,4-trichlorobenzene with a flow rate of 1.0 ml/min. The calibration was carried out by using PP with known molecular weight and distribution.

Results and Discussion

Quasiliving ATRP of styrene results in polymers with predetermined molecular weight, relatively low polydispersity and 2-chloro-2-phenylethyl chain end. The presence of this terminal functional group makes such polystyrene (PSt-Cl) suitable for further transformations in order to obtain well-defined macromolecular architectures. Numerous potential transformations of the 2-chloro-2-phenylethyl terminal group are offered by common and special chemical reactions. In the course of our recent investigations we have utilized PSt-Cl to initiate ATRP of divinylbenzene (DVB) for the preparation of star polymers, to carry out self-grafting by Friedel-Crafts alkylations for obtaining hyperbranched polystyrenes, and to perform carbocationic chain end

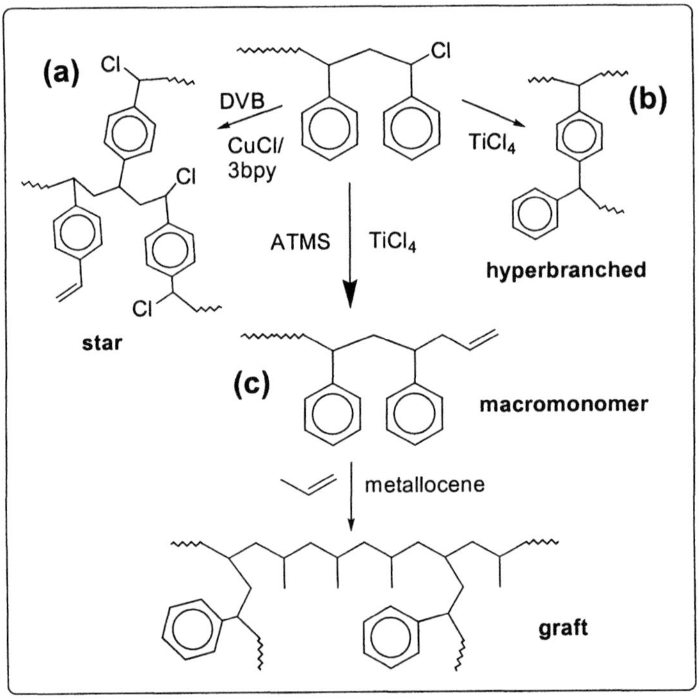

Scheme 1. Preparation of polystyrenes with various architectures from chlorine-terminated polystyrene obtained by quasiliving ATRP of styrene.

derivatization with ATMS followed by metallocene catalyzed copolymerization of the resulting PSt macromonomer with propylene. These reactions and the resulting new polymers are shown in Scheme 1.

PSt-Cl was used as macroinitiator for ATRP of DVB resulting in coupling of linear polystyrene segments by polymerization of the bifunctional monomer into a central polyDVB core (Scheme 1a). Figure 1 shows the molecular weight distribution (MWD) of the resulting polymers obtained at different polymerization times. As it is exhibited in this Figure the relative concentration of the starting PSt-Cl gradually decreases while the amount of higher molecular weight fraction increases with increasing reaction time. However, the presence of the starting linear polymer is clearly indicated after 8 hours reaction time. Similar findings were reported by Xia et al. (11) by using also pentamethyldiethylenetriamine (PMDETA) as ligand in catalyst of Cu-complex and several bifunctional monomers.

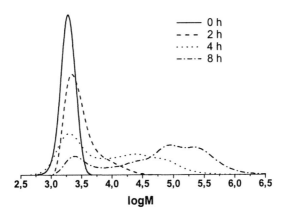

Figure 1. Molecular weight distributions of polystyrenes obtained by ATRP of DVB initiated by PSt-Cl in xylene. M_n(PSt-Cl)=1800, [PSt-Cl]=[CuCl]= 0.15 mol/dm^3, [bpy]= 0.45 mol/dm^3, [DVB]=1.11 mol/dm^3

In order to be able to characterize the higher molecular weight fraction the rest of the unreacted PSt-Cl had to be separated. Several solvents were tested as eluent for preparative chromatography of this polymer mixture. It was found that the most suitable process is passing the polymer in carbon tetrachloride solution through a chromatography column filled with neutral aluminum oxide followed by elution with tetrahydrofuran. As it is shown in Figure 2 the high molecular weight fraction passed through the column in the carbon tetrachloride solution, while the unreacted PSt-Cl was adsorbed to the neutral Al_2O_3 filler and could be eluted by tetrahydrofuran.

336

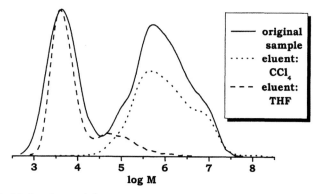

Figure 2. Molecular weight distribution of polystyrene samples before and after separation of the crude polymer obtained by ATRP of DVB initiated by PSt-Cl.

In the ^{1}H NMR spectrum of the high molecular weight polymer two broadened signals appeared in the range of 5-6 ppm. This finding clearly indicates the presence of unreacted double bonds of DVB units in the polymer, and in accordance with the increased molecular weight it confirms the efficient ATRP of DVB by the PSt-Cl macroinitiator. The shape of these macromolecules is expected to be star-like with polyDVB core and linear polystyrene arms. This was proved by the Mark–Houwink α coefficient (shape factor) calculated from GPC chromatograms. This value was found to be 0.11 indicating high branching degrees of the resulting star polymers.

PSt-Cl can be regarded as multivalent reagent in Friedel-Crafts alkylation since the 2-chloro-2-phenylethyl chain end can be easily transformed to a carbocation by a Lewis-acid while the pendant aromatic rings are excellent substrates of electrophilic substitution. Thus, after treatment of PSt-Cl in dichloromethane solution with titanium tetrachloride (TiCl$_4$) self-grafting of polystyrene by interalkylation was expected (Scheme 1b). The ^{1}H NMR spectrum of the resulting polymers (Figure 3) indicate that the chlorine end of PSt-Cl was completely transformed since the signal at 4.3-4.6 ppm, assigned to the methine proton next to the chlorine atom, disappeared. Simultaneously a new signal appeared at 3.3-3.5 ppm, which can be assigned to methine protons neighbouring with two aromatic rings, indicating the successful alkylation.

Figure 4 shows the results of GPC analysis. The average molecular weight and the polydispersity increased upon treatment with TiCl$_4$ indicating that the intermolecular alkylation, i.e. self-grafting, was dominant over intramolecular substitution, i.e. cyclization. In order to confirm the branched structure of the resulting polymers the α parameters of the Mark-Houwink equation were also

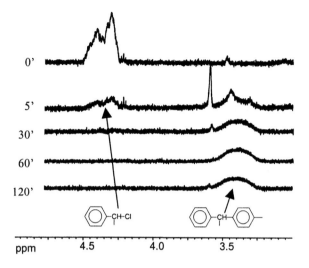

Figure 3. ^{1}H NMR spectra of polystyrene samples obtained by self-grafting of PSt-Cl. M_n(PSt-Cl)=2000, [PSt-Cl]=0.05 mol/dm^3, [TiCl$_4$]=0.4 mol/dm^3, solvent: dichloromethane and hexane (40/60), T= -78 °C

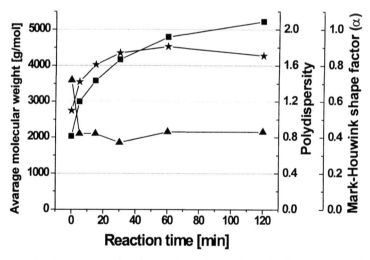

Figure 4. Average molecular weight (square), polydispersity (star) and Mark-Houwink shape factor (delta) of PSt-Cl treated by TiCl$_4$ as a function of time. M_n(PSt-Cl)=2000, [PSt-Cl]=0.05 mol/dm^3, [TiCl$_4$]=0.4 mol/dm^3, solvent: dichloromethane and hexane (40/60), T= -78 °C

calculated from the GPC chromatograms. It was found that the initial value of 0.72 decreased to 0.42 within 5 minutes reaction time and kept this value to the end of the investigated period. This value also signifies the branched structure as it is exhibited in Scheme 1.

The polystyryl carbocation generated by $TiCl_4$ can be reacted with a variety of other reagents. In the presence of ATMS the allylation of chain ends was expected in dichloromethane solution (Scheme 1c). Monitored by ^1H NMR spectroscopy the disappearance of methine protons neighboring with chlorine atom and the appearance of signals corresponding to allyl groups (at 4.8 and 5.5 ppm) were observed. The lack of signals at 3.3-3.5 ppm assigned to methine protons with two neighboring phenyl rings indicated the domination of the reaction with ATMS over the Friedel-Crafts alkylation. Thus, allyl-terminated polystyrene (PSt-allyl) without detectable side reactions was formed (8).

Since the PSt-allyl can be regarded as a substituted propylene, the copolymerization of this macromonomer with olefins by metallocene catalysis was expected to result in certain new graft copolymers. Thus a systematic investigation was carried out by us on the influence of synthesis conditions on the structure of poly(propylene-g-styrene) (PP-g-PSt) graft copolymer obtained by the copolymerization of propylene with PSt-allyl. The separation of the unreacted PSt-allyl was carried out by precipitation from hot xylene into cold acetone. A representative NMR spectrum of a purified copolymer sample is shown in Figure 5. This shows the presence of signals in the aromatic region corresponding to the PSt side chains in PP-g-PSt graft copolymer indicating that PSt-allyl incorporated into the polypropylene backbone. The molecular weights of PP-g-PSts were determined by GPC. The PSt content and the average number of side chains per molecule were calculated from the integral ratios of peaks in the ^1H NMR spectra and molecular weights of PSt-allyl and the resulting PP-g-PSts. These results are summarized in Table 1.

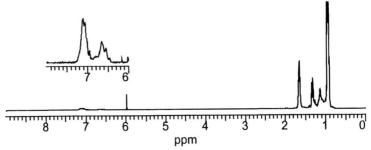

Figure 5. ^1H NMR spectrum of PP-g-PSt (copolymerization at 1.5 bar propylene pressure with PSt-allyl with MW of 2000, MBI catalyst at 50 °C).

Table 1. Effect of temperature (T), propylene pressure (p), molecular weight of PSt-allyl and type of catalyst on molecular weight, PSt content of PP-g-PSt, and average number of side chains per macromolecule in metallocene catalyzed copolymerization of propylene with PSt-allyl.

	$M_n(PP\text{-}g\text{-}PSt)$ g/mol	PSt content (w/w%)	Averag number of side chains
T (°C)	p=1.5 bar, catalyst: MBI, M_n(PSt-allyl)=18,000		
30	285,000	2.4	0.37
50	89,400	6.7	0.33
70	33,200	10.8	0.20
P (bar)	T=50 °C, catalyst: MBI, M_n(PSt-allyl)=18,000		
1	78,900	7.4	0.33
1.5	89,400	6.7	0.33
2	183,300	4.9	0.22
M_n(PSt-allyl)	T=50 °C, p=1.5 bar, catalyst: MBI		
2,000	54,000	8.7	2.1
11,000	80,300	7.0	0.56
18,000	89,400	6.7	0.33
Catalyst	T=50 °C, p=1.5 bar, M_n(PSt-allyl)=18,000		
EtInd	10,600	6.2	0.04
MBI	89,400	6.7	0.33
TiN	194,900	7.4	0.80

The PSt content of the resulting graft copolymers significantly increases with increasing polymerization temperature. This tendency can be explained by the more favorable concentration ratio of PSt-allyl to propylene. The change in the apparent reactivity ratio with temperature may also contribute to the increased PSt-allyl incorporation. Similar to homopolymerization of propylene, the M_n of the resulting graft copolymers decreases with increasing polymerization temperature. The PSt incorporation increases and the MW of copolymers decreases with decreasing propylene pressure. These changes are caused by the decreased concentration of propylene in the solutions. Some decrease in the PSt content and moderate increase in MW can be observed with increasing MW of PSt-allyl. This finding can be explained by the decreased apparent reactivity and decreased molar concentration (the molal concentration was kept constant) of longer chains while the increased MW is caused by the decreased PSt-allyl incorporation. As expected, the catalyst has considerable influence on the MW of copolymers. These catalysts provide the same tendency as in the case of propylene homopolymerizations. The average number of side chains per macromolecule was generally found to be lower than 1 with the

340

exception of PSt-allyl with MW of 2000 indicating the possibility of successful
synthesis of PP-*g*-PSt graft copolymers which might be efficient blending agents
for polypropylene and polystyrene (*12*).

Conclusions

Polystyrenes with 2-chloro-2-phenylethyl chain end (PSt-Cl) prepared by
quasiliving atom transfer radical polymerization are useful intermediates for the
synthesis of a variety of well-defined macromolecular architectures. In this
paper, we presented the preparation of star-shaped polystyrenes by quasiliving
ATRP of divinylbenzene initiated by PSt-Cl and hyperbranched polystyrene by
self-grafting of PSt-Cl mediated by TiCl₄. Allyl-terminated macromonomers
were also synthesized by reacting of PSt-Cl with allyltrimethylsilane in the
presence of TiCl₄. This PSt-allyl was successfully copolymerized with
propylene by metallocene catalyst yielding a variety of PP-*g*-PSt graft
copolymers.

Acknowledgements. The authors are grateful to D. Voit and E. Tyroler for GPC
and H. Komber for NMR measurements. The authors acknowledge loaning the
miniDawn® laser light scattering detector by Wyatt Technology Co. Financial
support by the Hungarian Scientific Research Fund (OTKA T29711, T25933,
T33107) and Sächsisches Ministerium für Wissenschaft and Kunst is also
acknowledged.

References
1. Iván, B. *Macromol. Chem. Phys.* **2000**, *201*, 2621-2628
2. Wang, J.-S.; Matyjaszewski, K. *J. Am. Chem. Soc.* **1995**, *117*, 5614-5615
3. Kato, M.; Kamigaito, M.; Sawamoto, M.; Higashimura, T. *Macromolecules*
   1995, *28*, 1721-1723
4. Georges, M. K.; Veregin, R. P. N.; Kazmaier, P. M.; Hamer, G. K.
   Macromolecules **1993**, *26*, 2987-2988
5. Chiefari, J.; Chong, Y. K.; Ercole, F.; Krstina, J.; Jeffery, J.; Le, T. P. T.;
   Mayadunne, R. T. A.; Meijs, G. F.; Moad, C. L.; Moad, G.; Rizzardo, E.;
   Thang, S. H. *Macromolecules* **1998**, *31*, 5559
6. Cocoa, S.; Matyjaszewski, K. *J. Polym. Sci., Part A:Polym. Chem.* **1997**, *35*,
   3595
7. Gao, B.; Chen, X.; Iván, B.; Kops, J.; Batsberg, W. *Polym. Bull.* **1997**, *39*,
   559

8. Iván, B.; Fónagy, T. *ACS Symp. Ser.* **2000**, *768*, 372
9. Coessens, V.; Matyjaszewski, K. *Macromol. Rapid Commun.* **1999**, *20*, 127
10. Matyjaszewski, K.; Nakagawa, Y.; Gaynor, S. G. *Macromol. Rapid Commun.* **1997**, *18*, 1057
11. Xia, J.; Zhang, X.; Matyjaszewski, K. *Macromolecules* **1999**, *32*, 4482
12. Schulze, U.; Fónagy, T.; Komber, H.; Pompe, G.; Pionteck, J.; Iván, B. *to be published*

Chapter 24

A Strategy to Prepare Anemone-Shaped Polymer Brush by Controlled/Living Radical Polymerization

Fu-Mian Li, Ming-Qiang Zhu, Xin Zhang, Liu-He Wei, Fu-Sheng Du, and Zi-Chen Li

Department of Polymer Science and Engineering, College of Chemistry, Peking University, Beijing 100871, China

A well-defined diblock copolymer containing a block of alternately structured maleic anhydride (MAn) with 4-vinylbenzyl chloride (VBC) and PVBC block, P(MAn-*alt*-VBC)-b-PVBC, was prepared by a one-pot reversible addition-fragmentation chain transfer polymerization (RAFT) of MAn and VBC. This block copolymer can form stable inverse micelles in tetrahydrofuran after the MAn moieties being reacted with 2-mercaptoethyl amine. Thus silver ions were embedded in the inner core of the micelle, which were *in-situ* reduced to silver in the existence of NaBH$_4$ to obtain stable silver nanoparticles. The benzyl chloride groups on the nanoparticle surfaces were able to initiate the atom transfer radical polymerization (ATRP) of St to achieve a new type of Anemone-shaped polymer brush.

There has been increasing interest in the development of new strategies to prepare a polymer brush on a solid surface such as silica or quartz wafer and noble metals or inorganic nanoscale particles(*1, 2*). There are two major currents to prepare polymer brush on the solid surface. One is to assemble a block copolymer whose segments possess distinct solubility in solution onto the solid surface (so called "grafting to") (*3,4*). Another way is *via* graft polymerization of vinyl or cyclic monomers on the solid surface (so called "grafting from") (*5-7*). The latter is capable of providing polymer hairs with high graft density on the solid surface. Furthermore, if a controlled radical polymerization technique including atom transfer radical polymerization (ATRP), nitroxide-mediated polymerization (NMP) and radical reversible addition-fragmentation chain transfer polymerization (RAFT) is employed, not only a well-defined polymer brush can be achieved, but also can be performed under conventional radical polymerization conditions even in aqueous milieu (*8*). Most of the surface-initiated controlled radical graft polymerizations reported are routinely carried out on the halomethylphenyl- (or α-haloester) and alkoxyamine-modified surface of silica, quartz wafers and particles as initiating sites for ATRP(*1, 9, 10*) and NMP (*11*), respectively.

In this paper, we describe a strategy to prepare a new type of anemone-shaped polymer brush (ASPB) whose hairs are tethered on the nanoscale silver particles. The whole process is outlined in Scheme 1. RAFT was used to synthesize a well-defined diblock copolymer composed by a block of alternately structured maleic anhydride (MAn) with 4-vinylbenzyl chloride (VBC) and a block of PVBC, P(MAn-*alt*-VBC)$_m$-*b*-PVBC$_n$ (I), in the first step, then, this block copolymer was allowed to react with 2-mercaptoethylamine to obtain a amphiphilic diblock copolymer (II) which can form stable inverse micelles in tetrahydrofuran with PVBC as the corona.

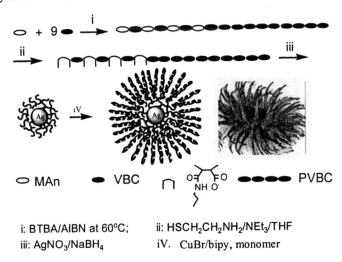

○ MAn   ● VBC   ∩ $O \overset{}{\underset{NH \ O^-}{\rceil} \overset{}{\underset{)}{} F O}$ ●●●● PVBC

i: BTBA/AIBN at 60°C;   ii: HSCH$_2$CH$_2$NH$_2$/NEt$_3$/THF
iii: AgNO$_3$/NaBH$_4$ iv. CuBr/bipy, monomer

Scheme 1

Thus silver ions can be embedded in the inner core of the micelle, and after the in-situ reduction with NaBH$_4$, stable silver nanoparticles were obtained. The ATRP of St was successively carried out on the surface of the Ag nanoparticles by using densely surface-anchored benzyl chloride groups in the PVBC segment to achieve an anemone-shaped polymer brush (ASPB).

Experimental

Materials. Azobisisobutylnitrile (AIBN, Aldrich) was recrystallized from methanol. Styrene (St, Beijing Chemicals Co.) and 4-vinylbenzyl chloride (VBC, Aldrich) were dried and redistilled before use. Maleic anhydride (MAn, Lushun Chemicals Co.) was recrystallized from benzene. S-benzyl dithiobenzoate (BTBA)was synthesized according to a literature method (12) CuBr was purified according to published procedure. 2-Mecaptoethylamine hydrochloride, 2,2'-bipyridine, AgNO$_3$, and NaBH$_4$ are used as received.

Characterization. The molecular weights and molecular weight distributions (M_w/M_n) were measured on a Waters 401 gel permeation chromatography at 35°C. THF was used as an eluent and polystyrene standards were used for calibration. The copolymer composition was determined by ^{1}H-NMR recorded on a Varian Gemini 200 MHz spectrometer in CDCl$_3$. UV-vis absorption spectra were recorded on a Schmadzu UV 2101PC spectrometer.

Dynamic Light Scattering (DLS). The average hydrodynamic radius (R$_h$) and R$_h$ distribution function, f(R$_h$) of these aggregates was determined using a self-regulating light scattering spectrometer (ALV/DLS/-5000) with a light wavelength of 514.5 nm. For each measurement, the solutions were allowed to stand at room temperature for 1 day to ensure that the solutions had reached equilibrium. The measurements of each solution sample were repeated for three times. The concentration of diblock copolymer in THF was 0.5 mg/mL.

Transmission Electron Microscopy (TEM). A solution of aggregates in THF (0.5mg/mL) was deposited onto the surface of 200 mesh Formvar-Carbon film-coated copper grids at 25°C. Excess solvent was removed with a filter paper. The samples were examined with a JEOL 1210TEM at a 100kv accelerating voltage.

Atomic Force Microscopy (AFM). Tapping-mode atomic force microscopy (AFM) observations were carried out in air with a commercial system (Seiko Instruments Inc., SPA300HV, Japan) operated under ambient conditions with SI-DF20 silicon cantilever. The samples for AFM analysis were prepared by depositing 1μL of the nanoparticle solution (0.5 mg/mL in THF) onto mica and allowing it to dry freely in air.

RAFT copolymerization of MAn with VBC. The RAFT polymerization was conducted in a sealed glass tube. S-Benzyl dithiobenzoate (BTBA) was used as one of the components of RAFT initiating system. In a typical run, MAn, VBC, BTBA and AIBN were charged in the glass tube in the molar ratio of 100: 900: 10: 1. After the mixture was degassed three times, the tube was sealed under vacuum, then was kept in an oil bath of 60 °C to conduct the polymerization. After 8 hr, the tube was broken and THF was added to cease the polymerization. The polymer was obtained by reprecipitation from methanol. The powdery pink polymer was dried under vacuum at 60 °C. The conversion of MAn is 100%, and the conversion of VBC is about 40%.

Preparation of polymer protected Ag nanoparticles. Block copolymer, $P(MAn-alt-VBC)_{10}-b-PVBC_{36}$ (20 mg) was dissolved in 40 mL of THF, and 10 mg of 2-mercaptoethylamine hydrochloride was added to the solution. The mixture was stirred at room temperature for 3 days to obtain a clear solution. To this solution was added 40 μl of 1 M $AgNO_3$ aqueous solution and again the mixture was stirred at room temperature for 2 days. Then, 40 mg of $NaBH_4$ was added, the clear solution change to dark blue after stirring at room temperature for 30 min. Stirring was continued for 2 days until a dark brown solution was obtained.

In-situ ATRP of St on the surface of Ag nanoparticles. The solvent in the above-prepared Ag nanoparticle solution was removed by evaporation. Then CuBr (4 mg), 2,2'-bipyridine (15 mg) and St (1 mL) was added to the dried Ag nanoparticles. The mixture was degassed for three times, the tube was sealed under vacuum, then was kept in an oil bath of 110 °C to conduct the polymerization. After 8 hr, the tube was broken and the polymer was obtained by precipitation from methanol. The ASPB sample was dried under vacuum at 60°C. The polymer yield is 80%.

Results and Discussion

Preparation of $P(MAn-alt-VBC)_m-b-PVBC_n$

It is known that the copolymerization of MAn with St via transition metal-mediated radical polymerization to directly prepare a well-defined alternating copolymer hardly takes place since the active anhydride residue would damage the catalytic system (13,14). Even though the NMP of MAn with St takes place at higher temperature, it does not provide an alternately structured copolymer due to high polymerization temperature performed (15) We have successfully synthesized a well-defined diblock copolymer of MAn with St composed by MAn-*alt*-St block and PSt block *via* RAFT copolymerization of MAn with excess amount of St initiating with AIBN and in the presence of S-benzyl dithiobenzoate (BTBA) at 60°C, and the resulting block copolymer self-assemble into well-defined nanoscale particles with narrow size distribution in aqueous medium after hydrolysis (*16*). Besides St in the MAn/St system, we demonstrated that this method could also be used for the preparation of P(MAn-*alt*-VBC)$_m$-*b*-PVBC through the RAFT copolymerization of MAn with an excess amount of VBC. As a representative run, the diblock copolymer (I) constituted by a block with alternately structured MAn/VBC copolymer block and PVBC block was also accomplished by simple one-pot procedure of MAn and VBC in the molar ratio of AIBN: BTBA: MAn: VBC=1:10:100:900 *via* RAFT at a moderate temperature of 60°C. SEC result indicated that the number average molecular weight of the polymer was 7,900 and the molecular weight distribution was 1.37. Based on the analysis of both gas chromatography (GC) and ^{1}H NMR, a diblock copolymer, P(MAn-*alt*-VBC)$_{10}$-*b*-PVBC$_{36}$, was successfully prepared, which was further hydrolyzed to form an amphiphilic diblock copolymer P(MA-*alt*-VBC)$_{10}$-*b*-PVBC$_{36}$ (II).

In situ formation of polymer-wrapped Ag nanoparticles

To improve the stability and interaction between metallic Ag and the diblock copolymer (II), mercapto groups were introduced to the diblock copolymer (I) as shown in Scheme 1. The mercaptonization of P(MAn-*alt*-VBC)$_{10}$-*b*-PVBC$_{36}$ was carried out in THF with 2-mercaptoethylamine. Then, AgNO$_3$ aqueous solution was added slowly to the above mercapto-modified amphiphilic diblock copolymer solution under agitation, yielding a nano-droplet of AgNO$_3$ aqueous solution shelled by the diblock copolymer via inverse micellization. The Ag$^+$ in the inner core of reverse micelles was *in situ* reduced by NaBH$_4$ to get black colored metallic Ag particles out-shelled by a corona with high density of benzyl chloride groups. Shown in Figure 1 are the UV-Vis absorption spectra of the AgNO$_3$ aqueous solution shelled by the diblock copolymer (II) via inverse micellization before and after the addition of NaBH$_4$. The absorption peak appeared at 415 nm after adding NaBH$_4$ to the solution indicated the formation of nanoscale Ag particles. The TEM photographs of the inverse micelles loading Ag$^+$ and Ag nanoparticles after reduction are shown in Figure 2. It can be seen that before reduction, the average diameters of the inverse micelles are in the range of 30 nm; when the loaded Ag$^+$ was reduced to metallic Ag, the apparent diameters of the Ag nanoparticle decreased to about 20 nm. This may be caused by the shrinkage of the inner core of the inverse micelle after the formation metallic Ag. Both of them are stable under the present conditions even when the solvent was removed. Therefore, they may be used for the subsequently surface initiated ATRP of St.

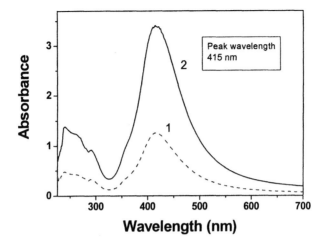

Figure 1 *UV-vis absorption spectra of the AgNO₃ aqueous solution shelled by the diblock copolymer via inverse micellization after the addition of NaBH₄. 1: 5min after adding NaBH₄; 2: 10min after adding NaBH₄.*

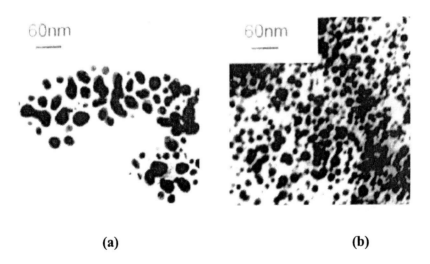

(a)   (b)

Figure 2 *TEM images of diblock copolymer inverse micelles loading Ag^+ (a) and Ag nanoparticles prepared (b).*

Preparation of ASPB *via* ATRP using surface-anchored benzyl chloride groups as the multi-initiators.

As mentioned hereinbefore, the benzyl chloride groups surface-anchored Ag nanoparticles can be used as the multi-initiators for surface-grafted ATRP of vinyl monomers such as St to prepare a polymer brush on the Ag nanoparticle. Since the ATRP should be conducted at higher temperature, whether the sphere-shaped Ag nanoparticles still keep their proper shapes during the ATRP is of our concern. Additionally, whether the amide and mercapto group embedded in the particle would damage the ATRP is also of our subsequent interest. In the present experiment, the surface-initiated graft ATRP of St was conducted at 110°C for 8 hours by using CuBr and 2,2'-bipyridine as the catalytic system and the benzyl chloride groups anchored onto the surface of Ag nanoparticles were acted as the multi-initiators. The monomer conversion was 80% as detected by GC, meaning that the amide and SH- groups did not affect the ATRP. This may be ascribed to the fact that amide and -SH groups inside the particle.

The ASPB thus obtained was characterized by means of atomic force microscopy (AFM), size-exclusion chromatography (SEC) and dynamic light scattering (DLS). AFM images are capable of providing the direct information about the shapes and sizes of particles in solid state. Thus, the changes of shape and size for (II)-shelled Ag nanoparticles before and after graft ATRP of St can be directly observed from the AFM images. It is seen from Figure 3 that the Ag nanoparticles still kept their regular spheres, there is almost no change in shape. However, the size of the grafted one became larger, that is, the average diameters of (II)-shelled Ag nanoparticles was ca. 19 nm, it was increased to ca. 30 nm

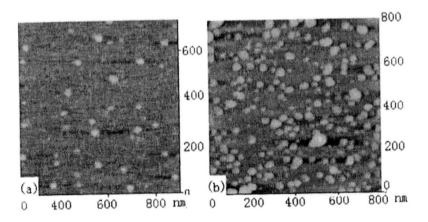

Figure 3 AFM images of Ag nanoparticles before (a) and after (b) grafting PSt.

after grafting. This indicates that the (II)-shelled Ag nanoparticles are stable enough to keep their proper spherical shape during the ATRP process. This may be attributed to the double protection, that is, Ag-S binding plus encapsulating by polymer layer (*17*).

The fluid average hydrodynamic radius (R_{av}) of started (II)-shelled Ag nanoparticles (a) and ASPB (b) were determined by SEC method. Figure 4 shows the SEC profiles for (a) and (b) by using THF as an eluent and linear PSt as a standard. The apparent number average molecular weight "M_n" of Ag nanoparticles-nucleated ASPB remarkably increased compared to that of the (a). The "M_n" of (a) was only about 4,000 with polydispersity of 1.40, whereas after grafting, the "M_n" of ASPB increased to as high as 180,000 with polydispersity of 1.42. This means that even though the size of polymer brush increased remarkably, the polydispersity did not change. Thus, the shapes of Ag nanoparticles and ASPB are still spherical. It should be pointed out that according to the particle size, the "M_n" of (II)-shelled Ag particle before and after grafted should be in the order of 10^5-10^6. However, we did not obtained such high "M_n", since the calibration standard was linear PSt not suitable for star-shaped PSt.

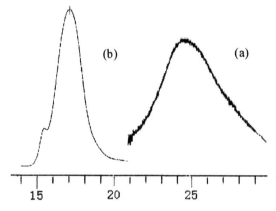

Figure 4 SEC profiles of Ag nanoparticles before (a) and after being grafted (b). (a) Ag-(II) $M_n=3950$, $M_w/M_n=1.40$; (b) ASPB $M_n=183900$, $M_w/M_n=1.42$.

Dynamic light scattering (DLS) was then used to measure the size and size distribution of (a) and ASPB (b) in solution state. Shown in Figure 5 are the distribution profiles of R_{av} of (a) and ASPB (b). DLS results confirmed that the surface-grafted ATRP of St proceeded undoubtedly and R_{av} increased after surface-grafted ATRP. The R_{av} of (a) was 9.5 nm. After surface-grafted ATRP of St, the R_{av} increased to 20.4 nm with a slight increase in size distribution width. The small peaks in the range of 1-5 nm may be from single PVBC chains or

350

PVBC-graft-PSt chains which did not assemble with Ag. This indicates that the grafting of St *via* ATRP on the surface of Ag nanoparticles resulted in the well-distributed increase of particles. In fact, the grafted (II)-shelled Ag particle can take its ASPB shape conformationally only in solution due to the flexibility of PSt hair. In solid state, the (II) forms an ultra thin film, consisting of a uniform sphere-shaped Ag nanoparticle core with polymer shell.

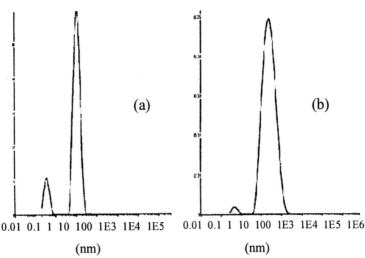

| 0.01 0.1 1 10 100 1E3 1E4 1E5 | 0.01 0.1 1 10 100 1E3 1E4 1E5 1E6 |
| (nm) | (nm) |

Figure 5 *Dynamic light scattering analysis of Ag nanoparticles before (a) and after grafted (b).*

In conclusion, ATRP and RAFT techniques were used to prepare well-defined ASPB in the successive steps. This new strategy may be extended to other noble metals such as gold, inorganic sulfides such as CdS, ZnS etc. The properties and detailed characterization of those ASPBs are under way.

Acknowledgement. This work was partially supported by the National Natural Science Foundation of China (under the contract No. 29992590-4).

References

1.   (a) Matyjaszewski, K.; Xia, J. H. *Chem. Rev.* **2001**, *101*, 2921. (b) Matyjaszewski, K.; Miller, P. J.; Shukla, N.; Immaraporn, B.; Gelman, A.; Luokala, B. B.; Siclovan, T. M.; Kickelbick, G.; Vallant, T.; Hoffmann, H.;

Pakula, T. *Macromolecules* **1999**, *32*, 8716.(c) Pyun, J.; Matyjaszewski, K.; Kowalewski, T.; Savin, D.; Patterson, G.; Kickelbick, G.; Huesing, N. *J. Am. Chem. Soc.* **2001**, 123, 9445.

2. Zhao, B.; Brittain, W. J. *Prog. Polym. Sci.* **2000**, *25*, 677.
3. Chang, Y. C.; Frank, C. W. *Langmuir* **1996**, *12*, 5824.
4. Zhao, W.; Krausch, G.; Rafailovich, M. H.; Sokolov, J. *Macromolecules* **1994**, *27*, 2933.
5. (a) Shah, R. R.; Merreceyes, D.; Husemann, M.; Rees, I.; Abbott, N. L.; Hawker, C. J.; Hedrick, J. L. *Macromolecules* **2000**, *33*, 597. (b) Huang, X.; Doneski, L. J.; Wirth, M. J. *Anal. Chem.* **1998**, *70*, 4023.
6. (a) Jordan, R.;, West, N.; Ulman, A.; Chou, Y. M.; Nuyken, O. *Macromolecules*, **2001**, *34*, 1606. (b) Zhao, B.; Brittain, W. *J. Am. Chem. Soc.* **1999**, *121*, 3557..
7. (a) Yamamoto, S.; Ejaz, M.; Tsujii, Y.; Fukuda, T. *Macromolecules*, **2000**, *33*, 5608. (b) von Werne, T.; Patten, T. E. *J. Am. Chem. Soc.* **2001**, *123*, 7497.
8. (a) Perruchot, C.; Khan, M. A.; Kamitsi, A.; Armes, S. P.; von Werne, T.; Patten, T. E. *Langmuir* **2001**, *17*, 4479. (b) Jones, D. M.; Huck, W. T. S. *Adv. Mater.* **2001**, *13*, 1256. (c) Kim, J. B.; Bruening, M. L.; Baker, G. L. *J. Am. Chem. Soc.* **2000**, *122*, 7616.
9. Zhao, B.; Brittain, W. J. *Macromolecules* **2000**, *33*, 8813.
10. Kong, X. X.; Kawai, T.; Abe, J.; Iyoda, T. *Macromolecules* **2001**, *34*, 1837.
11. Husseman, M.; Malmstrom, E. E.; McNamara, M.; Mate, M.; Mecerreyes, D.; Benoit, D. G.; Hedrick, J. L.; Mansky, P.; Huang, E.; Russell, T. P.; Hawker, C. J. *Macromolecules*, **1999** *32*, 1424.
12. Moad, G.; Chiefari, J.; Chong, Y. K.; Kristina, J.; Mayadunne, R. T. A.; Postma, A.; Rizzardo, E.; Thang, S. H. *Polym. Int.* **2000** *49*, 993.
13. Chen, G.-Q.; Wu, Z.-Q.; Wu, J.-R.; Li, Z.-C.; Li, F.-M. *Macromolecules* **2000**, *33*, 232.
14. Li, F.-M.; Chen, G.-Q.; Zhu M.-Q.; Zhou P.; Du, F.-S.; Li, Z.-C., In *Controlled Radical Polymerization*, ACS Symposium Series 768, Matyjaszewski, K., Ed.; American Chemical Society, Washington DC., 2000; p 384.
15. Benoit, D.; Hawker, C. J.; Huang, E. E.; Lin, Z. Q.; Russell, T. P. *Macromolecules* **2000**, *33*, 1505.
16. Zhu, M.-Q.; Wei, L.-H.; Li, M.; Jiang, L.; Du, F-S.; Li, Z-C.; Li F.-M. *Chem. Commun.* **2001**, 365.
17. Laibinis, P. E.; Bain, C. D.; Whitesides, G.. M. *J. Phys. Chem.* **1991**, *95*, 7017.
† The photo of Anemone Urticinopsis antarctica was downloaded from http://scilib.ucsd.edu/sio/nsf/fguide/cnidaria5.html.

Chapter 25

Surface Modification of Ethylene–Vinyl Alcohol Copolymer Films by Surface-Confined Atom Transfer Radical Polymerization

Ning Luo[1], Scott M. Husson[1,2,*], Douglas E. Hirt[1,2], and Dwight W. Schwark[3]

[1]Center for Advanced Engineering Fibers and Films and [2]Department of Chemical Engineering, Clemson University, 127 Earle Hall, Clemson, SC 29634
[3]Cryovac Division of Sealed Air Corporation, Duncan, SC 29334

Ethylene-vinyl alcohol (EVOH) copolymer films were used as substrates to grow surface-grafted poly(acrylamide) chains. In a first step, an initiator precursor, 2-bromoisobutyryl bromide, was used to functionalize surfaces of EVOH films. This step was performed from 2-bromoisobutyryl bromide solutions in a series of solvents to study solvent effects on initiator immobilization effectiveness and film integrity. In a second step, surface-confined atom-transfer radical polymerization (ATRP) was used to grow poly(acrylamide) from the initiator-functionalized EVOH films; $CuCl/Me_6TREN$ was used as a catalyst. For each step, changes in the physicochemical properties of the surface were monitored by ATR-FTIR spectroscopy and XPS. Using acetone as solvent, films of poly(acrylamide) were grown to about 8 nm thickness off of the EVOH films, with no distortion or visible changes in film transparency.

Ethylene-vinyl alcohol copolymers (EVOH) are prepared from the hydrolysis of copolymers of ethylene and vinyl acetate. The resulting vinyl alcohol units provide hydroxyl functionality along the polymer backbone, which makes EVOH different chemically from otherwise inert polyethylene(1). The existence of hydroxyl groups imparts two important features to the copolymers: high oxygen barrier properties and enhanced chemical reactivities. For the former, EVOH is used often in multi-layer films to improve barrier properties of polyethylene packaging films. In this case, the OH groups in EVOH copolymers can be used to react with tie layers so that the multi-layer films exhibit good adhesion(2-4). Regarding enhanced chemical reactivities, the OH groups serve as grafting points to incorporate additional chemical functionalities, including those used to initiate growth of other polymer chains from the EVOH surface(5-7). In the literature, this type of research generally concerns blood compatibility(6) and other medical applications(7-9). For example, Yao et al.(6) grafted poly(acrylamide) (hereafter, PA) on EVOH surfaces initiated by cerium (IV) ion. The permeability of urea through the PA-grafted EVOH film was improved compared to that of the original film, as was the blood compatibility. Kubota et al.(5) used a series of photo-initiators and cerium (IV) to graft PA on the surface of EVOH film. One difficulty that these researchers encountered was low graft efficiencies for the system studied. As reported(5), the graft efficiencies in their systems were less than 30%, indicating that a high percentage of monomer was consumed by solution-phase polymerization.

With the discovery of controlled radical polymerization in the mid-1990s(10,11), grafting polymers on surfaces with high density, high efficiency(12), and controllable polymer chain length (polymer brushes) has become an active topic in polymer science. As one of the controlled radical polymerization methods, atom-transfer radical polymerization (ATRP) shows great flexibility in producing surfaces with tailored physical and chemical properties. Roughly 100 journal articles and preprints have been written that describe research using ATRP to grow polymers from gold(13,14), silica(15,16), silicon(17,18), latex(19,20) and polymer surfaces(21). Substrates have included flat wafers, microspheres, glass filters(22), and fibers(21). By exploiting the living nature of ATRP, block copolymer surface structures have been prepared; using polymeric macroinitiators as the foundation, the "graft-from" approach has led to core-shell particles, wormlike micelles, and star polymers. However, only few described ATRP from polymer surfaces(19-21). Generally speaking, the chemical reactions performed on the surfaces of thin polymer films require mild conditions, relative to those used with gold, silica, or silicon substrates. Reaction temperatures close to room temperature are preferred, and good solvents for the polymer film should be avoided.

This contribution describes growth of surface-confined poly(acrylamide) (PA) by ATRP from the surfaces of transparent, ethylene-vinyl alcohol (EVOH) copolymer films. The existence of OH groups on the surface of EVOH film allows us to use typical ATRP chemistry in which bromoester groups serve as initiators for the surface grafting. Scheme 1 shows the reaction sequence used to prepare surface-confined polymers. The first reaction step immobilizes the initiator precursor, 2-bromoisobutyryl bromide, covalently to the EVOH surface. In a second step, the PA grows from the surface by ATRP.

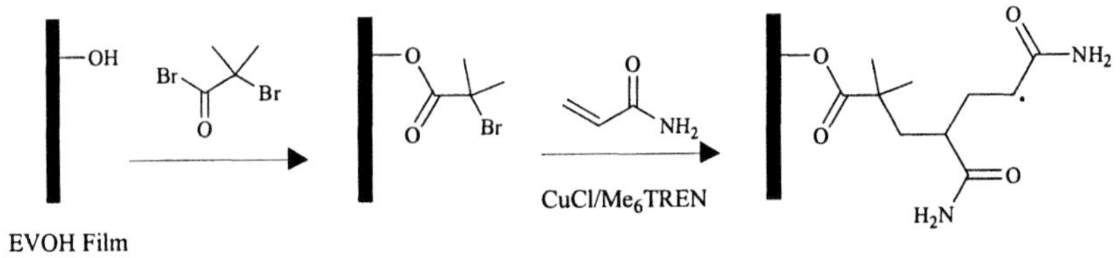

Scheme 1. Reaction steps of ATRP of PA from the surface of EVOH film.

Acrylamide was selected as the grafting monomer because of interest to impart hydrophilic surface character to hydrophobic ethylene-based polymer films. Reactions were performed at room temperature using the CuCl/Me$_6$TREN catalyst system. Data are presented that demonstrate the efficacy of using ATRP chemistry to grow surface-confined PA from EVOH films under conditions that preserve the starting properties of the film, including transparency.

Experimental Section

Materials

EVOH (62 mol% vinyl alcohol) copolymer films (thickness = 0.038 mm) were used as received from Cryovac Division of Sealed Air Corporation (Duncan, SC). 2-Bromoisobutyryl bromide (98%, Aldrich) was used in the initiator attachment reaction without further purification. Acrylamide (99+%, electrophoresis grade, Aldrich) was used as received. Toluene, tetrahydrofuran (THF), acetone, ethylene acetate, hexane, n-heptane, and methyl isobutyl ketone (MiBK) were purchased as ACS reagent grade and used as received.

Initiator Immobilization on EVOH Films

A ~1.5 cm × 5 cm piece of EVOH film was put in 200 mL of a solution of 3 % (w/w) 2-bromoisobutyryl bromide in an organic solvent to perform the initiator immobilization. The immobilization occurs by reaction of the acid bromide with hydroxyl groups on the EVOH surface; HBr is a by-product.

Various solvents were used to prepare the 2-bromoisobutyryl bromide solutions, including toluene, n-heptane, n-hexane, acetone, ethyl acetate, tetrahydrofuran (THF), methyl isobutyl ketone (MiBK), methylene chloride and mixtures thereof. Each reaction was performed for > 12 hours at room temperature (24 ± 3 °C) with stirring. The film was then removed, washed with 100 mL of ethyl acetate in an Aquasonic ultrasonicator, and then washed with a copious volume of water and then acetone. After drying in air, the initiator-functionalized film was characterized by ATR-FTIR spectroscopy and, in some cases, XPS. Hereafter, surfaces functionalized with initiator are referred to as EVOH-Br.

Polymerization Procedure

This reaction used an organometallic catalyst comprising Cu(I)Cl and ligand, hexamethyl tris(2-aminoethyl)amine (Me$_6$TREN), synthesized via methylation of tris(2-aminoethyl)amine(*23*). All polymerizations were carried out under the same conditions: The molar ratio of Cu(I)Cl to Me$_6$TREN was 1 to 1, and the molar concentration of acrylamide was 3M. Two solvents were used to perform the polymerization: water and acetone.

A typical polymerization run follows: 1.725 g of Me$_6$TREN was added to 25.0 mL of dimethylformamide (DMF). A piece of initiator-immobilized film, 0.29 g of Cu(I)Cl, and 4.27 g of acrylamide were put into a separate flask. To this flask, 20.0 mL of deionized water (18.2MΩ, Millipore) was added to dissolve the solid acrylamide. The solution was subjected to 3 freeze-thaw cycles with vacuum evacuation and nitrogen purging to remove oxygen. To begin polymerization, 1.0 mL of the Me$_6$TREN/DMF solution was transferred to the flask containing a piece of the initiator-immobilized film, Cu(I)Cl, and monomer. The reaction solution volume was enough to submerge the EVOH-Br film fully. Polymerization was performed at room temperature for a prescribed time up to 7 h. After removing the film from the polymerization system, it was sonicated in 400 mL of water for 15 min, and then washed with acetone.

Characterization Methods

Attenuated total reflectance (ATR)-FTIR spectra of the polymer films were obtained using a Nicolet Avatar 360 FTIR spectrometer equipped with a nitrogen-purged chamber. ATR was conducted with a horizontal multibounce attachment using a Germanium crystal and a 45-degree angle of incidence. All spectra were taken at 4 cm^{-1} resolution and reported as an average of 540 scans.

To measure the thickness of PA grown from an EVOH film, ATR-FTIR data were analyzed by a procedure similar to that described in Ref. 24; in this work, a 1 mm thick PA film was used as a reference. According to the principle of ATR-FTIR spectroscopy, the analysis penetration depth can be calculated as

$$d_p = \frac{\lambda}{2 \cdot \pi \cdot n_1 \left[\sin^2(\Theta) - \left(\dfrac{n_2}{n_1}\right)^2 \right]^{0.5}} \tag{1}$$

where n_1 and n_2 are the refractive indices of the crystal and sample, λ is the wavelength of interest, and Θ is the angle of incidence. For our ATR instrument, $\Theta = 45°$ and $n_1 = 4.0$. The average penetration depth was calculated to be 408 nm for pure PA film ($n_2 = 1.546$) over the wavenumber range of 1640 ~ 1660 cm^{-1}. The thickness of a grafted layer can be estimated from:

$$d(nm) = \frac{A_{grafted}}{A_{PAAm}} \cdot (408) \tag{2}$$

where $A_{grafted}$ is the area of the peak whose maximum absorbance is located ~ 1655 cm^{-1} for the surface grafted sample, and A_{PAAm} is the corresponding peak area of the amide peak (1655 cm^{-1}) of pure PA reference. There are several assumptions inherent to Eq. 2 as described in the Results and Discussion section.

All XPS data were obtained using a Kratos AXIS 165 XPS Spectrometer equipped with a monochromated Al Kα (1486.6 eV) X-ray source and hemispherical analyzer; this unit is housed at Cryovac.

Results and Discussion

Attachment of 2-Bromoisobutyryl Bromide to EVOH Surfaces

Figure 1 shows the ATR-FTIR spectra of plain EVOH film (Spectrum A) and initiator-functionalized EVOH film (Spectrum B). In this experiment, a mixture of toluene (60% by volume) and methylene chloride (40% by volume) was used to dissolve 2-bromoisobutyryl bromide for the initiator immobilization on EVOH. In the spectrum of plain EVOH film (A), there are no peaks in the wavenumber region between 2000 ~ 1600 cm^{-1}, which indicates that there is no measurable (by ATR-FTIR) concentration of carbonyl ester groups in the EVOH film. This finding is reasonable and supports that conversion of acetate ester groups to OH groups was efficient in the post-polymerization hydrolysis of ethylene-vinyl acetate copolymers to form EVOH. Comparing the spectra before (A) and after (B) the initiator immobilization, one sees the appearance of a carbonyl peak located at 1732 ± 12 cm^{-1}, which indicates the existence of the bromoester groups in the surface region of the EVOH film.

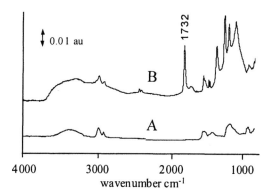

*Figure 1. Surface immobilization of 2-bromoisobutyryl bromide on EVOH film.
A: Blank EVOH film, as received. B: Initiator-functionalized EVOH film.*

XPS data further support the existence of surface-immobilized bromoester groups. Table I shows that Br exists in the EVOH-Br sample, and the carbonyl-group content is higher for the functionalized surface. Because of the reaction chemistry used and the extensive solvent washing of films prior to characterization, we conclude that the initiator groups were bonded covalently to the EVOH surfaces.

Table I. XPS Characterization of Initiator-Functionalized EVOH Films

	Atomic Composition (%)			*Percent of Carbon Atoms Present in Carbonyl Groups: O-C=O, C=O*
	C	*O*	*Br*	
EVOH plain film	86.9	13.1		1.7
EVOH-Br	73.2	21.4	5.4	9.7

Solvent Selection for the Initiator Immobilization Reaction

Any surface modification method applied to packaging films must, first and foremost, occur without distortion of the film properties (e.g., transparency). The above results demonstrate the efficacy of using chemical reaction to anchor initiator sites on EVOH film. However, when the reaction was conducted in a toluene/methylene chloride mixture, the EVOH film became distorted and non-transparent. Our hypothesis was that film distortion resulted from strong interaction between the solvents and the EVOH film. Therefore, various solvent systems were studied for the initiator immobilization reaction of 2-bromoisobutyryl bromide onto EVOH films. The aim was to identify a solvent system that provides high initiator density and maintains film integrity. To test this hypothesis, we selected solvents differing in polarity and hydrogen-bonding capacity, as measured by solubility parameter data.

Table II lists the pure-component solvents used, along with their solubility parameter data. While values of δ are similar for all solvents, differences exist in their abilities to form hydrogen bonds (δ_h) and in their polarities (δ_p). At one end, n-hexane and n-heptane are non-polar and have no capability to form hydrogen bonds. The δ_h and δ_p of toluene are much less than those of MiBK, methylene chloride, acetone, THF, and ethyl acetate, which are polar solvents. Of those, ethyl acetate has the highest hydrogen bonding capability.

Figure 2 shows experimental ATR-FTIR results for the initiator immobilization reaction in various pure solvents. The peak of interest occurs at about 1732 cm^{-1}, and corresponds to the carbonyl group of the surface-immobilized initiator. Clearly, toluene, n-hexane, and n-heptane provided a high content of surface-immobilized initiator; however, they also distorted the films. Acetone, THF, MiBK, and ethyl acetate showed lower contents of surface initiator than toluene but these solvents maintained film integrity. The increased content of surface-immobilized initiator seen with nonpolar solvents might result from higher positive deviations from ideal behavior in the liquid phase, from increased swelling of the films, or a combination of these effects. Because 2-bromoisobutyryl bromide is polar and capable of hydrogen bonding, it will have stronger intermolecular interactions with polar, hydrogen-bonding solvents than with nonpolar solvents. Thus, it should be more reactive toward the polar hydroxyl groups on the EVOH surface when placed in a nonpolar solvent. In this way, the concentrations of 2-bromoisobutyryl bromide in nonpolar solvents near a film surface are probably higher than that in polar solvents. Alternatively, high conversion of OH groups and interpenetration of 2-bromoisobutyryl bromide in nonpolar solvent systems might be attributed to increased swelling of the EVOH film and, correspondingly, improved access to hydroxyl groups. Previously inaccessible reaction sites might now be available for reaction. This might account for why the films were distorted while at the same time had high initiator content.

Among the polar solvents, it is not clear how, or whether, degree of solvent polarity (as measured by δ_p) affects the initiator immobilization reaction. All polar solvents, with the exception of ethyl acetate, gave fairly constant ATR-FTIR peak height at the wavenumber characteristic of carbonyl groups; this result indicates similar initiator content in the near-surface EVOH regions. As the exception, ethyl acetate has the highest hydrogen bonding capability and lowest peak height among these polar solvents.

Table II. Solubility Parameters of the Chemicals Used in this Study[a]

Solvents and polymers	δ $(J^{1/2}/cm^{3/2})$	δ_d $(J^{1/2}/cm^{3/2})$	δ_p $(J^{1/2}/cm^{3/2})$	δ_h $(J^{1/2}/cm^{3/2})$
Heptane	15.2	15.2	0	0
Hexane	14.8 ~ 14.9	14.8	0	0
Toluene	18.2 ~ 18.3	17.3 ~ 18.1	1.4	2.0
Methyl isobutyl ketone	17.2 ~ 17.5	15.3	6.1	4.1
Methylene chloride	19.9	17.4 ~ 18.2	6.4	6.1
Acetone	20.0 ~ 20.5	15.5	10.4	7.0
Tetrahydrofuran	19.5	16.8 ~ 18.9	5.7	8.0
Ethyl acetate	18.6	15.2	5.3	9.2
Poly(vinyl alcohol)	25.8 ~ 29.1	24.4 ~ 33.7	23.8 ~ 31.9	11.6 ~ 13.4
Polyethylene	15.8 ~ 17.1	15.8 ~ 17.1	0	0
EVOH (38/62)[b]	19.6 ~ 22.2	21.2 ~ 26.4	14.8 ~ 19.8	7.2 ~ 8.3

[a] Cited from Ref. 25. [b] Calculated according to Ref. 25.

Nonpolar-polar solvent mixtures were investigated for achieving high initiator content while maintaining the original properties of the films. Distorted films were produced in hexane/acetone (50/50) (v/v) and toluene/acetone (50/50) solvent systems. The films in ethyl acetate/hexane (50/50) and toluene/MiBK (50/50) were not distorted, but the initiator contents were low. Figure 3 shows ATR-FTIR results for the initiator immobilization reaction in various mixed solvents. Pure toluene and a 90/10 mixture of toluene/ethyl acetate distorted the films and turned them brown. However, in the composition range of 80/20 ~ 60/40 toluene/ethyl acetate, the initiator contents were similar, and the films showed no distortion. Unfortunately, these conditions produced films with the same initiator content of the pure, polar solvents. Therefore, initiator immobilization was done using pure acetone for all films prepared for surface-confined polymerization.

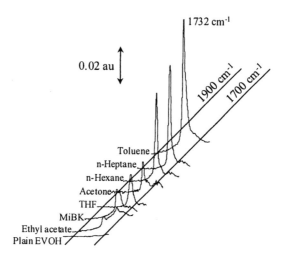

Figure 2. Comparison of the ATR-FTIR spectra supporting the initiator immobilization reactions of 2-bromoisobutyryl bromide on EVOH films from various pure solvents. For clarity, data are shown from only the wavenumber region used to identify the initiator.

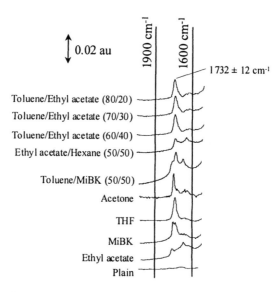

Figure 3. Comparison of the ATR-FTIR spectra supporting the initiator immobilization reaction on EVOH films from all solvent systems that produced non-distorted films.

Growth of Surface-Confined PA from Initiator-Functionalized EVOH Film

Figure 4 summarizes the observations and appearances of the PA-grafted EVOH films. The plain EVOH film begins as a transparent, wrinkled film (Figure 4 a). After growth of surface-confined PA by ATRP using water as the reaction solvent, the PA-grafted EVOH film became very soft and distorted after only 1 hour of polymerization. Subsequent washing with acetone hardened the film, but did not restore the film to its original size and shape (Figure 4 b). Grafting with acetone as the reaction solvent resulted in non-distorted, transparent films. The films kept their size and shape, and displayed no visible color (Figures 4 c, d). After polymerization, the reaction solutions were poured into 400 mL of methanol to determine whether any PA had formed in solution. In all cases, there was no detectable precipitation of PA, which demonstrated that only surface polymerization had occurred (i.e., high graft efficiency was obtained). This result contrasts the work of Kubota et al. (5), who reported that when ceric salt or photo-irradiation was used to graft PA on EVOH surface, graft efficiencies were less than 30%.

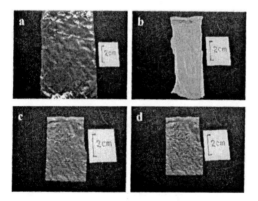

Figure 4. Photographs of PA-grafted EVOH films. (a): EVOH plain film.
(b): EVOH-PAAm, solvent water, polymerization for 1 h;
(c): EVOH-PAAm, solvent acetone, polymerization for 1 hr;
(d): EVOH-PAAm, solvent acetone, polymerization for 7 hr.

Figure 5 shows that the ATR-FTIR spectra of EVOH films following polymerization display amide character (~1655 cm⁻¹), which suggests that PA was grown from the EVOH surfaces. Also present in the spectra are peaks that correspond to the bromoester initiator (1739 ± 4 cm^{-1}). Figure 6 shows the time dependent thickness of PA grown from EVOH using acetone as the reaction solvent. Over the 7 hour period investigated, graft thickness, as estimated using

362

Eq. 2, increased slightly and then reached a plateau below 10 nm. *Based on these estimations*, the thickness appears to grow non-linearly with polymerization time; for flat, low surface area substrates, non-linear growth rate behavior indicates that growing chain concentration is non-constant(*18,26*). This result is consistent with findings from other ATRP studies that used acrylamide or methacrylamide(*27,28*). It also suggests that an insufficient concentration is generated of persistent, deactivating Cu^{2+} species. To ensure proper control for this system, one would need to use sacrificial initiator in solution, or, preferably, add Cu^{2+} prior to polymerization.(*29*)

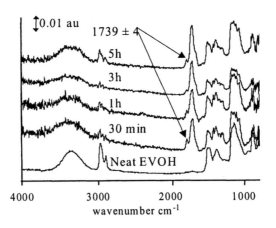

Figure 5. ATR-FTIR spectra of EVOH films as a function of polymerization time. Data are for growth of PA from EVOH films by ATRP.

It must be noted, however, that the thickness estimations (Eq. 2) assume that the refractive index of the ATRP-grown PA layer is the same as that of the PA reference film, which may not be exactly true; but the calculation of d_p is relatively insensitive to n_2 in the range $1.5 < n_2 < 1.6$. More importantly, the estimates are based on an approximation that the absorbance of the peak of interest increases linearly with the thickness of PA. This approximation is analogous to the Beer-Lambert law used for transmission measurements. The approximation is valid for bands with weak absorbances. Efforts are under way in our lab to develop a quantitative method for correlating IR peak absorbance to graft layer thickness using films of known graft thickness for calibration. These efforts will allow us to test the linear approximation between absorbance and layer thickness for this system.

The *apparent* steep increase in thickness seen in Figure 6 over the first 30 minutes may also be a result of the latter approximation used in Eq. 2, which assumes that the absorbance per unit thickness (β) remains constant throughout the film penetration depth. In reality, β decreases from the film surface into the bulk. For measurements on thin layers at the film surface, actual β values for these thin films will be higher than that used to calibrate Eq. 2. Thus, for a given measured absorbance, Eq. 2 provides an overestimate of thickness.

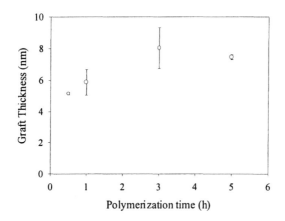

Figure 6. PA graft thickness on EVOH film as a function of polymerization time. The graft thickness was calculated using Eq. 2.

Conclusions

PA was grafted on the surface of EVOH films using ATRP chemistry. At the initiator immobilization step, non-polar solvents, toluene, hexane, and heptane, caused distortion of the films. Some polar solvents, like acetone and THF, provided non-distorted, transparent film, as did some non-polar/polar solvent mixtures, like toluene/ethyl acetate. Graft polymerization of PA can be done in acetone to produce non-distorted and transparent films.

Acknowledgments

This work was supported in part by the Cryovac Division of Sealed Air Corporation and the National Science Foundation under Grant Numbers CTS-9983737 and EEC-9731680.

References

1. Jenkins, W.A.; Harrington, J.P. *Packaging Foods with Plastics.* Technomic Press: Lancaster, PA, 1991; pp. 35-51.
2. Villalpando-Olmos, J.; Sanhez-Valdes, S.; Yanez-Flores, I. G. *Polymer Engineering and Science* **1999**, *39*, 1597-1603.
3. Demarquette, N. R.; Kamal, M. R. *J. Appl. Polym. Sci.* **1998**, *70*, 75-87.
4. Wen, J. Y.; Wilkes, G. L. *Polym. Bull.* **1996**, *37*, 51-57.
5. Kubota, H.; Sugiura, A.; Hata, Y. *Polym. Int.* **1994**, *34*, 313-317.
6. Yao, K. D.; Liu, Z. F.; Gu, H. Q.; Fan, T. Y. *J. Macromol. Sci., Chem.* **1987**, *A24*, 1191-1205.
7. Ikada, Y.; Iwata, H.; Mita, T.; NaGaoka, S. *J. Biomed. Mater. Res.* **1979**, *13*, 607-622.
8. Peng, C. Y.; Tsutsumi, S.; Matsumura, K.; Nakajima, N.; Hyon, S. H. *J. Biomed. Mater. Res.* **2001**, *54*, 241-246.
9. Matsumura, K.; Hyon, S. H.; Nakajima, N.; Peng, C.; Tsutsumi, S. *J. Biomed. Mater. Res.* **2000**, *50*, 512-517.
10. Georges, M. K.; Veregin, R. P. N.; Kazmaier, P. M.; Hamer, G. K.; Saban, M. *Macromolecules* **1994**, *27*, 7228-7229.
11. Wang, J. S.; Matyjaszewski, K. *J. Am. Chem. Soc.* **1995**, *117*, 5614-5615.
12. Wang, X. S.; Luo, N.; Ying, S. K. *Polymer* **1999**, *40*, 4515-4520.
13. Shah, R. R.; Merreceyes, D.; Husemann, M.; Rees, I.; Abbott, N. L.; Hawker, C. J.; Hedrick, J. L. *Macromolecules* **2000**, *33*, 597-605.
14. Huang, X. Y.; Doneski, L. J.; Wirth, M. J. *Chemtech* **1998**, *28*, 19-25.
15. Huang, X. Y.; Doneski, L. J.; Wirth, M. J. *Anal. Chem.* **1998**, *70*, 4023-4029.
16. Huang, X. Y.; Wirth, M. J. *Anal. Chem.* **1997**, *69*, 4577-4580.
17. Tsujii, Y.; Ejaz, M.; Yamamoto, S.; Fukuda, T.; Shigeto, K.; Mibu, K.; Shinjo, T. *Polymer* **2002**, *43*, 3837-3841.
18. Gopireddy, D.; Husson, S. M. *Macromolecules* **2002**, *35*, 4218-4221.
19. Min, K.; Hu, J. H.; Wang, C. C.; Elaissari, A. *J. Polym. Sci. Pol. Chem.* **2002**, *40*, 892-900.
20. Jayachandran, K. N.; Takacs-Cox, A.; Brooks, D. E. *Macromolecules* **2002**, *35*, 6070-6070.
21. Carlmark, A.; Malmstrom, E. *J. Am. Chem. Soc.* **2002**, *124*, 900-901.
22. Ejaz, M.; Tsujii, Y.; Fukuda, T. *Polymer* **2001**, *42*, 6811-6815.
23. Xia JH, Gaynor SG, Matyjaszewski K. *Macromolecules* **1998**, *31*, 5958-5959.
24. Xiao, D. Q.; Wirth, M. J. *Macromolecules* **2002**, *35*, 2919-2925.
25. Van Krevelen, D. W. *Properties of Polymers*, 3rd ed.; Elsevier: Amsterdam, 1990; pp. 189-224.

26. Husseman, M.; Malmstrom, E. E.; McNamara, M.; Mate, M.; Mecerreyes, D.; Benoit, D. G.; Hedrick, J. L.; Mansky, P.; Huang, E.; Russell, T. P.; Hawker, C. J. *Macromolecules* **1999**, *32*, 1424-1431.
27. Teodorescu, M.; Matyjaszewski, K. *Macromolecules* **1999**, *32*, 4826-4831.
28. Rademacher, J. T.; Baum, M.; Pallack, M. E.; Brittain, W. J. *Macromolecules* **2000**, *33*, 284-288.
29. Matyjaszewski, K.; Miller, P.J.; Shukla, N.; Immaraporn, B.; Gelman, A.; Luokala, B.B.; Siclovan, T.M.; Kickelbick, G.; Vallant, T.; Hoffmann, H.; Pakula, T. *Macromolecules* **1999**, *32*, 8716 – 8724.

Chapter 26

Polymers, Particles, and Surfaces with Hairy Coatings: Synthesis, Structure, Dynamics, and Resulting Properties

Tadeusz Pakula[1], Piotr Minkin[1], and Krzysztof Matyjaszewski[2]

[1]Max-Planck-Institute for Polymer Research, Postfach 3148, 55021 Mainz, Germany
[2]Center for Macromolecular Engineering, Department of Chemistry, Carnegie Mellon University, 4400 Fifth Avenue, Pittsburgh, PA 15213

Summary: In this chapter, we discuss problems related to the synthesis, structure and dynamic behavior of macromolecular systems, in which many linear chains are anchored to other macromolecules, to nanoparticles or, in a general sense, to surfaces forming "hairy" coatings at these objects. Computer simulated and real systems are considered. In the preparation of such systems, both by means of adsorption and by means of synthesis, strongly competitive situations on the molecular level take place, which result in various, not necessary desired effects, such as high polydispersity of surface layers or backbone extension in molecular brushes. It is demonstrated how the considered molecular structures can influence the system dynamics and consequently the mechanical properties. Examples of new super soft materials based on a specific class of macromolecular architectures are presented.

Introduction

Examples of systems we will discuss in this chapter are illustrated in Fig. 1. All of these macromolecular, colloidal and macroscopic objects contain a coating layer consisting of end-grafted macromolecular chains. Such a coating can strongly influence interactions of the considered objects with the surrounding because of a large variety of effects related to the new volume requirements, new chemical composition and new dynamics at the object peripheries. The hairy macromolecular structures can be considered as constituents of materials with new interesting properties.

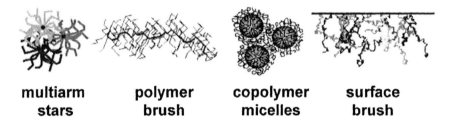

multiarm   polymer   copolymer   surface
stars   brush   micelles   brush

Figure 1. Examples of systems consisting of macromolecules, particles and surfaces with coating layers formed by end-grafted polymers

There is a considerable interest in controlled preparation of thin polymer layers with chains covalently bound to surfaces. Traditional methods consisting of adsorption of end-functionalized polymers and bounding them to the substrate show disadvantages, because an adsorption of the first fraction of chains is hindering access of additional macromolecules, which results in limited density of the layers. An alternative method has been proposed, in which surface initiated chain growth was expected to impose less restrictions on the formation of dense brushes (1,2). The latter method has been used for growth both from planar surfaces and from small particles with diameters down to the nanometer range. Similar technique has been applied to the synthesis of molecular brushes, in which a polymer chain is used as a macroinitiator, from which a large number of side chains grows (3).

Various kinds of brushes have been extensively studied theoretically as model systems, however, the most detailed information about the structure of brushes and the conformation of chains has been obtained from computer simulations (4). The "cooperative motion algorithm" (CMA), which is particularly suitable for simulation of dense polymer systems, has been very successful for simulation of both dry and wet layers of polymers consisting of chains grafted by one end at a surface. Various specific features of the layers such as orientation and extension of chains, concentration profiles and distributions of free ends have been established to primarily depend on the length of chains, the density of grafting and on the interaction with the surrounding medium.

In this chapter, we discuss some examples of recent advances in possibilities of control of synthetic procedures leading to these structures as well as the advances in characterization and understanding the properties of materials based on the complex constituents having their internal structures and dynamics.

To a large extent this discussion is based on the results of computer simulation, which is considered as a guide indicating reasonable directions for experimental studies.

Methods

Simulation: The method used was the cooperative motion algorithm (CMA) [5], in which macromolecules are approximated as beads connected by non breakable bonds in order to represent structures corresponding to macromolecular skeletons (see Fig. 1). The dynamics is simplified to strictly cooperative rearrangements taking place on a lattice system with all lattice sites occupied by the macromolecular elements or by single beads representing a solvent. The method has been described in the literature (4-7), nevertheless, it is useful to stress here some unique properties, which made it particularly effective for the studies of molecular and macromolecular systems within the size scale corresponding to the nanometer range. The cooperative rearrangements involve displacements of beads, which lead to changes of macromolecular conformations, however, in such a way that identities of macromolecules given by the specific architecture of the bond skeletons as well as by the sequences of beads within the skeletons remain preserved. A large variety of rearrangements are possible and therefore, it is not possible to specify all of them. The CMA is suitable for the studies of static properties of macromolecular models as well as for characterization of their dynamic behavior in a broad time range (5). Simulations are usually performed for three dimensional systems on the fcc lattice.

Motion of simulated molecules allows generating a large number of states, which can be averaged to get representative information about the structure of molecules in equilibrium. By monitoring displacements and orientations of the model molecules and of their elements in time, one gets information about the dynamics. The properties of the method important for simulation of complex macromolecular systems are: (i) the method is very flexible in representing various kinds of complex macromolecules with variable degree of simplification,

(ii) the model for the dynamics is the same for both small simple molecules and large macromolecules and does not require adjustments to particular structures, (iii) systems can be generated according to various polymerization processes taking place in space, which are good representations of reality with distributions of macromolecular masses, forms and constitutions and (iv) the simulation is very fast, which allows to consider large systems in a broad time range on small computers (i.e. PC).

Synthesis of brush polymers: The brush polymers have been obtained by grafting-from copolymerization using ATRP (*8,9*). The macroinitiator (poly(2-(2-bromopropionyl-oxy)ethyl methacrylate), pBPEM) was dissolved in chlorobenzene followed by three cycles of freeze-pump-thaw to exclude any oxygen from the reaction. Separately, a measured amount of catalyst (CuBr) and dNbpy were place in a Schlenk flask followed by degassing under vacuum and backfilling three times with N_2. Into the Schlenk flask was added a measured amount of *n*-butyl acrylate monomer (nBA) after degassing through bubbling of N_2 for 30 min. The catalyst and monomer solutions were transferred to the solution at room temperature under N_2. The polymerization reactor was immersed in an oil bath, which was preset to the specified reaction temperature. At timed intervals, samples were taken out from the flask via syringe.

Characterization: The structure of the bottle-brush polymer melts has been analyzed by means of the small angle X-ray scattering (SAXS) using a rotating anode X-ray source (Rigaku) with the position sensitive detector (Brucker). In order to characterize the dynamics, mechanical measurements have been performed using the mechanical spectrometer RMS 800 (Rheometric Scientific). Frequency dependencies of the complex shear modulus at a reference temperature (master curves) have been determined form frequency sweeps measured at various temperatures with a small amplitude sinusoidal deformation.

Results

Generation of systems: Off-space models for description of kinetics of polymerization processes can not account for a number of effects such as diffusion control of reaction rates, cyclization and steric or topological barriers. All of these effects can become very important steric or topological barriers. All of these effects can become very important when bulk polymerization of complex macromolecules is attempted. In order to demonstrate effects, which may be observed when polymerization is taking place in situations of a strong competition between growing sites, we consider here three cases with different

370

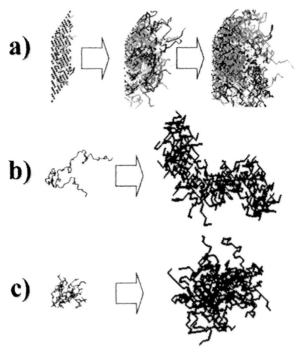

Figure 2. Illustration of various polymerizations initiating structures and resulting macromolecular "hairy" systems

initiator structures: (1) a planar surface, (2) a flexible linear chain and (3) a dendrimer from, which after polymerization surface brushes (Fig 2a), brush-like polymers (Fig. 2b) and multiarm stars (Fig. 2c) are obtained, respectively.

The first example describes the case of bulk polymerization initiated from a planar surface. Figure 3 shows concentration profiles of the growing polymers at various times after activation of the initiator fixed at the wall. Increase of the polymer layer thickness and layer density with time is well seen. With the progress of the reaction, the chains grow but not necessarily uniformly, which can lead to changes of chain length distributions, as illustrated in Figure 4.

The highlighted stages correspond to DP of $n = 12$, 54 and 92. In the case of relatively low initiation density ($\sigma=0.25$), the respective polydispersities of chains grown are $M_w/M_n = 1.11$, 1.06 and 1.06 and are remarkably larger than for unperturbed chains grown in solution (1.08, 1.02 and 1.01, correspondingly). Nevertheless, under such conditions, an almost linear dependence of the dry layer thickness on time has been observed, as in the reported experimental observations (*10*).

For other conditions of the layer growth, e.g. with a higher density of the initiator at the surface ($\sigma > 0.5$), much broader chain length distributions and a

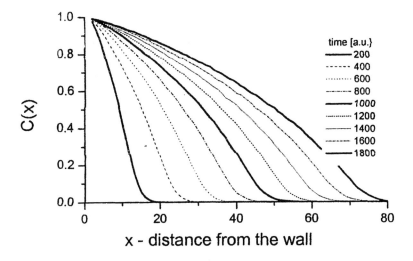

Figure 3. Evolution of polymer concentration distributions with time in a simulated system of polymers growing from the initiator attached to the wall. Initiation density σ=0.25.

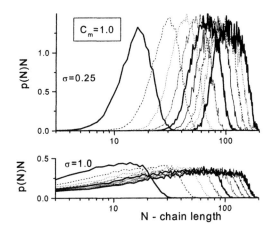

Figure 4. Effect of initiation density (σ) on distributions of length of linear polymers grown from the planar surface. The evolution of chain length distributions with time is shown for two initiation densities.

non-linear conversion versus time dependences have been observed, as illustrated in Figures 4 and 5, respectively.

The example shown illustrates that the films obtained by the surface initiated polymerization can exhibit a distribution of chain lengths, which can considerably change the layer structure in comparison with the structure of brushes consisting of uniform polymers, as usually considered. Distributions may become very broad when the initiation density is high and for systems with additional bimolecular activation-deactivation equilibria, which require diffusion of the activator and the deactivator to the chain end, as in ATRP systems.

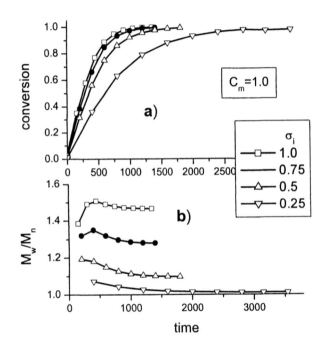

Figure 5. Effect of initiation density (σ) on dependencies of conversion and polydispersity vs. time for linear polymers grown from the planar surface. Bulk polymerization in a layer of fixed thickness is considered in this example.

In the case of polymerization initiated from a linear macroinitiator, brush-like macromolecules are obtained. Both the initiation density and the polymerization mechanism can influence the structure of such a complex macromolecular architecture. In the case of high density of the initiation the growth of side chains involves extension of the backbone. This is an important effect, which can strongly influence properties of materials containing such macromolecules. An example of the variation of the backbone end-to-end distance with the reaction progress is illustrated in Figure 6.

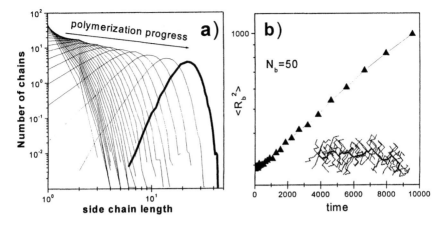

Figure 6. (a) Variation of side chain length distribution with the reaction progress in a system obtained by polymerization from a linear macroinitiator. (b) Extension of the backbone accompanying the side chain growth in the same system.

When the growing chains can fill the whole space around a small multifunctional initiator, as for example in the case of multiarm stars generated by polymerization initiated from end functionalized dendrimers, there is no effect on the arm size and arm polydispersity as long as the generation of the initiator is not higher than g=5 (64 arms). An example of star dimensions (mean squared radius-of-gyration of stars) vs. total mass distributed to various numbers of arms and various arm lengths is shown in Figure 7. The effect is observed in the scaling of star sizes with their arm length. Up to generation of g=5, the scaling exponents remain the same as for linear chains, whereas at higher initiator sizes the conditions of crowded growth are created and the arms can become increasingly polydisperse with the decrease of the initiator outer curvature, which shifts the system towards the situation closer to the case of the initiation from the wall.

The effects of the limited accessibility of space and limited accessibility of monomers to the reacting sites have been demonstrated here on the basis of simulated systems. There are limited possibilities for experimental verification of these observations in real systems. Attempts have been made to determine the distributions of chain lengths for the surface initiated polymerization. In the known cases, however, the conditions of polymerization were not yet corresponding to situations, which should be considered as the crowded growth conditions. Therefore, in the known experimental results no enhanced heterogeneities of the chain lengths were observed.

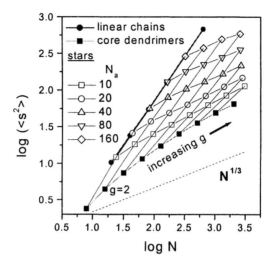

Figure 7. Mean squared radius of gyration of polymer stars with dendrimeric centers of various generations and with various arm lengths. The lines are plotted to guide the eye along dependencies with constant number of arms or constant arm length.

Structure and dynamics: Creating new macromolecular architectures can constitute a challenge for synthetic chemists but can additionally be justified if it can result in new properties of materials. We discuss here mechanical properties, which are related to the dynamics of systems with various macromolecular structures.

Joining monomeric units into linear polymer chains results in a dramatic change of properties. Whereas, a monomer in bulk can usually be only liquid-like or solid (e.g. glassy), the polymer can additionally exhibit a rubbery state with properties, which make these materials extraordinary in a large number of applications. This new state is manifested by the very slow relaxation of polymer chains in comparison with the fast motion of the monomers, especially when the chains become so long that they can entangle in a bulk melt. A mechanical manifestation of the relaxation, taking place in a polymer melt in comparison with the behavior of a low molecular system is illustrated in Figure 8. The dynamic mechanical characteristics of the two materials indicate a single relaxation in the monomer system in contrast to the two characteristic relaxations in the polymer. The characteristic rubbery state of a polymer extends between the segmental (monomer) and the chain relaxation frequencies and is controlled by a number of parameters related to the polymer structure. The most important among these parameters is the chain length, determining the ratio of the two relaxation rates. In the rubbery state the material is much softer than in the solid

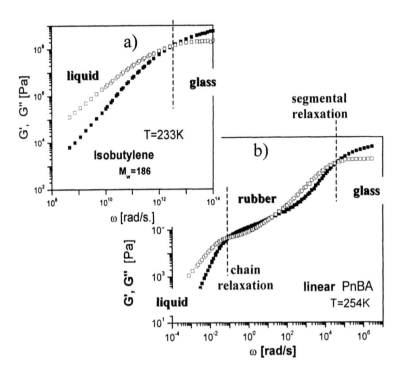

Figure 8. (a) Real (G') and imaginary (G'') part of the complex shear modulus vs. frequency for an isobutylene oligomer and (b) the same dependencies for a melt of linear p(n-butyl acrylate), p(nBA), chains.

376

state. If expressed by the real part of the modulus, the typical solid state elasticity is on the order of 10^9 Pa and higher, whereas the rubber like elasticity in bulk polymers is on the order 10^5-10^6 Pa.

The effect of polymerization on the relaxation behavior in the simulated systems can be simulated in the similar way. A single relaxation process can be detected in a non-polymeric simulated liquid and two dominant relaxations, the segmental and the chain relaxation, can be distinguished in the dynamic behavior of dense system of linear polymers. An example of such results is shown in Figure 9. A very good agreement in the behavior of simulated and experimental systems with the linear polymers has been documented elsewhere (6).

The next question, which arises, is related to possibilities of influence on the properties of materials by changing the polymer architecture to the one suitable for particular applications. We will demonstrate here that the highly branched macromolecular structures discussed in this chapter can lead to considerably different properties than those obtained by linking monomeric units into linear chains.

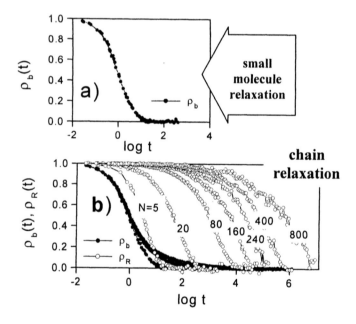

Figure 9. Dynamic behavior of (a) simulated liquid and (b) simulated melt of linear polymers as characterized by segmental and end-to-end vector correlation functions giving information about relaxation rates of the corresponding units. In the case of polymers, systems with various chain lengths are considered

The first example, which has been reported already earlier (5), concerns structures and behavior of multiarm stars in the melt. As illustrated in Figure 7, the stars have smaller dimensions than linear chains of the same molecular mass and the difference increases with increasing number of arms in the star. In the stars with a large number of arms, the space around the star center is predominantly occupied by the elements of considered macromolecule. This leads to ordering of the stars, which manifests itself in the X-ray scattering results. In Figure 10a, an example is shown of the scattered intensity distribution for the melt of multiarm polybutadiene stars synthesized by Roovers (5). The clear intensity peak indicates an ordering of star cores in such a system. The lower part of Figure 10 (b) shows the viscoelastic characteristics of this system, which indicates that besides the usual relaxation processes related to the segmental motion and to the motion of polymeric fragments (arms), there exists the third relaxation process with the longest relaxation time, which is interpreted as related to slow cooperative rearrangements in the structured system of stars.

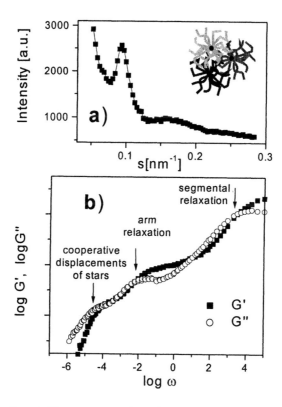

Figure 10. (a) Small angle X-ray scattering from the melt of polybutadiene stars with 128 arms of $M_w=7000$ and (b) the viscoelastic characteristics of this material by means of the frequency dependencies of the storage (G') and loss (G'') shear moduli (master curves at $T_{ref}=218K$).

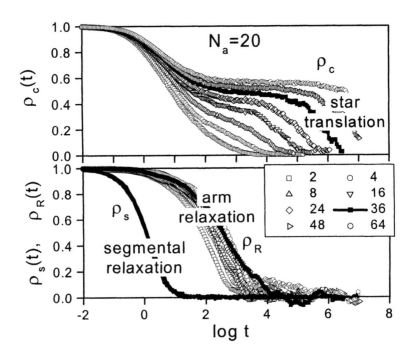

Figure 11. Dynamics in simulated multiarm star melt (5). The three correlation functions correspond to segmental ρs, arm ρR and global ρc relaxation. Effect of the number of arms on these relaxations is illustrated (arm length Na=20).

Simulated systems of the multiarm stars indicated the same relaxation processes as the real systems. The slowest relaxation has been identified as a translational motion of stars and has been observed as strongly dependent on star number, in a good agreement with the observations made for real systems. It was possible to observe in the simulation that the stars with arm numbers higher than 25 reach in the center the intramolecular density equal to the mean density of the system, which means that an impenetrable core of the star is created.

A kind of ordering and new slow relaxation processes can be observed also in melts of brush-like macromolecules. Figure 12 shows X-ray scattering results determined at small angles (SAXS), which indicate correlation distances between the backbones of these macromolecules. These results show that the correlation length can be controlled by the length of side chains. Figure 12 also presents an example of the viscoelastic properties determined for the melt of such brush-like macromolecules as a master curve of G' and G'' determined for the reference temperature of 254K. The results indicate again a presence of three relaxation

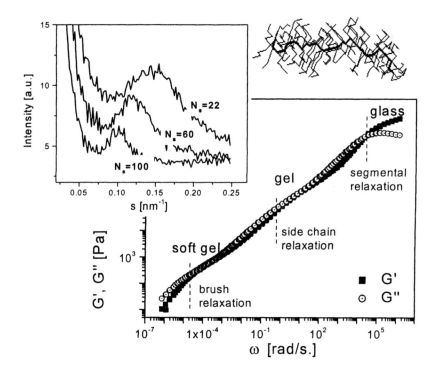

Figure 12. Viscoelastic characteristics of the p(nBA) brush polymer melt by means of frequency dependencies of the storage (G') and loss (G'') shear moduli (master curves at $T_{ref}=254K$), The insert: Small angle X-ray scattering from melts of p(nBA) brush like polymer with various side chain length but the same backbone i.e. the same number of side chains per macromolecule.

ranges: the high frequency relaxation corresponds to segmental motion the intermediate relaxation is attributed to the reorientation rates of the side chains and the slowest process is the global macromolecular relaxation in this system, which controls the zero shear flow and the corresponding viscosity. The rate and nature of this relaxation must be dependent on the length of the backbone. For short backbone chains the macromolecules can behave similarly to stars, for which translational motion dominates the slow dynamics but for longer backbone the reorientation possibilities should become slower than translation. This was observed clearly in the simulated systems as a crossover between translational and rotational relaxation times on the scale of backbone length.

It is important to notice that the described polymer architectures lead to relaxation spectra more complex than those of linear polymers. Depending on structural parameters of the systems, the proportions between relaxation rates corresponding to the three distinguished processes can be influenced. This should lead to various properties of materials in frequency ranges corresponding to modulus plateaus related to various unrelaxed states i.e. glassy, rubbery or the super soft state. For example, extension of side chain length to the range where they can entangle should extend the frequency range of the modulus plateau at the level typical for polymer rubber elasticity i.e. at 10^5-10^6 Pa (in detail depending on the chemical nature of the monomer). On the other hand, extension of the backbone or a slight crosslinking of the system with not yet entangled side chains, can lead to an extension of the frequency range with the super soft elastomeric plateau with the modulus level below 10^4 Pa. An example of this case is shown in Figure 13 where the mechanical properties of the lightly crosslinked system of brush polymers are characterized.

We have recently observed properties of this kind in a number of similar systems including brush copolymers, which can be considered as super soft thermoplastic elastomers. In the permanently crosslinked system illustrated in Figure 13, the modulus plateau related to the soft elastomeric state extends over a very broad temperature range (tested from room temperature up to 180°C) when detected isochronously.

Conclusions

Investigation of various systems modified by hairy coatings was presented. Modifications of macromolecular architectures can generate a new spectrum of relaxation, which leads to interesting mechanical properties of bulk systems. The examples described indicate that a broad variation of properties is possible by changes of the following molecular parameters: side chain length, grafting density, backbone length and also the crosslinking methodology and density. These parameters influence the extent and levels of the elastic plateaus in the complex viscoelastic spectrum and consequently should allow to generate materials with extremely different properties ranging between the hard glassy and super soft elastic, taken as examples.

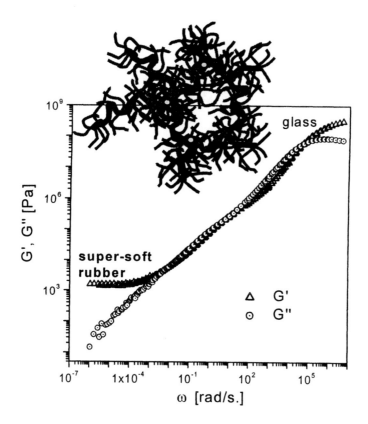

Figure 13. Viscoelastic behavior of a lightly cross-linked brush-like nBA polymer characterized by means of the frequency dependencies of the storage (G') and loss (G'') shear moduli (master curves at $T_{ref}=254K$).

Acknowledgments: Kathryn L. Beers and Shuhui Qin are gratefully acknowledged for their contribution to this work. The reported research has been partially supported by the Marie Curie Training Site at MPIP and by the National Science Foundation (DMR-0090409).

References

1. Halperin, A.; Tirrell, M.; Lodge, T. P. *Adv. Polym. Sci.* **1991**, *100*, 31.
2. Pakula, T. *Macromol. Symp.* **1999**, *139*, 49
3. Beers, K. L.; Gaynor, S. G.; Matyjaszewski, K.; Sheiko, S. S.; Moller, M. *Macromolecules* **1998**, *31*, 9413.

382

4. Pakula, T. *J. Chem. Phys.* **1991**, *95,* 4685; Pakula, T.; Zhulina, E. B. *J. Chem. Phys.* **1991**, *95,* 4691

5. Pakula, T; Vlassopoulos, D.; Fytas, G.; Roovers, J.; *Macromolecules* 1998,31,8931

6. Pakula, T.; Geyler, S.; Edling, T.; Boese, D.; *Rheol Acta* **1996**,33,631

7. Pakula, T.; *Comput. Theor. Polym. Sci.* **1998**, 8,21

8. Wang, J. S.; Matyjaszewski, K. *J. Am. Chem. Soc.* **1995**, *117,* 5614.

9. Matyjaszewski, K.; Xia, J. *Chem. Rev.* **2001**, *101,* 2921.

10. Matyjaszewski, K.; Miller, P. J.; Shukla, N.; Immaraporn, B.; Gelman, A.; Luokala, B. B.; Siclovan, T. M.; Kickelbick, G.; Vallant, T.; Hoffmann, H.; Pakula, T. *Macromolecules* **1999**, *32,* 8716.

Chapter 27

Photoinduced Free Radical Promoted Cationic Block Copolymerization by Using Macrophotoinitiators Prepared by ATRP and Ring-Opening Polymerization Methods

Yusuf Yagci and Mustafa Degirmenci

Science Faculty, Department of Chemistry, Istanbul Technical University, Maslak 80626, Istanbul, Turkey

The present work describes synthesis of well-defined polymeric photoinitiators via atom transfer radical polymerization (ATRP) and ring-opening polymerization (ROP) and their subsequent use in photoinduced block copolymerization. The first step of the process involves controlled polymerization of styrene and ε-caprolactone by using appropriate initiators possessing photolabile groups as well as functionality for ATRP and ROP, respectively. By selecting corresponding initiator and modifying the polymerization conditions, end-chain polymeric photoinitiators with various molecular weights and low polydispersities were obtained. These polymers were used in photoinduced free radical promoted cationic polymerization resulting in the formation of block copolymers.

Introduction

Macrophotoinitiators are polymers with photoinitiator functionality at side chains or in the end or middle of the chain (*1*). These materials are of great scientific and technological interest because of their application in UV-curable coatings (*2*) and as precursors for graft and block copolymers depending on the position of the photoinitiator moiety incorporated (*3*). Many macrophotoinitiators have been synthesized and their utilization in both applications has been studied. The major concern for their uses particularly in the latter application relates to the efficiency of functionalization, well-defined and predetermined structures, and low polydispersities. Obviously, upon irradiation non-functionalized chains will not be activated and consequently remain as homopolymers in the system if the all chains are not functionalized. Since the perfect functionalization is not achieved by conventional polymerization methods, controlled/living polymerization techniques are of interest for the preparation of macrophotoinitiators. This study describes the preparation of macrophotoinitiators by two controlled/living polymerization methods namely, atom transfer radical polymerization (ATRP) and ring-opening polymerization (ROP). The synthetic strategy (*4*) followed for the preparation of macrophotoinitiators is described in Scheme 1.

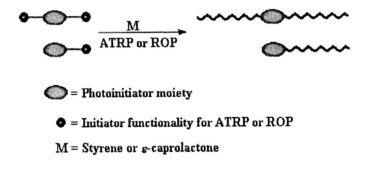

○ = Photoinitiator moiety

● = Initiator functionality for ATRP or ROP

M = Styrene or ε-caprolactone

Scheme 1. Preparation of macrophotoinitiators

Experimental

Materials

Preparation and purification of the initiators and the monomers, and polymerization procedures for ATRP (*5*) and ROP (*6*) were described elsewhere. Diphenyliodonium hexafluorophosphate (*7, 8*) ($Ph_2I^+PF_6^-$) and *N*-ethoxy-2-methylpyridinium hexafluorophosphate (*9*) ($EMP^+PF_6^-$) were prepared according to the published procedures. Cyclohexene oxide (CHO) (Aldrich) was vacuum distilled over calcium hydride just before use.

Photoinduced Free Radical Promoted Cationic Block Copolymerization

Appropriate solutions of polymeric photoinitiators, polystyrene (PSt) or poly ε–caprolactone (PCL), and monomer (CHO) in bulk containing onium salt (EMP$^+$or Ph$_2$I$^+$) in Pyrex tubes were degassed with nitrogen prior to irradiation. At the end of irradiation in a merry-go-round type photoreactor equipped with 15 Philips lamps emitting light nominally at 350 nm at room temperature, the solutions were poured into cold methanol. The precipitated copolymers were filtered off and dried in vacuo. Homopoly(cyclohexene oxide) was extracted with n-hexane.

Results and Discussion

Macrophotoinitiators Prepared by Atom Transfer Radical Polymerization

Among the newly developed controlled//living radical polymerization methods (*10*), atom transfer radical polymerization (ATRP) is the most versatile technique providing a variety of adjustable options (*11*). A series of photoinitiators possessing initiating sites i.e., alkyl halide for ATRP were prepared and listed below. The method for their preparation was described previously (*5*). Due to the better absorbance characteristics we have focused on the use of monofunctional initiator (B-Br).

The ATRP of styrene (St) in bulk at 110 °C by means of benzoin-linked initiator (B-Br) in conjunction with a Cu(I)Br/bipyridine yields end-chain functional macrophotoinitiator (Scheme2). The results are summarized in Table 1.

The number average molecular weight calculated by comparison of the integration ratio of aromatic proton resonances of benzoin group compares (Figure 1b) favorably with the molecular weight calculated by the monomer-to-initiator ratio and the one measured by GPC. Some discrepancy observed with the values calculated by spectral methods may be due to the inacuurancy of the molecular weight determination by end group analysis. However, the absorption and emission characteristics of the macrophotoinitiators closely resemble the precursor benzoin as measured by UV and fluorescence spectroscopy.

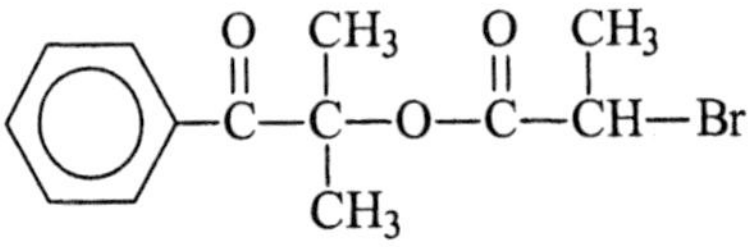

HMPP-Br

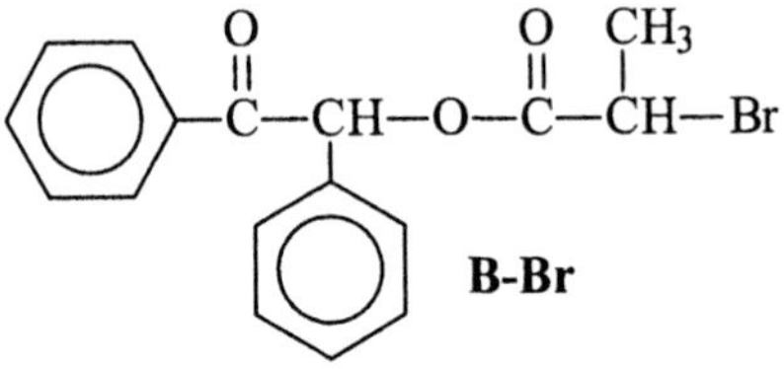

B-Br

Monofunctional ATRP initiators

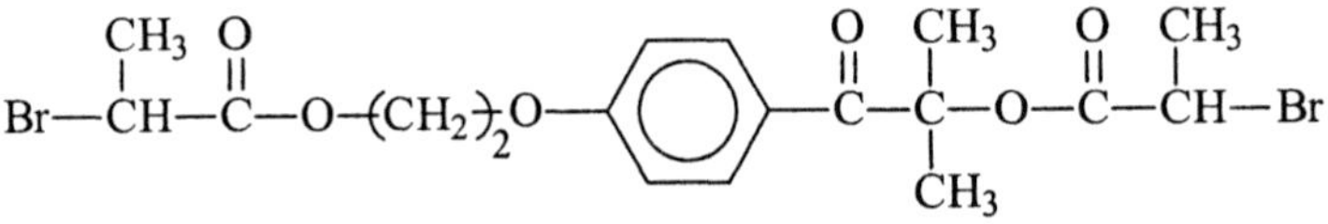

Br-HE-HMPP-Br

Bifunctional ATRP initiator

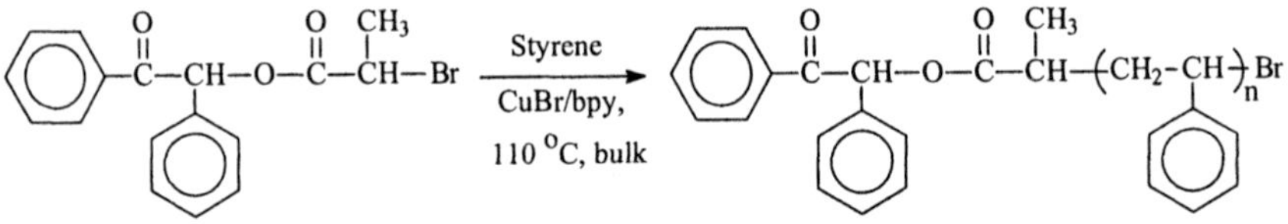

Scheme 2. Synthesis of polystyrene macrophotoinitiator

Table 1. Synthesis of Macrophotoinitiators by ATRP^a of Styrene

Run	$[I] \times 10^{-2}$ (mol.L^{-1})	Time (h)	Yield (%)	M_n (Theo.)	M_n (GPC)	M_w/M_n	M_n (H-NMR)	M_n (UV)
1^b	17.5	3	89	4962	4716	1.18	4462	4857
2^b	8.75	5	79	8521	8551	1.17	8991	8115
3^c	1.75	7	17	9291	8777	1.19	10900	9978

^{a} Temp.110°C, $[St]_o$=8,75 mol L^{-1} (in bulk),
^{b} [I]/[CuBr]/[Bpy]: 1/1/3,
^{c} [I]/[CuBr]/[Bpy]: 1/0.5/1.5

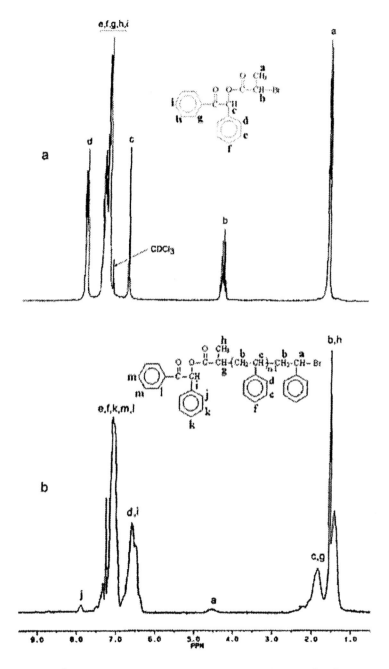

Figure 1. ¹*H-NMR spectra of ATRP initiator(B-Br)* **(a)** *and polystyrene macrophotoinitiator* **(b)** *in CDCl₃*

Macrophotoinitiators Prepared by Ring-opening Polymerization

Since several commercially available photoinitiators contain one or two hydroxyl groups, the ROP of ε–caprolactone could directly be applied to prepare structurally similar macrophotoinitiators (6). Thus, the ROP of ε-caprolactone was carried out in bulk at 110 °C using benzoin in the presence of catalytic amount of 2-ethyl-hexanoate (Sn(Oct)$_2$) (Scheme 3, Table 2). Although the polydispersities of the polymers are relatively large, the molecular weights are well controlled by the monomer-to-initiator molar ratio.

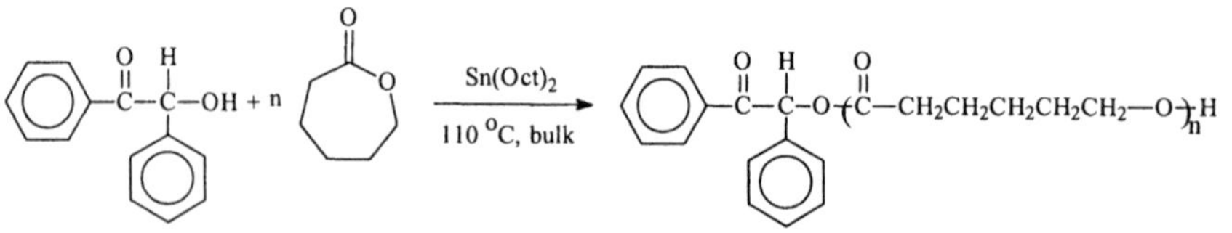

Scheme 3. Synthesis of poly(ε-caprolactone) macrophotoinitiator

Table 2. Synthesis of Macrophotoinitiators by ROP^a of ε-Caprolactone

Run	$[I]^b \times 10^2$ (mol L^{-1})	Time (h)	Yield (%)	Mn (Theo.)	Mn (GPC)	Mw/Mn	Mn (H-NMR)	Mn (UV)
4	45	69	100	2500	2100	1.16	2900	2800
5	45	72	100	2500	2550	1.42	3600	3400
6	22.5	103	85	4100	2500	1.18	4500	3900

^{a}In bulk; at 110 °C.
^{b}Initiator: Benzoin

Photoinduced Free Radical Promoted Cationic Block Copolymerization by Using Macrophotoinitiators

It was previously shown that macrophotoinitiators can be used in photoinduced radical polymerization to yield block copolymers (3,6,12). But, in the case of macrophotoinitiators prepared by ATRP, the resulting copolymers are eventually formed from radically polymerizable monomers. Obviously, this type of block copolymers can be synthesized directly by applying ATRP itself with the benefit of better control of polydispersity (*13*). However, the use of macrophotoinitiators in blocking reactions is not limited to free radical process. They can also be used in photoinduced radical-promoted cationic polymerization to yield block copolymers of monomers polymerizable with different mechanisms. The initial stage of the process is also radicalic. Upon irradiation benzoyl radicals and polymer bound radicals are formed via α-cleavage of the benzoin moiety incorporated to the polymer chains. It is known that certain onium salts such as $Ph_2I^+PF_6^-$ and $EMP^+PF_6^-$ can efficiently oxidize photochemically generated electron donating free radicals (*14*). If the photolysis is carried out in the presence of cyclohexene oxide (CHO) the polymer attached radical is converted to initiating cations to generate block copolymers (Scheme 4).

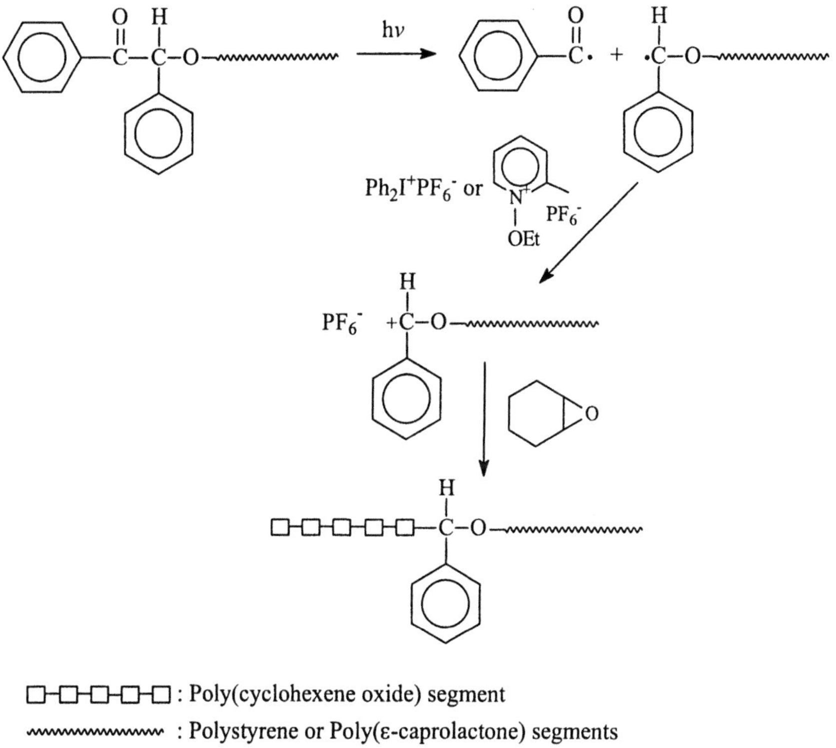

□-□-□-□-□ : Poly(cyclohexene oxide) segment

〜〜〜〜〜 : Polystyrene or Poly(ε-caprolactone) segments

Scheme 4. Photoinitiated Free Radical Promoted Cationic Block Copolymerization by Using Macrophotoinitiators

It should be pointed out that benzoyl radicals formed concomitantly do not participate in the oxidation process (*15*). Typical results concerning block copolymerization by using PSt and PCL macrophotoiniatitors are presented in Tables 3 and 4, respectively. Notably, no polymerization took place in the absence of onium salts since CHO is unreactive towards photochemically generated radicals.

Table 3. Photoinduced Free Radical Promoted Cationic Block Copolymerization[a] of CHO by Using PSt Macrophotoinitiator

Run	Onium salt	Time (min)	Yield[b] (%)	HomoPCHO[c] (%)	PSt-*b*-PCHO (%)	Mn^d	Mw/Mn
7	Ph_2I^+	35	42	5	95	6100	2.76
8	EMP^+	60	37	4	96	7900	2.34

[a][CHO] = 9.91 mol L^{-1}, [PSt] = 100 g L^{-1} (M_{nPSt} = 4716, M_w/M_n=1.18), [Onium salt] = 5 x 10^{-3} mol L^{-1} λ_{inc}= 350nm
[b]CHO conversion
[c]Extracted by n-hexane
[d]Determined by GPC according PSt standards

Table 4. Photoinduced Free Radical Promoted Cationic Block Copolymerization[a] of CHO Using PCL Macrophotoinitiator

Run	PCL	Onium salt	Time (min)	Yield[b] (%)	HomoPCHO[c] (%)	PCL-*b*-PCHO (%)	Mn^d	Mw/Mn
9	4	Ph_2I^+	15	56	7	93	6000	1.83
10	5	EMP^+	30	27	11	89	5500	1.64

[a][CHO] = 9.91 mol L^{-1}, [PCL] = 100 g L^{-1} (M_{nPCL} = 2500, M_w/M_n=1.42), [Onium salt] = 5 x 10^{-3} mol L^{-1}, λ_{inc}= 350nm
[b]CHO conversion
[c]Extracted by n-hexane
[d]Determined by GPC according PSt standards

Different reactivity of dipenyliodonium (Ph_2I^+) and N-ethoxy-2-methylpyrdinium (EMP^+) salts in initiating the free radical promoted cationic polymerization may be explained in terms of different redox potentials (relative to SCE) of the salts: -0.2 V (Ph_2I^+) (*16*) and -0.7 V (EMP^+) (*9*). Homopolymer formation, although at low rate, observed with both PSt and PCL macrophotoinitiators is due to the participation of the radicals formed from the decomposition of onium salts in further redox reactions. The block copolymer structures were assigned by NMR spectral measurements. The NMR spectra of the polystyrene-*b*-poly(cyclohexene oxide) and polycaprolactone-*b*-poly(cyclohexene oxide) display the signals characteristic of the corresponding segments (Figures 2 and 3).

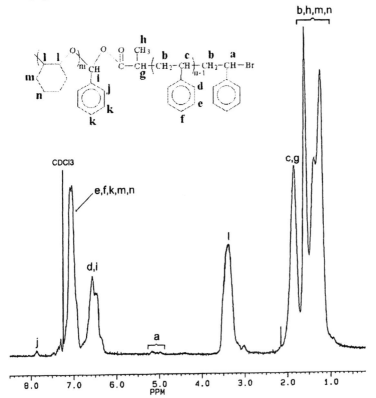

Figure 2. ¹H-NMR spectra of PSt-b-PCHO block copolymer (Table 3, Run 7)

In conclusion, the macrophotoinitiators with well-defined structures can be prepared by controlled/living polymerizations such as ATRP and ROP. Under UV irradiation, these photosensitive polymers generate polymeric radicals. In the presence of a radical polymerizable monomer block copolymers were formed. The type of macrophotoinitiator influences the polymerization products. While both homo and block copolymers are formed with the end-chain functional photoinitiators, the mid-chain functional photoinitiator yields purely block copolymers (*6*). On the other hand, transformation reaction was also carried out

392

in order to convert the polymeric radicals into initiating cations with the aid of oxidizing agents such as iodonium and pyridinium salts. This way block copolymer of monomers, which do not polymerize with the same mechanism, was prepared. Conceptually similar transformation reactions involving radical and cationic ring opening polymerizations were also reported by Kamachi (17) and Matyjaszewski (18). It is clear that the transformation reactions will continue to attract interest in near future because of the possibility of the various newly developed 'living'/controlled polymerization mechanisms such as ATRP. It would be possible to design and synthesize materials having precise structures with desired properties by combination of such mechanisms.

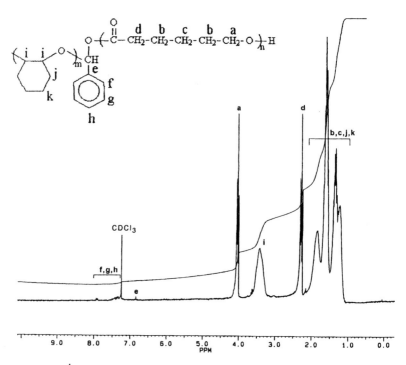

Figure 3. The ^{1}HNMR spectrum of PCL-b-PCHO block copolymer (Table 4, Run 10)

References

1. Carlini, C. ; Angiolini, L. *Adv. Polym. Sci.*, **1995**, 123, 12.
2. *Chemistry and Technology of UV&EB Formulation for Coatings, Inks&Paints*; Dietliker, K.; SITA Technology Ltd.: London, 1991; Vol.III
3. Yagci, Y. in *Macromolecular Engineering: Recent aspects*; Mishra, M. K.; Nuyken, O.; Kobayashi, S.; Yagci Y. Eds.; Plenum Press: New York, 1995; ch. 11.
4. Degirmenci, M.; Cianga, I.; Hizal, G.; Yagci, Y. *Polym. Prep.*, **2002**, 43 (2), 22
5. Degirmenci, M.; Cianga, I.; Yagci, Y. *Macromol. Chem., Phys.*, **2002**, 203, 1279
6. Degirmenci, M.; Hizal, G.; Yagci, Y. *Macromolecules*, **2002**, 22, 865
7. Crivello, J. V.; Lam, J. H. J. W. *Macromolecules*, **1977**, 10, 11307
8. Crivello, J. V.; Lam, J. H. J. W. *J. Polym. Sci., Polym. Chem. Ed.* **1980**, 18, 2677
9. Bottcher, A.; Hasebe, K.; Hizal, G.; Yagci, Y.; Stellberg, P.; Schnabel, W., *Polymer*, **1991**, 32, 2289
10. *Controlled Radical Polymerization*; Matyjaszewski, K., Ed.; ACS Symposium Series; American Chemical Society: Washington, DC, 1998; Vol. 685.
11. Matyjaszewski, K.; Xia, J. *Chem. Rev.* **2001**, 101, 2921
12. Onen, A.; Yagci, Y. *J. Macromol Sci., Chem.*, **1990**, *A27*, 743.
13. Matyjaszewski, K.; Teodorescu, M.; Acar, M. H.; Beers, K.L.; Coca, S.; Gaynor, S. G.; Miller, P.J.; Paik, H. *Macromol. Symp.* **2000**, 157, 183
14. Yagci, Y.; Reetz, I. *Prog.Polym.Sci.* **1998**, 23, 1465
15. Yagci, Y.; Borberly, J.; Schnabel, W. Eur. Polym. J. **1989**, 25, 129
16. Bachofner, H.E.; Beringer, F.M.; Meites, L. *J.Am.Chem.Soc,***1958**, 80, 4269
17. Guo, H-O.; Kajiwara, A.; Morishima, Y.; Kamachi, M. *Macromolecules*, **1996**, 29, 2354
18. Kajiwara, A., Matyjaszewski, K. *Macromolecules*, **1998**, 31, 3489

Chapter 28

Bioinspired Triblock Copolymers Prepared by Atom Transfer Radical Copolymerization

Jan C. M. van Hest*, Lee Ayres, Henri Spijker, Matthijn Vos, and Joost Opsteen

Department of Organic Chemistry, Nijmegen University, Toernooiveld 1, 6525 ED Nijmegen, The Netherlands
*Corresponding author: email: vanhest@sci.kun.nl

Well-defined block copolymers, comprising bio-related monomers could be conveniently synthesized via Atom Transfer Radical Polymerization. Elastin side chain triblock copolymers were prepared by macroinitiation from a bifunctional polyethylene glycol initiator. The obtained oligopeptide architecture exhibited similar thermoresponsive behavior as its main chain polymer analog. ATRP was performed on methacrylate functionalized nucleobases using an initiator equipped with a primary amine moiety. The resulting nucleobase polymers were successfully coupled to telechelic isocyanate-functional polymethylacrylate, resulting in triblock copolymers that could be interesting building blocks for supramolecular polymers.

Introduction

The development of methodologies that allow control over radical polymerization processes has greatly increased the possibilities of design and synthesis of polymer architectures (*1*). The tolerance of these controlled radical polymerizations toward many functional moieties has furthermore expanded the scope of monomer units that can be incorporated. Especially for the development of bio-related polymers this has been a major improvement, since monomers based on natural compounds are highly functional and therefore normally hard to polymerize via other controlled polymerization techniques (*2*). The construction of well-defined, bio-related polymers is of great interest for biomedical applications, where it is envisaged that these structures can function for example as drug delivery carriers or as scaffolds for tissue engineering.

Atom Transfer Radical Polymerization (ATRP) has recently been applied to construct polymers containing saccharide (*3,4,5*) and nucleobase moieties (*6,7,8*), and also polymers with activated ester units have been prepared that could be post modified with oligopeptide fragments (*9*). Polymer chemists have however only begun to explore ATRP for the preparation of bio-related polymers, and many opportunities are therefore still ahead to create new, highly functional polymers.

It is our aim to construct well-defined polymer architectures that are (partly) based on bio-inspired building blocks. In this chapter we describe our research activities for the preparation of two types of triblock copolymers. The first part deals with the construction of elastin side-chain block copolymers. In the second part we describe a modular approach for constructing nucleobase-containing triblock copolymers.

Elastin side chain polymers

Elastin is one of the most important classes of naturally occurring structural proteins (*10-14*). It is commonly found in ligaments, arteries, skin, and lung tissue of mammals where it functions as an elastomeric material (*15, 16*). There are many different types of elastin, but tropoelastin (the precursor protein of mammalian elastin) is one of the most studied. It was found to have VPGVG (V = valine, P = proline and G = glycine) as its most prominent amino acid repeat (*17*). The aggregation, conformational and mechanical properties of both chemically synthesized (*13,18,19*) and recombinantly prepared poly(VPGVG) (*20-22*) have been extensively investigated. It was discovered that poly(VPGVG) is soluble in water at room temperature but as the temperature is increased the solubility of this polypeptide is decreased. This remarkable lower critical solution temperature (LCST) behavior, which is completely reversible, is a result of a conformational change in elastin from random coil to β-spiral, which is caused by hydrophobic dehydration of the valine side chains (*23*).

Optimization of these hydrophobic interactions occurs when three of these long chain β spirals twist together into a larger aggregate.

This LCST behavior makes elastin like peptides, or ELP's, of interest for a wide range of applications, for example as thermally responsive hydrogels with an inverse temperature transition (21). It has also been shown that these peptide sequences are biologically compatible (24, 25), opening up the possibility of medical applications for ELP's (26, 27).

Until now all of the investigations that have been carried out have focused on linear polyVPGVG. However, it has been shown that this inverse temperature transition not only occurs in poly(VPGVG)$_n$ (where n > 10) but even in a single repeating unit of VPGVG (28). This has inspired us to investigate the possibility of synthesizing polymers with similar LCST behavior by introducing single repeats of VPGVG into the side chain. Connecting side chain elastin oligomers to the ends of a polymer chain could result in a versatile approach to producing thermally responsive triblock copolymers.

Although oligopeptide based monomers have been polymerized using a wide range of techniques (2, 29) to the best of our knowledge, however, there has been no example in which ATRP is used.

Results and Discussion

The preparation of VPGVG monomers was performed via solid phase peptide synthesis methods (figure 1). After constructing the pentapeptide via Fmoc chemistry, the Fmoc group of the last valine residue was removed and the free amine was allowed to react with an excess of a 2-(methacryloyloxy)ethyl isocyanate (30). This resulted in complete conversion to a methacrylate functional peptide. Because the high molecular weight (583 g/mol) of the oligopeptide did not allow a high concentration of polymerizable groups, a methacrylate moiety was preferred because of its high reactivity under ATRP conditions.

Two initiators were applied for the polymerization of the oligopeptide monomer: the standard initiator ethyl-α-bromo-isobutyrate (EBIB) and a bifunctional polyethylene glycol (PEG) initiator (di-α,ω-bromo-isobutyrate-PEG, PEG(BIB)$_2$), which could conveniently be synthesized via a DCC-mediated coupling between α-bromo isobutyric acid and PEG (M_n 1000 g/mol) (31).

Polymerizations were conducted with the catalytic system CuCl/bipy, in DMSO-$d6$. The use of a deuterated solvent allowed that the progress of reaction could conveniently be followed via ^{1}H-NMR spectroscopy.

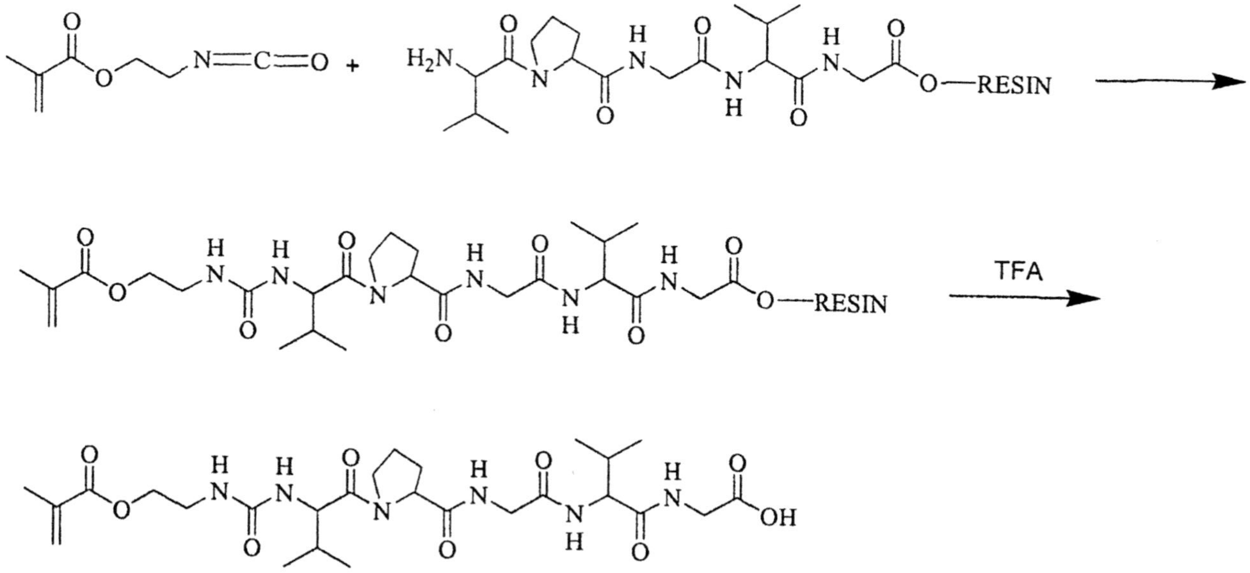

Figure 1: Reaction scheme for the synthesis of VPGVG monomers

Furthermore, DMSO assured the solubility of both the monomer and polymer species. However, still the temperature of the reaction (35°C) as well as the concentration of the monomer (0.25M) had to be kept low, in order to prevent precipitation during polymerization. The results of these polymerizations are depicted in table 1.

Table 1: ATRP of VPGVG methacrylate

Polymer	Initiator	Temp °C	Time min	Conv. %	M_n^a kg/mol	$M_{n,th}$ kg/mol	MWD
1	EBIB	36	240	80	53	5.8	1.25
2	PEG(BIB)₂	35	120	84	60	6.1	1.24

^{a} M_n determined by GPC, with 1-methyl-2-pyrrolidone (NMP) as mobile phase

The semi logarithmic plots of monomer concentration versus time were linear for both initiators, indicating that the polymerization of VPGVG methacrylate was a controlled process. The kinetics of polymerization experiment 2, leading to an elastin triblock copolymer are depicted in figure 2.

As shown in table 1, a deviation was observed when the M_n values obtained by GPC were compared to the theoretical M_n calculated from the initiator to monomer ratio. The degree of polymerization could also be determined by ^1H NMR spectroscopy of a purified sample of the triblock copolymer 2. These NMR calculations resulted in a value for M_n of 10kg/mol (DP = 14) which was closer to the theoretically predicted M_n, and indicated that the value obtained by GPC was anomalous. This can possibly be attributed to the difference in the hydrodynamic volume of the polystyrene calibration standards and our polymers, although the discrepancy in this case was particularly large.

Based on the abovementioned results, it can be concluded that ATRP is robust toward many functional groups such as urea, amides and even a free carboxylic acid end group. The application of DMSO as a solvent might suppress the interaction of the catalyst with these functional groups, allowing the polymerization to progress unimpeded.

As reported earlier by Reiersen and Rees (28) the thermoresponsive behavior in aqueous solution of single repeats of elastin could be determined by using circular dichroism (CD) spectroscopy. Upon increasing temperature, a change from a random coil type conformation, with a steep minimum at 200

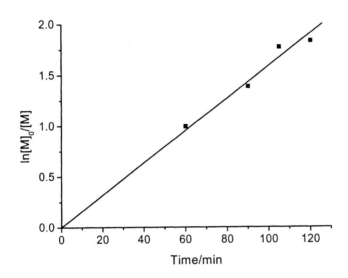

Figure 2 Semi-logarithmic plot of monomer conversion vs. time of polymerization 2.

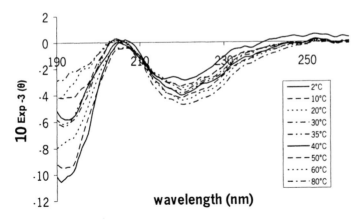

Figure 3 CD spectra of polymer 2, measured at different temperatures

nm, and a shallow minimum at 220 nm, to a type II β turn - characterized by a W-shaped spectrum with minima at 200 nm and 215 nm- could be observed. CD spectroscopy was therefore performed on polymers 1 and 2 to investigate the temperature dependent folding behavior of these two polymers. The CD measurements of polymer 2 are depicted in figure 3.

In both cases, we observed very similar spectra to those obtained by Reiersen and Rees. The same characteristic increase at 195 nm and a decrease at 220 nm was observed. There was only a slight shift in the wavelengths of the maxima and the minima, possibly due to the presence of the polymer backbone. Therefore it appeared that the incorporation of the peptide into the backbone of the polymer did not affect its inverse phase transition temperature behavior.

The thermoresponsive behavior was also studied by turbidity measurements at 480 nm. Only polymer 2 showed a transition from a clear solution to a turbid mixture (figure 4). The cloud point was determined to be 40.5°C. This behavior was completely reversible: upon cooling a clear solution was again obtained. This procedure could furthermore be repeated.

The phase transition temperature was in close range with the transition temperatures of other elastin-like polymers. The slight deviation in temperature from the reported values of 32-37°C could be attributed to the presence of the free acid groups at the end of the peptide unit. An increase in the number of acid functions would increase the temperature necessary for hydrophobic dehydration. Remarkably the effect of the polymer backbone seemed to be negligible.

The reason only polymer 2 seemed capable of reversible network formation can be explained by the fact that the insoluble elastin aggregates formed were interconnected by the PEG chains. In the case of polymer 1 it seems to be more likely that either only intramolecular aggregation occurred or that the clusters formed by intermolecular aggregation were too small to scatter light. For both

a   b

Figure 4 An aqueous solution of polymer 2, at room temperature (a)
and after heating (b)

polymers it has to be stated that the phase transition did not result in a visible change in viscosity. Therefore network formation was still a fragile process.

Nucleobase triblock copolymers

DNA is the source of inspiration for the second type of triblock copolymers described in this chapter. To achieve a high level of reproducibility in DNA replication and transcription, nature has developed with Watson-Crick base pairing a powerful tool to pass on information from a template chain to a daughter biopolymer. From a polymer chemist's point of view, the method can be regarded as the ideal template polymerization. Not only length of the template but also the exact composition is transferred. DNA template polymerization has therefore intrigued many polymer chemists for already a considerable amount of time (32-35).

The specific recognition between the complementary nucleobases is not only of interest for information transfer, but can also be readily used for directed self assembly processes. This feature of DNA has been extensively applied for nanotechnology purposes (36, 37, 38).

Another research area for which nucleobase self assembly could be very useful is supramolecular polymer chemistry. This field of research has recently become highly interesting by the demonstration by Meijer et al. that by using non-covalent interactions between monomeric building blocks polymers with real mechanical properties could be obtained (39, 40).

It is the aim of our research to construct well-defined ABA triblock copolymers, of which the A blocks consist of oligonucleobases. These block copolymers could make two contributions to the field of supramolecular polymer chemistry. First of all two complementary base pairs can be applied, which allows to introduce more information in the arrangement of supramolecular polymers, and therefore an easy access to well organized multi-block copolymers. Furthermore, by varying the length of the oligonucleobase fragments in a well-defined manner the association constant can be directed and more control over supramolecular organization is obtained.

Recently the advances of controlled radical polymerization have provided us with a tool to create the desired well defined block copolymers, and it has already been shown in literature that ATRP is compatible with nucleoside monomers (6-8). Our approach to synthesizing these nucleobase block copolymers is a modular one. The A block oligonucleobases will be polymerized with an amino-functional initiator. This will allow coupling of the A blocks with a telechelic isocyanate-functional B block. Compared to macroinitiation this is a more elaborate approach. However, the advantage is that we will be able to distinguish the ratio of formation of tri and diblock copolymers. Especially for non-covalent polymer chemistry this is of great importance, because the diblock polymers will function as chain stoppers and will have a pronounced effect on the degree of assembly of the supramolecular building blocks.

Results and discussion

The nucleobase functionalized monomers were synthesized using a simple and versatile synthetic route as depicted in figure 5 *(41)*.

Methacryloyloxypropyl bromide **4** was prepared using methacryloyl chloride and 3-bromopropanol with triethylamine base in dichloromethane. Subsequent alkylation of thymine with **4** using potassium carbonate and a catalytic amount of tetrabutylammonium iodide (TBAI) resulted in the desired thymine functionalized monomer **6**. Unfortunately, side reactions such as Michael addition and the formation of N-1,3-dialkylated product limited the yield of the desired product. Methacryloyloxypropyl adenine monomer **8** was obtained applying sodium hydride as base, followed by alkylation at the N-9 position with **4**.

Polymerization of both monomers was conducted using the same catalytic system as described for the elastin-based monomers. Again DMSO-*d6* was applied for solubility reasons and for the ability to monitor conversion by ^{1}H-NMR spectroscopy. As initiator both ethyl α-bromoisobutyrate (EBIB) and Boc protected aminobutyl α-bromoisobutyrate (**9**) were used (figure 6).

In figure 7, the results are depicted of the polymerization of thymine monomer **6** with EBIB. As can be observed the rate of polymerization was very fast when the reaction was performed at 76°C. Surprisingly, the polymerization still showed first order kinetics and the obtained polymer had a relative narrow PDI of 1.3. However, in order to have a high level of control over supramolecular assembly it was necessary to have a molecular weight distribution as narrow as possible. This could be accomplished by polymerizing at lower temperatures and at low monomer concentrations. Therefore, the reaction temperature was lowered to room temperature and the monomer concentration was reduced to 0.5 M, which resulted in a polymer with the desired narrow molecular weight distribution (PDI=1.18).

ATRP with initiator **9** resulted in an even faster polymerization than was observed for EBIB. Also in this case, sufficient control over livingness and PDI could be obtained by performing the reaction at room temperature. Similar results were obtained with ATRP of adenine monomer **8**.

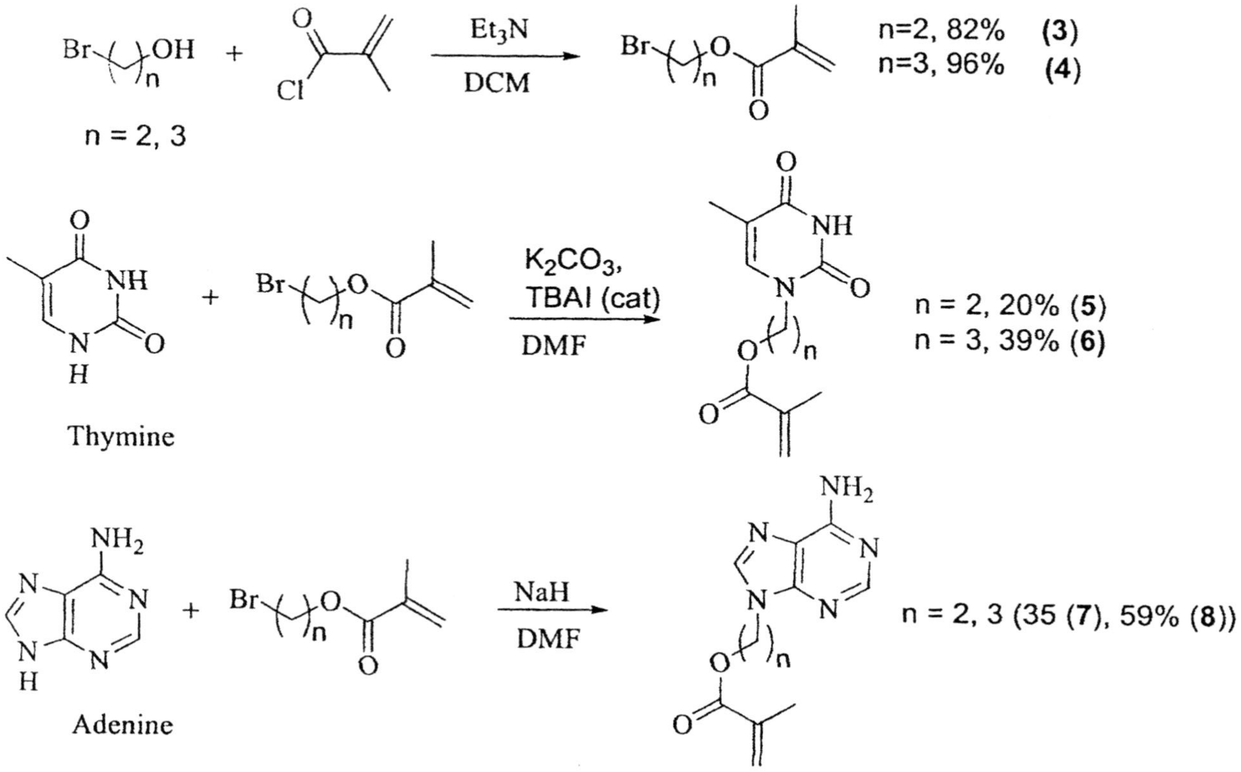

Figure 5 Synthesis of thymine and adenine nucleobase monomers.

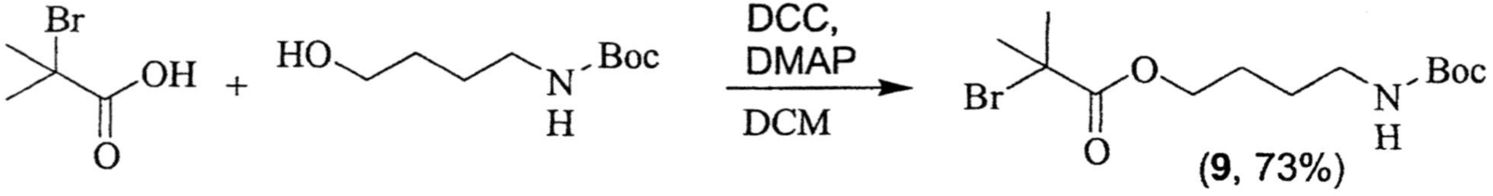

Figure 6 Preparation of amino functional ATRP initiator 9.

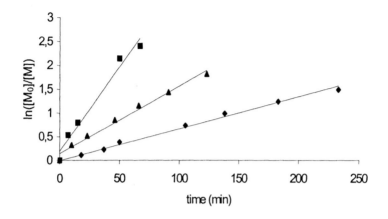

*Figure 7 Polymerization of thymine nucleobase monomer 6
at 76°C, 1.04 M (■), 41°C, 0.79 M (▲), 23°C, 0.53 M (◆)*

After having successfully synthesized the A blocks of our supramolecular module, the telechelic isocyanate functional middle block had to be prepared (figure 8). Again ATRP was chosen to construct a halide functional telechelic polymer. For ease of modification, poly methyl acrylate (PMA) was synthesized. The modification of the chloride end groups into isocyanates was partly adapted from the route developed by Matyjaszewski et al. (*42, 43*). Nucleophilic substitution of the chloride by NaN_3 was followed by hydrogenation, using H_2, Pd/C. The amines were subsequently modified into isocyanates by reaction with triphosgene. All modification steps could easily be followed by IR and ^1H NMR spectroscopy. To identify whether this procedure resulted in quantitative formation of telechelic PMA, benzyl amine was reacted with the isocyanate functional polymer. Based on NMR data the ratio of protons of the aromatic units of the end groups and the phenyl protons of the initiator moiety was 10:4, which showed that within the accuracy of the NMR measurement this modification route was successful. Currently we are optimizing the coupling procedure of the A and B blocks to construct our desired nucleobase functional triblock copolymers.

Experimental

ATRP of VPGVG methacrylate was carried out in solution using $PEG(BIB)_2$ as a bifunctional initiator. 154 mg (0.25 mmol) monomer, 10.0 mg (0.1mmol) CuCl, 34.0 mg (0.2 mmol) bipy were weighed into a Schlenk vessel. The vessel was evacuated and filled with N_2. This procedure was repeated three times to ascertain that O_2 was fully removed. 0.5 mL DMSO-d_6 was added to

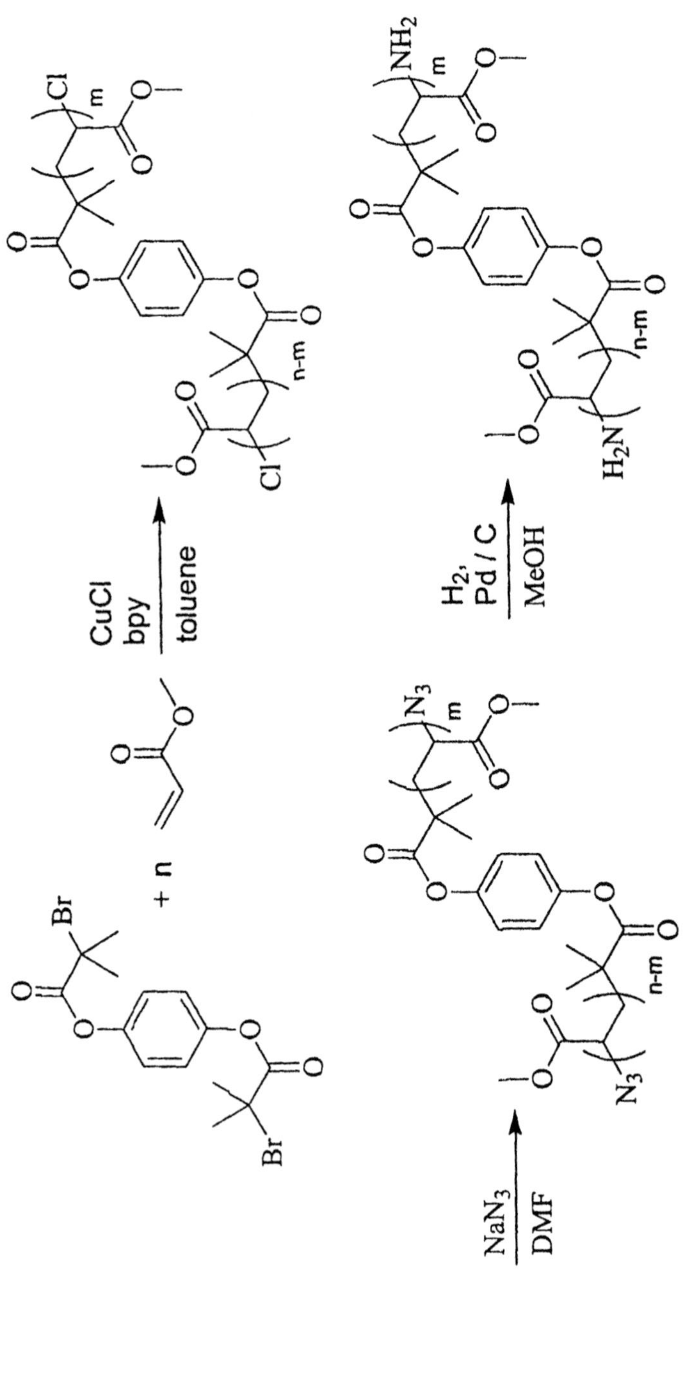
CuCl
bpy
toluene
H2, Pd / C
MeOH
NaN3
DMF

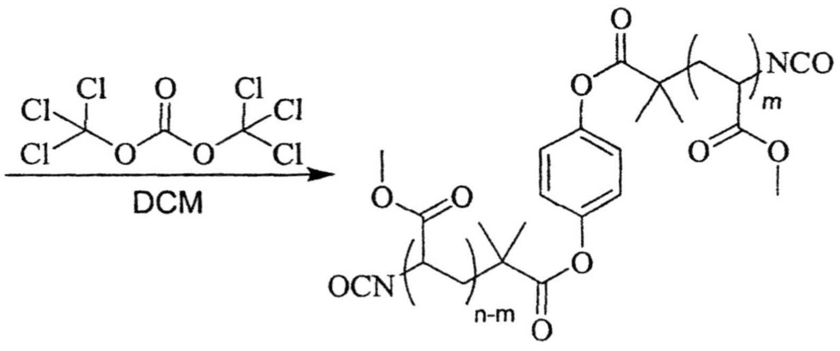

Figure 8 Synthesis of telechelic isocyanate functional PMA

dissolve reactants and 45 µL (0.5 mmol) xylene was introduced as internal standard. This mixture was then purged with N_2. 31 mg (0.025 mmol) bifunctional initiator was dissolved in 0.5 mL DMSO-d_6. The reaction mixture was heated to 35 °C and the initiator solution was added. After polymerization, the mixture was poured in ether. After decantation of the ether layer, the brown polymer precipitate was re-dissolved in demi-water (resulting solution pH 6.0). Acidification with 1M HCl solution to pH 1 resulted in precipitation of the polymer as a white powder. The polymer was filtered off and rinsed with ice cold demi-water. The product was then re-dissolved in a 1:1 water/THF mixture. After evaporation of THF the turbid water mixture was freeze dried yielding the desired ABA block copolymer. Characterization was performed by GPC with NMP as eluent and ^{1}H-NMR spectroscopy in DMSO-d_6.

ATRP of the nucleobase monomers was carried out according to the abovementioned procedure. Purification was performed by precipitation in an 0.060 M aqueous EDTA solution, followed by filtration, washing with distilled water and drying under vacuum. Molecular weight distributions were measured by GPC and the degree of polymerization was determined by ^{1}H-NMR spectroscopy.

Conclusion

With the results described in this chapter we have demonstrated that ATRP is a very versatile technique to construct functional, bio-related polymers. This allows us to introduce properties from natural polymers into synthetic analogues, leading to new, intelligent materials.

References

1. *Controlled/Living Radical Polymerization*; Editor Matyjaszewski, K.; ACS symposium series, ACS Washington, DC, 2000; Vol. 768.
2. Maynard, H. D.; Okada, S. Y.; Grubbs, R. H. *J. Am. Chem. Soc.* **2001**, *123*, 1275-1279.
3. Haddleton, D. M.; Edmonds, R.; Heming, A. M.; Kelly, E. J.; Kukulj, D. *New J. Chem.* **1999**, *23*, 477-479.
4. Haddleton, D. M.; Ohno, K. *Biomacromolecules* **2000**, *1*, 152-156.
5. Li, Z. C.; Liang, Y. Z.; Chen, G. Q.; Li, F. M. *Macromol. Rapid Commun.* **2000**, *21*, 375-380.
6. Khan, A.; Haddleton, D. M.; Hannon, M. J.; Kukulj, D.; Marsh, A. *Macromolecules* **1999**, *32*, 6560-6564.
7. Marsh, A.; Khan, A.; Haddleton, D. M.; Hannon, M. J. *Macromolecules* **1999**, *32*, 8725-8731.

8. Gross, R. A.; Kalra, B.; Kumar, A.; Gao, W.; Glausser, T.; Ranger, M.; Hedrick, J. L.; Hawker, C. J. *Abstr. Pap. Am. Chem. Soc.* **2002**, *224*, 27-BTEC.
9. Godwin, A.; Hartenstein, M.; Muller, A. H. E.; Brocchini, S. *Angew. Chem., Int. Ed. Engl.* **2001**, *40*, 594-597.
10. Urry, D. W.; Luan, C.-H.; Harris, C. M.; Parker, T. In *Protein-Based Materials*; McGath, K.; Kaplan, D., Eds.; Birkhäuser: Boston, 1997.
11. Manno, M.; Emanuele, A.; Martorana, V.; San Biagio, P. L.; Bulone, D.; Palma Vittorelli, M. B.; McPherson, D. T.; Xu, J.; Parker, T. M.; Urry, D. W. *Biopolymers* **2001**, *59*, 51-64.
12. van Hest, J. C. M.; Tirrell, D. A. *Chem. Commun.* **2001**, 1897-1904.
13. Urry, D. W. In *Methods in Enzymology*; Cunningham, L. W.; Frederiksen, D. W., Eds.; Academic Press: New York, 1982.
14. Urry, D. W. *Sci. Am.* **1995**, *272*, 64-69.
15. Urry, D. W.; Hugel, T.; Seitz, M.; Gaub, H. E.; Sheiba, L.; Dea, J.; Xu, J.; Parker, T. *Philos. Trans. R. Soc. London, Ser. B* **2002**, *357*, 169-184.
16. Gosline, J.; Lillie, M.; Carrington, E.; Guerette, P.; Ortlepp, C.; Savage, K. *Philos. Trans. R. Soc. London, Ser. B* **2002**, *357*, 121-132.
17. Gray, W. R.; Sandberg, L. B.; Foster, J. A. *Nature* **1973**, *246*, 461.
18. Lee, J.; Macosko, C. W.; Urry, D. W. *Biomacromolecules* **2001**, *2*, 170-179.
19. Lee, J.; Macosko, C. W.; Urry, D. W. *Macromolecules* **2001**, *34*, 5968-5974.
20. Nagapudi, K.; Brinkman, W. T.; Leisen, J. E.; Huang, L.; McMillan, R. A.; Apkarian, R. P.; Conticello, V. P.; Chaikof, E. L. *Macromolecules* **2002**, *35*, 1730-1737.
21. Wright, E. R.; McMillan, R. A.; Cooper, A.; Apkarian, R. P.; Conticello, V. P. *Adv. Func. Mater.* **2002**, *12*, 149-154.
22. Macmillan, R. A.; Lee, T. A. T.; Conticello, V. P. *Macromolecules* **1999**, *32*, 3643-3648.
23. Li, B.; Alonso, D. O. V.; Bennion, B. J.; Daggett, V. *J. Am. Chem. Soc.* **2001**, *123*, 11991-11998.
24. Nicole, A.; Gowda, D. C.; Parker, T. M.; Urry, D. W. *Biotechnology of bioactive polymers* **1994**.
25. Hoban, L. D.; Pierce, M.; Quance, J.; Hayward, I. *J. Surgical. Res* **1994**.
26. Rao, G. V. R.; Balamurugan, S.; Meyer, D. E.; Chilkoti, A.; Lopez, G. P. *Langmuir* **2002**, *18*, 1819-1824.
27. Okamura, A.; Hirai, T.; Tanihara, M.; Yamaoka, T. *Polymer* **2002**, *43*, 3549-3554.
28. Reiersen, H.; Clarke, A. R.; Rees, A. R. *J. Mol. Biol.* **1998**, *283*, 255-264.
29. Sanda, F.; Endo, T. *Macromol. Chem. Phys.* **1999**, *200*, 2651-2661.
30. Bamford, C.H.; Al-Lamee, K.G.; Middleton, I.P.; Paprotny, J.; Carr, R. *Bull.Soc.Chim.Belg.* **1990**, *99*, 919-930.
31. Tsarevsky, N.V.; Sarbu, T.; Goebelt, B.; Matyjaszewski, K. *Macromolecules* **2002**, *35*, 6142-6148.
32. Polowinski, S. *Prog. Polym. Sci.* **2002**, *27*, 537-577.

33. Van de Grampel, H. T.; Tan, Y. Y.; Challa, G. *Macromolecules* **1992**, *25*, 1041-1048.

34. Inaki, Y. *Prog. Polym. Sci.* **1992**, *17*, 515-570.

35. Akashi, M.; Takada, H.; Inaki, Y.; Takemoto, K. *J. Polym. Sci., Polym. Chem. Ed.* **1979**, *17*, 747-757.

36. Seeman, N. C. *Angew. Chem.-Int. Edit.* **1998**, *37*, 3220-3238

37. Watson, K. J.; Park, S. J.; Im, J. H.; Nguyen, S. T.; Mirkin, C. A. *Journal of the American Chemical Society* **2001**, *123*, 5592-5593.

38. Rowan, S. J. *Abstr. Pap. Am. Chem. Soc.* **2002**, *224*, 270-ORGN.

39. Brunsveld, L.; Folmer, B. J. B.; Meijer, E. W.; Sijbesma, R. P. *Chem. Rev.* **2001**, *101*, 4071-4097.

40. Folmer, B. J. B.; Sijbesma, R. P.; Versteegen, R. M.; van der Rijt, J. A. J.; Meijer, E. W. *Adv. Mater.* **2000**, *12*, 874-878.

41. Takemoto, K. *J. Polym Sci. : Symp Series* **1976**, *55*, 105-125

42. Coessens, V.; Pintauer, T.; Matyjaszewski, K. *Prog. Polym. Sci.* **2001**, *26*, 337-377.

43. Coessens, V.; Matyjaszewski, K. *J. Macromol. Sci.-Pure Appl. Chem.* **1999**, *A36*, 811-826.

Nitroxide-Mediated Polymerization and Stable Free Radical Polymerization

Chapter 29

Influence of Solvent and Polymer Chain Length on the Homolysis of SG1-Based Alkoxyamines

Olivier Guerret[1,2], Jean-Luc Couturier[1,3], Florence Chauvin[1], Hafid El-Bouazzy[1], Denis Bertin[1,*], Didier Gigmes[1], Sylvain Marque[1], Hanns Fischer[1,4], and Paul Tordo[1]

[1]UMR 6517 case 521, Université de Provence, Avenue Escadrille Normandie-Niemen, 13397 Marseille Cedex 20, France
[2]ATOFINA, Groupement de Recherche de Lacq, 64170 Lacq, France
[3]ATOFINA, Centre de Recherche de Rhônes Alpes, rue Henri Moissan, 69493 Pierre Bénite Cedex, France
[4]Institute of Physical Chemistry, University of Zuerich, Winterthurerstrasse 190, CH 8059 Zuerich, Switzerland

In nitroxide mediated radical polymerizations (NMP) the polymerization times and the polymer polydispersities decrease with the increasing homolysis rate of the C-ON bond between the polymer chain and the nitroxide end group. Therefore, the factors influencing the rate constants k_d of alkoxyamine cleavage are of considerable interest. Here, we describe the influence of the medium polarity and viscosity on k_d for the alkoxyamine 2-[N-*tertio*butyl-N-(1-diethoxy-phosphoryl-2,2-dimethylpropyl) aminoxy] methyl propionate, and the effect of the polymer chain length on k_d for polystyryl-SG1 and poly(n-butyl acrylate)-SG1 macro-alkoxyamines. In contrast to other alkoxyamines, k_d of 2-[N-*tertio*butyl-N-(1-diethoxyphosphoryl-2,2-dimethylpropyl) aminoxy] methyl propionate does not depend on the medium polarity. This points to the absence of polar contribution to the transition state of homolysis. The solvent viscosity does not influence the rate constant k_d, which means that cage effects are unimportant. Finally, the influence of the chain length depends on the type of polymer, it is very weak for polystyryl-SG1 but significant for poly(n-butyl acrylate)-SG1.

Introduction

A decade ago, Rizzardo*(1)* and Georges*(2)* first prepared well-defined polymers using nitroxyl radicals as controlling species. Nitroxide Mediated Polymerization (NMP) emerged,*(3)* and numerous studies were undertaken to elucidate the mechanism*(4)* and the polymerization kinetics.*(5)* These studies led to new polymers*(3,6)* and to more efficient initiators/controllers*(7)*. Scheme 1 displays the NMP mechanism, with k_d the rate constant for the homolysis of the C-ON bond of the alkoxyamine (the so-called dormant species), k_c the rate constant for its reformation, and k_p and k_t the rate constants for the propagation and the termination of the polymerization, respectively.

Scheme 1

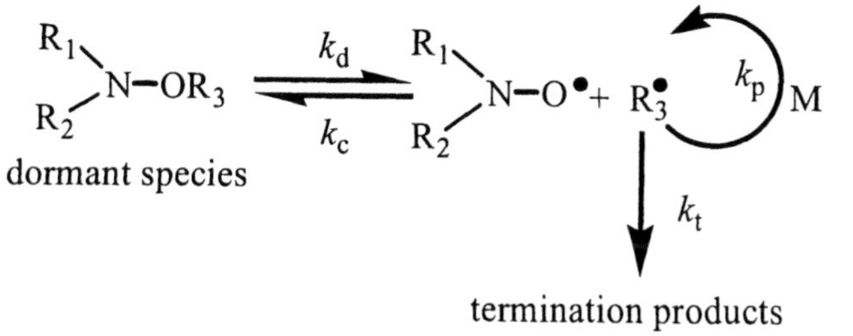

In principle, successful living and controlled polymerizations require that the rate constants k_d and k_c of the initiating and polymeric alkoxyamines ($R_1R_2NOR_3$) fall into specific and rather narrow ranges which depend on the monomer through k_p and k_t, the initial concentrations and the experimental conditions such as the temperature. In particular, a large ratio k_d/k_c often allows to reach high conversion in short time. However, in order to achieve a living and controlled polymerization, k_d/k_c must not exceed an upper limiting value. Furthermore, a low polydispersity requires that $1/k_d$ must be very small compared to the polymerization time.*(5)* Since k_d and k_c depend on the system, the choice of the nitroxide is crucial to obtain living and controlled polymerizations of a given monomer. In particular, k_d varies by many orders of magnitude with the substitution of the nitroxide moiety (R_1, R_2) and with the alkyl group R_3 of the dormant alkoxyamine.

In earlier work, we*(8a,9)* and others*(10,11)* have presented rate constants k_d and Arrhenius parameters for the cleavage of various alkoxyamines in order to find predictive rationalizations. The rate constants span eight orders of magnitude from 10^{-9} s^{-1} to 0.1 s^{-1} at 120 °C.*(8a)* This is mainly due to variations of the activation energy which is close to the bond dissociation energy, *BDE*, of the C-ON bond*(7a,8)*. This, in turn, depends on the stabilization of the leaving alkyl radical,*(7a,h,8a,9,11a-d,g)* that is, more stabilized alkyl radicals yield a faster homolysis, and of the nitroxide moiety*(11g)* which, for instance, can be stabilized by intramolecular*(11g)* and intermolecular*(12)* hydrogen bonding. Polar ground state effects*(9,11g)* and steric hindrance*(7a,h,8a,9,11a,f,g)* in both fragments also affect the *BDE*, and, hence, the activation energies and rate constants.

Although the effects of the alkoxyamine structure on the C-ON bond homolysis are now rather well established, data related to the effects of solvent polarity, solvent viscosity and chain length are scarce. Several years ago Rizzardo and Moad*(7a)* showed that k_d of alkoxyamine **1** (Scheme 2) increased by a factor of roughly two with increasing solvent polarity going from hexane (μ = 0.08 D, ε = 1.89) to methanol (μ = 2.87 D, ε = 32.66)*(13)*. This means that the transition state (TS) for the homolysis is more polar than the alkoxyamine ground state. Marque et al.*(11g)* observed the same effect for alkoxyamines **2** and **3** which carry the phenethyl group, and Ananchenko et al.*(10b)* confirmed this for **2**. On the other hand, Ananchenko et al.*(10c)* using *t*-BuPh (μ = 0.36 D, ε = 2.37)*(13)* and PhCl (μ = 1.62 D, ε = 5.62)*(13)* as solvents, did not observe a solvent polarity effect for the alkoxyamines **4** and **5** (Schemes 2 and 3).

Effects of chain length of the leaving groups have mainly been studied for polystyryl-TEMPO (PS-TEMPO). Thus, Bon et al.*(14)* showed that PS-TEMPO (M_n = 7600 g·mol^{-1}) cleaves 9 times faster than the low molecular weight analog **2**, while Fukuda et al.*(15)* found only a factor of 2 between PS-TEMPO (M_n = 1700 g·mol^{-1}) and **2**. Fukuda et al.*(16)* reported a two times larger k_d for polystyryl-SG1 **6** (Scheme 3, M_n = 1960 g·mol^{-1}) than for the analog **4**, and a slightly higher value for **6** (M_n = 3300 g·mol^{-1}) was also observed by Benoit et al..*(17)* We*(18)* recently reported that in a series of styryl based alkoxyamines **6** with 1400 < M_n < 51300 g·mol^{-1} chain length effects are absent. For the acrylate based alkoxyamines **7** (Scheme 3), Ananchenko et al.*(10c)* and Benoit et al.*(17)* observed rates of homolysis that were by factors between 2 and 6 (M_n = 4500 g·mol^{-1}), respectively, larger as compared to the analog **5**.

To the best of our knowledge, there are no studies on the influence of the solvent viscosity on k_d, and the reported effects of the solvent polarity and of the polymer chain length on k_d are in part controversial. This prompted this study on the influence of these factors using alkoxyamines **4-8** (Schemes 2 and 3).

Experimental Section

All chemicals and solvents (*t*-butylbenzene: *t*-BuPh, chlorobenzene: PhCl, *cis*-decaline and methyl caproate) were purchased from Aldrich and used as received. Alkoxyamine **5** was provided by Atofina as a mixture of two diastereoisomers and was used as received. Alkoxyamine **8** was prepared following known procedures.*(19)* NMR experiments were performed with CDCl₃

Scheme 2

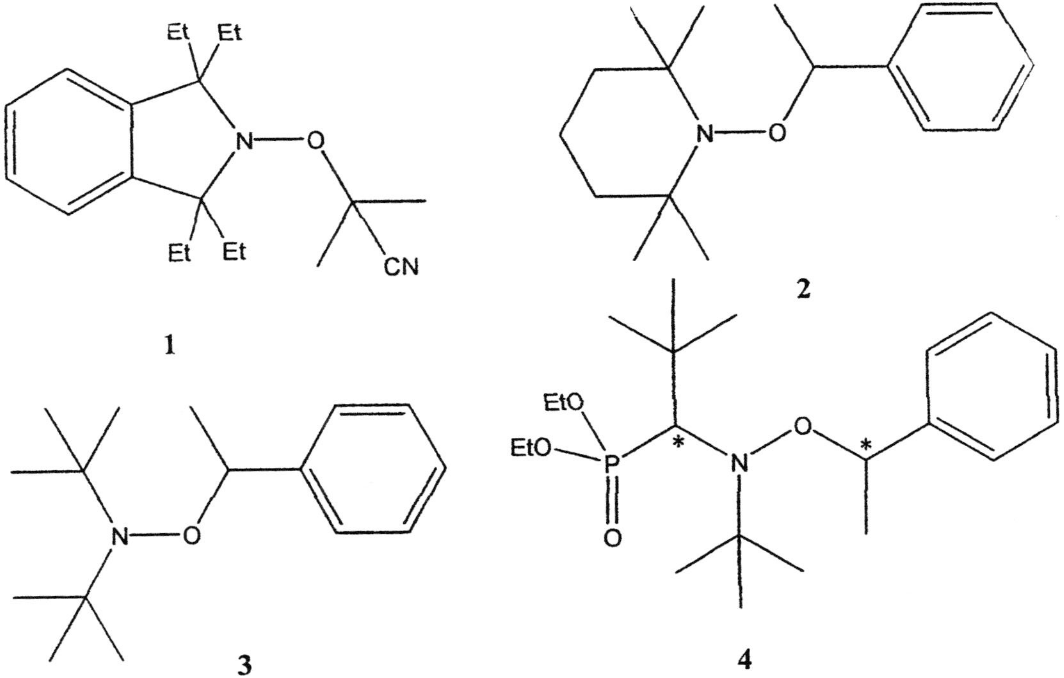

as solvent with a 300 MHz Bruker spectrometer (^{1}H 300MHz, ^{13}C 75.48 MHz and ^{31}P 121.59 MHz) in the Centre Regional de RMN at Marseilles.

Preparation of the styryl and *n*-butyl acrylate based polyalkoxyamines 6 and 7. A typical procedure is as follows: A solution of monomer (100 g), alkoxyamine **5** (0.379 g, for *n*-butyl acrylate) and free SG1 (7.6 mg) was degassed by three freeze-pump thaw cycles. Bulk polymerizations were then carried out in a round-bottom flask under nitrogen atmosphere at 115°C. At regular intervals samples were withdrawn and immediately cooled in an ice bath. Conversion was determined by ^{1}H NMR. M_n and *PDI* were determined by SEC chromatography. For kinetic measurements samples were purified by precipitation for polystyrene and stripping for poly(*n*-butyl acrylate) based macro-alkoxyamines.

Kinetic measurements. Measurements of k_d were carried out as descibed previously using CW-ESR Bruker*(8a,18,20)* and 300 MHz NMR Bruker*(21)* spectrometers. Viscosities were measured with a capillary viscometer.

Results and Discussion

Influence of the solvent polarity. As already mentioned, Ananchenko et al. did not observe an influence of the solvent polarity (*t*-BuPh vs PhCl) on the homolysis rate constant for the SG1 based alkoxyamines **4** and **5**. However, we*(21)* recently observed that k_d for **4** increased by a factor of 1.5 going from *t*-BuPh to the mixture *t*-BuPh/PhSH (1:1; PhSH : $\mu = 1.23$ D and $\varepsilon = 4.38$)*(13)*. Here, we determined k_d for **5** in solvents with different polarities. The results are collected in Table 1 and show that k_d of **5** does not depend on the solvent polarity. This also holds for the solvent methyl caproate which mimics the polarity of *n*-butyl acrylate or methyl methacrylate monomers. This result agrees with that of Ananchenko et al.*(10c)* The increase of k_d with increasing solvent polarity found for **4** could result from the contribution of a polar structure $[(SG1)^-(MeCHPh)^+]$ to the transition state of the homolysis. In the case of **5**, the contributions of $[(SG1)^-(MeCHCOOMe)^+]$ or $[(SG1)^+(MeCHCOOMe)^-]$ are very unlikely. This explanation should be checked with other low molecular weight and polymeric alkoxyamines with polar and unpolar residues.

Scheme 3

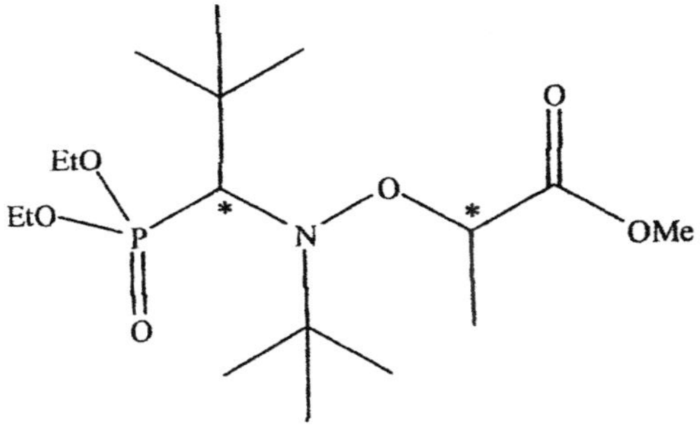

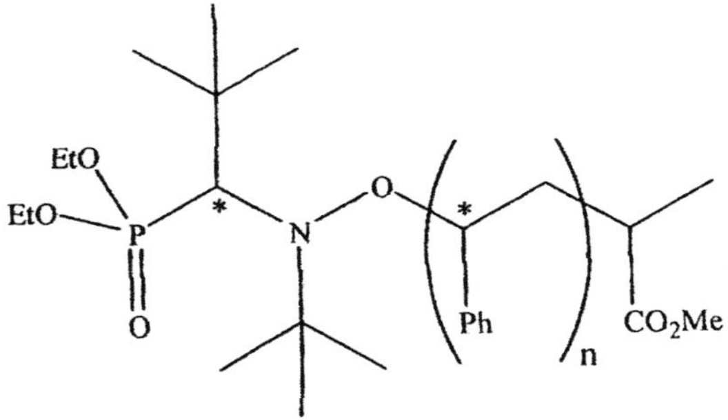

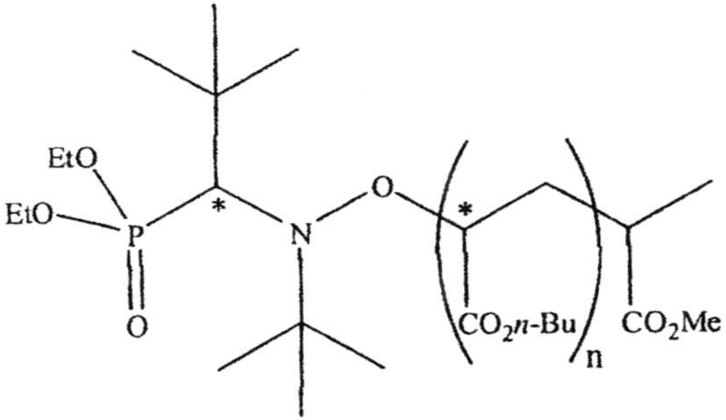

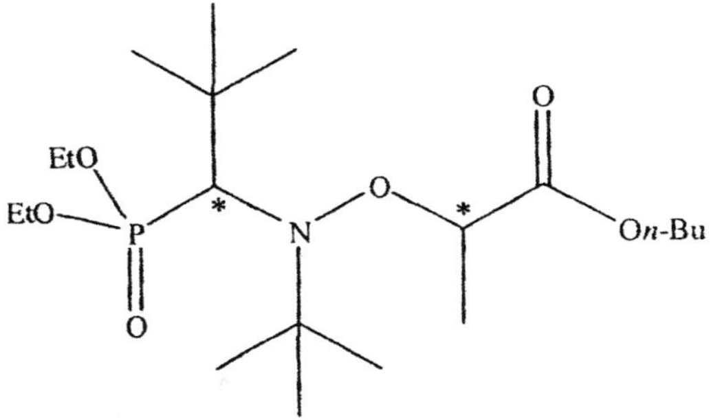

Table 1. Values of homolysis rate constants k_d of both diastereoisomers of 5 in solvents of various polarity and viscosity.

Entry	Solvent	η (cp)^c	μ (D)^d	E_a (kJ·mol^{-1})^a RR/SS	E_a (kJ·mol^{-1})^a RS/SR	$k_{d,120\,°C}$ $(10^{-3}\,s^{-1})^b$ RR/SS	$k_{d,120\,°C}$ $(10^{-3}\,s^{-1})^b$ RS/SR
1	cis-decaline	1.08	0.0	131.8	127.8	0.7	2.3
2	t-BuPh	0.5^{e,f}	0.36	131.3	127.5	1.0	3.0
3	PhCl	0.54^g	1.62	130.6	127.0	1.0	3.2
4	caproate	0.4	1.74	130.8	127.5	1.0	2.7
5	t-BuPh + PS	128.4^{e,h}	-	131.3	127.0	0.8	2.6

^{a} $E_a \pm 2$ kJ·mol^{-1}. E_a was estimated using a mean frequency factor A = $2.4 \cdot 10^{14}$ s^{-1}, see references (8a) and (11g). ^{b} Statistical error less than 10%. Each value is at least the average of two measurements. The whole set of rate constants was measured by ^{31}P NMR. ^{c} See reference (22) unless otherwise mentioned. Values given at 90°C. ^{d} See reference 13. ^{e}Measured at 90 °C using a capillary viscosimeter. ^{f} $\eta_{(t\text{-BuPh})}$ = 0.56 cp at 50°C from reference (22). ^{g} T = 55°C, See reference (23) ^{h} 1.5g of polystyrene (Mn = 182000 g·mol^{-1}, Mw = 340000 g mol^{-1}, PDI = 1.86) was solved in gauged flask filled up to 10 mL.

Influence of solvent viscosity. As it is well known, the rate constants of self-termination and propagation in radical polymerizations strongly depend on the solvent viscosity.(24) Furthermore, the efficiency and the rate of decomposition of most radical initiators depend on the solvent due to cage effects.(25) In the case of alkoxyamines a cage effect would decrease the homolysis rates. To our knowledge, no studies of the effect of solvent viscosity on k_d are available in the literature. Therefore, k_d of both diastereoisomers of 5 was measured in various solvents with different viscosity. The results are listed in Table 1. In methyl caproate, t-butylbenzene, chlorobenzene and cis-decaline (entries 1-4) k_d is the same within the error limits. To confirm that for a very viscous solution, a mixture of t-BuPh and polystyrene was prepared which mimics the experimental conditions of a polymerization. Even in that case k_d was not affected (Table 1, entry 5). Hence, appreciable cage effects are absent for 5, and the same is expected for alkoxyamine based macro-initiators. An explanation has already been given by Scaiano et al.(26) Cage effects arise from highly reactive reencounters of geminately formed radical pairs, and the rate constants k_c for coupling between alkyl and nitroxide radicals are orders of magnitude below the diffusion controlled limit.(27) Therefore, in alkyl/nitroxide radical pairs the number of reencounters before diffusive separation is not sufficient for cage product formation.

Chain Length Dependence. Polystyryl-SG1 alkoxyamines 6. As mentioned before, only a few results on the chain length dependence of k_d are available from the literature. Recently, we found that k_d does not depend on the length of the polystyryl chain.*(18)* The values for alkoxyamines **6** with different M_n were measured by EPR*(18)*, GPC*(16,17)* and for this work by ^{31}P NMR. The results of our work are listed in Table 2. It also displays literature data and the living fractions *LF* of the initiators as determined from the final radical concentration. In agreement with Fukuda*(16)* we find a small but distinct increase of k_d from the low molecular weight analog **4** to the polymeric alkoxyamines by a factor of at most 1.4. For M_n between 4100 and 51300 $g \cdot mol^{-1}$ k_d does not vary within the error limits (Figure 1).

Poly(*n*-butyl acrylate)-SG1 alkoxyamines 7. To our knowledge, values for k_d of poly(*n*-butyl acrylate)-SG1 alkoxyamines were only reported by Gnanou et al.*(17)* and Ananchenko et al.*(10c)* The latter observed a 3.5-fold higher value than for the low molecular weight analog **5**. Here, k_d was measured by either EPR or ^{31}P NMR for a series of alkoxyamines **7** with different M_n. The results are listed in Table 3 together with the value for **8**, literature data and the living fraction of the macro-initiators. The large difference between our data and the result of Gnanou et al.*(17)* may be due to the different techniques. According to our measurements k_d of **7** is up to about 6 times larger than k_d of **8**. This difference is larger than reported by Ananchenko et al.*(10c)* who compared the slowest isomer*(28)* of **7** with the diastereoisomeric mixture of **5**.*(29)* k_d also increases with the length of the polymer chain from $1.4 \cdot 10^{-3} s^{-1}$ for $M_n = 3000$ to $4.0 \cdot 10^{-3} s^{-1}$ for $M_n = 37000$ $g \cdot mol^{-1}$, that is by a factor of about 3.

Table 2: Rate constants[a] at 120 °C[b] for the C-ON bond homolysis of polystyryl-SG1 alkoxyamines.

M_n /g mol^{-1}	n^c	PDI^c	LF(%)	k_d $(10^{-3} s^{-1})$	References
398 (4)	0	-	100	5.2	18
1400	13	2.5	80	5.8	18
4100	41	1.4		6.6^{d,e}	this work
7400	74	1.2		6.4^{d,f}	this work
11300	113	1.1		7.4^{d,f}	this work
37200	358	1.1	83	6.4	18
51300	495	1.2	60	6.6	18
1960	19	1.1	95	11.0	16
3300	32	-	67	3.4	17

[a] Averages of several EPR or ^{31}P NMR measurements carried out with TEMPO or O_2 as scavengers. Statistical error at 15% [b] Values obtained at different temperatures were converted to 120 °C using a frequency factor A = $2.4 \cdot 10^{14}$ s^{-1}.*(8a, 11g)*. [c] Number of monomer units, *PDI*: polydispersity index. [d] Masured by ^{31}P NMR at 100 °C with [TEMPO] = 0.1 M. [e] **[6]** = 10^{-2} M. [f] **[6]** = $10^{-2} - 3 \cdot 10^{-2}$ M.

Several determinations were carried out in the two solvents *t*-BuPh and methyl caproate and with both EPR and ^{31}P NMR. The results did not depend on solvent and technique, in good agreement with the absence of a solvent polarity effect for **5**. Probably the poly(*n*-butyl acrylate) chains adopt similar conformations in both solvents.

For poly(*n*-butyl acrylate)-SG1 alkoxyamines, k_d is much larger than for their low molecular weight analog **8** (7 folds). For polystyryl-SG1 alkoxyamines, k_d is less sensitive to molecular weight. Two factors may contribute to the rate enhancements. Firstly, the polymeric leaving groups may impose additional steric strain on the dissociation energy of the C-ON bond of the polymeric species which lowers the activation energy. Secondly, the dissociation of the polymeric species may lead to a larger gain of motional freedom, that is, a larger activation entropy, which increases the frequency factor. These factors can be distinguished only by very precise measurements at different temperatures.

For poly(*n*-butyl acrylate)-SG1 alkoxyamines a significant increase of k_d with increasing chain length is observed (Figure 1). A reason why poly(*n*-butyl acrylate)-SG1 behaves differently from polystyryl-SG1 can not yet be given but it may be related to the elastomeric behaviour of the poly(*n*-butyl acrylate) homopolymer (Tg = - 54°C).

Table 3: Rate constants[a] at 120 °C[b] for the C-ON bond homolysis of poly(*n*-butyl acrylate)-SG1 alkoxyamines.

M_n (g mol^{-1})	n[c]	PDI[c]	LF (%)	k_d (10^{-3} s^{-1})
423[d] (**8**)	0			0.7
3000	30	1.4	97	1.4[e]
20800	195	1.2	96	1.9[e]
30700	307	1.2	99	2.4[e]
37000	370	1.2	99	4.0[e]
4500	45	1.4	97	7.0[f]

[a] Averages of several measurements with the scavengers TEMPO or O_2 and by EPR or ^{31}P NMR. Statistical error at 15%. [b] Values obtained at different temperatures were converted to 120 °C with a frequency factor A = $2.4 \cdot 10^{14}$ s^{-1}.*(8a,11g)* [c] Number of monomer units, *PDI*: polydispersity index. [d] Alkoxyamine **8** was prepared to have exactly the same ester group in the low molecular weight and the polymeric species. Full kinetic and synthetic data will be published elsewhere.*(19)* [e] [**7**] = 10^{-2} M for ^{31}P NMR experiments, [**7**] = 10^{-4} M for EPR experiments. [f] Ref. *(17)*.

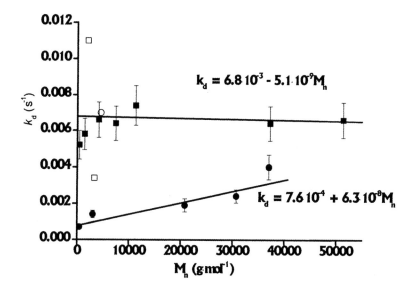

Figure 1. Evolution of k_d with M_n. ($\bullet$) Poly(n-butyl acrylate)-SG1 alkoxyamines, (o) data from Gnanou(17), ($\blacksquare$) polystyryl-SG1 alkoxyamines, ($\square$) data from Fukuda(16) and Gnanou(17).

Conclusion

We have shown that the solvent viscosity has no influence on k_d of low molecular weight and polymeric alkoxyamines. Effects of the solvent polarity depend on the structure of the alkoxyamines and amount to less than a factor of two. k_d is larger for polymeric alkoxyamines than for their low molecular weight analogs, and for poly(n-butyl acrylate)-SG1 an increase of the value of k_d with increasing chain length is observed which is absent or insignificant for poly(styryl)-SG1. Where solvent and chain length effects are found they are not large and this confirms that the homolysis of macro-initiator alkoxyamines is reasonably modeled by that of their low molecular weight analogs.

References

1. US Patent 4,581,429; Eur. Pat. Appl. 135280. Solomon, D. H.; Rizzardo, E.; Cacioli, P. *Chem. Abstr.* **1985**, *102*, 221335q.
2. Georges, M. K.; Veregin, R. P. N.; Kazmaier, P. M.; Hamer, G. K. *Macromolecules* **1993**, *26*, 2987.

3. Hawker, C. J. *Acc. Chem. Res.* **1997**, *30*, 373. Hawker, C. J.; Bosman, A. W.; Harth, E. *Chem. Rev.* **2001**, *101*, 3661 and references cited therein.

4. Greszta, D.; Matyjaszewski, K. *Macromolecules* **1996**, *29*, 7661 and references cited therein.

5. Fischer, H. *Chem. Rev.* **2001**, *101*, 3581 and references cited therein. Fukuda, T.; Goto, A.; Ohno, K. *Macromol. Rapid Commun.* **2000**, *21*, 151.

6. Matyjaszewski, K., Ed. *Controlled Radical Polymerization*; American Chemical Society: Washington, DC, 1998; Vol. 685. Matyjaszewski, K., Ed. *Controlled-Living Radical Polymerization: Progress in ATRP, NMP, and RAFT*; American Chemical Society: Washington, DC, 2000; Vol. 768. Matyjaszewski, K. *Macromol. Symp.* **2001**, *174*, 51.

7. (a) Moad, G.; Rizzardo, E. *Macromolecules* **1995**, *28*, 8722. (b) Grimaldi, S.; Le Moigne, F.; Finet, J.-P.; Tordo, P.; Nicol, P.; Plechot, M. *International Patent* WO 96/24620. 15 August **1996**. (c) Grimaldi, G.; Finet, J.-P.; Zeghdaoui, A.; Tordo, P.; Benoit, D.; Gnanou, Y.; Fontanille, M.; Nicol, P.; Pierson, J.-F. *Polym. Preprints (Am. Chem. Soc., Div. Polym. Chem.)* **1997**, *213*, 651. (d) Benoit, D.; Chaplinski, V.; Braslau, R.; Hawker, C. J. *J. Am.Chem. Soc.* **1999**, *121*, 3904. (e) Chong, B. Y. K.; Ercole, F.; Moad, G.; Rizzardo, E.; Thang, S. H. *Macromolecules* **1999**, *32*, 6895. (f) Grimaldi, S.; Finet, J.-P.; Le Moigne, F.; Zeghdaoui, A.; Tordo, P.; Benoit, D.; Fontanille, M.; Gnanou, Y. *Macromolecules* **2000**, *33*, 1141. (g) Zink, M.-O.; Kramer, A.; Nesvadba, P. *Macromolecules* **2000**, *33*, 8106. (h) Le Mercier, C.; Acerbis, S.; Bertin, D.; Chauvin, F.; Gigmes, D.; Guerret, O.; Lansalot, M.; Marque, S.; Le Moigne, F.; Fischer, H.; Tordo, P. *Macromol. Symp.* **2002**, *182*, 225. (i) Bertin, D.; Chauvin, F.; Couturier, J.-L., Gigmes, D.; Guerret, O.; Marque, S.; Tordo, P. *Eur. Patent submitted*.

8. (a) Marque, S.; Le Mercier, C.; Tordo, P.; Fischer, H. *Macromolecules* **2000**, *33*, 4403. (b) Marsal, P.; Roche, M.; Tordo, P.; de Sainte Claire, P. *J. Phys. Chem. A* **1999**, *103*, 2899.

9. Le Mercier, C.; Lutz, J.-F.; Marque, S.; Le Moigne, F.; Tordo, P.; Lacroix-Desmazes, P.; Boutevin, B.; Couturier, J. L.; Guerret, O.; Martschke, R.; Sobek, J.; Fischer, H. *ACS Symposium Series* **2000**, Chapter 8, 108.

10. (a) Anantchenko, G.; Matyjaszewski, K. *Macromolecules* **2002**, *35*, 8323. (b) Ananchenko, G. S.; Fischer, H. *J. Polym. Sci: Part A: Polym. Chem.* **2001**, *39*, 3604. (c) Anantchenko, G. S.; Souaille, M.; Fischer, H.; Le Mercier, C.; Tordo, P. *J. Polym. Sci.: Part A: Polym. Chem.* **2002**, *40*, 3264.

11. (a) Moad, C. L.; Moad, G.; Rizzardo, E.; Thang, S. H. *Macromolecules* **1996**, *29*, 7717. (b) Skene, W. G.; Belt, S. T.; Connolly, T. J.; Hahn, P.; Scaiano, J. C. *Macromolecules* **1998**, *31*, 9103. (c) Rizzardo, E.; Chieffari, J.; Chong, B. Y. K.; Ercole, F.; Krstina, J.; Jeffery, J.; Le, T. P. T.; Mayadunne, R. T. A.; Meijs, G. F.; Moad, C. L.; Moad, G.; Thang, S. H. *Macromol. Symp.* **1999**. (d) Ciriano, M. V.; Korth, H.-G.; Van Scheppingen, W. B.; Mulder, P. *J. Am. Chem. Soc.* **1999**, *121*, 6375. (e) Cameron, N. R.;

Reid, A. J.; Span, P.; Bon, S. A. F.; van Es, J. J. G. S.; German, A. L. *Macromol. Chem. Phys;* **2000**, *201*, 2510. (f) Goto, A.; Fukuda, T. *Macromol. Chem. Phys.* **2000**, *201*, 2138. (g) Marque, S.; Fischer, H.; Baier, E.; Studer, A. *J. Org. Chem.* **2001**, *66*, 1146.

12 Beckwith, A. L. J.; Bowry, V. W.; Ingold, K. U. *J. Am. Chem. Soc.* **1992**, *114*, 4983.

13. Herrenschmidt, Y.-L.; Guetté, J.-P. In *Techniques de l'Ingénieur*, 1988; pp K310-1.

14. Bon, S. A. F.; Chambard, G.; German, A. L. *Macromolecules* **1999**, *32*, 8269

15. (a) Goto, A.; Tomoya, T.; Fukuda, T.; Miyamoto, T. *Macromol. Rapid Commun.* **1997**, 673. (b) Goto, A.; Fukuda, T. *Macromol. Rapid Commun.* **1997**, 683.

16. Goto, A.; Fukuda, T. *Macromol. Chem. Phys;* **2000**, *201*, 2138.

17. Benoit, D.; Grimaldi, G.; Robin, S.; Finet, J.-P.; Tordo, P.; Gnanou, Y. *J. Am. Chem. Soc.* **2000**, *122*, 5929.

18. Bertin, D.; Chauvin, F.; Marque, S.; Tordo, P. *Macromolecules* **2002**, *35*, 3790.

19. Bertin, D.; Gigmes, D.; Marque, S.; Maurin, R.; S.; Tordo, P. to be published.

20. Kothe, T.; Marque, S.; Martschke, R.; Popov, M.; Fischer, H. *J. Chem. Soc. Perkin Trans. 2* **1998**, 1553.

21. Bertin, D.; Gigmes, D.; Marque,S.; Tordo, P. *e-Polymers.*2003, Paper No. 2

22. Viswanath, D. S.; Natarajan, G. *Databook on the Viscosity of Liquids;* Hemisphere Publishing Corporation. A member of the Taylor & Francis group: New York Washington Philadelphia London, 1989.

23. Heston, W. M.; Hennely, E. J.; Smith, C. P. *J. Am. Chem. Soc.* **1950**, *72*, 2071

24. Matyjaszeswki, K.; Davis, Th. P. *Handbook of Radical Polymerization.* Eds, Wiley Interscience, **2002**.

25. Solomon, D. H.; Moad, G. *The Chemistry of Free Radical Polymerization*, Pergamon, Elsevier Science Ltd, **1995**

26. Skene, W. G.; Scaiano, J. C.; Yap, G. P. A.; *Macromolecules* **2000**, *33*, 3536.

27. Sobek, J.; Martschke, R.; Fischer, H. *J. Am. Chem. Soc.* **2001**, *123*, 2849, and references therein.

28. Only one peak was observed and assumed to be RR/SS isomer from the ^{31}P NMR shift. Furthermore simulation with Predici program showed that the fastest isomer (RS/SR) was almostly isomerized in to RR/SS isomer in our time of polymerization.

29. We have shown that the values of k_d depend strongly on the alkyl group of the ester function, see reference *(19)*. Anantchenko and Matyjaszewski have already pointed out such behaviour in reference *(10a)*.

Chapter 30

Impact of Dilution on the Rate Constant of Termination $\langle k_t \rangle$ in Nitroxide-Mediated Polymerization

Céline Chevalier[1], Olivier Guerret[2], and Yves Gnanou[1,*]

[1]Laboratoire de Chimie des Polymères Organiques, UMR CNRS-ENSCPB-Université Bordeaux I, 16 Avenue Pey Berland, 33607 Pessac Cédex, France
[2]Groupe de Recherche de Lacq, ATOFINA, RN117, BP 34, 64170 Lacq, France

The variation of viscosity with conversion was determined for a series of five experiments carried out at 120°C and different dilutions. Styrene was the monomer chosen for this study and MONAMS, a monoalkoxyamine based on N-tert-butyl-N-(1-diethylphosphono-2,2-dimethylpropyl)-N-oxyl (SG1), served as initiator. Unexpectedly, the rate constant of termination $\langle k_t \rangle$ was found to vary with the initial dilution of the medium, but to remain unchanged with the viscosity build-up induced by monomer conversion.

Introduction

Nitroxide-mediated free radical polymerization affords polymers of controlled molar masses and narrow polydispersities, whose alkoxyamine end groups can be further used for chain extension and block copolymer formation (*1*). N-tert-butyl-N-(1-diethylphosphono-2,2-dimethylpropyl)-N-oxyl (DEPN), now available under the trademark of SG1, is an α-hydrogen containing nitroxide that meets almost ideally the above criteria for a number of monomers (*2,3*). These features were thus exploited to derive various complex architectures, including block copolymers (*4*), stars (*5*), etc...from di (*6*)- and plurialkoxyamines (*5*) based on SG1.

A full kinetic analysis that accounts for the evolution of such "living"/controlled systems was proposed by Fischer (*7,8,9,10,11,12*) and Fukuda (*13,14*). In polymerization under control of persistent radicals such as nitroxides, two specific events occur in addition to the usual propagation and termination steps: growing radicals (PS°) may indeed recombine with persistent radicals (SG1) to generate dormant species (PS-SG1) that can reversibly cleave into the previous reactive entities. Throughout polymerization irreversible self-termination occurs, causing the concentration of persistent radical to steadily build up. The probability for the transient radicals to be trapped by the persistent ones then increases with conversion and that of their self-termination inversely decreases but never disappears totally. These features form the basis of the so-called persistent radical effect (PRE) theorized by Fischer (*7,8,9*), whose kinetic analysis provides a comprehensive description of the time evolution of the various species present in the reaction medium. More precisely, this model predicts the existence of a quasi-equilibrium regime after an initial period, during which the vast majority of chains retain a dormant form and monomer is consumed under "living"/controlled conditions. According to this model $Ln([M_0]/[M])$ should vary linearly with $t^{2/3}$ during the quasi-equilibrium period:

$$Ln\left(\frac{[M_0]}{[M]}\right)=\frac{3}{2}\,k_p\left(\frac{k_d[I_0]}{3k_ck_t}\right)^{1/3} t^{2/3} \qquad (I)$$

$$\text{provided} \qquad K=\frac{k_d}{k_c}\leq\frac{[I_0]k_c}{4\,k_t}$$

In this model, the rate constants of recombination (k_c), activation (k_d), propagation (k_p) and self-termination (k_t) are assumed not to vary with the size of growing chains or the viscosity (*15,16*) of the reaction medium, two parameters that are known to build-up with conversion. In contrast to the propagation, recombination and activation steps, radical-radical self-termination is a diffusion-controlled reaction that heavily depends on chain length (*2,17,18*) and on the overall viscosity of the medium (*19,20*). $<k_t>$ values can decrease by several orders of magnitude with increasing conversion due to the gel effect (*21*). In an attempt to refine this model, Fischer showed through simulations that the chain length dependence of $<k_t>$ does not significantly alter the kinetics and course of polymerizations under control of PRE (9).

In this contribution, we investigate the effect of dilution on the overall course of polymerizations controlled by SG1. A series of polymerizations of styrene were thus carried out at different dilutions at 120°C in the presence of MONAMS (**scheme 1**), a SG1-based monoalkoxyamine used here as initiator.

Scheme 1: MONAMS, a SG1-based alkoxyamine

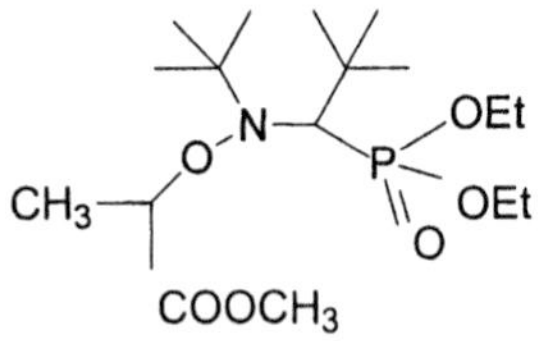

The $[\text{monomer}]_0/[\text{initiator}]_0$ ratio was taken constant in all experiments. One of our objectives is to determine the dependence of $<k_t>$ on the viscosity of the medium under our experimental conditions (T=120°C). Viscosity could be varied in a large range through dilution of the reaction medium by *tert*-butylbenzene, a solvent reputedly inert towards radicals. Using the Fischer equation (I), and rate constant values available in the literature, $<k_t>$ could be calculated from simulations carried out with the help of the PREDICI program. Do experimental conversions depart from simulated ones for a given experiment, which would mean that $<k_t>$ continuously varies with the increase of viscosity and chain length ? Do they fall in agreement instead, which would mean that $<k_t>$ remains constant throughout polymerization respecting the essential features of PRE? These are the questions we intend to settle in this contribution.

Experimental

Materials. Styrene and *tert*-butylbenzene (Aldrich) were purified by vacuum distillation over CaH_2. $CH_3CH(COOCH_3)$-SG1 (MONAMS), the alkoxyamine serving as initiator, was obtained as previously described (22).

Polymerizations. Polymerizations were performed in airproof glass tubes. The required amounts of monomer, solvent and alkoxyamine were introduced under an inert atmosphere. The solution was degassed by three pump-freeze-thaw cycles and heated at 120°C. Aliquots were withdrawn periodically to monitor the evolution of molar masses by SEC and that of conversion by gravimetry. Polymers were analysed without any precipitation. Viscosity measurements were carried out using a standardized parallel-plate viscometer (RS300 of Haake).

Characterization. The sample molar masses were determined using size exclusion chromatography (SEC), tetrahydrofuran as eluent and a flow rate of 1 mL.min^{-1}. The SEC apparatus was equipped with one TSK column (8×300 mm, 5μm) and a refractive index detector (Varian RI-4). SEC was calibrated using polystyrene standards.

Results and Discussion

Five polymerizations corresponding to five different dilutions of styrene were carried out at 120°C, using the same [styrene]$_0$/[MONAMS]$_0$ ratio. The initial monomer mole fraction ($f_{styrene}$) was varied from bulk to 0.12. *Tert*-butyl benzene was used as solvent and MONAMS as initiator. Aliquots were sampled out at various times and analyzed by SEC and gravimetry to determine the average molar mass of the sample and its conversion. Unlike polymerizations carried out in the presence of an initial excess of SG1 where Ln([M$_0$]/[M]) was shown to vary linearly with $k_P(K[SG1][P-SG1])t$ (*23,24*), in the present case the Ln([M$_0$]/[M]) vs time plots should not be linear, but exhibits a downward curvature. According to the Fischer-Fukuda (*9,13*) analysis, such evolution of the rate of polymerization merely mirrors the decreasing concentration of propagating radicals. As anticipated, Ln([M$_0$]/[M]) vs time plots **(Figure 1)** all exhibit a non-linear dependence independently of the experiment/dilution considered.

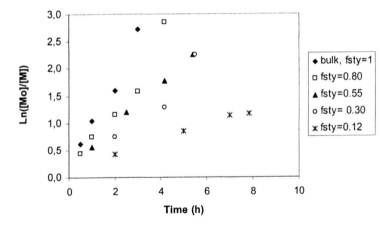

Figure 1 : Plot of Ln([M$_0$]/[M]) versus time at different dilutions in styrene; T=120°C, [styrene]/[MONAMS]=92.

After undergoing a strong downward curvature in the low conversion domain, Ln([M$_0$]/[M]) then adopts a close to linear evolution indicating that the probability for radical-radical self-termination decreases with the build-up of the

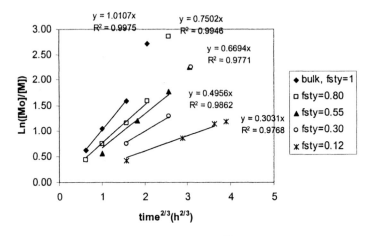

Figure 2 : Plot of Ln($[M_0]/[M]$) versus $t^{2/3}$ at different dilutions in styrene; T=120°C, [styrene]/[MONAMS]=92.

nitroxide concentration. In contrast, Ln($[M_0]/[M]$) vs $t^{2/3}$ **(Figure 2)** plots all show a linear dependence confirming the pertinence of the PRE model.
Apart from the points corresponding to the highest conversions ($\approx$90%), all others indeed fall in a same straight line for a given experiment, obeying the variation in $t^{2/3}$. As to the Mn of the samples analysed, they were found to increase linearly **(Figure 3)** with conversion and align on a same straight line independently of the experiment/dilution considered.

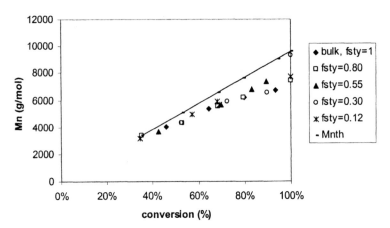

Figure 3 :Variations of Mn as a function of conversion for different dilutions in styrene; T=120°C, [styrene]/[MONAMS]=92.

However, one can note that this experimental variation of Mn departs significantly from the expected values as if additional chains were created throughout polymerization. From our point of view, these additional chains do not arise from styrene autopolymerization -which should be normally suppressed at high dilution- or from transfer to solvent -*tert*-butylbenzene being inert towards radicals- but certainly from another phenomenon that will be commented on later.

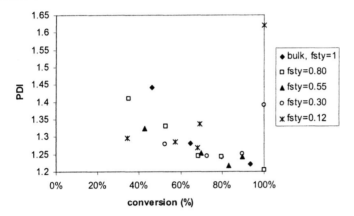

Figure 4 : Evolution of PDI as a function of conversion for different dilutions in styrene, T= 120°C, [styrene]/[MONAMS]=92.

As shown in **Figure 4**, the polydispersity indexes (PDI) of the samples decrease with increasing conversion, adopting for low conversion/short times the $1+\dfrac{1}{DP_n}+\dfrac{8}{3}(k_d t)^{-1}$ variation predicted by Fischer (*7*). Samples obtained from dilute solutions exhibit the lowest PDI's which is also conform to our expectations. The linear variation of $\ln([M_0]/[M])$ with $t^{2/3}$ suggests that k_t remains constant throughout polymerization and is unaffected by the build-up of viscosity. In spite of this observation, we felt essential to carry on with our initial idea and determine the scale of variation of η_0 with the dilution and the conversion.

Variation of viscosity (η_0) with conversion.

The viscosity (η_0) could not be measured directly on the reaction medium for practical reasons. Solutions corresponding to a particular conversion and dilutions were thus prepared using linear inactive PS standards of the same molar

430

mass as that of the corresponding sample. Viscosity measurements were carried out at 120°C with the help of a viscometer equipped with parallel plates. From the variation of the dynamic shear viscosity (η) with the shear rate ($\overset{\circ}{\gamma}$), the zero-shear viscosity (η_0) could be determined for a series of polymer solutions and melts representative of the reaction medium (**Figure 5**).

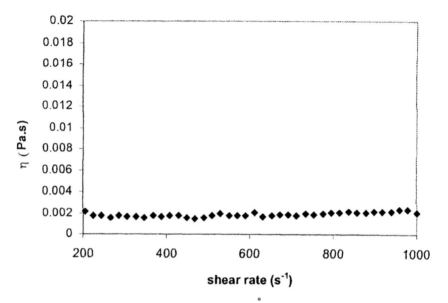

Figure 5: Evolution of η as a function of $\overset{\circ}{\gamma}$ at 120°C corresponding to $f_{sty}=0.75$ and 32.5% conversion.

All the solutions investigated even the most concentrated ones behave as newtonian fluids within the domain of shear rates tried. From these data, the conversion-dependent viscosity curves could be constructed (see **Figure 6**) for the five experiments.

C_η, the slope of the Lnη_0 vs conversion plot for each experiment/dilution, mirrors the increase of η_0 with conversion, which is even higher that the initial concentration in styrene is large. A linear dependence of C_η on $f_{styrene}$ is observed (**Figure 7**).

Effect of dilution on k_t and on the course of polymerization:

Knowing the conversion dependence of η_0 for the five experiments and the range -several orders of magnitude- of its variation, we checked how it influences $\langle k_t \rangle$ and to what extent. As previously stressed, the linear dependence of Ln($[M_0]/[M]$) on $t^{2/3}$ -with excellent coefficients of correlation- that was

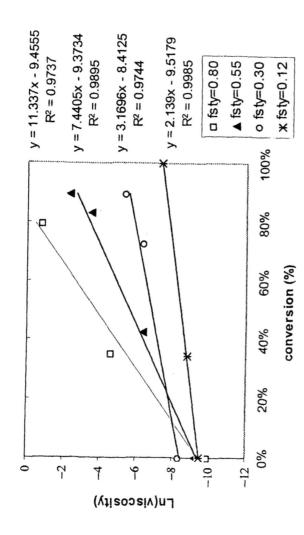

Figure 6 :Evolution of Lnη_0 as a function of conversion

observed independently of the experiment/dilution considered indicates that the

$\dfrac{3}{2} k_p \left(\dfrac{k_d [I_0]}{3 k_c k_t} \right)^{1/3}$ term -which represents the slope of these straight lines- is

constant. Assuming that k_p, k_d, and k_c remain constant throughout polymerization which seems reasonable according to literature data (**Table I**)-, we could easily extract the value of $<k_t>$ and thus infer that it is also constant for a given experiment. A series of simulations were subsequently carried out with the help of the PREDICI program. In these simulations, the thermal autopolymerization of styrene and the degradation of SG1 were taken into account. Using equation (I) and the values of k_p, k_d and k_c that previously served to deduce $<k_t>$, the variation of $Ln([M_0]/[M])$ was simulated for different $<k_t>$'s, the value retained for the latter rate constant corresponding to the best fit between simulated and experimental data (**Scheme 2**).

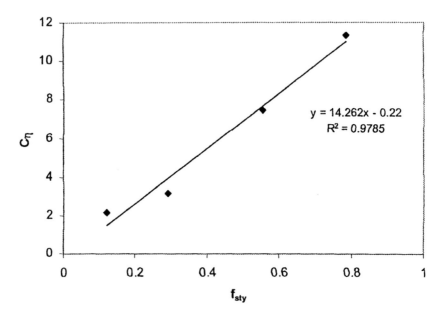

Figure 7 : variation of Cη versus f_{sty}

As shown in **Table II** and in **Figure 8**, the simulated values of $<k_t>$ fall in excellent agreement with those experimentally estimated, indicating again that the rate constant of termination is independent of the viscosity in our experimental conditions. This means that the build-up of viscosity and molar mass experienced by the reaction medium has little effect on $<k_t>$.

Scheme 2:Reactions taken into account for the simulations of SG1-mediated radical polymerization of styrene.

$$\text{Monams} \underset{\text{kcalkox}}{\overset{\text{kdalkox}}{\rightleftharpoons}} \text{I} + \text{SG}_1$$

$$3\,\text{M} \xrightarrow{\text{kth}} 2\,\text{I}\ (\text{self-initiation})$$

$$\text{I} + \text{M} \xrightarrow{\text{kams}} \text{PS}_{(1)}$$

initiation

$$\text{SG}_1 \xrightarrow{\text{kdeg}} \text{SG}_1\text{deg}$$

$$\text{PS(s)} + \text{M} \xrightarrow{\text{kp}} \text{PS(s+1)}$$

$$\text{PS(s)} + \text{SG}_1 \underset{\text{kd}}{\overset{\text{krec}}{\rightleftharpoons}} \text{PSdormant(s)}$$

propagation

$$\text{I} + \text{I} \xrightarrow{\text{kts}_0} \text{Idead}$$

$$\text{PS(s)} + \text{I} \xrightarrow{\text{kts}_4} \text{PSdead}$$

$$\text{PS(s)} + \text{PS(r)} \xrightarrow{\text{kt}} \text{PSdead(s+r)}$$

termination

Table I: Rate constants and their values

constant	value	ref
$k_{dalkox}\ (s^{-1})$	$9.88\ 10^{-4}$	32, 33
$k_{calkox}\ (L.mol^{-1}.s^{-1})$	$5.6\ 10^{6}$	33
$k_{th}\ (L^{2}\,mol^{-2}s^{-1})$	$1.2.10^{-10}$	34
$k_{ams}\ (L.mol^{-1}.s^{-1})$	$5.12\ 10^{5}$	2, 35
$k_{deg}\ (s^{-1})$	$2.79\ 10^{-5}$	24
$k_{p}\ (L.mol^{-1}\cdot s^{-1})$	2036	36
$k_{rec}\ (L.mol^{-1}\cdot s^{-1})$	$5.7\ 10^{5}$	2
$k_{d}\ (s^{-1})$	$8.26\ 10^{-3}$	2
$k_{ts0}\ (L.mol^{-1}\cdot s^{-1})$	$2.5\ 10^{9}$	18
$k_{ts4}\ (L.mol^{-1}\cdot s^{-1})$	1.10^{9}	18
k_{t}	To be determined	

Table II: experimentally estimated and simulated $<k_t>$'s.

f_{sty}	$<k_t>$ calculated by simulations $(L.mol^{-1}.s^{-1})$	$<k_t>$ experimentally estimated $(L.mol^{-1}.s^{-1})$
1-bulk	$1.75.10^8$	$1.80.10^8$
0.80	$3.22.10^8$	$3.09.10^8$
0.55	$2.96.10^8$	$2.90.10^8$
0.30	$3.41.10^8$	$3.57.10^8$
0.12	$4.56.10^8$	$6.25.10^8$

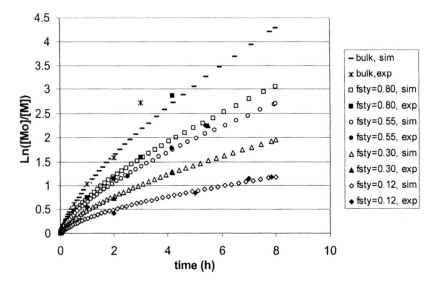

Figure 8 : Plot of Ln($[M_0]$/$[M]$) versus time for different dilutions in styrene; T=120°C, [styrene]/[MONAMS]=92; simulated and experimental points.

Two reasons may account for this experimental observation: first, the range of molar masses targeted is rather narrow (a few hundreds to 10^4 g/mol) and second, the medium heated at 120°C is only moderately viscous, far from those experiencing vitrification and the so-called "Trommsdorf-Norrish effect". In a recent study devoted to the determination of $<k_t>$ values in RAFT polymerizations, Vana et al.(37) also observed a weak chain-length dependence of $<k_t>$ for 10 to 100 monomer unit-long chains, but noted that $<k_t>$ can vary significantly in the $1 \le DP_n \le 10$ domain. Such a strong dependence of $<k_t>$ could not be evidenced in our case due to its little impact on the $Ln([M_0]/[M])$ vs time plots.

On the other hand, one point that is striking when looking closely at the values taken by $<k_t>$ is their dependence on the dilution of the reaction medium: the lower the initial concentration of styrene, the higher the value of $<k_t>$ measured and simulated. How to account for the fact that $<k_t>$ is independent of the conversion and viscosity but vary with the dilution of the medium? The explanation for this apparent contradiction can be found in the work of Buback and coworkers (25,26) who observed similar features. In several instances, these authors also found a dependence of $<k_t>$ on the dilution of the medium and its invariability with conversion. In toluene, the polymerization of methyl methacrylate (27) at T=30°C and P=1000 bar, and that of dodecyl acrylate (28,29) at T=40°C in CO_2 are two cases exhibiting such a behaviour. Buback (25,30,31) theorized that termination can be regarded as a consecutive three-step reaction, involving the translational diffusion of two radicals, their segmental diffusion to allow contact between the radical sites and the chemical reaction with:

$$\frac{1}{\langle k_t \rangle} = \frac{1}{k_{TD}} + \frac{1}{k_{SD}} + \frac{1}{k_{CR}}$$

where k_{TD}, k_{SD}, and k_{CR} are the corresponding rate constants. According to this approach, segmental mobility controls the termination in the initial period of polymerization (at low conversion), the contribution of translational diffusion becoming predominant only at the onset of the vitrification when the viscosity dramatically increases. To account for the fact that $<k_t>$ increases with dilution in some of their low conversion systems, the same team argued that segmental diffusion depends on intra-coil viscosity (28) and therefore increases with the decrease of the latter parameter due to dilution. Such an explanation may well also apply to our case. Even though viscosity increases by several orders of magnitude from one experiment to another it never reaches the regime typical of vitrified media. One can therefore suggest that termination is under control of segmental mobility until the end of polymerization. As the latter type of diffusion is sensitive to the internal friction of the polymer chain, it appears consistent that $<k_t>$ increases with the dilution -though in a modest range- in our experiments. Although a vast majority of experimental and simulated points agree almost ideally in the $Ln([M_0]/[M])$ vs $t^{2/3}$ plots, experimental points corresponding to highest conversions (>85%) significantly depart from simulated ones. This may well be due to the fact that in our simulations, we overlooked the possibility for the products of degradation of SG1 to initiate new chains. Lacroix-Desmazes (24) et al. showed in a recent paper on the controlled polymerization of n-butyl acrylate in the presence of SG1 that the additional chains were initiated by some of the products of decomposition of SG1. This may also explain why we systematically observed smaller molar masses than those targeted, independently of the experiment considered.

Conclusion

This contribution clearly establishes the influence of dilution of the reaction medium on the rate constant of termination $<k_t>$ and provides the scale of its variation. As one could have anticipated, the $k_p/<k_t>$ ratio decreases with dilution meaning that a higher proportion of dead chains are formed for a given conversion at higher dilutions. To our surprise, the build-up of viscosity induced by the monomer conversion was found not to affect $<k_t>$ which remains constant throughout polymerization even at high conversions.

Acknowledgements
The authors are thankful to Atofina for financial support.

References

1. Hawker, C.J.; Bosman, A.W.; Harth, E. *Chem. Rev.* **2001**, *101*, 3661.
2. Benoit, D.; Robin, S.; Grimaldi, S.; Tordo, P.; Gnanou, Y. *J. Am. Chem. Soc.* **2000**, *122*, 5929.
3. Benoit, D.; Grimaldi, S.; Finet, J.P.; Tordo, P.; Fontanille, M.; Gnanou, Y. in *Controlled Radical Polymerization*; Matyjaszewski, K., Ed.; ACS Symposium Series 685; American Chemical Society; Washington, DC, **1998**; Chapter 14.
4. Robin, S; Gnanou Y. in *Controlled/Living Radical Polymerization*; Matyjaszewski, K., Ed.; ACS Symposium Series 768; American Chemical Society; Washington, DC, **2000**; Chapter 23.
5. Robin, S.; Guerret, O.; Couturier, J-L.; Gnanou, Y. *Macromolecules* **2002**, *35*, 2481.
6. Robin, S.; Guerret, O.; Couturier, J-L.; Pirri, R.; Gnanou, Y. *Macromolecules* **2002**, *35*, 3844.
7. Fischer, H. *Macromolecules* **1997**, *30*, 5666.
8. Fischer, H. *J. Polym. Sci, Part A: Polym. Chem.* **1999**, *37*, 1885.
9. Souaille, M.; Fischer, H. *Macromolecules* **2000**, *33*, 7378.
10. Souaille, M.; Fischer, H. *Macromolecules* **2001**, *34*, 2830.
11. Souaille, M.; Fischer, H. *Macromolecules* **2002**, *35*, 248.

12. Fischer, H. *Chem. Rev.* **2001**, *101*, 3581.
13. Ohno, K.; Tsujii, Y.; Miyamoto, T.; Fukuda, T.; Goto, M.; Kobayashi, K.; Akaike, T. *Macromolecules* **1998**, *31*, 1064.
14. Fukuda, T.; Goto, A.; Ohno, K. *Macromol. Rapid. Commun.* **2000**, *21*, 151.
15. Bertin, D.; Chauvin, F.; Marque, S.; Tordo, P. *Macromolecules* **2002**, *35*, 3790.
16. Chauvin, F.; Gigmes, D.; Marque, S.; Bertin, D.; Tordo, P. *Polym. Prepr.* **2002**, *43*(2), 80.
17. Griffiths, M.C; Strauch, J.; Monteiro, M.J.; Gibert, R.G. *Macromolecules* **1998**, *31*, 7835.
18. Shipp, D.A.; Matyjaszewski, K. *Macromolecules* **1999**, *32*, 2948.
19. Yamada, B.; Kageoka, M.; Otsu, T. *Macromolecules* **1991**, *24*, 5234.
20. Zetterlund, P.B.; Yamazoe, H.; Yamada, B.; Hill, D.J.T.; Pomery, P.J. *Macromolecules* **2001**, *34*, 7686.
21. Yamazoe, H.; Zetterland, P.B.; Yamada, B.; Hill, J.T.; Pomery, P.J. *Macromol. Chem. Phys.* **2001**, *202*, 824.
22. Guerret, O. ; Couturier, J.L.; Vuillemain, B.; Lutz, J.F.; Le Mercier, C.; Robin, S.; FR. Patent 06329, 1999.
23. Veregin, R.P.N.; Odell, P.G.; Michalak, L.M.; Georges, M.K. *Macromolecules* **1996**, *29*, 2746.
24. Lacroix-Desmazes, P.; Lutz, J-F.; Boutevin, B. *Macromol. Chem. Phys.* **2000**, *201*, 662.
25. Beuermann, S.; Buback, M. *Prog. Polym. Sci.* **2002**, *27*, 191.
26. Beuermann, S.; Buback, M.; Schmaltz, C. *Ind. Eng. Chem. Res.* **1999**, *38*, 3338.
27. Beuermann, S.; Buback, M.; Russell, G.T. *Macromol. Chem. Phys.* **1995**, *196*, 2493.
28. Buback, M. in *Controlled/Living Radical Polymerization*; Matyjaszewski, K., Ed.; ACS Symposium Series 768; American Chemical Society; Washington, DC, **2000**; Chapter 3.
29. Beuermann, S.; Buback, M.; Nelke, D. In preparation.
30. Buback, M. *Macromol. Chem.* **1990**, *191*, 1575.
31. Buback, M.; Kuchta, F.D. *Macromol. Chem. Phys.* **1997**, *198*, 1455.
32. Le Mercier, C.; Acerbis, S.; Bertin, D.; Chauvin, F.; Gigmes, D.; Guerret, O.; Lansalot, M.; Marque, S.; Le Moigne, F.; Fischer, H.; Tordo, P. *Macromol.Symp.* **2002**, *182*, 225.
33. Marque, S.; Le Mercier, C.; Tordo, P.; Fischer, H. *Macromolecules* **2000**, *33*, 4403.
34. Hui, A.; Hamielec, A. *J. Appl. Polym. Sci.* **1972**, *16*, 749.
35. Kaszas, G.; Foldes-Berezsnick, T.; Tudos, F. *Eur. Polym. J.* **1984**, *20*, 395.
36. Buback, M.; Gilbert, R.G.; Hutchinson, R.A.; Klumperman, B.; Kuchta, F.D.; Manders, B.G.; O'Driscoll, K.F.; Russell, G.T.; Schweer, J. *Macromol. Chem. Phys.* **1995**, *196*, 3267.
37. Vana, P.; Davis, T.; Barner-Kowollick, C. *Macromol. Rapid. Comm.* **2002**, *23*, 952.

Chapter 31

Nitroxide-Mediated Polymerization in Miniemulsion: A Direct Way from Bulk to Aqueous Dispersed Systems

B. Charleux

Laboratoire de Chimie Macromoléculaire, Université Pierre et Marie Curie, T44, E1 4, Place Jussieu, 75252 Paris Cedex 05, France (email: charleux@ccr.jussieu.fr)

Owing to fundamental differences with bulk polymerization, the special features of nitroxide-mediated CRP in miniemulsion and the key for success are discussed in this article. The miniemulsion process, where polymerization is initiated by a SG1-based alkoxyamine, is examined in terms of initiation / nucleation, kinetics, monomer transport, and their consequences on control of molar mass and distribution.

Since the very first developments, controlled free-radical polymerization (CRP) methods proved to be extremely powerful for the synthesis of a variety of new macromolecules with increased complexity (*1,2*). When these chemically and architecturally well-defined polymers are prepared at the scale of the laboratory, homogeneous polymerizations like bulk or solution are particularly well suited. Additionnaly, these simple conditions are also the best choice for kinetic and mechanistic investigation. However, it is of industrial concern to develop CRP in aqueous dispersed systems (*3,4*). Emulsion polymerization is indeed a widespread process for the production of polymers via free-radical chemistry. Combination of the properties imparted by control of the polymer chains at the molecular level along with the advantages of this process will constitute a great achievement for the future developments.

The most significant progresses of CRP in aqueous dispersed systems were however not directly done in emulsion polymerization but in miniemulsion (*5,6*). Indeed, the miniemulsion process can be regarded as a simple model for emulsion polymerization (*7-9*). In miniemulsion, the initial monomer in water emulsion is strongly sheared in order to divide the organic phase into small droplets that remain stable throughout the reaction. In addition to classical surfactants that ensure stability against coalescence, the use of a hydrophobe (such as hexadecane and/or polymer) was shown to enhance droplet stability via inhibition of Oswald ripening. The complex nucleation step that exists in emulsion polymerization is replaced by droplet nucleation (the presence of micelles should be avoided). For this reason, droplets behave as individual bulk reactors with ideally no exchange between them. The process allows the use of oil-soluble initiators and is tolerant to thermal auto-initiation, which is not possible in emulsion because undesirable polymerization would take place in the large non-stabilized monomer droplets. This polymerization process is nowadays developing very fast for many applications: in addition to its use in CRP, it was shown to be very useful for the synthesis of organic/organic and organic/inorganic hybrid particles; moreover, other chemistries than free-radical can be applied, such as polyadditions (*8*) and anionic polymerization *(10)*.

The most direct pathway from bulk to miniemulsion CRP is to use the same reagents and particularly the same initiator and control agent. In the simplest case, as far as nitroxide-mediated polymerization in miniemulsion is concerned, the initiator is a monomer-soluble alkoxyamine, either a low molar mass one or a polymeric one, and the control agent is an oil-soluble nitroxide (TEMPO or SG1 for instance, see Figure 1) with low water-solubility (*11,12*). A classical oil-soluble radical initiator can also be selected in conjunction with added free nitroxide; such bicomponent initiating system forms oil-soluble alkoxyamines and hence, behaves very similarly to the previous system. Another way to perform nitroxide-mediated CRP in miniemulsion is to use a water-soluble initiator in conjunction with an oil-soluble nitroxide. In this case, the initiating system can be either a bicomponent one with classical initiator and free nitroxide, or a monocomponent one with specially designed water-soluble alkoxyamine.

Owing to fundamental differences with bulk polymerization, the special features of nitroxide-mediated CRP in miniemulsion and the key for success are discussed in this article. The miniemulsion process is examined here in terms of initiation / nucleation, kinetics, monomer transport and their consequences on control of molar mass and distribution. We will focus on SG1-mediated polymerization and examine more particularly the use of alkoxyamine initiators that allow the best control over molar mass and architecture (*13*).

A Brief Description of Nitroxide-Mediated CRP and its Application to Aqueous Dispersed Systems

Nitroxides are stable radicals that are able to trap carbon centered radicals at a nearly diffusion controlled rate (*1,2*). At low temperatures, the formed alkoxyamine is stable and therefore the nitroxide behaves as an inhibitor. However, at elevated temperature, the C-O bond may undergo homolytic cleavage, leading back to the propagating radical and to the nitroxide. This equilibrium between propagating radical and inactive alkoxyamine is the key step in nitroxide-mediated polymerization. Moreover, owing to the stability of their alkoxyamine end-group, the dormant macromolecules can be isolated and further used as macroinitiators for the polymerization of the same or a different monomer. The application of nitroxides to control macromolecular architecture has been recently reviewed by Hawker et al. (*13*).

Initially, TEMPO (2,2,6,6-tetramethylpiperidinyl-1-oxy) (Figure 1) was the most widely used and studied nitroxide for CRP of styrene and derivatives (*14-16*), enabling the synthesis of well-defined block copolymers and star-shaped structures. However, the application of this nitroxide to other monomers appeared to be less straightforward (*13*). A new class of acyclic nitroxides was more recently proposed (*17-20*). One of them is the N-*tert*-butyl-N-(1-diethylphosphono-2,2-dimethylpropyl) nitroxide (also called SG1) (Figure 1) (*18-20*).

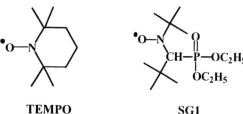

TEMPO   SG1

Figure 1. Structure of TEMPO (2,2,6,6-tetramethylpiperidinyl-1-oxy) and SG1 (N-tert-butyl-N-(1-diethylphosphono-2,2-dimethylpropyl) nitroxide).

Faster kinetics than with TEMPO were observed for styrene polymerization and this nitroxide was also shown to be particularly well suited for the controlled polymerization of acrylic esters such as n-butyl acrylate. This feature opened the way to the synthesis of complex copolymer architectures using nitroxide-mediated polymerization (*13,21-23*), as it was also the case with the other CRP techniques, namely atom transfer radical polymerization (ATRP) (*24,25*) and reversible addition-fragmentation transfer (RAFT) (*26*).

TEMPO and derivatives have been used as mediators in aqueous dispersed systems such as suspension (*27-30*), seeded emulsion (*31*), ab initio batch emulsion (*32,33*) and miniemulsion polymerizations (*34-42*). Styrene was the most studied monomer, but more recently, the CRP of n-butyl acrylate was also made possible in a miniemulsion system (*37,38*). Nevertheless, TEMPO presents many drawbacks, which are not in favor of its use in aqueous dispersed systems and the more recent progresses have been done with SG1 as a mediator (*12,43-48*). For this reason, the examples presented in this article will be based on SG1-mediated polymerization of n-butyl acrylate (BA) and styrene (S) in miniemulsion.

Initiation and Nucleation

Oil-Soluble Alkoxyamine Initiator

When this type of initiator is used, transposition of bulk polymerization to miniemulsion process is quite straightforward. The oil-soluble alkoxyamines that have been applied as initiators in miniemulsion are TEMPO-capped polystyrene (PS-TEMPO) (*36-38*), SG1-capped poly(n-butyl acrylate) (PBA-SG1)(*47*) and SG1-based low molar mass alkoxyamines such as MONAMS (*12,44,45,47*) and DIAMS (*48*) (Figure 2).

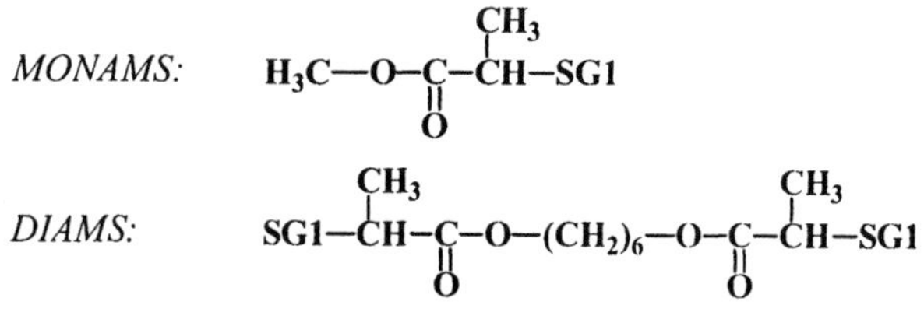

$$\text{MONAMS:} \qquad H_3C-O-\underset{\underset{O}{\|}}{C}-\overset{\overset{CH_3}{|}}{C}H-SG1$$

$$\text{DIAMS:} \qquad SG1-\overset{\overset{CH_3}{|}}{C}H-\underset{\underset{O}{\|}}{C}-O-(CH_2)_6-O-\underset{\underset{O}{\|}}{C}-\overset{\overset{CH_3}{|}}{C}H-SG1$$

Figure 2. Structrure of MONAMS and DIAMS SG1-based alkoxyamines.

An advantage of the polymeric macroinitiators is that they act as reactive hydrophobic agents against Oswald ripening, avoiding the use of other molecules that can be considered as volatile organic compounds. However, to reduce the number of steps and avoid a preliminary bulk polymerization, low molar mass alkoxyamines are of great interest too.

When an oil-soluble alkoxyamine is used, exhibiting fast initiation because of fast dissociation rate, droplet nucleation is necessarily a fast process too. However, this ideal situation can be complicated by the dispersed state of the system with possible partitioning of the alkoxyamine or/and of the initiating carbon-centered radical. This actually does not apply for polymeric macroinitiators because neither the alkoxyamine nor the initiating radical are water-soluble; hence they are trapped in the droplets/particles and cannot undergo exit. Low molar mass alkoxyamines behave however quite differently. For instance, MONAMS has been used to initiate n-butyl acrylate polymerization in miniemulsion at 112 and 125 °C, in the presence of a small concentration of free SG1 (44-47). Whereas partition coefficient of the alkoxyamine in water is low (49), it is not the same for the 1-(methoxycarbonyl)eth-1-yl (MCE) primary radical formed upon homolytic cleavage. Indeed, MCE is known to be quite hydrophilic: for instance, water solubility of methyl acrylate at saturation at 50 °C is 0.6 mol.L^{-1} and partition coefficient, i.e. ratio of concentration in water over concentration in the organic phase, is 0.05 (50). Therefore, after dissociation of the alkoxyamine, the initiating radical has three possibilities: it can either exit, initiate, or recombine with free SG1 (to simplify, termination with another carbon-centered radical is not considered; such hypothesis actually holds if sufficient concentration of free nitroxide is initially introduced in the reaction medium). If one compares first, initiation (rate = $k_p.[BA]_0.[MCE]$) and recombination (rate = $k_c.[SG1]_0.[MCE]$), the ratio of initiation rate over recombination rate, $(k_p.[BA]_0)/(k_c.[SG1]_0)$, is $0.017/[SG1]_0$. For the calculation, the rate constant of recombination k_c is 4×10^7 L.mol^{-1}.s^{-1} (19,20); the rate constant of addition of MCE to BA is assumed to equal the rate constant of propagation of n-butyl acrylate at 125 °C, i.e. $k_p = 94000$ L.mol^{-1}.s^{-1} (51) and $[BA]_0 = 7$ mol.L^{-1}. As mentioned above, a small concentration of free SG1 is usually added to ensure a better control of the polymerization of BA (20,45,47) and a value close to $[SG1]_0 = 8 \times 10^{-4}$ mol.L^{-1} in the monomer phase is a common situation (45,47). In this case, $(k_p.[BA]_0)/(k_c.[SG1]_0) = 21$. Not only is initiation favored, but also the initial kinetic chain length is large enough to lead to oligomers with relatively large molar mass, preventing further exit. Actually, to ensure fast initiation, $[SG1]_0$ should not be larger than 0.017 mol.L^{-1} in this case, which is quite a large upper limit. For styrene initiated with MONAMS, the initiation rate would be very fast too because of large cross-propagation rate constant (see reactivity ratios in ref. 52).

It is then interesting to compare initiation rate with exit rate. A mathematical model for exit of monomeric radicals from latex particles has been proposed by Ugelstadt et al. (53) and was also described later by Gilbert (4). The first-order rate coefficient for desorption of a monomeric radical, k_{dM}, can be expressed as written in Eq. 1 with D_w, the diffusion coefficient for the desorbing species in the water phase (taken as $1.6\ 10^{-7}\ dm^2.s^{-1}$)(4), r_s, the radius of the swollen particles, and q, the ratio of monomeric radical concentration in water over concentration in the organic phase (assumed to be the same as for methyl acrylate monomer, i.e. q = 0.05).

$$k_{dM} = \frac{3\,D_w}{r_s^{\ 2}} \cdot q \qquad (1)$$

From this, the probability of exit as a function of r_s can be calculated as also done previously by Monteiro et al. for CRP in emulsion, using reversible transfer technique (Eq. 2) (54).

$$P(exit) = \frac{k_{dM}}{k_{dM} + k_p \cdot [BA]_0} \qquad (2)$$

$[BA]_0$ is the initial concentration of n-butyl acrylate in the monomer droplets (7 mol.L^{-1}) and k_p is, like above, the rate of addition of the MCE radical to n-butyl acrylate (at 125 °C, k_p = 94000 L.mol^{-1}.s^{-1} (51)). The plot of P(exit) as a function of the particle diameter is shown in Figure 3.

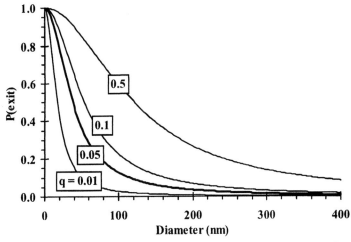

Figure 3. Probability of exit of the initiating radical with partition coefficient q as a function of particle size in miniemulsion polymerization of n-butyl acrylate at 125 °C.

Probability of exit for MCE with q = 0.05 is very large for small particles, but strongly decreases to negligible values for large particles. Therefore, miniemulsion polymerization of BA, in which particle diameter is usually larger than 100 nm, should not be affected by extensive exit of initiating radicals, unless an initiating radical much more water-soluble than MCE is employed (q >> 0.05). The same conclusion can be drawn for styrene initiated by MONAMS. Indeed because of fast initiation rate due to very large cross-propagation rate constant (*52*), the probability of exit should be very small too. Moreover, the addition of the first monomer unit would lead to a drop in the partition coefficient q, and hence in the value of k_{dM}, making the further exit quite unfavorable. Nevertheless, in all cases a great attention should be paid to the value of q, especially when using an alkoxyamine with low addition rate constant.

With fast initiator dissociation and fast addition of the primary radical to monomer along with negligible exit, all droplets become monomer swollen polymer particles within a short time span. Therefore, the nucleation step, which is determined by the initiation step, is short with respect to propagation and all monomer droplets undergo polymerization simultaneously.

Water-Soluble Initiator

One considers here alkoxyamine initiators and initiating radicals exhibiting very large partition coefficient q. An example is the SG1-adduct of sodium acrylate (*55*) (Figure 4). In this case, the dissociation and initiation steps take place in the water-phase, where monomer and nitroxide concentrations are significantly different from those in the droplets.

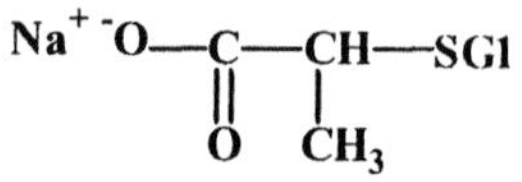

$$Na^+ \; {}^-O—\underset{\underset{O}{\|}}{C}—\underset{\underset{CH_3}{|}}{CH}—SG1$$

Figure 4. SG1 adduct of sodium acrylate.

For instance, polymerization of n-butyl acrylate is considered. Water-solubility of BA at saturation at 125 °C can be estimated at approximately 0.015 mol.L^{-1} (it is 6.4 × 10^{-3} mol.L^{-1} at 50 °C (*4*)). Using the rate constant of recombination with SG1 k_c = 4×10^7 L.mol^{-1}.s^{-1} (*19,20*) and the rate constant of propagation k_p = 94000 L.mol^{-1}.s^{-1} (*51*), a ratio (k_p.[BA]$_{aq}$/(k_c.[SG1]$_{aq}$) = 3.5×10^{-5}/[SG1]$_{aq}$ can be calculated. If [SG1]$_{aq}$ > 3.5×10^{-5} mol.L^{-1} (saturation concentration is 0.05 mol.L^{-1}), this ratio is then lower than 1, i.e. much lower than in the monomer phase. For styrene, because water-solubility is slightly lower than that of BA, the propagation step is even less favored. When deactivation of the water-soluble oligoradicals is faster than propagation (here again self-termination of carbon centered radicals is considered as negligible), i.e. when p, the probability of propagation with respect to recombination (Eq. 3), is lower than 1, then initial chain growth in the aqueous phase is strongly slowed down.

$$p = \frac{k_p \cdot [M]_{aq}}{k_p \cdot [M]_{aq} + k_c \cdot [X]_{aq}} \approx \frac{k_p \cdot [M]_{aq}}{k_c \cdot [X]_{aq}} \tag{3}$$

From now on, X represents the nitroxide deactivator, RX is the water-soluble alkoxyamine initiator (initial concentration $[RX]_{aq0}$), M is the monomer, RM_iX is an oligomeric alkoxyamine with i monomer units, and the concentrations are expressed per volume unit of water-phase. According to the activation-deactivation equilibrium, k_d is the rate constant of alkoxyamine dissociation and k_c is the rate constant of coupling between a propagating radical and X (both are supposed to be chain length independent).

With the assumptions that p < 1 and that the $RM_i^\bullet$ radicals are totally water-soluble until they reach a critical degree of polymerization z+1, it is possible to calculate the rate of entry of these radicals in the monomer phase: $Re_{(z)} = k_p.[M]_{aq}.[RM_z^\bullet]_{aq}$. This means that the radicals with i = z + 1 irreversibly enter the oil-phase, while those with i ≤ z propagate in the aqueous phase only. The step of entry itself is not rate determining. The critical value z is a function of the nature of the initiator and of the monomer. Such a theoretical approach was proposed by Maxwell et al. (56) for the rate of entry in classical emulsion polymerization. To simplify, the RM_iX alkoxyamines will be considered as also water-soluble for i ≤ z, although a partition coefficient should be applied for every chain length.

The concentration of each species RM_iX and $RM_i^\bullet$ in the water-phase can be calculated as a function of time as given in Eqs 4-6 (they follow the Poisson distribution). As all the considered events take place at low conversion, the monomer concentration remains constant in the water-phase, and equal to the saturation concentration (thus p is a constant too).

$$[RX]_{aq} = [RX]_{aq0} \cdot \exp(-p \cdot k_d \cdot t) \tag{4}$$

$$[RM_iX]_{aq} = \frac{(p \cdot k_d \cdot t)^i}{i!} \cdot [RX]_{aq0} \cdot \exp(-p \cdot k_d \cdot t) \tag{5}$$

$$[RM_i^\bullet]_{aq} = \frac{k_d \cdot [RM_iX]_{aq}}{k_c \cdot [X]_{aq}} \tag{6}$$

For a given monomer with critical value z, the rate of entry can then be calculated as given in Eq.7.

$$Re_{(z)} = p \cdot k_d \cdot \frac{(p \cdot k_d \cdot t)^z}{z!} \cdot [RX]_{aq0} \cdot \exp(-p \cdot k_d \cdot t) \qquad (7)$$

The integrated form of this equation represents the amount of alkoxyamine that has entered the monomer phase at a given time. An example is given in Figure 5, for various values of z (between 0 and 4, corresponding to usual hydrophobic monomers (4)), using $k_d = 3 \times 10^{-3}$ s^{-1}, that is a common value for SG1-based alkoxyamines, and p = 0.3.

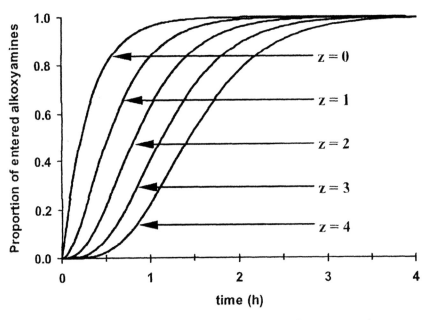

Figure 5. Proportion of alkoxyamine that has entered the monomer phase at a given time, for various values of z. $k_d = 3 \times 10^{-3}$ s^{-1} and p = 0.3.

It appears very clearly that entry of the alkoxyamines, which corresponds to the true initiation step (i.e. the real start of chain growth), is significantly retarded when z increases. Such situation would be very unfavorable to good control over molar mass and distribution, due to an effect of apparent slow initiation. For a given monomer and a given water-soluble initiator, i.e. a given value of z, very few experimental parameters can be modified. To favor water-phase propagation and fast entry, the concentration of free SG1 should be reduced. Additionnally, if possible, an alkoxyamine initiator exhibiting fast dissociation rate should be selected.

Polymerization Kinetics

Because of radical segregation, conventional emulsion and miniemulsion polymerizations usually exhibit slow termination reactions resulting in much faster polymerization than in the homogeneous systems under similar conditions. However, the controlled radical polymerizations operating via a reversible termination reaction (nitroxide-mediated polymerization and ATRP) do not follow the same kinetics as a conventional radical polymerization. They are governed instead by the activation-deactivation equilibrium. The transient and persistent radical concentrations are regulated by the persistent radical effect (PRE) (*57,58*) rather than by a steady state resulting from an initiation/termination balance. As a consequence, reversible termination is favored due to the large concentration of deactivator, which continuously increases with conversion.

A theoretical kinetic analysis based on the persistent radical effect (no other source of radicals than the activation reaction) was applied to miniemulsion systems (*59*). To simplify the model, an oil-soluble alkoxyamine initiator was considered, the propagating radicals were supposed to be compartmentalized (true for macroradicals), whereas free nitroxide could exchange between the particles due to molecular diffusion through the water-phase. The instantaneous rate of polymerization is proportional to $[P^\bullet]$, the concentration of propagating radicals, and inversely proportional to the concentration of nitroxide deactivator at a given time. The latter depends on both the initial concentration, and the concentration that is released owing to the PRE (equal to the concentration of irreversibly terminated chains). It can then be calculated for a miniemulsion system, as a function of monomer conversion, x, according to Eq. 8.

$$[X]_{(ME)} = -A + \sqrt{(A+[X]_0)^2 - B.\ln(1-x)} \tag{8}$$

$$\text{with } A = \frac{k_t}{N_A.v_p.k_c} \text{ and } B = \frac{4.K.[RX]_0.k_t}{k_p}$$

The concentrations are expressed per volume unit of the overall organic phase. $[X]_0$ represents the initial concentration of nitroxide in the system; k_t, k_c, and k_p are the rate constants of termination, recombination with nitroxide and propagation respectively; K is the activation-deactivation equilibrium constant, $[RX]_0$ is the initial concentration of alkoxyamine, v_p is the particle volume and N_A is the Avogadro's number.

In a bulk polymerization the same relationship leads to:

$$[X]_{(BULK)} = \sqrt{[X]_0^2 - B.\ln(1-x)} \tag{9}$$

From simple mathematics, $[X]^2_{(BULK)} = [X]^2_{(ME)} + 2.A.([X]_{(ME)} - [X]_0)$, so that $[X]_{(BULK)} \geq [X]_{(ME)}$. Therefore, the difference between the concentrations of free nitroxide released at a given conversion in bulk and miniemulsion is to a large extent determined by the value of parameter A, which is dependent on the size of the particles for a given monomer and a nitroxide. When the particles are very small, A is large and $[X]$ is larger in bulk than in miniemulsion, indicating the greater degree of irreversible termination. As a consequence, the polymerization should be faster in miniemulsion than in bulk, the proportion of dead chains should be smaller, but the molar mass distribution should be broader. The polydispersity index can be reduced by addition of free nitroxide, at the expense however of the polymerization rate. In contrast, when particle volume is large, A becomes small and the concentration of released nitroxide is nearly the same in bulk and in miniemulsion. Consequently, the compartmentalization effect does not operate any longer, and the kinetics should be the same in both systems. In other words, the overall concentration of propagating radicals (relative to the organic phase volume) is not much larger in a miniemulsion system than in the corresponding bulk polymerization. Therefore, the average number of radicals per particle is simply inversely proportional to the number of particles in the system, N_p, and usually far below 1, while the average number of deactivator molecules per particle is much larger than 1 (typically from a few tens to a few hundreds). In other words, a pseudo-bulk kinetics operates and polymerization rate is independent of N_p.

Consequences on Control of Molar Mass and Molar Mass Distribution

With monomer-soluble alkoxyamine initiator, because of low extent of exit and fast nucleation, all monomer droplets become particles within a short time. With similar concentrations of chains in all droplets, the rate of monomer consumption in those droplets should be the same. Thus, they should behave ideally as independent nanoreactors, exhibiting the same kinetics. This situation does not require monomer transport. All the chains start their growth at approximately the same time. With high initiator efficiency, the experimental and target molar mass should match, i.e. M_n, the number average molar mass, should increase linearly with monomer conversion. This trend was actually observed in most of the miniemulsion examples (5,6). Consequently, the molar mass distribution should not be affected with respect to bulk polymerization (59). The issue is quite different when a water-soluble initiator is used. As shown previously in this article, if the rate of entry is too slow with respect to propagation, then dramatic effects should be observed. Polymer chains would not start their growth (i.e. growth in monomer rich particles) at the same time, the consequence of which would be a broadening of the molar mass distribution (*43*). Additionally, a slow rate of entry might lead to differences in alkoxyamine concentration in the particles, and hence to differences in polymerization rates leading to monomer transport and broadening of the particle size distribution. As a consequence, type of initiator and experimental conditions should be very carefully selected to maintain a good control over polymerization kinetics and polymer characteristics.

Conclusion

The use of controlled free-radical polymerization in conjunction with miniemulsion process has opened a rather easy pathway towards the synthesis of well-defined polymers in aqueous dispersed systems. Nitroxide-mediated polymerization was quite successful since many examples of controlled homopolymers and copolymers have been reported so far.

References

1. "Controlled Radical Polymerization", K. Matyjaszewski Ed., *ACS Symp. Series* **1998**, 685.
2. "Controlled/Living Radical Polymerization: Progress in ATRP, NMP, and RAFT", K. Matyjaszewski Ed., *ACS Symp. Series* **2000**, *768*.
3. Lovell, P. A.; El-Aasser, M. S. "Emulsion Polymerization and Emulsion Polymer" John Wiley & Sons, Chichester (England) **1997**.
4. Gilbert, R. G. "Emulsion Polymerization. A Mechanistic Approach." Academic Press, London **1995**.
5. Qiu, J.; Charleux, B.; Matyjaszewski, K. *Prog. Polym. Sci.* **2001**, *26*, 2083.
6. Cunningham, M.F. *Prog. Polym. Sci.* **2002**, *27*, 1039.
7. Miller, C. M.; Sudol, E. D.; Silebi, C. A.; El-Aasser, M. S. *Macromolecules* **1995**, *28*, 2754 ; 2765 ; 2772
8. Landfester, K. *Macromol. Rapid Commun.* **2001**, *22*, 896.
9. Asua, J.M. *Prog. Polym. Sci.* **2002**, 27, 1283.
10. Barrère, M.; Ganachaud, F.; Bendejacq, D.; Dourges, M.-A.; Maitre, C.; Hémery, P. *Polymer* **2001**, *42*, 7239.
11. Ma, J.W.; Cunningham, M.F.; McAuley, K.B.; Keoskherian, B.; Georges, M.K. *J. Polym. Sci. Polym. Chem.* **2001**, *39*, 1081.

12. Farcet, C.; Lansalot, M.; Charleux, B.; Pirri, R.; Vairon J.P. *Macromolecules* **2000**, *33*, 8559.
13. Hawker, C.J.; Bosman, A.W.; Harth, E. *Chem. Rev.* **2001**, *101*, 3661.
14. Solomon, D.H.; Rizzardo, E.; Cacioli, P.; *U.S. Patent 4,581,429*, **1985**.
15. Georges, M.K.; Veregin, R.P.N.; Kazmaier, P.M.; Hamer, G.K. *Macromolecules* **1993**, *26*, 2987.
16. Hawker, C.J. *J. Am. Chem. Soc.* **1994**, *116*, 11185.
17. Benoit, D.; Chaplinski, V.; Braslau, R.; Hawker, C.J. *J. Am. Chem. Soc.* **1999**, *16(121)*, 3904.
18. Grimaldi, S.; Finet, J.P., Le Moigne, F.; Zeghdaoui, A.; Tordo. P.; Benoit, D.; Fontanille, M., Gnanou, Y. *Macromolecules* **2000**, *33*, 1141.
19. Benoit, D.; Grimaldi, S.; Robin, S.; Finet, J.P.; Tordo, P.; Gnanou, Y. *J. Am. Chem. Soc.* **2000**, *122*, 5929.
20. Lacroix-Desmazes, P.; Lutz, J.F.; Chauvin, F.; Severac R.; Boutevin, B. *Macromolecules* **2001**, *34*, 8866.
21. Robin, S.; Gnanou, Y. *ACS Symp. Series.* **2000**, *768*, 334.
22. Robin, S.; Gnanou, Y. *Macromol. Symp.* **2001**, *165*, 43.
23. Robin, S.; Guerret, O.; Couturier, J.L.; Pirri, R.; Gnanou, Y. *Macromolecules*, **2002**, *35*, 2481.
24. Matyjaszewski, K.; Xia, J. *Chem. Rev.* **2001**, *101*, 2921.
25. Kamigaito, M.; Ando, T.; Sawamoto, M. *Chem. Rev.* **2001**, *101*, 3689.
26. Chong, Y.K.; Le, T.P.T.; Moad, G.; Rizzardo, E.; Thang, S.H. *Macromolecules* **1999**, *32*, 2071.
27. Schmidt-Naake, G.; Drache, M.; Taube, C. *Angew. Makromol. Chem.* **1999**, *265*, 62.
28. Taube, C.; Schmidt-Naake, G., *Macromol. Mater. Eng.* **2000**, *279*, 26.
29. Taube, C.; Schmidt-Naake, G. *Chem. Eng. Tech.* **2001**, *24*, 1013.
30. Taube, C.; Schmidt-Naake, G. *Chem. Eng. Tech.* **2001**, *73*, 241.
31. Bon, S.A.F.; Bosveld, M.; Klumperman, B.; German, A.L. *Macromolecules* **1997**, *30*, 324.
32. Marestin, C.; Noël, C.; Guyot, A.; Claverie, J. *Macromolecules* **1998**, *31*, 4041.
33. Cao, J.; He, J.; Li, C.; Yang, Y. *Polym. J.* **2001**, *33*, 75.
34. Prodpan, T.; Dimonie, V.L.; Sudol, E.D.; El-Aasser, M.S., *Macromol. Symp.* **2000**, *155*, 1
35. MacLeod, P.J.; Barber, R.; Odell, P.; Keoshkerian, B.; Georges, M.K. *Macromol. Symp.* **2000**, *155*, 31.
36. Pan, G.; Sudol, E.D.; Dimonie V.L.; El-Aasser M. *Macromolecules* **2001**, *34*, 481.
37. Keoshkerian, B.; MacLeod, P.J.; Georges, M.K. *Macromolecules* **2001**, *34*, 3594.

451

38. Keoshkerian, B.; Szkurhan, A.R. ; Georges, M.K. *Macromolecules* **2001**, *34*, 6531.
39. Tortosa, K.; Smith, J.-A.; Cunningham, M.F. *Macromol. Rapid Commun.* **2001**, *22*, 957.
40. Cunningham, M.F.; Xie, M.; McAuley, K.B.; Keoshkerian, B.; Georges, M.K. *Macromolecules* **2002**, *35*, 59.
41. Cunningham, M.F.; Tortosa, K.; Ma, J.W.; McAuley, K.B.; Keoshkerian, B.; Georges, M.K. *Macromol. Symp.* **2002**, *183*, 273.
42. Pan, G.; Sudol, E. D.; Dimonie, V. L.; El-Aasser, M. S. *Macromolecules* **2002**, *35*, 6915.
43. Lansalot, M.; Farcet, C.; Charleux, B.; Vairon, J.P.; Pirri, R.; Tordo, P. *ACS Symp. Series* "Controlled/Living Radical Polymerization: Progress in ATRP, NMP, and RAFT", K. Matyjaszewski Ed., **2000**. *768*, 138.
44. Farcet, C.; Charleux, B.; Pirri, R. *Macromolecules* **2001**, *34*, 3823.
45. Farcet, C.; Charleux, B.; Pirri, R., *Macromol. Symp.* **2002**, *182*, 249.
46. Farcet, C.; Belleney, J.; Charleux, B.; Pirri, R. *Macromolecules* **2002**, *35*, 4912.
47. Farcet, C.; Nicolas, J.; Charleux, B. *J. Polym. Sci. Polym. Chem.* **2002**, *40*, 4410.
48. Farcet, C.; Charleux, B.; Pirri, R.; Guerret, O. *Am. Chem. Soc., Polym. Prepr.* **2002**, *43*(2), 98.
49. Farcet, C. *PhD dissertation*, University of Paris 6 (France) **2002**.
50. Van Doremaele, G.H.J.; Geerts, F.H.J.M., Schoonbrood, H.A.S.; Kurja, J.; German, A.L. *Polymer* **1992**, *33*, 1914.
51. Beuermann, S.; Buback, M. *Prog. Polym. Sci.* **2002**, *27*, 191. The value of k_p extrapolated at high temperature from the Arrhenius plot should not be considered as accurate, but as a rough estimation.
52. Chambard, G.; Klumperman, G.; German, A.L. *Polymer* **1999**, *40*, 4459.
53. Ugelstadt, J.; Hansen, F.K. *Rubber Chem. Technol.* **1976**, *49*, 536.
54. Monteiro, M.J.; Hodgson, M.; de Brouwer, H. *J. Polym. Sci. Polym. Chem.* **2000**, *38*, 3864.
55. Le Mercier, C. *PhD dissertation*, University of Aix-Marseille (France) **2000**.
56. Maxwell, I.A.; Morrison, B.R.; Napper, D.H.; Gilbert R.G. *Macromolecules* **1991**, *24*, 1629.
57. Fischer, H. *J. Polym. Sci. Polym. Chem.* **1999**, *37*, 1885.
58. Souaille, M.; Fischer, H. *Macromolecules* **2000**, *33*, 7378.
59. Charleux, B. *Macromolecules* **2000**, *33*, 5358.

Chapter 32

The Use of PROXYL Nitroxides in Nitroxide-Mediated Polymerization

Neil R. Cameron, Catherine A. Bacon, and Alistair J. Reid

Department of Chemistry, University of Durham, South Road, Durham DH1 3LE, United Kingdom

The use of alkoxyamines derived from substituted PROXYL nitroxides is studied and performance is compared to analogous species obtained from TEMPO. PROXYLs are found to have significant differences relative to TEMPO: more rapid styrene polymerization; the ability to bring about the living polymerization of *n*-butyl acrylate; and a lower propensity to undergo disproportionation. The latter is suggested to be the key parameter producing the different behaviour of PROXYL nitroxides.

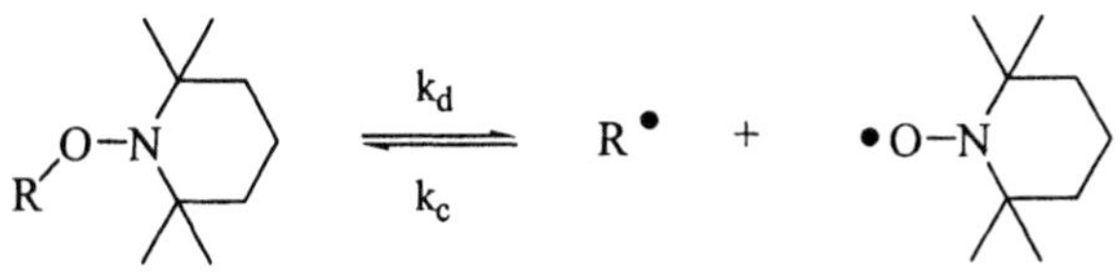

Figure 1. Equilibrium between dormant and active chains in NMP.

Nitroxide-mediated polymerization (NMP) is one of three main controlled radical polymerization (CRP) techniques(*1*). At the heart of NMP is the equilibrium between dormant and active centres set up when radical polymerizations are conducted in the presence of sufficient amounts of nitroxide (Figure 1).

NMP has been the subject of intense study for several years, following initial work by Rizzardo, Solomon, Moad et al.(*2*) Subsequent work carried out by Georges and coworkers(*3,4*) employing TEMPO (species on right hand side of Figure 1) as the mediator indicated that this species had limitations, most seriously an inability to polymerize monomers other than styrenes. More recent work has produced a number of acyclic nitroxides that have much greater monomer scope than TEMPO(*5,6*). Differences between nitroxides have been shown to be due to a fine balance between k_d, k_c and k_p for the monomer in question(*7*). Our work has focused on PROXYLs, which are the 5-membered ring analogues of TEMPO. Derivatives of these substituted adjacent to the ring N atom are much more accessible synthetically than TEMPO analogues, allowing us to investigate the influence of structure on activity. Other groups have also studied PROXYLs and importantly have observed differences in behaviour compared to TEMPO. Veregin et al.(*4*) demonstrated that 3-carboxy PROXYL mediated styrene polymerization more rapidly than TEMPO. A 2,5-diphenyl-subsituted derivative was also found to result in a significantly faster polymerization of styrene than TEMPO (*8*). Hawker, Braslau et al. also employed this mediator and similarly observed a faster rate of styrene polymerization compared to TEMPO(*6*). Furthermore, *n*-butyl acrylate could be polymerized in its presence, although broad polydispersities (~ 2) were obtained. Yamada and coworkers(*9*) found that ring-substituted PROXYL species led to rate enhancements over TEMPO or PROXYL itself.

Here we describe our studies into the use of alkoxyamines derived from substituted PROXYLs in NMP.

Experimental

Toluene (solvent for ESR studies) was distilled before use. TEMPO was purified by vacuum sublimation. *n*-Butyl acrylate and styrene were freed of inhibitor by passing through basic alumina and were distilled under N_2 immediately prior to use. THF and diethyl ether were purified by heating at reflux over and distilling from Na/benzophenone under N_2. CH_3CN was heated at reflux over and distilled from CaH_2 and stored over 4Å molecular seives. All other chemicals and solvents were used as received from commercial suppliers

454

(mainly Aldrich). Size exclusion chromatography (SEC) with refractive index, viscosity and light scattering detectors, using THF as a solvent, was employed to determine molecular weights and molecular weight distributions. ESR spectra were recorded on a Bruker EMX spectrometer fitted with a variable temperature probe. NMR spectra were obtained with a Varian Innova 400 fitted with a variable temperature probe, operating at either 400 MHz (^{1}H) or 100 MHz (^{13}C), using either CDCl$_3$ or d$_8$-toluene as the solvent and with tetramethylsilane (TMS) as an internal standard. Mass spectroscopy was performed with a Micromass Auto Spec.

The nitroxides and alkoxyamines employed were prepared according to procedures described elsewhere(*10,11*). The species used in the present work are shown in Figure 2.

Di-*tert*-butyl peroxalate (DTBPO) was prepared by a literature method(*12*) (**Caution: DTBPO when dry is known to be shock sensitive and to detonate in contact with metal objects**). A typical procedure for polymerizations is as follows: a solution of monomer (39 mmol.) and alkoxyamine (0.125 mmol.) in a round-bottomed flask was degassed by 3 freeze-pump-thaw cycles. The flask was back-filled with N$_2$ and divided between several previously purged GC vials with PTFE seals, by syringe transfer. The vials were heated at 125 °C in an oil bath, withdrawn periodically and immediately cooled in an ice bath. Conversion was determined by ^{1}H NMR spectroscopy, molecular weights and polydispersity were determined without further purification by SEC.

Results and Discussion

Nitroxide and Alkoxyamine Synthesis

Nitroxides **1** to **4** were generally obtained in low yields (16 – 36%), however usable quantities were produced and the starting materials are cheap. Alkoxyamines **5** to **9** were prepared by trapping of the product of addition of *tert*-butoxy radicals to styrene with the appropriate nitroxide. For these reactions, yields were higher (typically 70 – 80%) apart from the synthesis of **9**, where it is presumed that **4** and DTBPO, the source of *tert*-butoxy radicals, participate in a redox reaction(*13*). **10** and **11** were prepared from the corresponding nitroxide and styrene in the presence of Jacobsen's catalyst(*14*).

Styrene Polymerization

Alkoxyamines **5** to **9** were tested for their ability to mediate the polymerization of styrene. The kinetic plots from heating styrene in their presence at 125°C are shown in Figure 3.

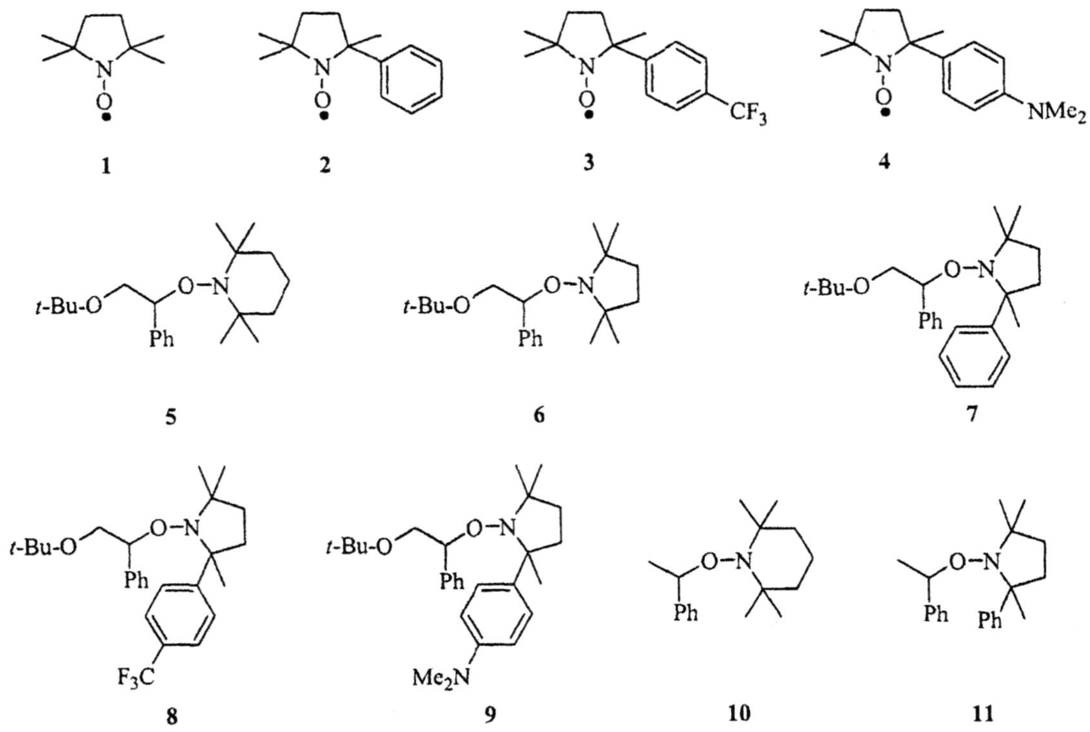

Figure 2. Nitroxides and alkoxyamines used in the present work

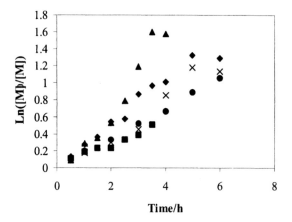

Figure 3. Kinetic plots for the polymerization of styrene at 125 °C with: **5** (●); **6** (■); **7** (▲); **8** (♦); and **9** (X). (Adapted with permission from reference(11). Copyright 2000 Wiley-VCH).

The alkoxyamines can be seen to fall broadly into one of two classes: those that are 'TEMPO-like' (**5**, **6** and **9**); and those that mediate styrene polymerization at an appreciably higher rate (**7** and **8**). Polymerization with **7** is in fact around 2.5 times faster than with **5**. The cause of these differences could be either steric or electronic (or both). Comparing **5** with **7** suggests that steric effects are important, as has been found by other groups(5,6,15,16). The similarity in polymerization rate between **7** and **8** implies that an electron withdrawing group does not increase the rate of polymerization further, however an electron donating (by resonance) species (**9**) appears to counteract the steric effect of the phenyl substituent and decrease the polymerization rate to that observed in the presence of the TEMPO-derived species. Thus, the relationship between sterics and electronics and how these influence polymerization rate is a complex one. It should be pointed out that these results conflict somewhat with semiempirical molecular orbital calculations of Moad and Rizzardo, who predicted that nitroxides with electron donating subsituents would result in alkoxyamines with lower C-O bond homolysis energies(16).

Molecular weight data for polymerizations conducted in the presence of **5** to **9** are depicted in Figure 4.

For all apart from **8**, M_n is seen to grow linearly with conversion and there is good agreement with predicted values (dotted line). This suggests that the polymerization of styrene is well controlled in the presence of PROXYL derived alkoxyamines. The evolution of polydispersity with conversion was also studied. In all cases, polydispersity decreases with time to around 1.3, in agreement with the Persistent Radical Effect(17). The molecular weight distribution for the polymerization in the presence of **8** is broader than the others (1.6 at high conversion), which agrees with the conclusion that this alkoxyamine leads to poorer control.

The polymerization with **7** was repeated in the presence of small quantities of added nitroxide **2**; the results are presented in Figure 5.

An excess nitroxide concentration of 0.10 equivalents leads to an induction period of around 1 hour, whereas lower added quantities of **2** result in shorter induction periods. In each case, once started the polymerization proceeds at a constant rate with a linear increase in $\ln([M]_0/[M])$. The reason for the induction period is unknown, but it suggests that the position of the equilibrium is changing during the early stages of the polymerization. This could be due to nitroxide decomposition(*18*), however we have no reason to believe that our PROXYL nitroxides are unstable on the timescale of polymerization. An alternative explanation is that there is a source of external radicals, e.g. from trace levels of impurities capable of initiation such as peroxides.

The molecular weights of the resulting poly(*n*-butyl acrylate)s were investigated by SEC. M_n was found to increase very rapidly to around 50,000 at low conversion and then did not rise much on increasing conversion. On the other hand, M_w and especially M_z increase to very high levels with increasing conversion, as does polydispersity (Figure 6). These data indicate that control over the polymerization mediated by **7** is poor. However, it is also probable that branching is occurring, which is well known for *n*-butyl acrylate(*19*) and has

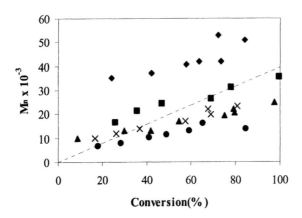

Figure 4. M_n against conversion for the polymerization of styrene in the presence of alkoxyamines (key as in Figure 3). (Adapted with permission from reference(11). Copyright 2000 Wiley-VCH)

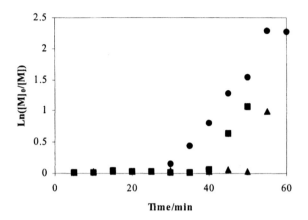

Figure 5. Kinetic plot of the polymerization of n-butyl acrylate at 125°C in the presence of excess quantities of 2: 0.03 eq. (●); 0.05 eq. (■); 0.10 eq. (▲).

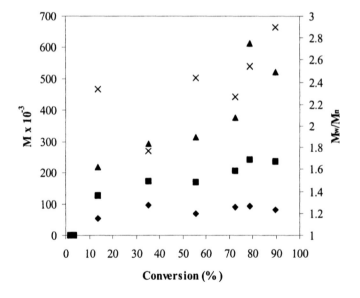

Figure 6. Molecular weight and polydispersity against conversion for the polymerization of n-butyl acrylate in the presence of 7 and 0.05 eq. of 2: M_n (♦); M_w (■); M_z (▲); M_w/M_n (X)

been observed during both its atom transfer radical polymerization(*20*) and NMP in bulk and miniemulsion(*21*). Quantitative ^{13}C NMR spectroscopy was used to demonstrate that branching was indeed occurring in our polymerizations (data not shown).

Despite the poor control over the polymerization of *n*-butyl acrylate by 7, we decided to investigate the ability of the resulting polymer to act as a macroinitiator for subsequent polymerizations. The polymerization of a charge of styrene was seen to proceed linearly, indicating that the PBA macroinitiator is indeed capable of initiation. The molecular weight data of the resulting p(BA-S) copolymers were obtained, again by SEC. Interestingly, M_n was seen to increase with conversion, and the polydispersity to decrease (Figure 7).

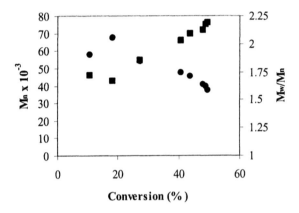

Figure 7. Molecular weight and polydispersity data for the polymerization of styrene initiated by a PBA macroinitiator: M_n (■); M_w/M_n (●)..

This indicates a controlled polymerization, suggesting that the PBA macroinitiator is capable of producing a block copolymer and that the polymerization of styrene proceeds in a controlled manner. SEC data obtained with a light scattering detector indicated that the molar mass of the PBA macroinitiator increases on reinitiation with little or no evidence of bimodality in the block copolymer trace, showing that the macroinitiator contained little or no dead PBA material (Figure 8).

The ability of PBA to initiate and induce some control in the polymerization of styrene with little evidence of dead material indicates that the vast majority of PBA chains possess a nitroxide moiety. However, the polymerizations are characterized by poor control and broad polydispersities. Fischer(*7,17*) has pointed out that living character and polymerization control are separate phenomena and that it is indeed possible to have a living polymerization, producing little or no dead material, that gives a broad polydispersity product. Livingness is determined by K (see Figure 1), which has a limiting value, dependent on k_p and k_t for the monomer in question, above which large fractions of dead material are produced at high conversion. On the other hand, control is

460

determined by the product $k_d k_c$, which should be larger than a threshold value, again peculiar to each monomer, required to give a low polydispersity. Thus, for a given set of values of k_p, k_t and initial initiator concentration there is a range of values of k_d and k_c that gives rise to a polymerization that is both living *and* controlled. It may be the case that the product $k_d k_c$ for **7** in *n*-butyl acrylate is relatively small, leading to poor control, while $K(=k_d/k_c)$ is still sufficiently low to ensure that little dead material is produced. The differences in performance between **7** and TEMPO could also be due to differences in k_d and k_c. **7** has been found to give a higher rate of styrene polymerization than **5**, which suggests a higher value of K. However, it is known that TEMPO is prone to causing hydrogen abstraction(*22-24*) and this may in fact be the crucial difference between **7** and **5**.

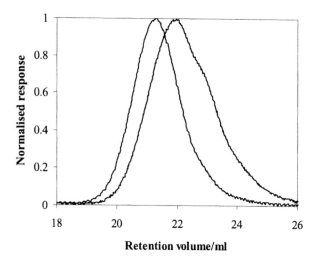

Figure 8. SEC traces for a poly(n-butyl acrylate-b-styrene) copolymer and the original poly(n-butyl acrylate) macroinitiator (right-hand trace).

Alkoxyamine Homolysis and Disproportionation

Studying the fundamental reactions of alkoxyamines, including homolysis and disproportionation, is key to understanding their behaviour in NMP. Therefore, we were keen to investigate the properties of our alkoxyamines to elucidate the reasons for their different performances.

The most direct method of studying alkoxyamine homolysis is quantitative electron spin resonance (ESR). ESR spectra were collected from alkoxyamines

5, 6, 7, 10 and 11 cleaved in the presence of oxygen as a radical scavenger at different temperatures(25). A linear fit to the double integral with time data allows the calculation of alkoxyamine bond homolysis rate constants (k_d), which leads to an Arrhenius plot ($\ln(k_d)$ vs. $1/T$) for each alkoxyamine system. In each case, a nonlinear least squares fitting algorithm was used to obtain the Arrhenius parameters, namely activation energy E_a and the pre-exponential factor A, from the slope and intercept respectively.

The Arrhenius parameters for 5, 6, 7, 10 and 11 are shown in Table 1. It can be seen that the alkoxyamines fall into two classes. Those with 1-phenethyl residues (10 and 11) have significantly higher values of E_a than alkoxyamines derived from *tert*-butoxy radicals (5 to 7). The influence of the nitroxide fragment on alkoxyamine homolysis appears less significant. For the series 5 to 7 there is hardly any effect and indeed E_a appears to increase slightly on changing from a TEMPO to a substituted PROXYL substituent. With the 1-phenethyl derived alkoxyamines there is a drop in E_a when the nitroxide residue is varied and this result is in line with our expectations based on previous work.

Table 1. Arrhenius Parameters for Alkoxyamines Investigated

Alkoxyamine	A/s^{-1}	$E_a/(kJmol^{-1})$
5	3.2×10^{10}	102.5
6	2.0×10^{9}	104.3
7	2.05×10^{11}	109.8
10	3.23×10^{14}	129.9
11	1.61×10^{13}	121.8

Fischer et al.(26) performed quantitative ESR measurements on a series of 27 alkoxyamines based on six nitroxides with different carbon centred radical fragments. It was suggested that the observed variation of rate constants was not due to changes in Arrhenius frequency factor but was the result of bond strength differences. The frequency factors observed were similar to those obtained previously by Scaiano and coworkers(22); the majority of systems evaluated had an average value of 2.6×10^{14} s^{-1} with a 2.5 fold variation about this average. The difficulty in discussing variations in frequency factor is a consequence of a statistical enthalpy-entropy compensation effect(27) caused by measurement and fitting errors, which masks any real effect when the variation is small. For this reason it was difficult to draw conclusions from values of the frequency factor, as no general trend with nitroxide or transient fragments was evident and reported values vary. The larger variation in activation energies (which more or less equates to bond dissociation energy) was more reliably assigned to chemical differences (mainly steric).

Homolysis activation energies do not explain the observed difference in behaviour of the alkoxyamines described here (E_a for **5** is lower than that for **7**). In an attempt to rationalize this, we examined their decomposition via hydrogen abstraction to generate an unsaturated species an d hydroxylamine (Figure 9).

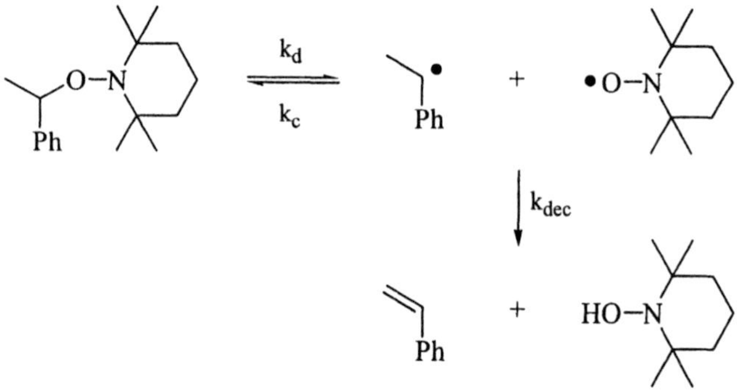

Figure 9. Disproportionation of alkoxyamine 10.

Heating **10** at 125°C for a period of time resulted in the formation of significant quantities styrene, as determined by ^{1}H NMR spectroscopy (spectrum not shown). The method of Fukuda et al.(*24*) was used to determine the disproportionation rate constant k_{dec} for **10**, giving a value of $k_{dec} = 4.8 \times 10^{-6}$ s^{-1}. This compares with the value of 4.5×10^{-5} s^{-1} determined by Fukuda and coworkers for the same alkoxyamine at 140°C. In contrast, heating related alkoxyamine **5** for 16 hours resulted in the production of very small quantities of vinyl containing species at a level that was too low to obtain a reliable integration (spectrum not shown). This indicates that the introduction of a β substituent greatly reduces the propensity for TEMPO-derived alkoxyamines to undergo disproportionation, in agreement with findings by other workers(*22,24,28*). Alkoxyamines **7** and **11**, based on 2,2',5-trimethyl-5'-phenylpyrrolidinyl-*N*-oxyl, showed much less formation of unsaturated species over long periods of time than corresponding TEMPO-derived analogues. **11** leads to the production of observable quantities of styrene after 16 hours at 125°C, however not enough to be quantified reliably and certainly much less than analogous alkoxyamine **10**. This demonstrates the powerful influence of nitroxide residue structure on the tendency of alkoxyamines to undergo disproportionation. On the other hand, β-*tert*-butoxy substituted derivative **7** under the same conditions does not generate unsaturated decomposition products at NMR-detectable levels. These results tell us two things: β-substitution reduces the tendency for disproportionation of PROXYL containing alkoxyamines; and PROXYL-derived alkoxyamines are less prone to disproportionation than those based on TEMPO. We can relate these results to our observations of the polymerization behaviour of PROXYL-derived alkoxyamines. **7** gave rise to a more rapid styrene polymerization than **5**. Since their homolysis activation energies are similar (in fact, that of **5** is slightly lower; Table 1), this suggests that disproportionation is the dominant factor. We have also observed that the polymerization of styrene mediated by **11** (results not shown) is significantly slower than that with **7**. From Table 1 it can be seen that **11** has a higher E_a but is also more prone to β-hydrogen abstraction than **7**. Thus, the difference in polymerization rate could be due either to differences in k_d or tendency to disproportionate (or both). Differences in ability to polymerize n-butyl acrylate have also been found. **7** gives rise to a rapid, living but uncontrolled polymerization whereas both **5** and **11** are inactive. Since **5** has a similar homolysis rate constant to **7** we conclude again that the greater tendency of the former species to undergo disproportionation is governing its behaviour. The performance of **11** could once again be due either to its higher E_a or larger extent of β-hydrogen abstraction than **7** (or a combination of both). In our opinion, based on the comparison of **5** and **7**, disproportionation is the dominant factor. This is particularly important when considering the polymerization of monomers such as acrylates, in which the backbone hydrogens are significantly labile and have been shown to undergo more readily this type of decomposition.

Conclusions

The synthesis and use of alkoxyamines derived from PROXYL nitroxides in NMP has been described. Importantly, crucial differences between these mediators and TEMPO have been observed. The rate of styrene polymerization in the presence of **2** is more than twice as fast as that with TEMPO, and alkoxyamine **7** obtained from **2** is able to mediate the living (but not controlled) polymerization of n-butyl acrylate. Investigations of the alkoxyamines themselves reveal that those bearing β-substituents (**5** to **7**) have a lower homolysis activation energy than 1-phenethyl analogues (**10** and **11**). Furthermore, the latter are more prone to disproportionation, as are alkoxyamines possessing TEMPO residues. These results imply that differences in polymerization behaviour between TEMPO and PROXYL based alkoxyamines are due to the greater tendency of the former to disproportionate.

Acknowledgement

The authors would like to thank the EPSRC and Schlumberger for funding, Dr S. Marque (CNRS-Université d'Aix-Marseille 1 et 3) and Dr G. Tustin (Schlumberger Cambridge Research) for helpful discussions, Dr A. Royston (University of Durham) for assistance with ESR experiments and Mr D. Carswell (University of Durham) for performing the SEC analyses. We are grateful to the

464

EPSRC for granting a Joint Infrastructure Fund award (GR/M87917) which enabled the purchase of the ESR spectrometer.

References

1   *Controlled Radical Polymerization*; Matyjaszewski, K., Ed.; American Chemical Society: Washington DC, 1998; Vol. 685; *Controlled / Living Radical Polymerization*; Matyjaszewski, K., Ed.; American Chemical Society: Washington, 2000; Vol. 768.

2   Johnson, C. H. J.; Moad, G.; Solomon, D. H.; Spurling, T. H.; Vearing, D. J. *Aust. J. Chem.* **1990**, *43*, 1215-1230; Rizzardo, E. *Chem. Aust.* **1987**, *54*, 32; Rizzardo, E.; Chong, B. Y. K., *2nd Pacific Polymer Conference* Tokyo, 1991, p 26-27; Solomon, D. H.; Rizzardo, E.; Cacioli, P., US 4,581,429, 1986

3   Georges, M. K.; Veregin, R. P. N.; Kazmaier, P. M.; Hamer, G. K.; Saban, M. *Macromolecules* **1994**, *27*, 7228-7229; Georges, M. K.; Veregin, R. P. N.; Kazmaier, P. M.; Hamer, G. K. *Macromolecules* **1993**,*2 6*, 2987-2988.

4   Veregin, R. P. N.; Georges, M. K.; Hamer, G. K.; Kazmaier, P. M. *Macromolecules* **1995**,*2 8*, 4391-4398.

5   Benoit, D.; Grimaldi, S.; Finet, J. P.; Tordo, P.; Fontanille, M.; Gnanou, Y. In *Controlled Radical Polymerization*; Matyjaszewski, K., Ed.; American Chemical Society: Washington DC, 1998; Vol. 685, Chapter 14.

6   Benoit, D.; Chaplinski, V.; Braslau, R.; Hawker, C. J. *J. Am. Chem. Soc.* **1999**, *121*, 3904-3920.

7   Souaille, M.; Fischer, H. *Macromolecules* **2000**,*3 3*, 7378-7394.

8   Puts, R. D.; Sogah, D. Y. *Macromolecules* **1996**,*2 9*, 3323-3325.

9   Yamada, B.; Miura, Y.; Nobukane, Y.; Aota, M. In *Controlled Radical Polymerization*; Matyjaszewski, K., Ed.; American Chemical Society: Washington DC, 1998; Vol. 685, Chapter 12.

10 Hideg, K.; Hankovszky, H. O.; Halasz, H. A.; Sohar, P. *J. Chem. Soc.-Perkin Trans. 1* **1988**, 2905-2911; Hideg, K.; Lex, L. *J. Chem. Soc.-Chem. Commun.* **1984**, 1263-1265; Keana, J. F. W.; Seyedrezai, S. E.; Gaughan, G. *J. Org. Chem.* **1983**,*4 8*, 2644-2647.

11 Cameron, N. R.; Reid, A. J.; Span, P.; Bon, S. A. F.; van Es, J. J. G. S.; German, A. L. *Macromol. Chem. Phys.* **2000**, *201*, 2510-2518.

12 Bartlett, P. D.; Benzing, E. P.; Pincock, R. E. *J. Am. Chem. Soc.* **1960**, *82*, 1762-1768.

13 Moad, G.; Solomon, D. H. *The Chemistry of Free Radical Polymerization*; Elsevier Science Ltd.: Oxford, 1995.

14 Dao, J.; Benoit, D.; Hawker, C. J. *J. Polym. Sci. Pol. Chem.* **1998**,*3 6*, 2161-2167.

15 Puts, R. D.; Sogah, D. Y. *Macromolecules* **1997**,*3 0*, 7050-7055.

16 Moad, G.; Rizzardo, E. *Macromolecules* **1995**,*2 8*, 8722-8728.

17 Fischer, H. *Chem. Rev.* **2001**,*1 01*, 3581-3610.

18 Lacroix-Desmazes, P.; Lutz, J. F.; Chauvin, F.; Severac, R.; Boutevin, B. *Macromolecules* **2001**,*3 4*, 8866-8871.

19 Ahmad, N. M.; Heatley, F.; Lovell, P. A. *Macromolecules* **1998**, *31*, 2822-2827.

20 Roos, S. G.; Muller, A. H. E. *Macromol. Rapid Commun.* **2000**, *21*, 864-867.

21 Farcet, C.; Belleney, J.; Charleux, B.; Pirri, R. *Macromolecules* **2002**, *35*, 4912-4918.

22 Skene, W. G.; Scaiano, J. C.; Yap, P. A. *Macromolecules* **2000**, *33*, 3536-3542.

23 Souaille, M.; Fischer, H. *Macromolecules* **2001**,*3 4*, 2830-2838.

24 Goto, A.; Kwak, Y.; Yoshikawa, C.; Tsujii, Y.; Sugiura, Y.; Fukuda, T. *Macromolecules* **2002**,*3 5*, 3520-3525.

25 Bon, S. A. F.; Chambard, G.; German, A. L. *Macromolecules* **1999**, *32*, 8269-8276.

26 Marque, S.; Le Mercier, C.; Tordo, P.; Fischer, H. *Macromolecules* **2000**, *33*, 4403-4410.

27 Krug, R. R.; Hunter, W. G.; Grieger, R. A. *J. Phys. Chem.* **1976**, *80*, 2335-2341.

28 Ohno, K.; Tsujii, Y.; Fukuda, T. *Macromolecules* **1997**,*3 0*, 2503-2506.

Chapter 33

Nitroxide-Mediated Semibatch Polymerization for the Production of Low-Molecular Weight Solvent-Borne Coating Resins

Yanxiang Wang, Frederic Naulet, Michael F. Cunningham, and Robin A. Hutchinson[*]

Department of Chemical Engineering, Queen's University, Kingston, Ontario K7L 3N6, Canada

Nitroxide-mediated polymerization is being explored as an alternative to free-radical chemistry for production of solvent-borne polymers ($M_n < 6000$) used in automotive coatings. Through the production of polymer with decreased polydispersity and more defined composition and chain structure, it may be possible to increase polymer content in solution and decrease the required levels of higher-cost functional monomers. Results for styrene polymerization mediated by 4-hydroxy-TEMPO, combined with insight gained through computer simulation, illustrate some of the technical challenges that must be overcome to maintain the current semibatch industrial process technology, including efficient nitroxide usage at the high concentrations required during startup, and maintaining sufficient reaction rate to match current batch times.

North American regulations are a dominant force driving change in the basic nature of coatings resins: the volatile organic content (VOC) must be reduced from 1990 levels of 480 g/L of paint to below 300 g/L by 2010. Low molecular weight highly functionalized polymer and oligomer solutions at 60 to 80 weight percent solids have replaced high molecular weight, non-functional polymer solutions at 30 to 40 weight percent solids as key components in acrylic coatings formulas (1). The oligomeric chains form a high-MW polymer network on the surface to be coated via reaction of the functional groups (e.g.; hydroxyl or epoxy) with an added cross-linking agent. The base resins are made via high-temperature solution free-radical semibatch polymerization, with monomers (styrene, acrylates, methacrylates) and initiator fed continuously over several hours. Semibatch operation is adopted for heat removal and safety reasons as well as to minimize composition drift. Composition control is especially important during production of the new generation of low-MW materials; with an average chain-length of less than 50 monomeric units, it is essential that *all* chains contain sufficient functionality to participate in the crosslinking reactions needed to form a durable and tough coating.

Living radical polymerization (LRP) offers the potential for major advances in the manufacture of polymeric materials through its control of polymer microstructure – narrowing of molecular weight distributions (MWD), controlled composition distribution along the chain, and targeted placement of functional groups and branchpoints. In the coatings industry, this could translate to a variety of advantages such as higher solids contents (and therefore lower VOC levels) and lower required levels of high-cost functional monomers. These advantages are more likely to be realized commercially if LRP chemistry can be employed using semibatch process technology. Thus in this work we compare the production of low MW polystyrene via nitroxide mediated polymerization (NMP) in a semibatch reactor system to current free-radical technology. The experimental study is supported and guided by insights gained through computer simulation.

Experimental

Styrene and mixed xylenes from the Aldrich Chemical Co. were used as received, as were 4-hydroxy-TEMPO (TEMPO-OH, 4-hydroxy-2,2,6,6,-tetramethylpiperidinyloxy) and initiators. The initiators used in the study include BPO (Benzoyl Peroxide, 97%, Aldrich Chemical Co.), Vazo® 67 (2,2'-Azobis(2-methylbutyronitrile)), Luperox® 231 (1,1-Bis(tert-butylperoxy)-3,3,5-trimethylcyclohexane 92%), and TBPA solution (tert-Butyl peroxyacetate, 75 wt. % solution in aliphatic hydrocarbons); in the following text TBPA refers to pure initiator. PE-T (1-(2,2,6,6-tetramethylpiperidinyloxy)-1-phenylethane) was synthesized using procedures from the literature (2). The polymerizations were performed in a 1 L automated lab reactor (Mettler Toledo LabMax™), with temperature and component addition rates controlled by Camile TG software.

Semibatch Free-Radical Polymerization (FRP). 210 g xylenes were first added to the reactor. After the temperature stabilized at 138°C, 510 g styrene/TBPA mix (58:1 by w/w or 73:1 by mole/mole) was concurrently fed to the reactor at a constant rate over a 6 hour period, followed by a 30 minute hold at reactor temperature. Samples were withdrawn into preweighed sample bottles containing ca. 1-1.5% inhibitor 4-methoxyphenol at regular intervals throughout the experiment, and analyzed as described below.

Semibatch Nitroxide-Mediated Polymerizations. In the basic recipe with TBPA initiator, 210 g xylenes were brought to a temperature of 138°C (heat-up time of approximately 30 minutes), followed by addition of 21.55 g TEMPO-OH, 33.3 g styrene, and 11.82 g TBPA (styrene:nitroxide:initiator molar ratio of 2:1:0.7) to the reactor. After a one hour hold period at 138°C to form the initiating nitroxide-capped chains, 467 g styrene was added at a constant rate over a 6 hour period, and then the solution was held at temperature for another 8 hours. The basic procedure is illustrated schematically in **Figure 1**. In a variation of this technique, 3.1 g of additional TBPA was mixed with the 467 g styrene and fed over the 6 hour addition period. The startup procedure for the other initiators examined in this study was the same as for the basic TBPA recipe, with minor variations summarized in Table I. With TBPA, addition temperature was found to have a negligible effect on polymerization behaviour.

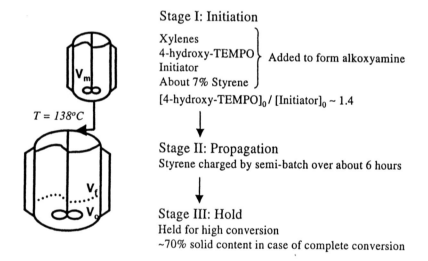

Stage I: Initiation

Xylenes
4-hydroxy-TEMPO
Initiator   } Added to form alkoxyamine
About 7% Styrene

$[\text{4-hydroxy-TEMPO}]_0 / [\text{Initiator}]_0 \sim 1.4$

$T = 138°C$

Stage II: Propagation
Styrene charged by semi-batch over about 6 hours

Stage III: Hold
Held for high conversion
~70% solid content in case of complete conversion

Figure 1. Schematic representation of the semibatch polymerization procedure using TBPA and 4-hydroxy-TEMPO.

Table I. Variation in Semibatch Startup Procedures with Various Initiators

	TBPA/ TEMPO-OH	Luperox®231/ TEMPO-OH	Vazo®67/ TEMPO-OH	BPO/ TEMPO-OH	PE-T
Nitroxide/ Initiator	1.4	1.4	1.4	1.4	-
Addition Temp. (°C)	138	20	20	20	20
Hold time (min) at 138 °C	60	15	15	60	0

Characterization. Monomer conversion was determined from the concentration of residual styrene using a Varian CP-3800 GC installed with a Model 8410 autosampler and a flame ionization detector set at 250°C. A 30M Chrompack Capillary Column (CP-Sil 8 CB) was used for the separation and the injector temperature was held at 200°C. The samples were diluted in acetone and external calibration was made before the measurements. Conversion data were checked on several samples via gravimetry measurements. For GPC measurement, the dry samples were dissolved in THF, which was also used as eluant at a flow rate of 1 mL/min. Before injection, the solutions were filtered through Chromspec syringe filter (25 mm Nylon, 0.2 μm Non-Sterile). The measurements were performed using a Waters 2690 Separations Module equipped with Waters Styragel HR columns (a five column set of HR0.5, HR1, HR3, HR4, HR5) in THF at 35°C and a Waters 410 differential refractometer, calibrated using narrow MW PS standards and analyzed using Waters Millenium software.

Results and Discussion

The production of low-MW polymer by living-radical chemistry in semibatch procedures introduces some unique challenges. With a target M_n of 2000-5000 g/mol in a solution containing 70 wt% polymer, the concentration of chains (and therefore controlling agent) will be 0.1-0.25 mol/L at the end of the semibatch reaction. For a perfectly controlled system, all of these chains must be initiated at the start of the semibatch process when the volume is much lower (cf. Figure 1): thus, the concentration of chains in the system at startup should be in the range of 0.3-0.8 mol/L. These concentrations are 1-2 orders of magnitude higher than those typically used in living radical investigations.

Nitroxide-mediated styrene polymerization was chosen as the starting point for this investigation due to previous experience with the system in miniemulsion studies (*3a*). Unimolecular alkoxyamine is often chosen for stable-free radical polymerization studies since it provides a 1:1 ratio of living chains to nitroxide controlling agent at the start of polymerization (*4*). However, the majority of this study involves a two-component initiator/nitroxide system due to the large quantities of reagents involved (~20 g TEMPO-OH per experiment), and to eliminate the extra synthesis step that would add to process cost. As will be presented later, the complexities of the two-component system at high nitroxide and initiator concentrations make achieving a target MW in the semibatch reactor a difficult task.

FRP vs NMP in a Semibatch System

Typical experimental results for monomer fractional conversion and number-average MW (M_n) data for a free-radical semibatch polymerization are shown in **Figure 2**; conversion is defined as the fraction of monomer converted to polymer relative to the total amount of monomer fed until time *t*. The conversion profile is typical of "starved-feed" semibatch reactor operation; styrene polymerizes more slowly than methacrylates and acrylates, which reach 80% conversion in less than an hour (*1*). The M_n values are slightly higher than many coatings formulations, which can range to values as low as 2000-3000 g/mol; the final polydispersity of 1.9 is a typical value for these products.

In addition to providing a benchmark for comparison with NMP experiments, these data are valuable for establishing the validity of the free-radical mechanisms and rate coefficients used in the model. The simulations were performed with Predici® using a standard set of mechanisms: initiation, propagation, termination by combination and disproportionation, and transfer to solvent and monomer. Rate coefficients were taken from recent literature (*5,6*), and used without adjustment. Thermal initiation of styrene was modeled as third order in monomer, with the rate coefficient estimated from a fit to experimental data. The curves in Figure 2 were generated assuming an initiator efficiency of 0.5. The discrepancy between simulation and experiment towards the end of the batch is attributed to a slight decrease in the termination rate coefficient (k_t) as polymer content in the reactor increases. No attempt was made to include this effect in the model since the focus of this work is the nitroxide-mediated systems. Full details of the simulations will be presented in more detail in a future publication.

The conversion and MW profiles for a typical NMP experiment with TBPA and TEMPO-OH (monomer fed between 60 and 420 minutes after an initial 60 min startup period) are shown as **Figure 3**. As described earlier, the low target MW of the final product requires a high nitroxide concentration, about 0.1-0.15 mol/L in the final reaction mixture. In an attempt to initiate all of the chains prior to feeding the bulk of the monomer, TEMPO-OH and TBPA (1.4:1 molar ratio) were held at 138 °C in the presence of a small amount of styrene (~2 moles per mole of TEMPO-OH) in the xylene solvent for one hour. During this startup period, the concentration of the nitroxide is about 0.32 mol/L, much higher than we have seen reported in literature.

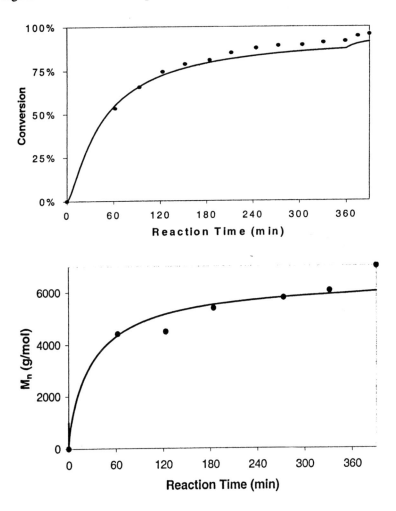

Figure 2. Conversion (top) and number-average MW (bottom) profiles for semibatch styrene FRP. Data points indicate experimental measurements, the line is model predictions generated via computer simulation.

The conversion profile for the NMP is very close to that of a thermally initiated experiment (no initiator, no nitroxide) run under an identical semibatch feeding schedule, as expected for this system (7,8). However, the value of 30% after 7 hours is much lower than the 95% level achieved by FRP; even after 15 hours conversion has only reached 65%. The final polydispersity of 1.4 is much lower than the value of 1.9-2.0 achieved in the FRP system.

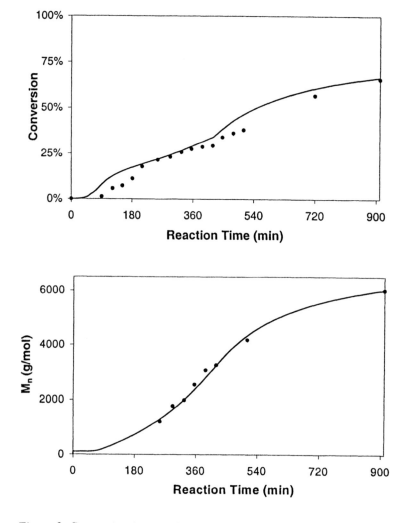

Figure 3. Conversion (top) and number-average MW (bottom) profiles for semibatch styrene NMP with TBPA initiator. Data points indicate experimental measurements, the line is model predictions generated via computer simulation.

The M_n of 6000 g/mol at 65% conversion is much higher than expected; the total number of chains calculated from this value (0.05 moles) is only 40% that of the initial TEMPO-OH charged to the system (0.125 moles). This is a strong indication that a significant fraction of the nitroxide is irreversibly deactivated during the startup procedure. This deactivation is believed to result from reactions involving initiator-derived primary radicals and nitroxide during startup, accentuated by their high concentrations, as discussed in more detail in the following section.

A mechanistic model of the NMP has been constructed, with kinetic parameters taken from the recent simulation study by Ma et al. (*3b*). A good match to the experimental results (predicted polydispersity of 1.5; conversion and M_n profiles plotted in Figures 3) could be achieved by decreasing the initial TEMPO-OH concentration in the system to 40% of its actual value. In addition, it was necessary to decrease TBPA initiator efficiency to 0.32 from 0.5. These parameter changes are consistent with the conclusion drawn from analysis of the experimental results: initiator and nitroxide are both consumed by side-reactions during the startup procedure.

A comparison of Figures 2 and 3 indicates that, as expected, the NMP system is significantly slower than the corresponding FRP process. To mitigate this situation, an experiment was performed where a small amount of additional TBPA (3.1 g, compared to the initial charge of 11.8 g) was fed concurrently with the styrene monomer. As shown in **Figure 4**, this small addition had a significant effect: a conversion of 85% was achieved after 8 hours, comparable to the FRP result in Figure 2. This increased rate was achieved without a significant broadening of the polymer MWD – final Mn was 6400 and polydispersity was 1.42. The simulated conversion profile and MW (predicted Mn of 6000, polydispersity of 1.48) show remarkable agreement with experiment; this result was achieved by keeping the parameters identical to the simulation of the NMP case (initial TEMPO-OH charge set to 40% of the experimental value; TBPA efficiency set to 0.32 during the startup period). The efficiency of the fed TBPA was set at 0.50, identical to the value used in the FRP simulation; this higher value is reasonable due to the low concentration of the continuously added initiator.

The results shown in Figure 4 are promising. Low MW polymer can be produced by controlled-radical polymerization chemistry using the semibatch operating policy currently used in the coatings industry. Product polydispersity is significantly narrowed, while batch time is held comparable to that of FRP. Yet considerable challenges remain. The next section examines in more detail the complexities of the initiator-nitroxide chemistry during startup.

474

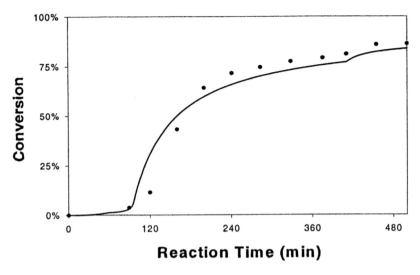

Reaction Time (min)

Figure 4. Conversion profile for semibatch styrene NMP with additional TBPA fed over time. Data points indicate experimental measurements, the line is model predictions generated via computer simulation.

Chain Initiation and Control in the NMP Semibatch System

The NMP semibatch results with TBPA indicate that the system suffers from a significant loss of nitroxide and initiating radicals during startup. To further define and understand the reactions responsible for this loss, four additional initiators have been employed in the semibatch system. The bimolecular initiators were TEMPO-OH combined with peroxides (TBPA, Luperox® 231, BPO), and an azonitrile (Vazo® 67). The fifth system is the unimolecular alkoxyamine (PE-T).

Figure 5 shows the semibatch monomer concentration profiles for the different initiating systems. Data from the conventional FRP run and the thermally-initiated polymerization experiment are also plotted, as well as the monomer concentration profile calculated assuming no reaction (0% conversion). For the recipes using BPO and Vazo® 67 initiators, as well as that with PE-T unimer, the polymerization rate was very slow and essentially no conversion was observed during the 6 hour semibatch feeding stage. For recipes with TBPA and Luperox® 231, monomer is consumed at a rate similar to that of thermally-initiated polymerization.

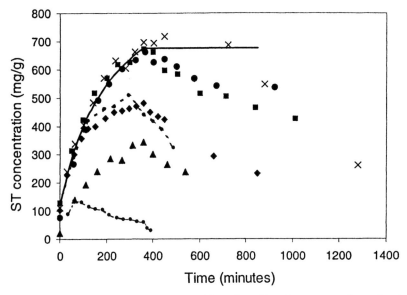

Figure 5. Monomer concentration profiles of semibatch styrene polymerizations using different initiation systems. From bottom to top: Conventional radical polymerization (– - –), Luperox® 231/TEMPO-OH (▲), TBPA/TEMPO-OH (♦), thermal initiation (---), Vazo® 67/TEMPO-OH (•), unimer PE-T (■), BPO/TEMPO-OH (×), without polymerization (——).

Figure 6 plots the molecular weight and polydispersity results for the systems that achieved appreciable conversion. In controlled radical polymerization without side-reactions, the expected molecular weight is equal to the mass of monomer consumed divided by the moles of alkoxyamine, and the molecular weight should increase linearly with conversion. The living nature of these systems is evident when compared to conventional radical polymerization. However, the observed deviation from linearity, especially at high conversion, indicates the importance of side-reactions in this system. Table II summarizes the total moles of polymer chains at the end of reaction, as calculated by using experimental conversion and molecular weight; this value is much lower in all cases than the moles of nitroxide added to the system.

The characteristics of semibatch nitroxide mediated polymerization with the five different initiators can be interpreted as follows. First, the recipes using TBPA and Luperox® 231 as initiators give a reasonable polymerization rate, although the number of polymer chains is much lower than the number of nitroxide added in the startup stage (Table II). We hypothesize that the low

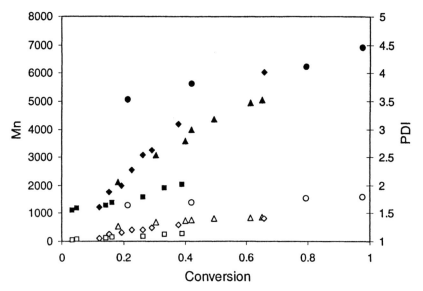

Figure 6. Number-average molecular weight (solid symbols) and polydispersity index (open symbols) plotted against global conversion. Conventional radical polymerization (●), Luperox® 231/TEMPO-OH (▲), TBPA/TEMPO-OH (♦), unimer PE-T (■).

Table II. Nitroxide Utilization in Semibatch Polymerizations

	TBPA/ TEMPO-OH	Luperox® 231/ TEMPO-OH	Vazo® 67/ TEMPO-OH	BPO/ TEMPO-OH	PE-T
Nitroxide (moles)	0.125	0.107	0.107	0.125	0.106
Chains (moles)	0.04	0.04	0.05	0.03	0.06
Chains /Nitroxide	0.32	0.37	0.47	0.24	0.57

number of polymer chains is caused by the formation of inactive alkoxyamine during reactor startup, according to the following: TBPA decomposes by one-bond homolysis to an acyloxy and an alkoxy radical [5], which cannot be directly capped by the TEMPO-OH. In addition to possible hydrogen abstraction from xylene, reaction with the low concentration of styrene present, and primary radical termination, these oxygen-centred radicals also have a high probability to decompose further to methyl radicals (5), as shown in **Figure 7**. Luperox® 231 decomposes to give two tert-butoxy radicals, which then follow the processes described for TBPA. We hypothesize that, due to the high concentrations present during startup, a significant fraction of the TEMPO-OH reacts with methyl radical. The resulting alkoxyamine is very stable (9), and thus a substantial fraction of the TEMPO-OH (and the primary radicals from initiator decomposition) is irreversibly removed from the system.

It is also worth commenting on the fast rate of polymerization for the Luperox® 231 system. In nitroxide-mediated styrene polymerization, it is known the thermal initiation plays an important role and polymerization rate is independent on the concentration of alkoxyamine (7,8). In Figure 5, the polymerization rate for the TBPA/TEMPO-OH system is comparable to the corresponding thermal initiation, as expected. The rate for the Luperox® 231/TEMPO-OH is significantly faster. At this point, we have no clear explanation for this behaviour. However, it may be related to the known accelerating effect of free nitroxide on the rate of styrene thermal initiation (10).

The other three initiating systems – Vazo® 67/TEMPO-OH, BPO/TEMPO-OH, and PE-T unimer – led to negligible conversion during the semibatch styrene polymerization. The recent work of Georges et al. (11) provides an explanation for the BPO results: promoted initiator dissociation at low temperatures leads to the destruction of a significant fraction of the nitroxide in the system, especially for the high reagent concentrations used in this study. For the other two systems, we believe that a high occurrence of primary-radical termination leads to an excess of free nitroxide in the system, creating the observed induction period.

Experimental and simulation work is underway to further elucidate the results obtained with the different initiating systems. What can be unequivocally stated at this point is that the high initiator and nitroxide concentrations required to produce low-MW chains via semibatch polymerization leads to side reactions that can consume not only the primary radicals but also the nitroxide in the system. These reactions make it a challenge to achieve a robust polymerization in which one is able to achieve appreciable reaction rates as well as control the total number of chains (and thus final polymer MW) in the system.

$$\begin{array}{c}
\text{CH}_3-\overset{\overset{\displaystyle\text{CH}_3}{|}}{\underset{\underset{\displaystyle\text{O}^\bullet}{|}}{\text{C}}}-\text{CH}_3 \;+\; \text{H}_3\text{C}-\overset{\overset{\displaystyle\text{O}}{\|}}{\text{C}}-\text{O}^\bullet
\end{array}$$

$$\Big\updownarrow$$

$$\text{H}_3\text{C}-\overset{\overset{\displaystyle\text{O}}{\|}}{\text{C}}-\text{O}-\text{O}-\overset{\overset{\displaystyle\text{CH}_3}{|}}{\underset{\underset{\displaystyle\text{CH}_3}{|}}{\text{C}}}-\text{CH}_3$$

$$\text{H}_3\text{C}-\overset{\overset{\displaystyle\text{CH}_3}{|}}{\underset{\underset{\displaystyle\text{CH}_3}{|}}{\text{C}}}-\text{O}-\text{CH}_2-\overset{\bullet}{\text{CH}}-\text{C}_6\text{H}_5$$

$$\text{CH}_3-\text{CH}_2-\text{O}-\text{CH}_2-\overset{\bullet}{\text{CH}}-\text{C}_6\text{H}_5$$

479

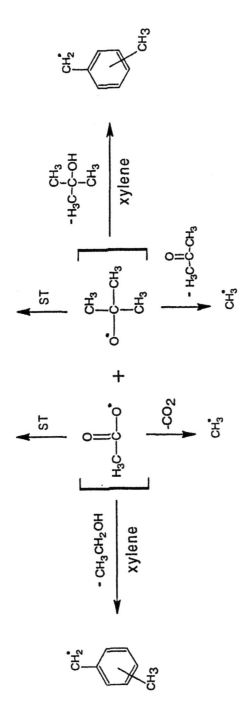

Figure 7. TBPA decomposition pathways.

Conclusions

This work demonstrates that low-MW polymer can be produced by controlled-radical polymerization chemistry using the semibatch operating policy currently used in the coatings industry. Product polydispersity is significantly narrowed, while batch time is held comparable to that of FRP. Yet considerable challenges remain. In particular, the rates of nitroxide and primary radical consuming reactions are promoted by the high concentrations of initiating reagents during startup of the semibatch system. Further study is underway to better understand these mechanisms, and to design a system in which their effect is minimized. In addition to these efforts to improve the startup procedure, we are starting work on copolymer systems. Other living radical techniques will also be explored, to identify and quantify product and process improvements that may be achieved via controlled radical polymerization for production of coatings resins.

References

1. Grady, M.C.; Simonsick, W.J.; Hutchinson, R.A. *Macromol. Symp.* **2002**, *182*, 149-162.
2. Matyjaszewski, K.; Woodworth, B.E.; Zhang, X.; Gaynor, S.G.; Metzner, Z. *Macromolecules* **1998**, *31*, 5955-5957.
3. (*a*) Cunningham, M.F.; Xie, M.; McAuley, K.B.; Georges, M.K.; Keoshkerian, B. *Macromolecules* **2002**, *35*, 59-66. (*b*) Ma J.; Smith J.-A.; McAuley K.B.; Keoshkerian B.; Georges M.K.; Cunningham M.F. *Chem. Eng. Sci.*, in press.
4. Benoit, D.; Chaplinski, V.; Braslau, R.; Hawker, C.J. *J. Am. Chem. Soc.* **1999**, *121*, 3904-3920. Hawker, C.J.; Bosman, A.W.; Harth, E. *Chem. Rev.* **2001**, *101*, 3661-3688.
5. Buback, M.; Klingbeil, S.; Sandmann, J.; Sderra, M.-B.; Vogele, H.P.; Wackerbarth, H.; Wittkowski, L. *Z. Phys. Chem.* **1999**, *210*, 199-221. Buback, M.; Sandmann, J. *Z. Phys. Chem.* **2000**, *214*, 583-607.
6. Beuermann, S.; Buback, M. *Prog. Polym. Sci.* **2002**, *27*, 191-254.
7. Goto, A.; Fukuda, T. *Macromolecules* **1997**, *30*, 4272-4277. Fukuda, T.; Terauchi, T.; Goto, A.; Ohno, K.; Tsujii, Y.; Miyamoto, T.; Kobatake, S.; Yamada, B. *Macromolecules* **1996**, *29*, 6393-6398.
8. Greszta, D.; Matyjaszewski, K. *Macromolecules* **1996**, *29*, 7661-7670. Shipp, D.A.; Matyjaszewski, K. *Macromolecules* **1999**, *32*, 2948-2955.
9. Marque, S.; Le Mercier, C.; Tordo, P.; Fischer, H. *Macromolecules* **2000**, *33*, 4403-4410.
10. Boutevin, B.; Bertin, D. *Eur. Polym. J.* **1999**, *35*, 815-825.
11. Georges, M. K.; Hamer, G.; Szkurhan, A. R.; Kazemedah A; Li, J. *Polym. Prepr.* **2002**, *43(2)*, 78-79.

Chapter 34

Boroxyl-Based Radical Initiators and Polymerization

T. C. Chung and H. Hong

Department of Materials Science and Engineering, The Pennsylvania State University, University Park, PA 16802

This paper discusses a new family of living free radical initiators, alkylperoxy-dialkylborane ($C-O-O-BR_2$), which show living polymerization of acrylate and methacrylate monomers at ambient temperature. The $C-O-O-BR_2$ species can be prepared by in situ selective mono-oxidation of an asymmetrical trialkylborane ($-C-BR_2$) with a control amount of oxygen. Apparently, in the presence of polar monomers the $C-O-O-BR_2$ engages spontaneously hemolytic cleavage to form active alkoxyl radical ($C-O*$) and "stable" boroxyl radical ($*O-BR_2$), due to the delocalization of the free radical with the empty p-orbital of boron. The alkoxyl radical is active in initiating the polymerization of vinyl monomers. On the other hand, the stable borinate radical may form a reversible bond with the propagating radical site to prevent undesirable termination reactions. The living polymerization was characterized by predictable polymer molecular weight, narrow molecular weight distribution, and the formation of telechelic polymers and block copolymers by sequential monomer addition.

The control of polymer structure has been an important issue in polymer synthesis, both for scientific interests and industrial applications. Living polymerization provides an optimal means of control that can result in polymers with well-defined molecular structures; i.e. desirable molecular weight and polymer chain ends, narrow molecular weight distribution, and the formation of block and graft polymers.

Free radical polymerization is particularly interesting due to its compatibility with a wide range of functional groups. Early attempts to realize a living free radical polymerization involved the concept of reversible termination of the growing polymer chains by iniferters (*1,2*), such as N,N-diethyldithiocarbamate derivatives. However, this approach suffered from poor control of the polymerization reaction and the formation of polymer with high polydispersity. The first living radical polymerization was observed in reactions involving a stable nitroxyl radical, such as 2,2,6,6-tetramethylpiperidinyl-1-oxy (TEMPO) (*3-5*), which does not react with monomers but forms a reversible end-capped propagating chain end. Usually, the reactions have to be carried out at an elevated temperature (>100 °C) to obtain a sufficient concentration of propagating radicals for monomer insertion. Subsequently, several research groups have replaced the stable nitroxyl radical with transition metal species or reversible chain transfer agents as the capping agents to mediate living free radical systems. These polymerization reactions follow the mechanisms of atom transfer radical polymerization (ATRP) (*6-8*) or reversible addition-fragmentation chain transfer (RAFT) (*9*), respectively. Overall, these systems have a central theme-reversible termination via equilibrium between the active and dormant chain ends at an elevated temperature.

electron withdrawing   electron donating

In the past decade, we have been studying a new free radical initiation system based on the oxidation adducts of organoborane and oxygen, which may contain boroxyl radical - a mirror-image of the stable nitroxyl radicals (shown above). Our early interest in the borane/oxygen radical initiator stemmed from the desire to develop a new effective route in the functionalization of polyolefins (i.e. PE, PP, EP, etc.) (*10*), which was a long-standing scientific challenge with great potential for industrial applications. The unexpected good control in the incorporation of borane groups to polyolefin and the subsequent radical chain extension by the incorporated borane groups promoted us to examine this free radical polymerization mechanism in greater details.

Borane Containing Polyolefins and Polyolefin Graft Copolymers

We firstly studied the copolymerization of α-olefin with borane containing α-olefin monomers (*11, 12*) by Ziegler-Natta and metallocene catalysts. The incorporated borane groups pending along the polymer chain were interconverted to polar functional groups, such as OH, NH_2, etc., as illustrated in Equation 1. Due to the unique combination of good stability of trialkylborane to the transition metal catalysts, good solubility of borane in hydrocarbon medium, and the improved copolymerization ability of single-site catalyst systems, a broad range of functional polyolefin copolymers have been prepared with an relatively well-defined molecular structure, i.e. narrow molecular weight and composition distributions and a controlled amount of functional groups.

The major drawback in this functionalization process is the cost of borane compound and the undesirable reduction of crystallinity and melting temperature of PE and PP due to the side chains. To lessen these concerns, our research turned to converting the borane groups to the radical initiators that can carry out graft-from radical polymerization, as illustrated in Equation 1, which incorporate hundreds of functional groups (via monomers) to polyolefin by each borane group. Such a process not only dramatically increased the efficiency of borane groups, but also reduced the number of side chains. The resulting graft copolymers (*13, 14*) were found to be very effective compatibilizers in polyolefin blends and composites.

With the proper choice of borane moieties, such as alkyl-9-BBN having a stable bicyclononane double chair-form structure and a linear alkyl group to the polymer chain, the mono-oxidation by oxygen is selective at the linear alkyl group to form C-O-O-BR_2 species (*15*). Next, these moieties initiated free radical graft-from polymerization of functional monomers (such as acrylic and methacrylic monomers) at ambient temperature to form polyolefin graft copolymers containing polyolefin and functional polymer segments. The unexpected good control in the chain extension process (almost no homopolymer or crosslinked material) prompted us to further examine this free radical polymerization mechanism.

Borane-terminated Polyolefins and Diblock Copolymers

One of the ideal reactions to examine borane/oxygen initiators was to prepare borane-terminated polyolefin (*16-18*) containing only one borane group. The efficiency of the borane group in the radical polymerization can be easily determined by the formation of diblock copolymer, and the polymerization profile (molecular weight and molecular weight distribution) of the forming diblock copolymers provides a wealth of detail about the radical polymerization process.

Several methods were developed for the preparation of borane-terminated polyolefins. The most effective and convenient route involves H-BR_2 chain

484

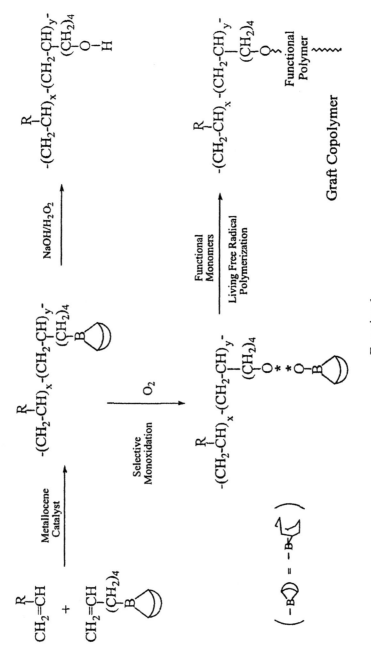

Equation 1

transfer agent during metallocene-mediated α-olefin polymerization. The resulting borane-terminated polyolefin usually has a relatively narrow molecular weight distribution (Mw/Mn~2), and its molecular weight is basically inversely proportional to the molar ratio of [borane]/[α-olefin].

The terminal borane group was used in the radical chain extension. One example is 9-BBN terminated polyethylene (PE-t-9-BBN) (18). Upon oxidation of PE-t-9-BBN with a control quantity of oxygen in the presence of MMA, the free radical chain extension took place to form PE-b-PMMA diblock copolymer. The resulting reaction mixture was carefully fractionated by Soxlet extraction using boiling THF for 24 hours to remove any PMMA homopolymer. In most cases, almost no PMMA homopolymer was isolated. Figure 1 compares the GPC curves of two PE-b-PMMA diblock copolymers that were sampled during the chain extension process and the starting PE-t-B polymer (Mn= 43,000 g/mole and Mw/Mn= 2.2). It is clear that the polymer continuously increased its molecular weight through the entire polymerization process. The polymer's molecular weight distribution was maintained at very constant and narrow levels (Mw/Mn= 2.0-2.4). The inset shows the linear plot of polymer molecular weight vs. the monomer conversion, and compares the results with a theoretical line based on the polymer molecular weight estimated from [g of monomer consumed]/[mole of initiator]. A good match with the straight line through the origin strongly supports the presence of living polymerization in the reaction.

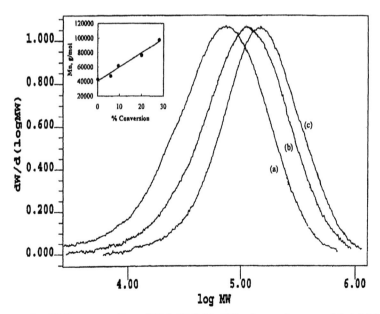

Figure 1. GPC curves of two PE-b-PMMA diblock copolymers with (a) Mn= 98,000 g/mole and (b) Mn= 62,000 g/mole and (c) the starting PE-t-B polymer (Mn= 43,000 g/mole and Mw/Mn= 2.2). (Redrawn from Macromolecules 2001,34, 8040. Copyright 2001 Am. Chem. Soc.)

These molecular weight results were further confirmed by the ^{1}H NMR probe. Figure 2 shows ^{1}H NMR spectra of three PE-b-PMMA copolymers that were sampled at different reaction times during the same radical chain extension process. The reaction was started by using a PE-t-B polymer with Mn= 20,000 g/mole and Mw/Mn= 2.7. The new peak at 3.58 ppm, corresponding to methoxyl groups (CH$_3$O) in PMMA, increased its intensity with the reaction time. Apparently, the PMMA segment in PE-b-PMMA grows with the reaction time, and high molecular weight diblock copolymer with up to 85 mole% of PMMA copolymer (Figure 2, c) has been prepared. Considering there is only one terminal borane group in each PE chain, these experimental results imply a very effective living radical polymerization chain extension process.

The similar results were also observed in poly(ethylene-co-1-octene), poly(ethylene-co-styrene), polypropylene, and syndiotactic polystyrene (16-18). Overall, the process resembles a transformation reaction from metallocene to living radical polymerization via a borane terminal group, which employs the best polymerization mechanisms (metallocene and living radical) for preparaing polyolefin and functional (polar) polymer blocks, respectively.

Alkyl-9-BBN/Oxygen Initiator

The living radical polymerization observed in the borane-containing polymer cases prompted us to study the corresponding small borane molecules, such as 1-octyl-9-borabicyclononane (1-octyl-9-BBN). However, the experimental results (19) showed a very different picture. Although it also exhibits the general trend of increased polymer molecular weight with monomer conversion, indicating the longevity of the active site, the overall catalyst efficiency is very poor. Only < 15% of the borane molecules engage in the polymerization process. In addition, the polymers formed during the polymerization show a steadily increasing molecular weight and broadening molecular weight distribution. Figure 3 shows the GPC curves of poly(ethyl methacrylate) (PEMA) sampled during the polymerization. This initiator is very difficult to use for the preparation of block copolymer.

The major difference between the two C-BR$_2$ moieties (one bonded to PE solid and the other in a soluble molecule) is intriguing, and may be associated with the prompt intermolecular reactions in a small molecule case, as illustrated in Equation 2. In the PE-t-9-BBN case, the isolated borane terminal groups in the PE matrix prevent any significant intermolecular reaction, and the selectively formed C-O-O-BR$_2$ species only engage in living radical polymerization. On the other hand, the continuous evolution of the borane group in the small molecule via intermolecular reactions creates many inactive species. In several control reactions, alkoxyl-9-BBN (C-O-BR$_2$) showed no reactivity to acrylic and methacrylic monomers. The oxidation adducts of C-O-BR$_2$ also exhibited extremely low reactivitiy for the polymerization.

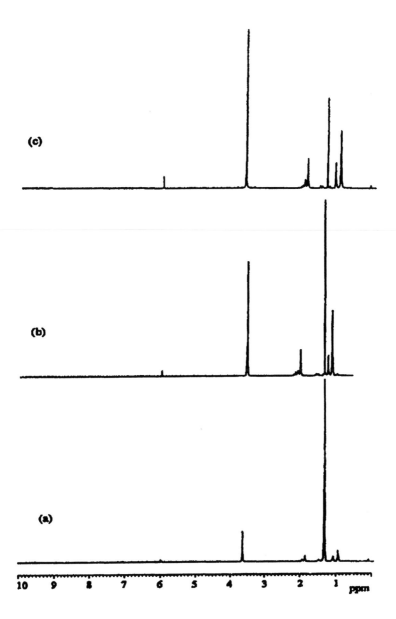

Figure 2. ^1H *NMR spectra of three PE-b-PMMA copolymers, including (a) I-1, (b) I-2, and (c) I-3 shown in Table 5. (solvent: C2D2Cl4; temp.: 110 °C). (Redrawn from Macromolecules 2001,34, 8040. Copyright 2001 Am. Chem. Soc.)*

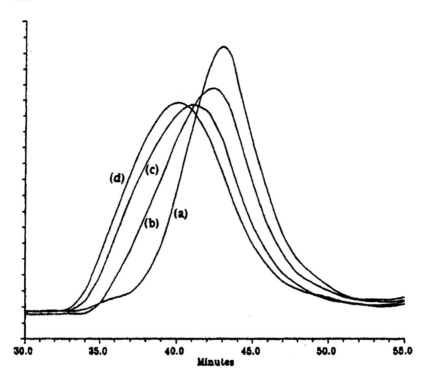

Figure 3, The comparison of GPC curves between PEMA homopolymers with various reaction times (a) 1 hour (PDI=2.2), (b) 3 hours (PDI=2.5), (c) 5 hours (PDI=2.6) and (d) 24 hours (PDI=2.5). (Redrawn from J. Am. Chem. Soc.1996, 118, 705. Copyright 1996 Am. Chem. Soc.)

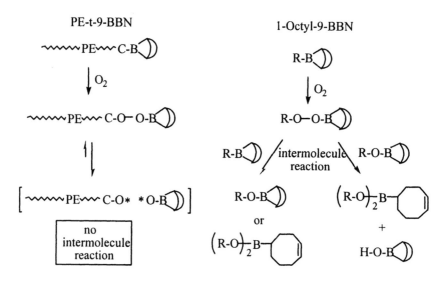

Equation 2

8-Bora-indane/O$_2$ Initiator

The ideal trialkylborane (BR$_3$) initiators shall perform selective mono-oxidation reaction to form peroxide (C-O-O-BR$_2$) species that shall not react with other borane species before polymerization. With this idea in mind, we recently studied several asymmetric trialkylborane systems that contain a roboric framework and a reactive B-C bond. In addition, the selective mono-oxidation reaction was carried out in the presence of monomers, so that the *in situ* formed peroxide (C-O-O-BR$_2$) immediately initiated the radical polymerization.

8-Bora-indane (I) (*20*) is a very interesting example, which has a boron atom bridging between a robust 6-member ring and a reactive 5-member ring, as illustrated in Equation 3. Upon contacting with a stoichmetric amount of oxygen, the B-C bond in the 5-member ring is promptly oxidized to form B-O-O-C speices in a newly expanded 7-member ring, which may release a small ring stress. With the presence of MMA monomers, the resulting C-O-O-BR$_2$ (II) immediately initiated the radical polymerization of MMA to create the polymer (III) that has the residual boroxyl group located at the beginning of the polymer chain. The two unreacted B-C bonds in the borane residue can be quantitatively converted to OH groups by NaOH/H$_2$O$_2$ reagent to form a telechelic PMMA polymer (IV) with two OH groups located at the beginning of polymer chain.

Equation 3

In a typical polymerization reaction, a 100 ml glass reactor equipped with a magnetic stirrer was introduced with desirable quantities of solvent (benzene, THF, etc.), monomer (MMA, EMA, etc.) and 8-bora-indane by syringe under a nitrogen atmosphere. The reactor was then placed in a water bath (0-20° C) for 10 minutes before injecting in a specific quantity of dry oxygen to the solution. The mixture was stirred at constant temperature under nitrogen with periodic removal of samples from the reactor. The polymer solution was quenched with NaOH aqueous solution, and then treated with H_2O_2 at 45 °C to oxidize the residue terminal borane group to two hydroxyl groups. The precipitated polymer was collected, washed with water and methanol, and dried in a vacuum oven at 60° C. The resulting polymers were weighted and analyized by NMR and GPC.

Figure 4 shows the GPC curves of the poly(methyl methacrylate) polymers sampled during the radical polymerization in benzene by 8-bora-indane/O_2 initiator. It is clear that the polymer continuously increased its molecular weight during the entire polymerization process. The polymer's molecular weight distribution was narrow (Mw/Mn= ~1.3-1.6) and slightly increased with the long reaction time, which may due to the fact that a trace amount of oxygen always presents in the system to continue the oxidation of the boroxyl moieties in the propagating site (III). Inset plots the polymer molecular weight vs. the monomer conversion, which is almost a straight line through the origin and consistent with the theoretical values based on the [g of monomer consumed]/[mole of initiator]. The good match strongly supports the presence of living radical polymerization in the reaction.

The same experimental results were also observed in many acrylic and methacrylic polymers, including poly(t-butyl acrylate) and poly(trifluorethyl acylate). However, this initiator is less reactive to pure hydrocarbon monomers, such as styrene, as well as the monomers causing strong acid-base interaction with borane, such as vinylpyridine.

As discussed in Equation 3, each polymer chain shall have two terminal OH groups, a primany and a secondary one. Figure 5 shows ^1H and ^{13}C NMR spectra of a PMMA sample (Mn=20,100 g/mol; Mw/Mn= 1.3). The ^{13}C (DEPT-135) NMR spectrum exhibits two minor peaks at 63.02 and 77.75 ppm corresponding to methylene alcohol and methine alcohol, respectively. For quantitative ^1H NMR analysis, the OH terminal groups were converted to $O(SiCH_3)_3$ groups. In addition to all the peaks for PMMA backbone, there is a chemical shift (at 0.05 ppm) corresponding to $SiCH_3$. The terminal OH group concentration is calculated by the peak intensity ratio between 0.05 PPM and 3.56 ppm ($O-CH_3$ in PMMA backbone). The PMMA molecular weight estimated from the chain end and GPC curve is in good agreement.

It is very interesting to study the effect of oxygen to the polymerization, since oxygen is essential for activating borane initiator (as illustrated in Equation 3) but also known as a termination agent in radical polymerization. Figure 6 shows the plots of monomer conversion versus reaction time using two different oxygen concentrations, [oxygen]/[8-bora-indane] mole ratio of 2/3 and 1/3. In general, the polymer yields are almost double by doubling the oxygen

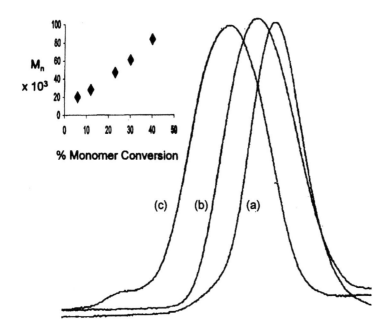

Figure 4, GPC curves of poly(methyl methacrylate) polymers sampled during the polymerization reaction (a) 5, (b) 10, and (c) 20 hours, respectively. Inset is a plot of polymer molecular weight vs. monomer conversion

492

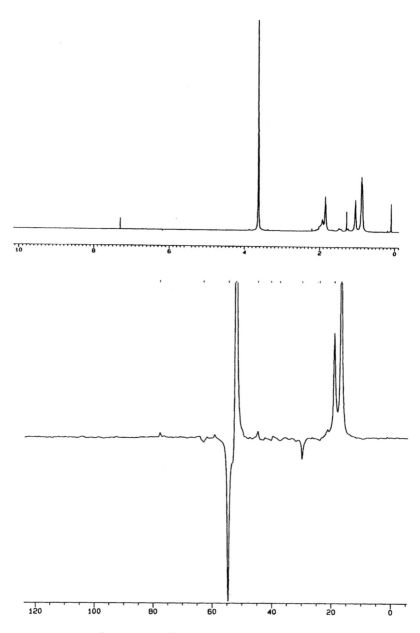

Figure 5, ¹H (top) and ¹³C (bottom) NMR spectra of a PMMA sample (Mn=20,100 g/mlo; Mw/Mn= 1.3) prepared by 8-bora-indane/O₂ initiator.

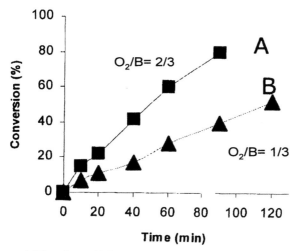

Figure6, The plots of MMA conversion versus reaction time by varying [oxygen]/[8-bora-indane] mole ratio, (a) 2/3 and (b) 1/3.

concentration, and the monomer conversion (with polymer molecular weight) contineously increases with the reaction time. It is clear that oxygen was consumed in the beginning of reaction, and the formed peroxyl borane adducts provide living polymerization of MMA. In addition, the propagating polymer chain ends appear to be stable with the unreacted 8-bora-indane.

This living radical polymerization was also evidenced in the preparation of diblock copolymers by sequential monomer addition. Figure 7 compares ^{1}H NMR and GPC curves of poly(methyl methacrylate-b-trifluoroethyl acrylate) diblock copolymer and the corresponding poly(methyl methacrylate) homopolymer. The molecular weight almost doubles from the homopolymer to the diblock copolymer without significantly changing the narrow molecular weight distribution (Mw/Mn= ~1.3-1.4). Basically, the copolymer composition is controlled by the monomer feed ratio and reaction time.

Overall, it is remarkable to think the simplicity of this living radical initiator and its polymerization process occurring at ambient temperature with the injection of oxygen to 1-octyl-9-borafluorene in the presence of monomers. The diminished termination reactions imply the *in situ* formation of a stable boroxyl radical, which serves as the reversible capping agent with the propagating radical during the living radical polymerization. Both the homo- and di-block copolymers formed are white solids with well-defined molecular structures.

Acknowledgment

The authors would like to thank the Office of Naval Research for the financial support.

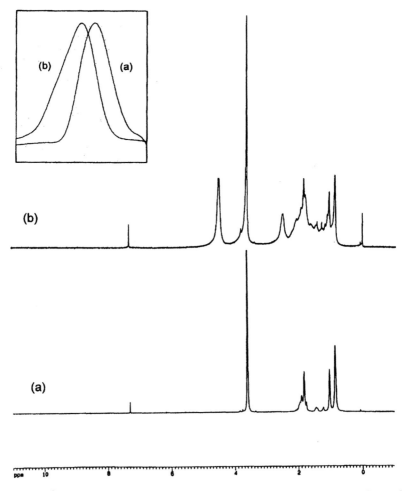

Figure 7, ^1H NMR spectra of (a) poly(methyl methacrylate) and (b) poly(methyl methacrylate-b-trifluoroethyl acrylate). Inset shows their GPC curves.

References

1. Otsu, T.; Yoshida, M. *Makromol. Chem., Rapid Commun.*, **1982**, *3*, 127.
2. Otsu, T.; Kuriyama, A. *J. Macromol. Sci., Chem.*, **1984**, *A21*, 921.
3. Georges, M. K.; Veregin, P. R. N.; Kazmaier, P. M.; Hamer, G. K. *Macromolecules,* **1993**, *26*, 2987.
4. Hawker, G. J. *J. Am. Chem. Soc.*, **1994**, *116*, 11185.
5. Hawker, C. J.; Bosman, A. W.; Harth, E. *Chem Rev.*, **2001**, *101*, 3661.
6. Wang, J. S.; Matyjaszewski, K. *J. Am. Chem. Soc.*, **1995**, *117*, 5614.

7. Wang, J. S.; Matyjaszewski, K. *Macromolecules,* **1995**, *28*, 7901.
8. Kato, M.; Kamigaito, M.; Sawamoto, M.; Higashimura, T. *Macromolecules,* **1995**, *28*, 1721.
9. Chiefari, J; Chong, Y. K.; Ercole, F.; Krstina, J.; Jeffery, J.; Le, T.; Mayadunne, R.; Meijs, G. F.; Moad, C. L.; Moad, G.; Rizzardo, E.; Thang, S. H. *Macromolecules,* **1998**, *31*, 5559.
10. Chung, T. C. *"Functionalization of Polyolefins"* Academic Press, London, **2002**.
11. Chung, T. C. *Macromolecules,* **1988**, *21*, 865.
12. Chung, T. C.; D. Rhubright, D. *Macromolecules,* **1993**, *26*, 3019.
13. Chung, T. C.; Jiang, G. J. *Macromolecules,* **1992**, *25*, 4816.
14. Chung, T. C.; Jiang, G. J.; Rhubright, D. *Macromolecules,* **1993**, *26*, 3467.
15. Chung, T. C.; Janvikul, W.;Bernard R.; Jiang, G. J. *Macromolecules,* **1994**, *27*, 26.
16. Xu, G.; Chung, T. C. *J. Am. Chem. Soc.,* **1999**, *121*, 6763.
17. Xu, G.; Chung, T. C. *Macromolecules,* **1999**, *32*, 8689.
18. Chung, T. C.; Xu, G.; Lu, Yingying; Hu, Youliang *Macromolecules,* **2001**, *34*, 8040.
19. Chung, T. C.; Janvikul, W.; Lu, H. L. *J. Am. Chem. Soc.,* **1996**, *118*, 705.
20. Chung, T. C.; Han, H.; Xu, G. *Polymer Preprint,* **2002**, *43(2)*, 82.

Chapter 35

Controlled Radical Polymerization of Alkyl Methacrylates in the Presence of NO/ NO₂ Mixtures

Christophe Detrembleur[1,2], Michael Claes[1], and Robert Jerome[1,2]

[1]Center for Educational and Research on Macromolecules (CERM), University of Liege, Sart-Tilman, B6, 4000 Liege, Belgium
[2]Current address: Bayer AG, Leverkusen, Germany

Radical polymerization of alkyl methacrylates initiated by AIBN is controlled when conducted in the presence of a mixture of NO/NO_2. Reaction of alkyl methacrylates with NO/NO_2 leads indeed to the monomer adduct and parent α-nitro, ω-nitroso oligomers, which are precursors of nitroxides, known to control the radical polymerization of alkyl methacrylates, according to a "Nitroxide-Mediated Polymerization" (NMP) mechanism. Although some side reactions may occur with time (increasing polydispersity), polymerization of MMA initiated by AIBN at low temperature (60°C) after bubbling of NO/NO_2 is relatively fast, and the molecular weight is dictated by the amount of NO/NO_2. Finally, for the first time, copolymerization of MMA with HEMA (10/1; v/v) has been controlled by this mixture of NO and NO_2 although the reaction remains very fast (ca. 65% monomer conversion after 5 h at 60°C), which is of prime importance for coating applications.

Introduction

Nitroxide-Mediated Polymerization (NMP) is one of the three main strategies commonly used to control the radical polymerization (CRP) of a large range of monomers *(1-3)*. The basic mechanism consists of the capture of the propagating radicals by nitroxides with formation of thermally labile alkoxyamines (Scheme 1). The dormant species are fragmented at high temperature (T > 100°C), and the released polymeric radicals add a limited number of monomers before recombination with nitroxides. Repetition of this homolysis-monomer addition-recombination cycle allows macromolecules to grow without significant interruption by irreversible termination.

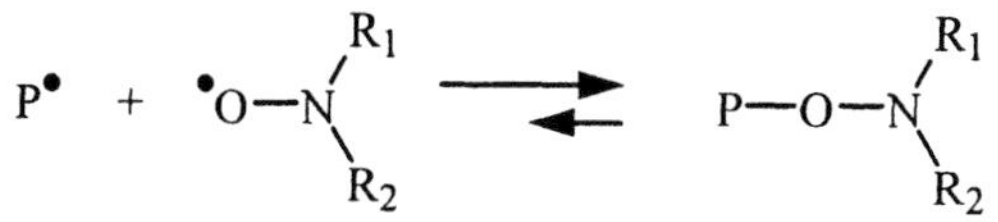

$$P^{\bullet} \ + \ {}^{\bullet}O-N\!\!\diagup\!\!\!\!\diagdown{}^{R_1}_{R_2} \ \rightleftharpoons \ P-O-N\!\!\diagup\!\!\!\!\diagdown{}^{R_1}_{R_2}$$

where $P^{\bullet}$ is the propagating chain

Scheme 1

Instead of using preformed nitroxides and alkoxyamines, that may be expensive, several research groups have contemplated the "in situ" formation of nitroxides from readily available and cheap precursors. In 1995, Matyjaszweski et al. *(4)* observed that radicals were trapped by nitroso compounds with formation of alkoxyamines. Recently, Nesvadba et al. patented the use of nitrones and nitroso compounds as control agents for the radical polymerization of vinyl monomers *(5)*. The addition of these compounds to initiating and/or propagating radicals (Scheme 2) actually forms nitroxides able to mediate radical polymerization. Only one monomer, i.e., n-butyl acrylate (n-BuA), was considered by the authors, and low molecular weight poly(n-BuA) (M_n < 10000) with polydispersity ranging from 1.4 to 2.0 was formed. Later on, Grishin et al. showed that the polymerization of methyl methacrylate (MMA) initiated by azobisisobutyronitrile (AIBN) was controlled by the addition of either a nitrone, the C-phenyl-N-tert-butylnitrone (PBN) *(6)*, or a nitroso compound, the 2-methyl-2-nitrosopropane *(7)*. The polydispersity was ca. 1.6 at low monomer conversion (< 10%) and increased up to 2.0 to 2.5 with the monomer conversion. Very high molecular weight (Mw) was also reported (10^5 to $3 \ 10^6$).

In 2001, Catala at al. reported on the CRP of styrene in the presence of nitroso-tert-octane *(8)*. In all these investigations, only one monomer was tested and no experiment of block copolymerization was considered. Finally, Detrembleur et al. successfully used an easily available nitrone (N-tert-butyl-α-isopropylnitrone) for the controlled radical polymerization of styrene, styrene/acrylonitrile mixtures, and dienes *(9)*. For the first time, well-defined

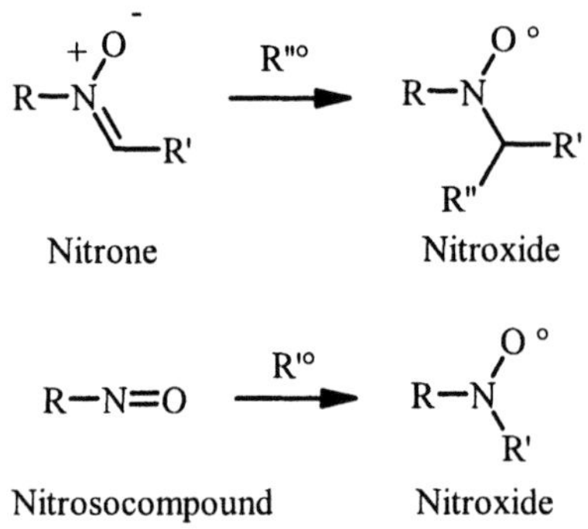

Nitrone   Nitroxide

Nitrosocompound   Nitroxide

Scheme 2

poly(styrene)-b-poly(styrene-co-acrylonitrile),   poly(styrene)-b-poly(n-butyl acrylate), and poly(styrene)-b-poly(isoprene) copolymers were synthesized in the presence of this nitrone used as a nitroxide precursor.

In an effort to make the NMP process still more attractive, some of us proposed to use sodium nitrite as precursor of nitroxides (10,11). The radical polymerization of tert-butylmethacrylate (tBMA) in water was controlled by combining sodium nitrite with a reducing agent, e.g. iron(II) sulfate and ascorbic acid, at a relatively low temperature (80°C). Because tBMA is not miscible with water, droplets of the organic phase (tBMA, growing polymer chains and initiator) were dispersed in the water phase under vigorous stirring, and no gross precipitation of PtBMA was observed during polymerization. Nitroxides are thought to be formed by a three-step process (scheme 3): (i) formation of nitric oxide (NO) by reduction of sodium nitrite by iron(II) sulfate or ascorbic acid; (ii) trapping of radicals (initiating or propagating species) by NO with formation of the parent nitroso compounds (R-NO); (iii) these compounds are efficient spin-traps, that react with radicals to form nitroxides (R_2-NO°).

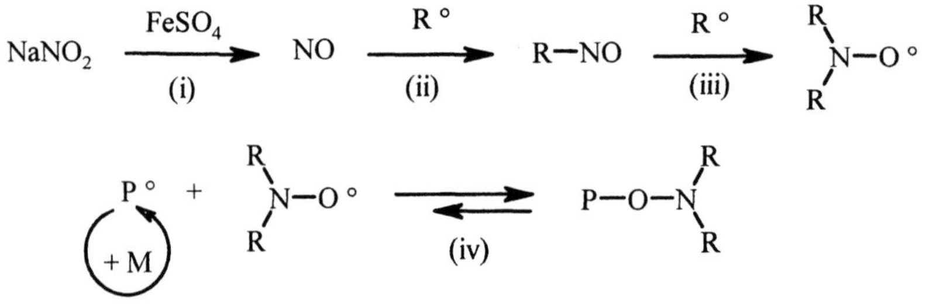

where R ° is the propagating chain or the initiating radical;
P ° is the propagating chain and M the monomer.

Scheme 3

The radical polymerization of tBMA is thus controlled by NMP from readily available and very cheap sodium nitrite and iron(II) sulfate or ascorbic acid. Whenever the radicals $R°$ that participate to steps (ii) and (iii) are propagating radicals $P°$, the final polymer is three-arm star-shaped. A mixture of linear and star-shaped chains are thus expected to be formed in the polymerization medium.

It must be noted that, in addition to NO, NO_2 is also formed *(10)*. tBMA reacts rapidly with the NO/NO_2 mixture with formation of α-nitro, ω-nitroso poly(tBMA) oligomers (Scheme 4; with $n \geq 1$) *(10)*, which are expected to have a beneficial effect on the control of the radical polymerization.

$$NO + NO_2 + n \quad \underset{CO_2R}{\diagup\!\!=\!\!\diagdown} \quad \longrightarrow \quad O_2N\left[\!\!\underset{CO_2R}{\diagup\!\diagdown}\!\!\right]_n\!\!-N{=}O$$

α-nitro ω-**nitroso polymethacrylate**

dimerization

$$O_2N\left[\!\!\underset{CO_2R}{}\!\!\right]_n\!\!-\underset{\underset{O_-}{+}}{\overset{\overset{O^-}{|}}{N}}{=}\underset{+}{N}\!\!-\!\!\left[\!\!\underset{CO_2R}{}\!\!\right]_n\!\!NO_2$$

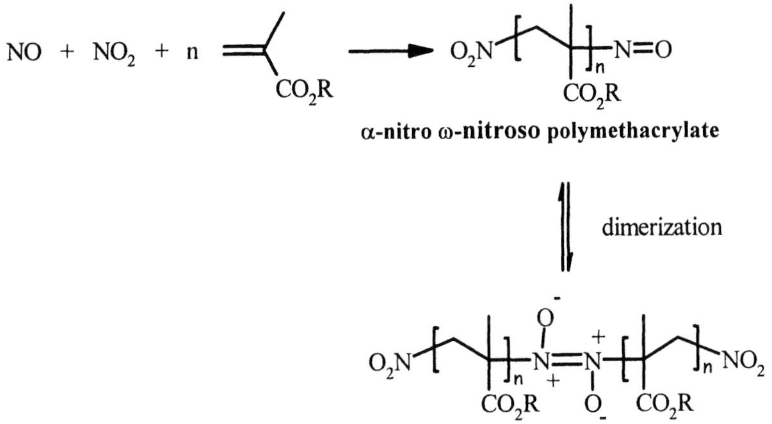

Scheme 4

This paper will report on the role of the "in-situ" formed adducts, i.e., α-nitro, ω-nitroso monomer and oligomers, on the control of the radical polymerization of tBMA, MMA and HEMA conducted in the presence of sodium nitrite. Moreover, formation of nitroso compounds by reaction of a preformed NO/NO_2 mixture with alkyl methacrylates prior to polymerization will be investigated. These nitroso compounds together with the nitroso derivatives formed by the addition of alkyl radicals to NO are thought to be regulators for the radical polymerization. In order to assess this hypothesis, a mixture of NO and NO_2 was bubbled into the monomer at 25°C prior to polymerization conducted in the presence of a conventional free-radical initiator at a relatively low temperature (60°C).

Experimental Section

Materials. Methyl methacrylate (MMA), 2-hydroxyethyl methacrylate (HEMA; Aldrich) and tert-butyl methacrylate (tBMA; BASF) were distilled just before use in order to remove the stabilizer. Toluene was refluxed over calcium hydride and distilled before use. Acetonitrile (Aldrich) and diethylether (Vel) were used as received, and water was deionized prior to use. Nitrogen was bubbled through the solvents in order to eliminate molecular oxygen. NaNO$_2$ (Backer), AIBN (Aldrich and Fluka), sulfuric acid (Analar) were used as received. NO and NO$_2$ were provided by Messer. Liquids were transferred under nitrogen by means of syringes or stainless steel capillaries.

Synthesis of methyl 2-methyl-3-nitro-2-nitrosopropionate (NMMA; Scheme 4 with n=1 and R=CH$_3$). The synthesis was carried out as reported by Shechter et al. (12). A solution of 12.903 g of NaNO$_2$ (0.187 moles) in 30 ml of water was added to a solution of MMA in diethylether (20 ml MMA in 80 ml diethylether) at 0°C. This solution was carefully added with a sulfuric acid solution (10 ml concentrated sulfuric acid in 30 ml water), so leading to the formation of the nitrosocompound as testified by an intense blue color. After 2h of reaction at 0°C, the organic solution was washed several times with water, dried with MgSO$_4$, filtered, and the residual monomer and solvent were eliminated under vacuum at room temperature. A white solid (NMMA) was formed and recovered by filtration. The viscous light-green oily filtrate was dissolved in diethylether and added with hexane until a white solid was formed (NMMA) and collected by filtration. This treatment was repeated several times. The white solid was purified by recrystallization from a hexane/diethylether mixture, and dried under vacuum at room temperature. The yield (15-20%) was not optimized.

- Elemental analysis: theoretical composition: 15.9 % N, 34.1 % C, 4.6 % H; found 15.4 % N, 34.0 % C, 4.7 % H.
- IR: 676, 868, 993, 1134, 1263, 1299.53, 1381, 1436, 1561 (nitro group), 1746 (ester group), 2959, 3041, and 3439cm^{-1}.

Formation of α-nitro, ω-nitroso poly(tBMA) and poly(MMA) oligomers. Nitric oxide was synthesized as reported in the scientific literature (13). Briefly, a degassed solution of sodium nitrite (1M in water) was added dropwise to a degassed acidic solution of FeSO$_4$ (1M). The non-purified nitric oxide was swept by a nitrogen flow through 30 ml of degassed monomer (MMA or tBMA) at room temperature. A blue (tBMA) or green (MMA) color was rapidly observed as result of the nitrosocompound formation. After 2-3h, the residual monomer was eliminated under vacuum, and a viscous light-green residue was left. The use of non degassed monomer and solutions did not change these observations.

Polymerization in the presence of α-nitro, ω-nitroso oligomers. In a typical procedure, a mixture of oligomer (0.1 g), AIBN (0.0555 g; 3.38 10^{-4} mol), toluene (5 ml) and tBMA (5 ml; 0.03 mol) was added into the reaction flask, that was evacuated by three nitrogen-vacuum cycles and heated at 60°C. Samples were regularly withdrawn, and monomer conversion was determined gravimetrically after drying at 80°C under vacuum, taking into account the original amounts of initiator and oligomer.

Polymerization in the presence of NO and NO₂. MMA (10 ml; 93.5 mmoles) was added into a 50 ml one-necked round bottom flask fitted with a three-way stopcock connected to either a nitrogen line or a vacuum pump. Oxygen was removed from the reaction flask by repeated vacuum-nitrogen cycles and finally filled with nitrogen. NO (85 mg/h; 2.36 10^{-4} mol within 5 minutes) and NO₂ (100 mg/h; 1.81 10^{-4} mol within 5 minutes) were then bubbled at the same time into MMA. The bubbling was stopped after 5 min., a solution of AIBN in toluene (7.5 ml; 0.015 M; 1.12 10^{-4} mol) was added, and polymerization was conducted at 60°C. Samples were regularly withdrawn from the polymerization medium, and the monomer conversion was determined as aforementioned.

This recipe was unchanged when MMA was copolymerized with HEMA. This comonomer (1 ml) was actually added to the reaction medium together with AIBN.

ESR analysis. α-nitro, ω-nitroso oligomers (57 mg) were mixed with AIBN (57 mg; 3.47 10^{-4} mol) in 5 ml of toluene. Oxygen was removed by repeated vacuum-nitrogen cycles and replaced by nitrogen. A sample of this solution was introduced into a NMR tube filled with nitrogen. This tube was heated at 60°C, and the ESR spectra were regularly recorded at this temperature.

Characterization. Size exclusion chromatography (SEC) was performed in THF (neat or added with 5% NEt₃), at 40°C, with either a Hewlett-Packard 1090 liquid chromatograph equipped with a Hewlett-Packard 1037A refractive index detector and HP PL gel 5μ columns (10^5 Å, 10^4 Å, 10^3 Å, 100 Å), or a Waters 600 liquid chromatograph equipped with a Waters 410 refractive index detector and Styragel HR columns (M_w = HR 1: 100-5000; HR 2: 500-20000; HR 4: 5000-600000). The columns were calibrated with PMMA standards and the flow rate was 1 ml/min. ¹H NMR spectra were recorded in CDCl₃ with a Brucker AN 400 (400MHz) apparatus at 25°C. IR spectra were recorded with a 1720X Perkin Elmer FT-IR spectrometer and a Nexus 670 FT-IR from Thermonicolet. The ESR experiments were performed at 9.46 GHz with a Brucker ESP-300E spectrometer equipped with a temperature controller. Spectra were recorded with a 20 mW microwave power and an 1-G modulation amplitude.

502

Results and Discussion

Contribution of α-nitro, ω-nitroso poly(tBMA) (10) and α-nitro, ω-nitroso poly(MMA) to the CRP of tBMA.
Whenever non purified NO (synthesized by mixing a solution of sodium nitrite with an acidic solution of iron (II) sulfate) is bubbled into tBMA at room temperature, the mixture becomes rapidly blue as result of formation of α-nitro, ω-nitroso poly(tBMA) (scheme 4). After 2 h of reaction, a viscous light-green solid is collected upon elimination of the residual monomer under vacuum. The elemental analysis confirms that ca. 9 to 10 w% of this residue is nitrogen, consistent with the addition of the nitric oxides to tBMA and formation of α-nitro, ω-nitroso poly(tBMA), in agreement with the scientific literature (10,14). As reported in a previous paper (10), the IR spectrum confirms that poly(tBMA) is end-capped by a nitro group (sharp and intense peaks at 1561 and 1371 cm⁻¹; Figure 1).

When the nitroso end-group is concerned, its assignment is a problem because these compounds have a strong tendency to form dimers (15) (Scheme 4) and the absorption of the NO group depends on the unimer vs dimer structure. Formation of the nitroso group is however supported by the Liebermann Test, which is commonly used to identify this type of compounds (16). The same observations are reported when MMA is substituted for tBMA.

Both α-nitro ω-nitroso adducts (α-nitro, ω-nitroso poly(tBMA) and α-nitro, ω-nitroso poly(MMA)) have a very favourable effect on the radical polymerization of tBMA initiated by AIBN in toluene at 60°C. First of all, the characteristic color of the nitroso compounds is slowly fading as result of reaction with free radicals. More importantly, the dependence of M_n on the monomer conversion (Figure 2) and the $\ln[M]_o/[M]$ vs time plot (Figure 3) tends to linearity (eventhough the experimental data are more scattered in Fig. 3A) in agreement with a controlled process. In parallel, the polymerization rate drops down (e.g. 72 % tBMA converted after 24h of reaction) and the polydispersity index which is low at low monomer conversion, increases with time.

Comparison of figures 2A and 2B shows that the regulating PMMA adducts are of a higher molecular weight than the tBMA ones. Moreover, an induction period of time (ca. 3h) is observed when the α-nitro, ω-nitroso poly(MMA) adduct is used rather than the α-nitro, ω-nitroso poly(tBMA) adduct (Figs. 3A and 3B).

Theoretical molecular weight (Mn,th) for poly(tBMA) formed in the presence of α-nitro, ω-nitroso poly(tBMA) has been calculated as the product of

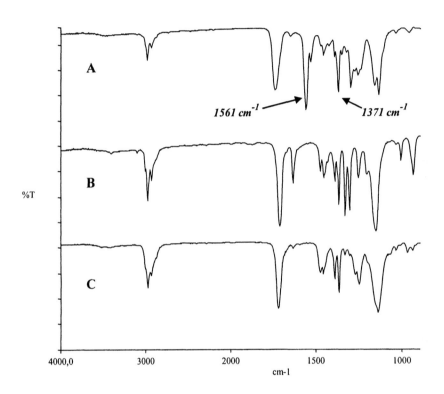

Figure 1: IR spectra of (A) the residue collected after the bubbling of NO and
NO₂ in tBMA, (B) tBMA and (C) poly(tBMA).

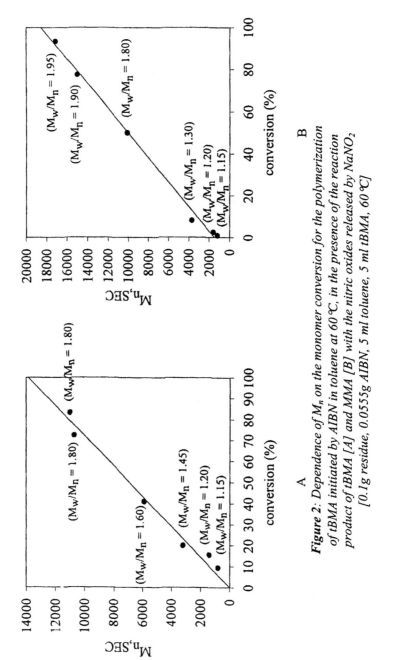

Figure 2: *Dependence of M_n on the monomer conversion for the polymerization of tBMA initiated by AIBN in toluene at 60 °C, in the presence of the reaction product of tBMA [A] and MMA [B] with the nitric oxides released by $NaNO_2$ [0.1g residue, 0.0555g AIBN, 5 ml toluene, 5 ml tBMA, 60 °C]*

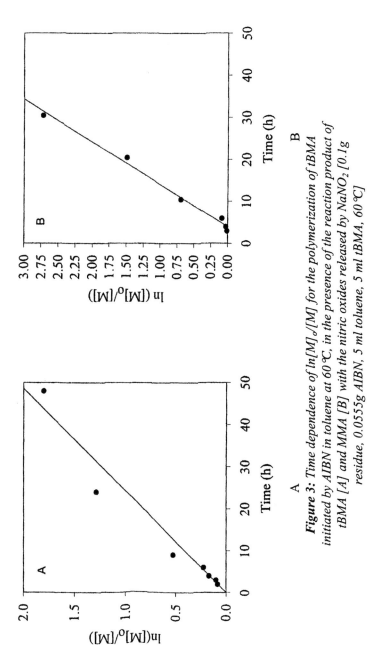

Figure 3: *Time dependence of ln[M]$_o$/[M] for the polymerization of tBMA initiated by AIBN in toluene at 60 °C, in the presence of the reaction product of tBMA [A] and MMA [B] with the nitric oxides released by NaNO$_2$ [0.1g residue, 0.0555g AIBN, 5 ml toluene, 5 ml tBMA, 60 °C]*

the tBMA mass by the tBMA conversion, divided by the number of moles of nitroxides formed in-situ (approximated to the number of moles of nitroso groups of the α-nitro, ω-nitroso poly(tBMA) adduct). On the basis that 10 wt% of this adduct is nitrogen (see elemental analysis), that each adduct molecule contains one nitroso group and one nitro group, and that one nitroso group forms one nitroxide, the theoretical molecular weight of polytBMA at 100% monomer conversion is 12200. Figure 2A shows a good agreement with this prediction, in spite of SEC calibration by polyMMA and not by polytBMA standards.

Therefore, whenever the radical polymerization of tBMA is conducted in the presence of $NaNO_2$ and a reducing agent, the $NO(NO_2)$ gas which is released has a key role because it leads to the "in situ" formation of nitroso compounds (and thus nitroxides) which can impart control to the polyaddition. Indeed, nitroso compounds result from the addition of the nitric oxides (NO and NO_2) to the monomer and from the addition of NO to the initiating and/or propagating radicals (Scheme 5).

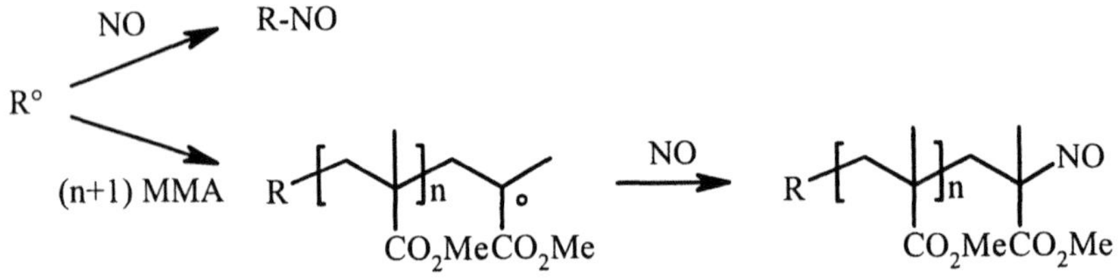

Scheme 5

Although $NaNO_2$ is a very common and cheap precursor of regulating nitroxides, the variety of structure of the nitroxides formed "in situ" is source of complexity which makes the mechanistic analysis difficult.

From this preliminary analysis of the origin of the control of the radical polymerization of tBMA and MMA in the presence of $NaNO_2$, it may be inferred that bubbling a NO/NO_2 mixture into the monomer(s) prior to polymerization should be another way to control the free-radical polymerization of alkyl methacrylates.

Controlled radical polymerization of methyl methacrylate as result of a preliminary reaction with a NO/NO_2 mixture.

Homopolymerization

When NO (85 mg/h) and NO_2 (100 mg/h) are bubbled into MMA at room temperature (NO/NO_2 molar ratio = 1.3), the solution becomes rapidly green, consistent with the formation of nitroso compounds (Scheme 4). After 5 minutes of reaction, temperature is increased up to 60°C and AIBN is added to initiate the polymerization. The free-radicals which are accordingly formed slowly react with the nitroso compounds as attested by the fading of the green color. The progress of the MMA polymerization is such that the molecular weight increases linearly with the monomer conversion (Figure 4). In parallel, the polydispersity index increases with time (Figures 4 and 5). SEC chromatograms illustrate the increase in both the molecular weight and polydispersity with monomer conversion (Figure 5). Moreover, the time dependence of $\ln[M]_o/[M]$ is also linear (Figure 6), which indicates that the concentration of the propagating chains remains constant all along the polymerization, in agreement with a controlled process. An induction period of ca. 1h is observed, whereas Mn,exp is higher than Mn,th. The molecular weight is actually controlled by the amount of nitroxides in a classical NMP process. Because the nitroxides are formed "in situ" by reaction of radicals with the nitrosocompound formed in the pre-reaction (scheme 4), the amount of NO/NO_2 should basically control M_n. In this work, Mn,th has been calculated on the basis of the NO_2 amount, which is used in lower amount than NO [Mn,th = [MMA(g) x conversion]/NO_2 (mol)].

According to fig. 4, the initiator efficiency is lower than 1 (ca. 0.6), which indicates that ca. 60% of NO_2 (and thus 47 % of NO) contribute to the formation of nitroxide active in CRP. Several reasons might explain this observation : (i) reaction of NO and NO_2 with MMA (scheme 4) is not complete, (ii) reaction of the nitroso compounds with alkyl radicals does not form nitroxides quantitatively, (iii) the nitroso compounds and/or the nitroxides are unstable and partly lost with time.

In order to confirm that the PMMA molecular weight is determined by the amount of the nitric oxides, the bubbling time of NO and NO_2 (at constant bubbling rate) in the pre-reaction (thus before addition of AIBN) has been increased from 5 to 10 min. As expected, the molecular weight is decreased by a factor of two (Figure 7) because two times more nitroso compounds and thus nitroxides are formed in-situ. Moreover, the polymerization rate is independent of the amount of NO and NO_2, and $\ln([M]_0/[M])$ remains linear, consistent with a constant concentration of propagating chains (Figure 8). Whatever the bubbling time of NO/NO_2, the polydispersity increases from 1.1(low monomer conversion) up to 1.7 at ca. 70% conversion.

Although the polymerization control is not perfect (broadening of the molecular weight distribution with monomer conversion), a mixture of NO/NO_2 is a valuable substitute for the nitroxides (or alkoxyamines) commonly used in the controlled radical polymerization of vinyl monomers.

In an effort to elucidate the structure of the reaction product formed by bubbling NO/NO_2 into MMA, the residual monomer has been eliminated under vacuum after 5 min. of reaction at room temperature. It is well-known that methyl 2-methyl-3-nitro-2-nitrosopropionate (NMMA) is formed and mainly

508

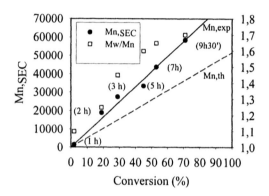

Figure 4: *Dependence of M_n,SEC on the monomer conversion for the MMA polymerization initiated by AIBN after bubbling of NO/NO_2 into the monomer [Bubbling of NO (85 mg/h) and NO_2 (100 mg/h) in MMA (10 ml) for 5 min. at 25°C, followed by the addition of AIBN (7.5 ml; 0.015 M) at 60°C]*

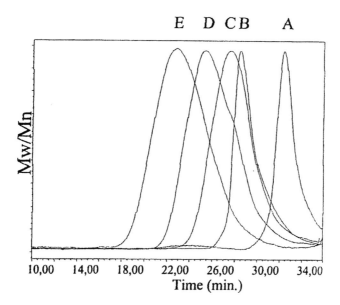

Figure 5: SEC chromatograms for the polyMMA samples analyzed in Figure 4. Polymerization time: (A) 1h; (B) 2h; (C) 3h; (D) 5h; (E) 9.5h; (Figure 4)

510

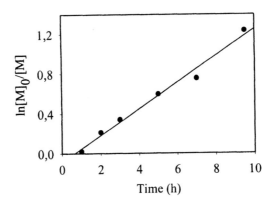

Figure 6: *Time dependence of ln[M]₀/[M] for the MMA polymerization initiated by AIBN after bubbling of NO/NO₂ into the monomer [Bubbling of NO (85 mg/h) and NO₂ (100 mg/h) in MMA (10 ml) for 5 min. at 25°C, followed by the addition of AIBN (7.5 ml; 0.015 M at 60°C]*

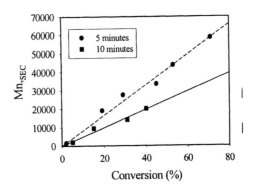

Figure 7: *Effect of the bubbling time of No/NO₂ into MMA prior to polymerization, on the PMMA molecular weight [Bubbling of NO (85 mg/h) and NO₂ (100 mg/h) in MMA (10 ml) at 25°C for 5 min. and 10 min., respectively, followed by the addition of AIBN (7.5 ml; 0.015 M) at 60°C]*

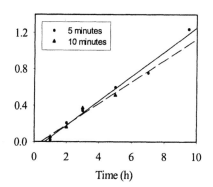

Figure 8: *Effect of the bubbling time of No/NO₂ into MMA prior to polymerization, on the polymerization rate [Bubbling of NO (85 mg/h) and NO₂ (100 mg/h) in MMA (10 ml) at 25°C for 5 min. and 10 min., respectively, followed by the addition of AIBN (7.5 ml; 0.015 M) at 60°C]*

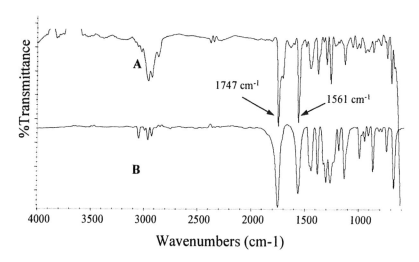

Figure 9: *IR spectra of the reaction product of NO and NO₂ with MMA at 25°C [A] and NMMA [B]*

collected as a dimer whenever NO (contaminated with NO_2) is reacted with MMA in a water/diethylether mixture (Scheme 4, with n = 1 and R = CH_3) *(12)*. It is therefore reasonable to assume that NMMA is formed in the experiments carried out in this work, together with oligomeric nitroso compounds (n $\geq$ 1 and R = CH_3 in scheme 4). The IR-spectra for NMMA (synthesized according to the scientific literature) and for the product of reaction between MMA and NO/NO_2 at 25°C are compared in fig. 9. They show that the strong absorptions of both the ester at 1747 cm^{-1} and the nitro group at 1561 cm^{-1} of NMMA are also observed for the reaction product. As previously discussed, the assignment of the absorption peak of the NO group is quite uncertain because of dimer formation. However, nitroso groups are detected, at least qualitatively, by the Liebermann Test *(16)*.

Formation of nitroxides in the polymerization medium at 60°C has also been mimicked by reaction of the aforementioned mixture of NMMA and oligomeric α-nitro ω-nitroso PMMA with AIBN at 60°C. A triplet with A_N = 14 G and g = 2.01 is observed (Figure 10), which is characteristic of nitroxides *(17)* and confirms their formation in the polymerization medium. The three-line ESR spectrum results from the hyperfine coupling of the radical to the nitrogen atom which is bonded to two tertiary carbons. Indeed, a more complex ESR spectrum should be observed in case of bonding of the nitrogen atom to a secondary or primary carbon, as result of the radical coupling to the hydrogen atom(s) of these carbons. This spectrum remains unchanged when a twofold molar excess of the mixture of NMMA and oligomeric α-nitro ω-nitroso PMMA is used with respect to AIBN.

Figure 11 shows the time dependence of the concentration of the nitroxides formed by reaction of AIBN with the mixture of NMMA and reactive oligomers. This concentration increases very rapidly for the first 5 min. and then decreases slowly with time. The 2-cyano isopropyl radicals formed by thermolysis of AIBN are thus rapidly trapped by the nitroso compounds with formation of nitroxides, which are then at least partly converted into alkoxyamines by reaction with additional 2-cyano isopropyl radicals. In addition to alkoxyamine formation, the decreasing nitroxide concentration could also result from their thermolysis, which should however be low at 60°C (otherwise no polymerization control could be observed).

Random copolymerization of MMA and HEMA.
For a successful copolymerization of MMA with HEMA, NO and NO_2 must be bubbled into MMA for 5 min. Then, HEMA and AIBN are added to the pre-treated MMA and the copolymerization is initiated by increasing the temperature up to 60°C. Under these conditions, the copolymerization is controlled as supported by the linear increase of Mn with the comonomer conversion (Figure 12) and ln([M]$_0$/[M]) with time (Figure 13). In contrast to the MMA

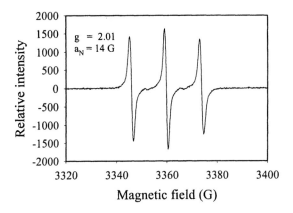

Figure 10: *ESR spectrum for the reaction product of AIBN with the product formed by bubbling NO and NO₂ into MMA in toluene at 60 °C*

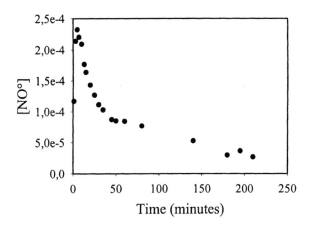

Figure 11: *Time dependence of the concentration of nitroxides formed by reaction of AIBN with the adduct of NO and NO₂ to MMA (scheme 4)*

514

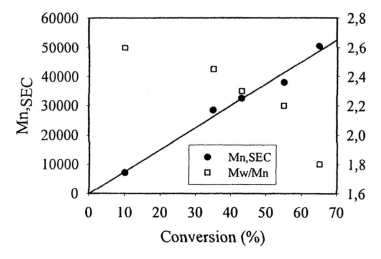

Figure 12: Dependence of M_n,SEC on the monomer conversion for the MMA/HEMA (10/1) copolymerization initiated by AIBN after bubbling of NO/NO₂ into MMA [Bubbling of NO (85 mg/h) and NO₂ (100 mg/h) in MMA (10 ml) for 5 min. at 25°C, followed by the addition of AIBN (7.5 ml; 0.015 M) and HEMA (1 ml) at 60°C]

homopolymerization, the polydispersity index is very high at low monomer conversion and it decreases rapidly with the copolymerization progress, in total agreement with a controlled process. Moreover, this well-controlled copolymerization is very fast as indicated by 65% monomer conversion after only 5h at 60°C, which makes this process attractive.

As an additional evidence of control, the MMA/HEMA copolymerization has been repeated under the same experimental conditions, except for the bubbling of NO and NO₂. In absence of NO and NO₂, the molecular weight is rapidly very high and does not increase further with the monomer conversion (Figure 14) consistent with conventional free-radical polymerization. Moreover, the polydispersity is very broad all along the reaction (2.25 ≤ Mw/Mn ≤ 3.0). Comparison of figures 12 and 14 is clear evidence for the control imparted to the MMA/HEMA copolymerization by NO/NO₂.

Conclusions

Methyl 2-methyl-3-nitro-2-nitrosopropionate (NMMA) and derived α-nitro, ω-nitroso oligomers are formed when a mixture of NO and NO₂ is reacted with MMA at 25°C. The same reaction occurs when tBMA is substituted for MMA. These nitro/nitroso compounds impart control to the radical polymerization of the parent monomers at relatively low temperature (60°C). This observation

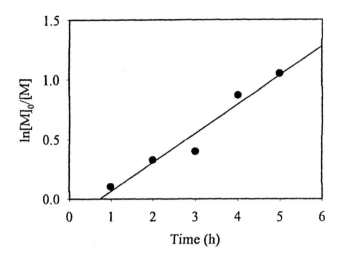

Figure 13: *Time dependence of ln([M]$_0$/[M]) for the MMA/HEMA (10/1) copolymerization initiated by AIBN after bubbling of NO/NO$_2$ into MMA [Bubbling of NO (85 mg/h) and NO$_2$ (100 mg/h) in MMA (10 ml) for 5 min. at 25°C, followed by the addition of AIBN (7.5 ml; 0.015 M) and HEMA (1 ml) at 60°C]*

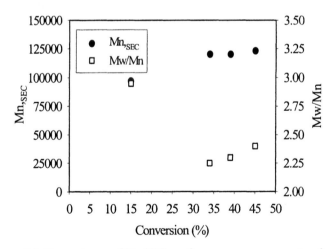

Figure 14: *Dependence of M$_n$ SEC on the monomer conversion for the MMA/HEMA (10/1) copolymerization initiated by AIBN [MMA (10 ml); AIBN (7.5 ml; 0.015 M); HEMA (1 ml) at 60°C]*

explains why $NaNO_2$ added with a reducing agent controls this type of radical polymerization. Indeed, nitrosocompounds are formed "in situ" by the addition of the released nitric oxides (NO and NO_2) to the monomer and the addition of NO to the initiating radicals and propagating radicals (Scheme 5). For the first time, a mixture of NO and NO_2 has been successfully used to provide the radical polymerization of alkyl methacrylates with control. Actually, a pre-reaction of NO and NO_2 with the monomer is carried out at 25°C in order to form in-situ the nitroso compounds, i.e. the precursors of nitroxides. Upon release of free-radicals in the pretreated monomer, controlled polymerization of MMA and tBMA is observed at low temperature (60°C). When MMA is copolymerized with HEMA (10/1) under these conditions (pre-reaction of MMA), the control is maintained and the reaction is fast. So, this non-conventional NMP process makes the radical polymerization of alkyl methacrylates controlled until high monomer conversion, which is impossible by "classical" NMP because of the dominating β-elimination of the terminal methacrylic proton of the growing chains. More investigation is needed to improve our understanding of this system and to improve it by restricting the extent of side reactions at the origin of the polydispersity that increases with the monomer conversion.

Although the polydispersity of polyMMA and polytBMA is high compared to traditional NMP, the new sodium nitrite/reducing agent and NO/NO_2 are highly valuable for industrial applications, because of low price, mild polymerization conditions (water, low temperature, ...), control imparted to copolymerization of alkylmethacrylates with functional monomers, such as HEMA, in short reaction time.

Acknowledgments

The authors are much indebted to the "Services Fédéraux des Affaires Scientifiques, Techniques et Culturelles" for general support to CERM in the frame of the "Pôles d'Attraction Interuniversitaires: PAI 5/03".

References

1   (a) Salomon, D.H.; Rizzardo, E.; Cacioli, P. **1986**, *US Patent* 4,581,429. (b) Rizzardo, E. *Chem. Austr.* **1987**, *54*, 32.

2   (a) Georges, M.K.; Veregin, R.P.N.; Kazmaier, P.M.; Hamer, G.K. *Trends Polym. Sci.* **1993**, *2*, 66. (b) Georges, M.K.; Veregin, R.P.N.; Kazmaier, P.M.; Hamer, G.K. *Macromolecules* **1993**, *26*, 2987.

3   (a) Grimaldi, S.; Lemoigne, F.; Tordo, P.; Nicol, P.; Plechot, M. *WO 96/24620* **1996**. (b) Benoit, D. Ph.D. Thesis, *Polymérisation radicalaire contrôlée en présence de radicaux nitroxydes*, Université de Bordeaux,

France, **1997**. (c) Benoit, D.; Grimaldi, S.; Finet, J.P.; Tordo, P.; Fontanille, M.; Gnanou, Y. *ACS Symp. Ser.* **1998**, *685*, 225. (d) Benoit, D.; Grimaldi, S.; Robin, S.; Finet, J.P.; Tordo, P.; Gnanou, Y. *J. Am. Chem. Soc.* **2000**, *122*, 5929. (e)Le Mercier, C.; Lutz, J.F.; Marque, S.; Le Moine, F.; Tordo, P.; Lacroix-Desmazes, P.; Boutevin, B.; Couturier, J.L.; Guerret, O.; Martschke, R.; Sobek, J.; Fischer, H. *ACS Symp. Ser.* **2000**, *768*, 108. (f) Benoit, D.; Chaplinski, V.; Braslau, R.; Hawker, C.J. *J. Am. Chem. Soc.* **1999**, *121*, 3904. (g) Benoit, D.; Harth, E.; Helms, B.; Rees, I.; Vestberg, R.; Rodlert, M.; Hawker, C.J. *ACS Symp. Ser.* **2000**, *768*, 123. (h) Benoit, D.; Harth, E.; Fox, P.; Waymouth, R.M.; Hawker, C.J. *Macromolecules* **2000**, *33*, 363.(i) Hawker, C.J.; Bosman, A.W.; Harth, E. *Chem. Rev.* **2001**, *101(12)*, 3661. (j) Harth, E.; Van Horn, B.; Lee, V.Y.; Germack, D.S.; Gonzales, C.P.; Miller, R.D.; Hawker, C.J. *J. Am. Chem. Soc.* **2002**, *124*, 8653.

4 Matyjaszewski, K.; Gaynor, S.; Grestzta, D.; Mardare, D.; Shigemoto, T. *Macromol. Symp.* **1995**, *98(73)*, 83.

5 (a) Nesvadba, P. ; Kramer, A. ; Steinmann, A. ; Stauffer, W. **1999** *PCT Inter. Applic.* WO 99/03894.
(b) Nesvadba, P. ; Kramer, A. ; Steinmann, A. ; Stauffer, W. **2001** *US Patent* US 6,262,206.

6 (a) Grishin, D.F. ; Semyonycheva, L.L. ; Kolyakina, E.V. *Vysokomol. Svedin., Ser. A* **1999**, *41*, 609 [*Polym. Sci. (Engl. Transl.)* **1999**, *41*, 401].
(b) Grishin, D.F.; Semenycheva, L.L.; Kolyakina, E.V. *Russian Journal of Applied Chemistry* **2001**, *74(3)*, 494.

7 (a) Grishin, D.F.; Semyonycheva, L.L.; Kolyakina, E.V. *Mendeleev Commun.* **1999**, 250. (b) Grishin, D.F.; Ignatov, S.K.; Razuvaev, A.G.; Kolyakina, E.V.; Shchepalov, A.A.; Pavlovskaya, M.V.; Semenycheva, L.L. *Polymer Science, Ser. A* **2001**, *43(10)*, 989.

8 Catala, J.-M.; Jousset, S.; Lamps, J.-P. *Macromolecules* **2001**, *34*, 8654.

9 Detrembleur, C.; Teyssie, Ph.; Jerome, R. *Macromolecules* **2002**, *35*, 7214.

10 (a) Vanhoorne, P.; Meyer, R.V.; Detrembleur, C.; Jerome, R. *Eur. Pat. Appl. EP 1236742* **2002**. (b) Detrembleur, C.; Teyssie, Ph.; Jerome, R. *Macromolecules* **2002**, *35*, 1611.

11 (a) Detrembleur, C.; Teyssie, Ph.; Jerome, R. *e-Polymers* **2002**, 004. (b) Detrembleur, C. *Ph.D. Thesis: Developments of new regulators for controlled radical polymerization*, University of Liege **2001**.

12 Shechter, H.; Ley, D.E. *Chem. and Ind.* **1955**, 535.

13 Blanchard, A.A. *Inorg. Syn.* **1946**, *2*, 126.

14 (a) Brown, J.F. *J. Am. Chem. Soc.* **1957**, *79*, 2480. (b) Park, J.S.B.; Walton, J.C. *J. Chem. Soc., Perkin Trans. 2* **1997**, 2579.

15 (a) Lagercrantz, C. *J. Phys. Chem.* **1971**, *75*, 3466. (b) Terabe, S.; Kuruma, K.; Konaka, R. *J. Chem. Soc. Perkin Trans. II* **1973**, 1252.

16 Feigl, F.; Anger, V.; Oesper, R.E. *Spot Tests in organic analysis*, Elsevier Publishing Company, **1966**, 290.

17 (a) Knowles, P.F.; Marsh, D.; Rattle, H.W.E. Magnetic Resonance of Biomolecules, John Wiley & Sons, **1976**, 181-197. (b) Yoshioka, H. J. Colloid and Interface Sci. **1978**, 66, 3527.

RAFT and Other Degenerative Transfer Processes

Chapter 36

Kinetics and Mechanism of RAFT Polymerization

Graeme Moad[1], Roshan T. A. Mayadunne[2], Ezio Rizzardo[1], Melissa Skidmore[2], and San H. Thang[1]

[1]CSIRO Molecular Science and [2]CRC for Polymers, CSIRO Molecular Science Bag 10, Clayton South, Victoria 3169, Australia

RAFT polymerization has emerged as one of the more versatile methods of living radical polymerization. In this paper aspects of the kinetics and mechanism of RAFT polymerization are discussed with a view to pointing out some of the advantages and limitations of various RAFT agents and providing some guidance on how to select a RAFT agent for a particular polymerization. Factors discussed include: transfer constants (C_{tr}, C_{-tr}) of RAFT agents - (measurement, substituent effects, prediction with MO calculations, reversibility), retardation (examples, dependence on RAFT agent and monomer, possible mechanisms) and formation of multimodal distributions (examples, contributing mechanisms).

Over the last 10 years, a considerable effort has been expended to develop free radical processes that display the essential characteristics of living polymerizations *(1-6)*. These radical polymerizations can provide molecular weights that are predetermined by reagent concentrations and conversion, yield narrow molecular weight distributions and, most importantly, polymer products that can be reactivated for chain extension or block copolymer synthesis and enable the construction of complex architectures.

RAFT Polymerization (Radical Polymerization with Reversible Addition-Fragmentation chain Transfer) is one of the most recent entrants and arguably one of the more effective methods in this field *(7-11)*. Some of the advantages of RAFT polymerization, over competing technologies [atom transfer radical polymerization (ATRP) *(5,6)*, nitroxide mediated polymerization (NMP) *(4)*], stem from the fact that it is tolerant of a very wide range of functionality in monomer and solvent (*e.g.* -OH, -COOH, $CONR_2$, $-NR_2$, SO_3Na). This means that it is applicable to a vast range of monomer types and that polymerizations and copolymerizations can be successfully carried out under a wide range of reaction conditions (bulk, solution, emulsion, suspension). The RAFT process has been shown to be effective over a wide temperature range (polymerizations have been successfully performed over the range 20-150°C). Indeed, with some limitations imposed by the need to limit termination reactions, the reaction conditions employed in RAFT polymerization are typical of those used for conventional free radical polymerization.

The RAFT process is extremely versatile. However, it is important to recognize that not all RAFT agents work with equal efficiency in all circumstances. In this paper we consider aspects of the kinetics and mechanism of RAFT polymerization with a view to understanding how to choose RAFT agents and polymerization conditions to maximize living characteristics and minimize side reactions and retardation.

Results and Discussion

The mechanism originally proposed for RAFT polymerization is shown in **Figure 1** *(11)*. In RAFT polymerization, the chain equilibration process is a chain transfer reaction. Radicals are neither formed nor destroyed in this step. In principle, if the RAFT agent behaves as an ideal chain transfer agent, the reaction kinetics should be similar to those of conventional radical polymerization. The rate of polymerization should be half-order in initiator and zero order in RAFT agent. This behavior is observed with, for example, methyl methacrylate (MMA) over a wide range of RAFT agent concentrations *(12)*. However, departures from this ideal are evident with certain RAFT agents, particularly when used at high concentrations *(8)*, and can be pronounced for monomers with high k_p, for example, acrylate esters *(8)* and vinyl acetate *(9)* - see below.

Reaction conditions for RAFT polymerization should usually be chosen such that the fraction of initiator-derived chains is negligible. The degree of polymerization (DP) can then be estimated using the relationship (1) *(12)*.

$$DP \text{ (calc)} \sim [\text{monomer consumed}]/[\text{RAFT agent}] \qquad (1)$$

Positive deviations from equation (1) indicate incomplete usage of RAFT agent. Negative deviations indicate that other sources of polymer chains are significant. These will include initiator-derived chains. With due attention to reaction conditions it is possible to achieve and maintain a high degree of livingness. It is possible to prepare narrow polydispersity block copolymers with undetectable levels of homopolymer impurities *(13)*.

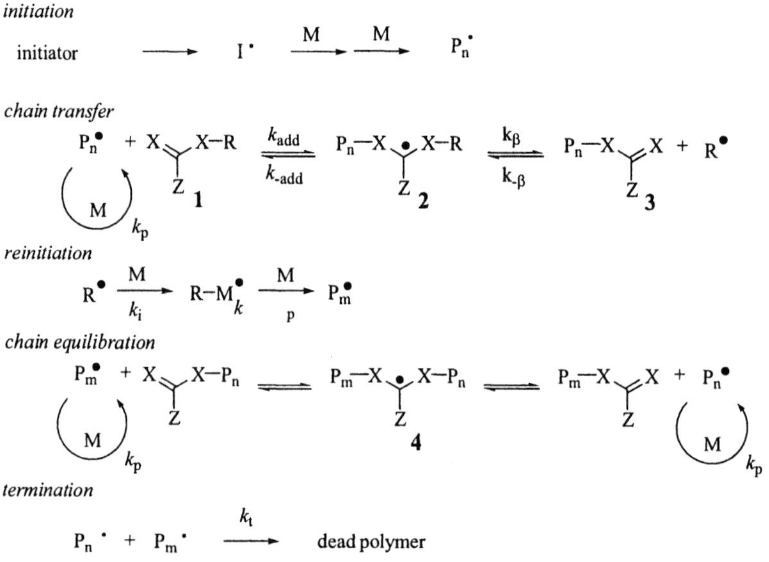

initiation

initiator $\longrightarrow$ $I^{\bullet}$ $\xrightarrow{\quad M \quad M \quad}$ $P_n^{\bullet}$

chain transfer

$P_n^{\bullet}$ + $X \diagdown X{-}R$ (Z) $\underset{k_{-add}}{\overset{k_{add}}{\rightleftharpoons}}$ $P_n{-}X_{\bullet} X{-}R$ (Z) $\mathbf{2}$ $\underset{k_{-\beta}}{\overset{k_{\beta}}{\rightleftharpoons}}$ $P_n{-}X \diagup X$ (Z) $\mathbf{3}$ + $R^{\bullet}$

$\mathbf{1}$, k_p

reinitiation

$R^{\bullet}$ $\xrightarrow[k_i]{M}$ $R{-}M^{\bullet}$ $\xrightarrow[p]{M}$ $P_m^{\bullet}$

chain equilibration

$P_m^{\bullet}$ + $X \diagdown X{-}P_n$ (Z) k_p $\rightleftharpoons$ $P_m{-}X_{\bullet} X{-}P_n$ (Z) $\mathbf{4}$ $\rightleftharpoons$ $P_m{-}X \diagup X$ (Z) + $P_n^{\bullet}$ k_p

termination

$P_n^{\bullet}$ + $P_m^{\bullet}$ $\xrightarrow{k_t}$ dead polymer

Figure 1. Mechanism of RAFT polymerization

Polydispersities also depend on the properties of both the initial (**1**) and the polymeric RAFT agent (**3**). In order to obtain narrow polydispersities, the initial RAFT agent (**1**) and reaction conditions need to be chosen such that **1** is rapidly consumed during the initial stages of the polymerization. We can show that the rate of consumption of **1** depends on two transfer constants as shown in equation (2) *(12,14,15)*.

$$-\frac{d[1]}{d[M]} \approx C_{tr} \frac{[1]}{[M] + C_{tr}[1] + C_{-tr}[3]} \qquad (2)$$

where $C_{tr} = k_{tr}/k_p$, $C_{-tr} = k_{-tr}/k_i$, $k_{tr} = k_{add}[k_\beta/(k_{-add} + k_\beta)]$ and $k_{-tr} = k_{-\beta}[k_{-add}/(k_{-add} + k_\beta)]$. Other parameters are defined in Figure 1. The value of C_{-tr} depends on properties of the radical R• and how it partitions between adding monomer and adding to the polymeric RAFT agent. Depending on the value of C_{-tr}, the rate of consumption of 1 will be slower when high RAFT agent concentrations are used and may reduce with conversion.

The generic features common to all RAFT agents are summarized in Figure 2. The RAFT agent should to be chosen with attention to the particular polymerization process (the monomers and the reaction conditions). Particular design features to take into account are:

• The RAFT agents (1 and 3) should have a high C_{tr} in the monomers being polymerized. This requires a high rate of addition (k_{add}) and a favorable partition coefficient ($k_\beta/(k_\beta + k_{-add})$). The value of k_{add} is determined mainly by X and Z while the partition coefficient depends on the relative leaving group abilities of R• and the propagating radical. For 3 the partition coefficient will be ~ 0.5.

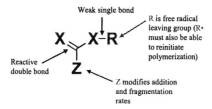

Figure 2. Generic RAFT agent structure

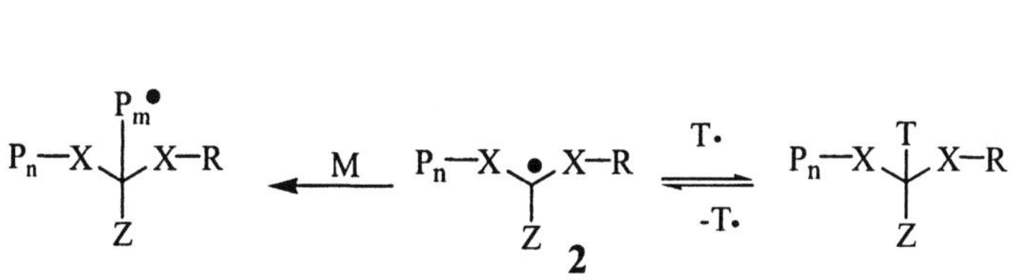

Figure 3. Possible side reactions in RAFT polymerization

- The intermediates (**2** and **4**) should fragment rapidly and give no side reactions such as coupling (where T· might be **2**, **4**, R·, an initiator derived radical, a propagating radical or oxygen - as might be present in poorly degassed samples) or propagation (Figure 3).
- The radical R· should efficiently reinitiate polymerization. This requires that $k_i > k_p$ and it is also desirable that $k_i > k_{-\beta}$. The value of C_{-tr} should be small.

Transfer Constants of RAFT Agents

Transfer constants of RAFT agents have been estimated using the Mayo method *(16,17)* or by fitting the evolution of the molecular weight distribution with conversion *(18,19)*. We have advocated determining transfer constants by analyzing the rate of consumption of RAFT agent with monomer conversion *(8,12,20)* using the integrated form of rate equation. This avoids some of the difficulties associated with measuring high transfer constants. It is often assumed that chain transfer to the initial RAFT agent is irreversible. In this case equation (2) simplifies as follows,

$$-\frac{d[1]}{d[M]} \approx C_{tr}\frac{[1]}{[M]} \qquad \text{which suggests} \quad C_{tr} \sim \frac{d\ln[1]}{d\ln[M]}$$

and a plot of $\ln[1]$ *vs* $\ln[M]$ should provide a straight line with slope C_{tr}.

However, in the case of the more active RAFT agents, we have shown that this assumption is not justified. Transfer constants obtained with disregard of C_{-tr} should be regarded as apparent transfer constants applicable only to the specific reaction conditions. By conducting polymerization for a range of [**1**] and conversions it is possible to use equation (2) to obtain both C_{tr} and C_{-tr} *(12)*.

Another method, which allows estimation of both C_{tr} and C_{-tr} of **1** and is appropriate when the transfer constants are high, is to fit the evolution of the polydispersity with reaction time or conversion. In figure 4 we demonstrate the application of this method to determine C_{tr} and C_{-tr} for benzyl and cumyl dithiobenzoates in styrene polymerization at 60 °C *(14)*. The weight average molecular weight is more sensitive to C_{tr} and C_{-tr} than the number average molecular weight. If it is assumed C_{-tr} (**1**)=0 *(14)* only a poor fit to the experimental data can be obtained for low conversions.

RAFT agents with $X=CH_2$ (*e.g.* methacrylic acid macromonomer **5**) are most suited to the synthesis of methacrylic and similar polymers and copolymers. However, transfer constants are generally <1.0 and very narrow polydispersities can only be achieved using feed addition protocol *(21-23)*.

Carbon-sulfur bonds are weaker than analogous carbon-carbon bonds. The reactivity of $>C=S$ is substantially greater than $>C=CH_2$ towards radical addition. Dithioesters *(11)* and trithiocarbonates *(9,25)* and certain dithiocarbamates (where the nitrogen lone pair is delocalized) *(26,27)* are preferred with (meth)acrylic and styrenic monomers in that their use affords narrow polydispersity polymers in a batch polymerization process. For styrene polymerization, rates of addition decrease (and rate of fragmentation increase) in the series Z is aryl > S-alkyl ~ alkyl ~ *N*-pyrrolo >> OC_6F_5 > *N*-lactam >

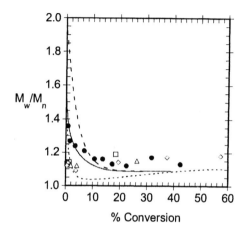

Figure 4 (14). Evolution of polydispersity with conversion for bulk polymerization of styrene at 60°C (0.0017 M AIBN initiator) in the presence of dithiobenzoates PhC(=S)SR, where R = -CH₂Ph (0.0093 M) (•); R = -C(Me)₂Ph (0.0083 M) (✧); -C(Me)₂CO₂C₂H₅ (0.0083 M) (Δ); -C(CH₃)₂CH₂C(CH₃)₃ (0.0083 M) (□). Kinetic simulation with initial RAFT agent C_{tr} = 50 and C_{-tr} = 0 (– – –), C_{tr} = 400 and C_{-tr} = 11600 (———), or C_{tr} = 2000 and C_{-tr} = 10000 (- - - -). C_{tr} (4) = 6000 .(24)

5

6 R'=Ph
7 R'=CN

526

OC_6H_5 > O-alkyl >> N(alkyl)$_2$ *(8,15)*. Only the first four of this series provide narrow polydispersities (M_w/M_n<1.2) in batch polymerization.

We have indicated previously that the relative activities of RAFT agents can be predicted using molecular orbital calculations *(8)*. The calculations indicate that there is a strong correlation of k_{add} in styrene polymerization with calculated heats of reaction, of LUMO energies of the RAFT agent, and of partial charge on the =S (Figure 5). The energy differences are large and, while absolute values differ the same general trends in LUMO energies are seen in the results of higher order *ab initio* and density functional calculations *(15)*. Similar calculations indicate that the influence of the free radical leaving group R on LUMO energy and thus on k_{add} should be comparatively small *(14)*.

Rate constants for fragmentation (k_{-add} and k_β) and for readdition ($k_{-\beta}$ and thus C_{-tr}), however, may be strongly affected by the nature of R *(12,14)*. More sterically hindered, more electrophilic, more stable R add slower and fragment faster. Nucleophilic radicals may prefer to add RAFT agent rather than monomer. A very high C_{tr} explains why cumyl dithiobenzoate (**6**; C_{tr}~56, C_{-tr}~2500), even though it has a two-fold higher C_{tr} in MMA polymerization, may be a less effective RAFT agent than cyanoisopropyl dithiobenzoate (**7**; C_{tr}~25 C_{-tr}~450) *(12,14)*.

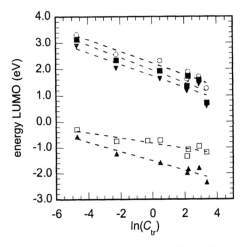

Figure 5 (15). Plot of logarithm of apparent transfer coefficient (of ZC(=S)S-CH$_2$Ph) in MMA polymerization vs calculated LUMO energy for methyl RAFT agents (ZC(=S)S-CH$_3$). Values from AM1 calculations (□), ab initio calculations with Gaussian 98 and 3/21G* (○), 6/31G* (▲), MP2/D95 (▼), B3LYP/6-31G* (■) basis sets.

Side Reactions during RAFT

It has been proposed that the intermediate **2** (or **4**) may react either reversibly or irreversibly with another radical species (T•·) by coupling (Figure 3) or by disproportionation. The reaction provides one explanation for the retardation sometimes observed when high concentrations of RAFT agents are used *(12,19,28,29)*.

There is some evidence that the coupling product may be isolated in styrene polymerization with dithiobenzoate RAFT agents under conditions of high radical flux *(28)* or at low polymerization temperatures with γ–irradiation *(30)*. The intermediate **2** (or **4**) may also react with oxygen. Fragmentation will be facilitated by higher reaction temperatures, low RAFT agent concentrations, and with 1,1-disubstituted monomers.

When the monomer is vinyl acetate, N-vinyl pyrrolidone *(9,27)*, the problem of inhibition can be alleviated by the use of RAFT agents where Z is such as to give a less stable intermediate. Dithioesters and trithiocarbonates generally give inhibition. Xanthates and dithiocarbamates are preferred.

In general, faster fragmentation means shorter lifetimes for the intermediates, less side reactions and less retardation. Faster fragmentation is also correlated with slower addition and a lower transfer constant. Thus an appropriate balance needs to be achieved between these requirements.

Another side reaction is the that of the initial RAFT agent with expelled radicals R•, initiator-derived radicals (I•) or other radicals formed during polymerization to give a new adduct species. These may fragment to give new RAFT agents or react further by the pathways already discussed. If I• (or other radical) is a poor free radical leaving group the adduct may be relatively stable (increasing the likelihood of coupling or other reactions) and the derived RAFT agent relatively inert.

For X=CH$_2$, the species **2** may add to monomers - *i.e.* the RAFT agent may copolymerize to give a graft copolymer *(22)* (see Figure 3). The reaction is currently unknown when X=S. In the former case the reaction is disfavored at high reaction temperatures and with 1,1-disubstituted monomers (*e.g.* methacrylates) *(22)*.

Retardation Mechanisms

Experimental Findings.

Rates of polymerization of acrylate esters are significantly retarded in the presence of dithiobenzoate RAFT agents. We have shown that for acrylate polymerization aliphatic dithioesters (*e.g.* dithioacetate) and trithiocarbonates

528

give less retardation *(8,12)*. It has recently been reported that dithiophenylacetate RAFT agents enable polymerization of acrylates at ambient temperature whereas cumyl dithiobenzoate gives inhibition *(31)*. The form of retardation is illustrated in Figure 6 with conversion time profiles for polymerizations of methyl acrylate in benzene solution at 60 °C with benzyl dithiobenzoate, cyanoisopropyl dithiobenzoate and benzyl dithioacetate as RAFT agents. The rate of polymerization in the presence of dithiobenzoate derivatives does not depend on R and is strongly retarded with respect to that observed in the absence of RAFT agent. While the dithioacetate also slows the rate of polymerization, the extent of retardation is substantially less than that observed with the dithiobenzoates, even when a 10-fold higher RAFT agent concentration is used. All polymerizations provide narrow polydispersity products. All polymerizations ultimately give high conversions.

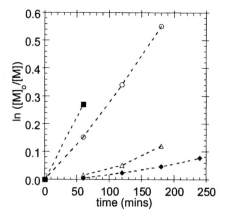

Figure 6. Pseudo first order rate plot for bulk polymerization of methyl acrylate (4.45 M in benzene) at 60°C with ~3.3 x10⁻⁴ M AIBN in the absence (■) or presence of MeC(=S)CH₂Ph (0.00306 M) (○); MeC(=S)SCH₂Ph (0.0306 M) (Δ); PhC(=S)SCH₂Ph or PhC(=S)SC(Me)₂CN (0.00366 M) (♦).

There is some controversy regarding retardation of styrene polymerization. We have reported that styrene polymerization is strongly retarded by high concentrations of cumyl dithiobenzoate (**6**). We have also shown that the retardation is alleviated with use of a dithiobenzoate RAFT agent with a different R group (*e.g.* cyanoisopropyl dithiobenzoate **7**) or by use of a RAFT agent with a different Z group (*e.g.* cyanoisopropyl dithioacetate). Both changes would be expected to make fragmentation of adduct (**1**) more facile. We also reported that retardation is small with lower concentrations of cumyl dithiobenzoate (bulk, 0.003 M, 60°C). Under these conditions, the rate of polymerization is reduced, but is consistent with that expected in the absence of a gel effect. Monteiro *et al.(29)* and Kwak *et al.(28)* have also reported on retardation with **6** but only provide data for high concentrations. Barner-Kowollik *et al.(18)* report that there is strong retardation even in the presence of low concentration of dithiobenzoate esters and irrespective of R.

With methacrylates we find behavior that is qualitatively similar to that with styrene. Severe retardation is observed when using high concentrations of, in particular, cumyl dithiobenzoate (**6**)*(12)*. Little retardation is observed with lower concentrations of **6**, with other dithiobenzoates (*e.g.* cyanoisopropyl dithiobenzoate **7**), or with other RAFT agents (dithioacetates, trithiocarbonates).

A variety of factors may cause the rate of polymerization in the presence of RAFT agents to be less than that seen in a conventional polymerization under the same reaction conditions. These are discussed below.

Reduced Gel Effect

It is known that the magnitude of the gel effect in radical polymerization depends strongly on molecular weight and molecular weight distribution *(32)* and typically is less in polymerizations providing lower molecular weight polymers. We have found that for bulk MMA and styrene polymerization in the presence of low concentrations (<0.003 M) of cumyl dithiobenzoate, while the rate of polymerization is lower than that seen in the absence of RAFT agent, it is similar to or greater than that predicted by kinetic simulation assuming a continuation of low conversion kinetics (*i.e.* a k_t that is independent of conversion). This is consistent with the slower rate of polymerization under these conditions being associated with a substantially reduced gel effect.

Slow Fragmentation

A fraction of radicals in RAFT polymerization are present as the adducts **2** and **4**. ESR experiments show that for polymerizations of acrylates and styrenes in the presence of dithiobenzoate RAFT agents these species are present in concentrations more than an order of magnitude higher than the total concentration of propagating species *(28,33)*. It has been suggested that slow fragmentation in itself might be responsible for retardation observed in styrene polymerization with dithiobenzoate RAFT agents *(8,12,18,19)*. Kinetic simulation of styrene polymerization shows that the effect of slow fragmentation to give radical concentrations of the magnitude observed by ESR on the rate of polymerization is very small and only discernable as a short inhibition period. A steady state is quickly established such that the concentration of propagating radicals is only slightly reduced. We conclude that slow fragmentation, by itself, cannot account for retardation in this system.

Reaction of Adducts 2 or 4 with Other Radicals

The reaction of the adducts (*e.g.* **2** and/or **4**) with propagating, initiator or RAFT agent derived radicals by combination (or disproportionation) has already been mentioned. Monteiro *et al.(29)* and Kwak *et al.(28)* have shown by kinetic simulation that this reaction, if it occurs, will cause retardation. One can also envisage that, dependent on the radical species involved, combination could be reversible *(12,19)*. There is no definitive evidence that this process occurs or is significant for the reaction conditions discussed above (styrene, bulk, 60 °C).

Reaction of Adducts 2 or 4 with Oxygen

We have observed that RAFT polymerization with dithiobenzoates appears to be more oxygen sensitive than conventional polymerization. This is particularly noticeable in the polymerization of acrylate esters. This may be a consequence of adducts (*e.g.* 2 and/or 4) being present in high concentration relative to the propagating radicals. These species are likely to react with oxygen at diffusion-controlled rates and their consumption will cause retardation. For successful and reproducible RAFT polymerization, it is essential to efficiently degas the reaction media. Oxygen sensitivity is reduced with RAFT agents (aliphatic dithioesters, trithiocarbonates) which give less stable adducts.

Multimodal Molecular Weight Distributions from RAFT Polymerization

Bimodal or multimodal molecular weight distributions are sometimes observed in RAFT polymerization. In some cases, these are easily rationalized in terms of by-products by radical-radical termination involving the propagating species.

This issue of dead chain is more important in star polymer synthesis where the size of higher molecular weight peaks increases according to the number of arms (Figure 7). It should be pointed out that the problem of star-star coupling can be avoided by selection of RAFT agents such as 9 where the radical center is never attached to the core of the star *(8,34)*.

In polymerization of acrylic monomers to high conversion bimodal distributions have been reported *(9,35)*. In Figure 8 we illustrate this with GPC traces for a polymerization of methyl acrylate to high conversion. In this case, the amount of by-product is too large to be readily explained by radical-radical coupling reactions. Analysis by GPC with UV detection shows that the higher molecular weight polymer is substantially alive (it retains the thiocarbonylthio chromophore - Figure 8). Note that at 304 nm there is no poly(methyl acrylate) absorption and only chains with the dithioacetate end groups are observed. The high molecular weight shoulder appears smaller because intensity is proportional to M_n (vs. M_n^2 for the refractive index trace). Similar findings as regards bimodal peaks have been reported by McCormick *et al.* for N,N-dimethylacrylamide polymers *(36)*. Extended polymerization time does not result in any marked change in the molecular weight distribution though the formation of some oligomeric products containing the dithioacetate chromophore is evident in the UV trace.

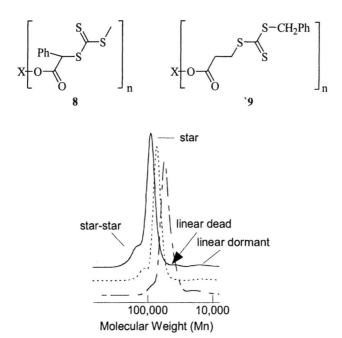

Figure 7. Molecular weight distributions after 16 hours during synthesis of star polymers by thermal polymerization of styrene at 110 °C in the presence of RAFT agents 8 (34). The concentration of RAFT agent was 0.00296/(no. arms). From top to bottom are: 8 arm M_n 114000, PD 1.07, 52 % Conv.; 6 arm M_n 92000, PD 1.04, 50 % Conv; 3 arm M_n 55000, PD 1.11, 59 % Conv. Molecular weights are absolute molecular weights based on the use of a multi-angle light scattering detector.

532

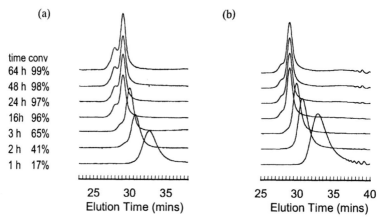

(a)   (b)

time conv
64 h 99%
48 h 98%
24 h 97%
16h   96%
3 h   65%
2 h   41%
1 h   17%

25   30   35   25   30   35   40
Elution Time (mins)   Elution Time (mins)

Figure 8. GPC traces at various reaction times/conversions for polymerization of methyl acrylate (4.43 M in benzene, 0.0033 M AIBN) in presence of benzyl dithioacetate(0.0306 M) (a) RI detection and (b) UV detection at 304 nm.

We have previously reported that multimodal peaks in acrylate polymerization may be eliminated or reduced by choice of RAFT agent *(9)*. Figure 9 shows molecular weight distributions for poly(methyl acrylate) prepared at high conversion with three different RAFT agents **(10-12)** under similar experimental conditions.

10   **11**   **12**

Polymerization with RAFT agent **(10)** gives a monomodal distribution with a narrow molecular weight distribution (M_w/M_n =1.19) The RAFT agent **(11)** gives a narrower distribution (M_w/M_n =1.08) and a small shoulder is apparent on the high molecular weight side of the distribution. The trithiocarbonate RAFT agent **(12)** gives a similar polydispersity (also M_w/M_n =1.08) but a peak, which is distinctly bimodal. The differences in polydispersity reflect the transfer constants of the **(10-12)** and the corresponding polymeric RAFT agents. In the case of **10** and **11** the distribution is sufficiently broad to completely or partially hide the high molecular weight peak. A polydispersity of 1.2, while narrow, is sufficient to obscure bimodality.

The origin of the high molecular weight peak in these and similar polymerizations has not been fully elucidated. For the examples shown, the high molecular weight peaks are too large to be fully explained by radical coupling processes involving propagating species and/or the adducts **2** or **4**. The finding that the higher molecular weight peak retains the thiocarbonylthio chromophore also argues against this. The size of the peak does depend on the molecular weight of the polymer (more important with higher molecular weight polymers, $M_n > 50000$). It does depend on conversion (usually only observed for >50% conversion).   One further process that may lead to multimodal

533

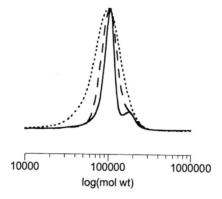

```
10000          100000        1000000
            log(mol wt)
```

Figure 9. GPC traces of high conversion poly (methyl acrylate) prepared in the presence of various RAFT agents. Molecular weights distributions shifted to correct for differences in conversion to facilitate comparison (X axis correct for sample made with 12). Samples prepared with 10 (0.0019 M) M_n 87000, M_w/M_n 1.19, 72% conv. (······); 11 (0.0036M) M_n 110100, M_w/M_n 1.08, 89% conv. (- - - -); 12 (0.0037 M) M_n 123700, M_w/M_n 1.08, 92% conv. (——). Molecular weights are in polystyrene equivalents. Initiator AIBN (0.00033 M).

distributions is long chain branching caused by intermolecular transfer to polymer. In this case, such bimodal distributions should not only be independent of RAFT agent they should also be observed in other living radical polymerizations (ATRP, NMP) carried to high conversion. Bimodal distributions have been reported (37) for high conversion, high molecular weight products from ATRP of methyl acrylate but were attributed to radical-radical termination.

Conclusions

RAFT Polymerization is a versatile method for providing narrow polydispersity polymers of controlled architecture. With attention to kinetics and mechanism, it is possible to select of RAFT agents and reaction conditions to achieve optimal results with most monomers.

Acknowledgement

We are grateful to DuPont Performance Coatings for supporting this work.

References

1. Matyjaszewski, K. *ACS Symp. Ser.* **1998**, *685*, 2-30.
2. Matyjaszewski, K. *ACS Symp. Ser.* **2000**, *768*, 2-26.

534

3. Moad, G.; Solomon, D.H. *The Chemistry of Free Radical Polymerization* Pergamon: Oxford, 1995.
4. Hawker, C.J.; Bosman, A.W.; Harth, E. *Chem. Rev.* **2001**, *101*, 3661-88.
5. Kamigaito, M.; Ando, T.; Sawamoto, M. *Chem. Rev.* **2001**, *101*, 3689-745.
6. Matyjaszewski, K.; Xia, J. *Chem. Rev.* **2001**,*101*, 2921-90.
7. Rizzardo, E.; Chiefari, J.; Chong, Y.K.; Ercole, F.; Krstina, J.; Jeffery, J.; Le, T.P.T.; Mayadunne, R.T.A.; Meijs, G.F.; Moad, C.L.; Moad, G.; Thang, S.H. *Macromol. Symp.* **1999**, *143*, 291-307.
8. Moad, G.; Chiefari, J.; Krstina, J.; Postma, A.; Mayadunne, R.T.A.; Rizzardo, E.; Thang, S.H. *Polym. Int.* **2000**,*49*, 933-1001.
9. Rizzardo, E.; Chiefari, J.; Mayadunne, R.T.A.; Moad, G.; Thang, S.H. *ACS Symp. Ser.* **2000**,*768*, 278-96.
10. Le, T.P.; Moad, G.; Rizzardo, E.; Thang, S.H. Int. Patent Appl. WO 9801478 (*Chem. Abs.* (1997) 128: 115390)
11. Chiefari, J.; Chong, Y.K.; Ercole, F.; Krstina, J.; Jeffery, J.; Le, T.P.T.; Mayadunne, R.T.A.; Meijs, G.F.; Moad, C.L.; Moad, G.; Rizzardo, E.; Thang, S.H. *Macromolecules* **1998**, *31*, 5559-62.
12. Moad, G.; Chiefari, J.; Moad, C.L.; Postma, A.; Mayadunne, R.T.A.; Rizzardo, E.; Thang, S.H. *Macromol. Symp.* **2002**, *182*, 65-80.
13. Pasch, H.; Mequanint, K.; Adrian, J. *e-Polymers* **2002**,*00 5*,
14. Chong, Y.K.; Krstina, J.; Le, T.P.T.; Moad, G.; Rizzardo, E.; Thang, S.H. *Macromolecules* **2003**, in press.
15. Chiefari, J.; Mayadunne, R.T.A.; Moad, C.L.; Moad, G.; Postma, A.; Rizzardo, E.; Thang, S.H. *Macromolecules* **2003**, in press.
16. Ladaviere, C.; Doerr, N.; Claverie, J.P. *Macromolecules*, **2001**,*34*, 5370-2.
17. Moad, G.; Moad, C.L.; Rizzardo, E.; Thang, S.H. *Macromolecules* **1996**, *29*, 7717-26.
18. Barner-Kowollik, C.; Quinn, J.F.; Morsley, D.R.; Davis, T.P. *J. Polym. Sci., Part A: Polym. Chem.* **2001**,*39*, 1353-65.
19. Barner-Kowollik, C.; Quinn, J.F.; Nguyen, T.L.U.; Heuts, J.P.A.; Davis, T.P. *Macromolecules* **2001**, *34*, 7849-57.
20. Goto, A.; Sato, K.; Tsujii, Y.; Fukuda, T.; Moad, G.; Rizzardo, E.; Thang, S.H. *Macromolecules* **2001**, *34*, 402-8.
21. Moad, G.; Ercole, F.; Johnson, C.H.; Krstina, J.; Moad, C.L.; Rizzardo, E.; Spurling, T.H.; Thang, S.H.; Anderson, A.G. *ACS Symp. Ser* **1998**,*685*, 332-60.
22. Krstina, J.; Moad, C.L.; Moad, G.; Rizzardo, E.; Berge, C.T.; Fryd, M. *Macromol. Symp.* **1996**,*111*, 13-23.
23. Krstina, J.; Moad, G.; Rizzardo, E.; Winzor, C.L.; Berge, C.T.; Fryd, M. *Macromolecules* **1995**, *28*, 5381-5.
24. Goto, A.; Sato, K.; Fukuda, T.; Moad, G.; Rizzardo, E.; Thang, S.H. *Polym. Prepr.* **1999**,*40(2)*, 397-8.
25. Mayadunne, R.T.A.; Rizzardo, E.; Chiefari, J.; Krstina, J.; Moad, G.; Postma, A.; Thang, S.H. *Macromolecules* **2000**,*33*, 243-5.
26. Mayadunne, R.T.A.; Rizzardo, E.; Chiefari, J.; Chong, Y.K.; Moad, G.; Thang, S.H. *Macromolecules* **1999**, *32*, 6977-80.
27. Destarac, M.; Charmot, D.; Franck, X.; Zard, S.Z. *Macromol. Rapid Commun.* **2000**,*21*, 1035-9.
28. Kwak, Y.; Goto, A.; Tsujii, Y.; Murata, Y.; Komatsu, K.; Fukuda, T. *Macromolecules* **2002**, *38*, 3026-9.
29. Monteiro, M.J.; de Brouwer, H. *Macromolecules* **2001**, *34*, 349-52.

30. Barner-Kowollik, C.; Vana, P.; Quinn, J.F.; Davis, T.P. *J. Polym. Sci., Part A: Polym. Chem.* **2002**, *40*, 1058-63.
31. Quinn, J.F.; Rizzardo, E.; Davis, T.P. *Chem. Commun.* **2001**, 1044-5.
32. O'Neil, G.A.; Wisnudel, M.B.; Torkelson, J.M. *Aiche Journal* **1998**, *44*, 1226-31.
33. Hawthorne, D.G.; Moad, G.; Rizzardo, E.; Thang, S.H. *Macromolecules* **1999**,*32* , 5457-9.
34. Mayadunne, R.A.; Moad, G.; Rizzardo, E. *Tetrahedron Lett.* **2002**, *43*, 6811-4.
35. Donovan, M.S.; Sanford, T.A.; Lowe, A.B.; Sumerlin, B.S.; Mitsukami, Y.; McCormick, C.L. *Macromolecules* **2002**, *35*, 4123-32.
36. Donovan, M.S.; Lowe, A.B.; Sumerlin, B.S.; McCormick, C.L. *Macromolecules* **2002**, *35*, 4123-32.
37. Davis, K.A.; Paik, H.J.; Matyjaszewski, K. *Macromolecules* **1999**,*32* , 1767-76.

Chapter 37

On the Importance of Xanthate Substituents in the MADIX Process

Mathias Destarac[1,*], Daniel Taton[1,2], Samir Z. Zard[3,*], Twana Saleh[3], and Yvan Six[4]

[1]Rhodia Recherches, Centre de Recherches d'Aubervilliers, 52, rue de la Haie Coq, 93308 Aubervilliers Cedex, France
[2]Laboratoire de Chimie des Polymères Organiques, ENSCPB, 16, Avenue Pey Berland, 33 607 Pessac Cedex, France
[3]Département de Synthèse Organique, Ecole Polytechnique, 91128 Palaiseau Cedex, France
[4]Institut de Chimie des Substances Naturelles, CNRS, 91198 Gif sur Yvette Cedex, France

The structural effect of miscellaneous xanthates RS(C=S)OZ' on the level of control of free radical polymerizations was investigated. Such polymerizations referred as to the MADIX process involve the use of xanthates as reversible chain transfer agents (CTAs). In the case of S-alkyl-O-ethyl xanthates, RS(C=S)OEt, the process involves slow degenerative transfer of xanthate end-groups between polymer chains. This leads to styrene and acrylate-derived polymers with predetermined molar masses at high conversion and polydispersity index (PDI) between 1.5 and 2. In this series, the more substituted and stabilized the R leaving group, the better the control over molar masses as a function of the monomer conversion. Excellent control can be achieved in the polymerization of vinyl esters, acrylic acid and acrylamide. The reactivity of the C=S double bond of xanthates can be dramatically increased towards polystyryl and polyacrylyl radicals by incorporating electron-withdrawing substituents in the activating Z=OZ' group of S-(1-ethoxycarbonyl)ethyl-O-alkyl xanthates, $C_2H_5OCO(CH_3)CH$-$S(C=S)OZ'$. This allows the preparation of polymers with controlled molar masses and PDI close to unity.

An important development in the area of controlled/living radical polymerization is based on the use of thiocarbonyl thio compounds, Z-C(=S)SR, such as dithioesters, dithiocarbamates, trithiocarbonates or xanthates as reversible chain transfer agents (CTA) in two independently discovered processes. These have been designated Reversible Addition Fragmentation Chain Transfer (the RAFT process) (*1*) and Macromolecular Design *via* Interchange of Xanthates (the MADIX process) (*2*). Both MADIX and RAFT follow a reaction mechanism that was reported in 1988 for the formation of 1:1 adducts (*3*). Although at the time such radical additions were performed with the more readily accessible xanthates, their obvious extension to other related classes was noted. The addition reaction is most simply accomplished by heating an *O*-ethyl xanthate RS(C=S)OEt and an alkene in the presence of a peroxide as initiator. Of particular interest in such a process is the great variety of functional groups and complex substrates which are tolerated (*4*).

The concept of xanthate group transfer radical addition has been successfully extrapolated by the Rhodia group to a new C/LRP system (*2,5*). At that time, the discovery of the RAFT process by the CSIRO with dithioesters as controlling agents was still kept under secret (*1,6*) and the acronym *MADIX* was proposed for this new fashion of controlling free radical polymerizations involving xanthates. Since then, significant developments have been directed towards RAFT and MADIX processes, including the use of new CTAs such as dithiocarbamates (*7,8*) or trithiocarbonates (*9*), kinetical and mechanistic investigations (*1,10-15*), direct synthesis of water-soluble materials (*16-18*), or applications of RAFT and MADIX to dispersed media (*1,2,19,20*).

From a mechanistic viewpoint, the MADIX and RAFT processes are thus strictly identical and eventually only differ by the nature of the CTA: the RAFT terminology prevails for CTAs Z-C(=S)-S-R in general whereas MADIX refers to xanthates exclusively (with Z = OZ', see Scheme 1) (*1,2*). A fast equilibration relative to the rate of propagation is needed for an optimal control of the polymerization. This can be achieved through variation of the Z and/or the R groups which dramatically affect the chain transfer constant ($C_{tr} = k_{tr}/k_p$) (*1*).

In this contribution, we wish to demonstrate how both R and Z = OZ' groups of xanthates can be finely tuned to reach optimal control of the polymerization. Miscellaneous xanthates have thus been designed and tested in the radical polymerization of various monomers.

Experimental

Materials. Monomers were distilled over CaH_2 under vacuum prior to use. The synthesis of X_1, X_2, X_4, and X_6 to X_{10} is described in ref. *2* and *15*. Xanthates X_3 and X_5 were synthesized following a procedure similar to that employed for X_4, except that 4-cyanobenzaldehyde and 4-methoxybenzaldehyde were used as starting materials, respectively. Xanthates X_{11}, X_{12}, X_{14}, X_{15} **and** X_{16} were.prepared by essentially the same procedure to that detailed below for xanthate X_{17}.

Transfer to xanthate (X) $C_{tr}(X)=(k_a/k_p)[k_\beta/(k_{-a}+k_\beta)]=k_{tr}/k_p$

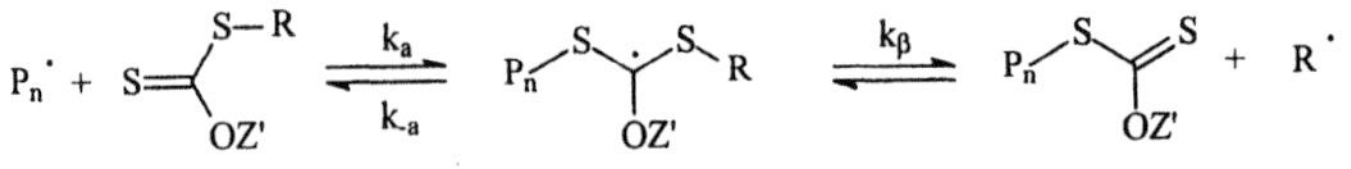

Chain-to-chain transfer (P_nX) $C_{tr}(P_nX)=k_a/2k_p$

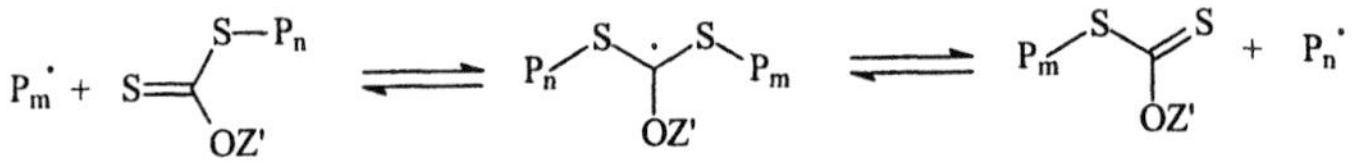

Scheme 1. Mechanism of the MADIX process

Synthesis of xanthate X_{17}.
A 75% aqueous solution of fluoral hydrate (10.0g, 64.6 mmol) and diethylphosphite (64.6 mmol) in triethylamine (9.0 mL, 64.6 mmol) was stirred at room temperature for 15 hrs. The mixture was concentrated and the residue was purified by chromatography (petroleum ether /acétone 10 : 1 then ether, then ether/methanol 10 :1) to give diethyl 2,2,2-trifluoro-1-hydroxy-ethylphosphonate in 82% yield. To an ice-cooled solution of this alcohol (1.44g; 6.1 mmol) in DMF (12 mL) was added carbon disulfide (1.5 ml; 24.4 mmol) followed by the portion-wise addition of sodium hydride as a 60% suspension in oil (244 mg; 6.1 mmol). Stirring was continued for a further 15 minutes and ethyl 2-bromopropionate (0.87 ml; 6.7 mmol) was added, and the resulting mixture was stirred for one hour at 0°C then one hour at room temperature. The red-colored solution was poured into a 1:1 mixture of ether and a saturated aqueous solution of ammonium chloride (200 ml). The organic layer was separated and the aqueous layer further extracted with ether (2 x 200 ml). The combined organic phases were dried over magnesium sulfate and concentrated under vacuum. Purification by chromatography over silica gel (ethyl acetate: petroleum ether 1:4) afforded the desired xanthate X_{17} as a 1:1 mixture of two diastereoisomers (690 mg; 58%). ^{1}H NMR (CDCl$_3$) δ 4.03-4.37 (m, 7H), 1.75 (d, J=6.8 Hz, 3H), 1.20-1.56 (m, 9H) ;^{13}CMR (50 MHz, CDCl$_3$) δ 198.0, 170.3, 64.8, 64.7, 61.9, 40.2, 28.4, 15.9, 15.8, 13.8.
Polymerizations. All polymerizations were carried out in sealed tubes after degassing by three pump-freeze-thaw cycles. Transfer constants (C_{tr}) were determined using the Mayo method (conversion < 5 %). Conversion was determined by gravimetry.
Characterization. Molar masses and polydispersities were measured by size exclusion chromatography, using Phenogel columns: Guard, linear, 1000 Å and 100 Å (eluent: THF (1 mL.mn^{-1}). M_n values were calculated based on PS standards.

Results and Discussion

Xanthates can be synthesized from well established organic synthetic routes (*3,4*). Some of the *O*-alkyl xanthates described here (Figure 1) are entirely new compounds and have been purposely designed for this work.

Structural Effect of MADIX Agents Z'OC(=S)SR

The effect of xanthates containing the same Z = OEt activating group (X_1 to X_{10}) was first investigated for the polymerization of styrene and ethyl acrylate. These results are summarized in Tables 1 to 3 and Figure 1; complementary data can be found in ref. *15*. In this first series of experiments, it is assumed that the rate constant of addition k_a (Scheme 1) is mainly influenced by the Z = OEt group. Therefore, the results obtained are directly related to the leaving R group ability of these xanthates, that is to the probability ($k_\beta/(k_{-a}+k_\beta)$) for the transient radical to undergo a β-scission (fragmentation step). At low monomer and xanthate conversion, the transfer constant to xanthate, $C_{tr}(X)$, is calculated by the Mayo equation (*1,15*)

$$(DP_n)_0=[M]_0/(C_{tr}(X)*[X]_0) \qquad (1)$$

where $[M]_0$ and $[X]_0$ are the initial concentration of monomer and xanthate respectively. Before the entire consumption of X, $DP_n > \Delta[M]_t/[X]_0$. Consequently, a non linear evolution of molar masses (M_n) with monomer conversion is observed, though experimental M_n values eventually match the theoretical ones at high conversion. In other words, the higher the transfer constant, the closer the M_n vs monomer conversion profile to the theoretical straight line which corresponds to an instantaneous xanthate consumption. Generally speaking, these features apply for "living polymerization processes exhibiting slow equilibria" (*21,22*). In these polymerizations also, the polydispersity index (PDI) of the polymers is generally much higher than unity (PDI around 2) (*21*).

The R group of *O*-ethyl xanthates has a marked influence on the M_n evolution profile due to a change of the C_{tr} value (see Tables 1 and 2). For instance, the M_n value obtained at high conversion with X_1 is slightly higher than that predicted by [Styrene]/[X_1] feed ratio. This is ascribed to a slow and incomplete consumption of X_1 over the course of the polymerization. This is supported by the C_{tr} value of X_1 lower than unity ($C_{tr}(X_1)$ = 0.89). The chain transfer activity increases in the following order: $X_1 \sim X_2 < X_3 < X_4 \sim X_6 < X_5 \sim X_7 < X_8 < X_9 < X_{10}$. It turns out that incorporation of electron-withdrawing groups increases xanthate reactivity: $C_{tr}(X_6)$ = 1.65 > $C_{tr}(X_2)$ = 0.82 and $C_{tr}(X_7)$ = 2 > $C_{tr}(X_1)$ = 0.89. Also, variation of the electron density of the 1,3-diphenyl-3-oxo-propyl group has a slight effect on the C_{tr}. Indeed, the cyano group of X_5 increases the reactivity as opposed to the methoxy group of X_3: $C_{tr}(X_5)$ = 2 > $C_{tr}(X_4)$ = 1.65 > $C_{tr}(X_3)$ = 1.

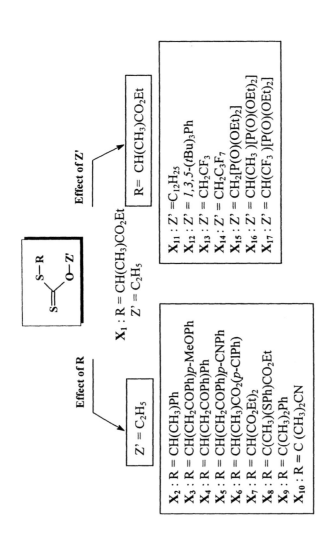

Effect of R

Effect of Z'

X_1 : R = CH(CH$_3$)CO$_2$Et
      Z' = C$_2$H$_5$

Z' = C$_2$H$_5$

R = CH(CH$_3$)CO$_2$Et

X_2 : R = CH(CH$_3$)Ph
X_3 : R = CH(CH$_2$COPh)p-MeOPh
X_4 : R = CH(CH$_2$COPh)Ph
X_5 : R = CH(CH$_2$COPh)p-CNPh
X_6 : R = CH(CH$_3$)CO$_2$(p-ClPh)
X_7 : R = CH(CO$_2$Et)$_2$
X_8 : R = C(CH$_3$)(SPh)CO$_2$Et
X_9 : R = C(CH$_3$)$_2$Ph
X_{10} : R = C (CH$_3$)$_2$CN

X_{11} : Z' = C$_{12}$H$_{25}$
X_{12} : Z' = *1,3,5-(tBu)$_3$Ph*
X_{13} : Z' = CH$_2$CF$_3$
X_{14} : Z' = CH$_2$C$_3$F$_7$
X_{15} : Z' = CH$_2$[P(O)(OEt)$_2$]
X_{16} : Z' = CH(CH$_3$)[P(O)(OEt)$_2$]
X_{17} : Z' = CH(CF$_3$)[P(O)(OEt)$_2$]

Figure 1. Series of xanthates studied in this work.

Table 1. Polymerization of Styrene in the Presence of Xanthates.
$[St]_0/[X]_0=80.$ T=110 °C.[a]

Entry	X	t (h)	$M_{n\ th}$[b]	M_n[c]	M_w/M_n	Conversion[d] (%)
1	X_1	2	1050	8000	2.03	10.2
2	X_1	90	6550	8150	2.01	76.3
3	X_3	0.5	630	8100	2.04	3.2
4	X_4	0.5	650	6050	2.06	3.8
5	X_5	0.5	530	4650	2.07	2.1
6	X_{10}	1	1050	1700	1.85	10.2
7	X_{10}	48	7600	7600	2.41	88.9
8	X_{13}	2	900	2650	1.93	7.8
9	X_{13}	90	5750	5600	1.57	65.8
10	X_{15}	1	1150	5250	1.98	9.8
11	X_{15}	24	6750	6850	1.90	77
12	X_{16}	1	950	6500	2.05	7.2
13	X_{17}	5	900	1100	1.10	7.4
14	X_{17}	116	6100	6200	1.15	73.7

[a] Entries 1 to 7: Bulk polymerization. Entries 8 to 14: Polymerizations in 50% toluene.
[b] $M_{n\ th}$=($[M]_0/[Xanthate]_0$)*(monomer conversion)*$(M_W)_{Monomer}$ + M_W of the xanthate (assuming a complete consumption of X and a negligible contribution of the initiator-derived chains).
[c] measured by SEC in THF. Molar masses measurements were based on PS standards.
[d] conversion was determined gravimetrically.

542

Table 2. Polymerization of Ethyl Acrylate (EA) in the Presence of
Xanthates. $[EA]_0$=4.6 M, $[X]_0$=5.75 .10^{-2}M, $[AIBN]_0$=1.72.10^{-3} M. T=80 °C.
Solvent: Toluene.

Entry	X	t (h)	$M_{n\ th}$ [a]	M_n [b]	M_w/M_n	Conversion[c] (%)
1	X_1	0.17	1150	5700	1.81	11.6
2	X_1	3.5	7940	7290	1.77	96.5
3	X_9	1	2160	1990	2.11	24
4	X_9	5.5	8200	7500	1.57	98.8
5	X_{10}	0.5	1100	1400	2.05	11.4
6	X_{10}	4	7320	8100	1.53	89.1
7	X_{11}	0.4	1450	5640	1.76	13.7
8	X_{11}	2	7760	7320	1.73	92.5
9	X_{12}	1.5	4530	60500	1.63	51
10	X_{13}	0.33	1150	2950	1.90	11.0
11	X_{13}	2.33	7650	7950	1.42	91.8
12	X_{14}	1.5	8400	7100	1.41	81.8
13	X_{17}	0.25	1100	900	1.26	8.6
14	X_{17}	0.42	4400	4400	1.12	49.6
15	X_{17}	1	7000	6900	1.14	82.6

[a] $M_{n\ th}$=$([M]_0/[Xanthate]_0)$* (monomer conversion)* $(M_W)_{Monomer}$+ M_W of the xanthate (assuming a complete consumption of X and a negligible contribution of the initiator-derived chains)

[b] measured by GPC in THF. Molar mass measurements were based on PS standards.

[c] conversion was determined gravimetrically

Finally, O-ethyl xanthates with tertiary leaving groups further improve the control, the cyanoisopropyl group of X_{10} proving the best leaving group in this series: $C_{tr}(X_8)$ = 3 < $C_{tr}(X_9)$ = 3.8 < $C_{tr}(X_{10})$ = 6.8. These results are consistent with the findings for dithioesters employed in the RAFT process: the more substituted and stabilized the R° leaving group, the higher the transfer constant (1). Noteworthy, an excellent correlation between the experimental M_n evolution profiles and those predicted by Eq (1) is observed with X_1 to X_{10}, taking into account the C_{tr} values determined by the Mayo method (15).

On the other hand, little influence of the R group on the molar mass distributions is observed: PDIs are typical of those obtained in a xanthate-free polymerization (1.9-2.4). This is ascribed to a slow interchange of the xanthate end-groups between polymer chains, that is to a low $C_{tr}(P_nX)$. Catala and coll. (23) have recently calculated the chain transfer constant value of a S-polystyryl-O-ethyl xanthate and found $C_{tr}(P_nX) = 0.8$, that is a very close value to $C_{tr}(X_2)$ which is equal to 0.82 (15). This verifies, in this particular case, that the phenylethyl group exhibits the same leaving ability as the polystyryl chains. The observation of PDIs around 2 are consistent with the statement by Müller and coll. who have predicted that the PDI of "living" polymerization involving slow equilibria evolves according to the following formula (21).

$$PDI = 1 + 1/C_{tr}(P_nX) \qquad (2)$$

Despite the rather high PDI values, excellent control of chain structures is achieved, as evidenced by NMR analysis and MALDI-TOF mass spectroscopy (15): no chains derived from thermally generated radicals have been detected. This high "end group fidelity" eventually allowed block copolymer synthesis from xanthate-capped homopolymers serving as macro-CTA in chain extension experiments (2,5).

Importantly, the nature of the R group has no significant influence on the overall rate of polymerization for this series of xanthates. Significant retardation is only observed with X_9, presumably due to degradative transfer (15).

Next, the effect of the activating group $Z = OZ'$ of the xanthates on the quality of control of MADIX was investigated. This was achieved using a series of xanthates carrying the same R leaving group, namely a (1-ethoxycarbonyl)ethyl group. Both electron density and steric hindrance of the $Z = OZ'$ group were varied. In this case, the differences observed can be related to a difference of reactivity of the C=S double bond towards growing radicals. The results are summarized in Tables 1 and 2 and Figure 2. First of all, one can note that there is no influence of O-alkyl chain length on the quality of control, X_1 and X_{11} giving roughly the same results. Introduction of a bulky group such as tris-(tert-butylphenyl) group in α-position to the oxygen prevents the growing radicals from accessing the C=S double bond of xanthate X_{12}, which proves ineffective as MADIX agent. For MADIX polymerization of styrene and ethyl acrylate, the chain transfer activity decreases in the following order: $X_{17} > X_{13} \sim X_{14} > X_{15} > X_1 \sim X_{11} > X_{16} >> X_{12}$. It appears that the perfluoroalkyl chain length has no influence on the activity of xanthates, X_{13} and X_{14} giving similar results. As for X_{17}, it is fully consumed before 10% of the monomer is converted, resulting in a linear increase of the molar masses as a function of monomer conversion from the early stages of the polymerization, with Mn values perfectly matching the theoretical ones based on the [styrene]/[X_{11}] ratio.

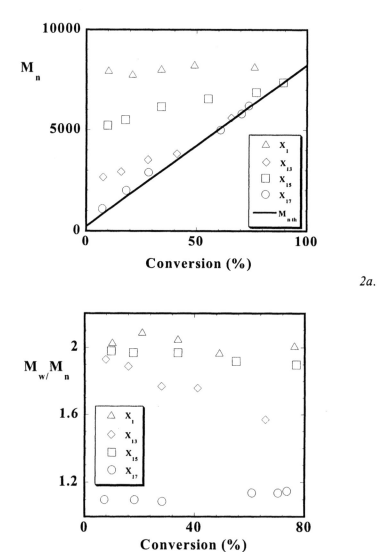

2a.

2b.

Figure 2. Evolution of (2a) M_n and (2b) M_w/M_n during polymerization of styrene in the presence of xanthates with various Z' groups. $[M]_0/[X]_0=80$. Thermal polymerization at 110°C, in 50% toluene.

Figure 2b shows that the O-alkyl group has also a dramatic impact on PDIs. The moderate reactivity of O-ethyl xanthates is attributed to the conjugation of the lone pairs of electrons on the oxygen atom with the C=S bond, resulting in low k_a values (1,2). It is interesting to note in this respect that dithiocarbamates with no electron-withdrawing groups on the nitrogen atom are poor transfer agent since conjugation with the electron pair on the nitrogen has an even greater stabilizing effect on the thiocarbonyl group. However, in the case of xanthates, the enhanced capability for transfer (increase of both $C_{tr}(X)$ and $C_{tr}(P_nX)$- Scheme 1) leading to a significant decrease in the PDI is achieved through the use of a fluoro-alkyl substituents in the Z' moiety (1,24). This can be rationalized by the fact that the conjugation effect mentioned above is considerably reduced with such electron-withdrawing substituents.

Of particular interest, the 1-diethoxyphosphonyl and 2,2',2''-trifluoromethyl groups on the α-carbon bonded to the oxygen atom have a cooperative effect since X_{17} further activates the chain transfer process, as compared to X_{13} and X_{14}. It appears, however, that the introduction of the diethoxyphosphonyl group alone is not sufficient to enhance the reactivity of these xanthates since the substitution of the CF_3 group for a methyl group (X_{16}) or a hydrogen (X_{15}) results in a moderate control. The fact that X_{15} is slightly better than X_{16} is explained by the donating inductive effect of the methyl group which stabilizes the C=S double bond.

The scope of the MADIX process

As emphasized above, xanthates afford variable control over PDIs and molar masses depending on their substituents as well as on the monomer undergoing polymerization. In other words, there is no MADIX agent (neither RAFT agent) that would exhibit a universal character. It appears that the higher the reactivity of the propagating radicals (e.g. those deriving from vinyl acetate), the better the control of the polymerization with O-ethyl xanthates. In contrast, the reactivity of these MADIX agents is moderate towards polystyryl radicals, slightly improved with polyacrylyl radicals and they are ineffective towards poorly reactive radicals such as polymethacrylyl radicals (1-2,15).

Case of Alkyl Acrylates

Polyacrylyl radicals exhibit relatively low steric hindrance and are highly reactive. With properly selected xanthates efficient control of the polymerization of acrylates is achieved. For instance, O-ethyl xanthates are mildly effective towards acrylates, although they can provide excellent control over molar masses vs monomer conversion profiles, in particular with a tertiary R leaving group. This is illustrated in the Figure 3 below. Also, O-ethyl xanthates produce polyacrylates with lower PDIs (entries 1 to 6 in Table 2) as compared to polystyrene samples (Table 1), suggesting more rapid interchange of the dithiocarbonate moieties -higher $C_{tr}(P_nX)$- during polymerization of acrylates, as

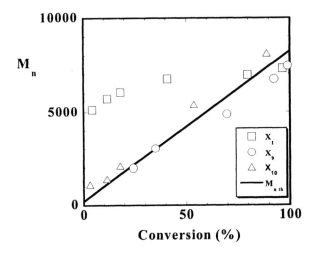

Figure 3. Mn vs conversion in MADIX polymerization of EA in the presence of O-ethyl xanthates with different leaving groups. Conditions of Table 2.

compared to styrene polymerization. As discussed above, however, control is highly improved by fine tuning the Z = OZ' of xanthates (entries 7 to 15 in Table 2).

Case of Vinyl Esters

Whereas the polymerization of vinyl esters has been shown to be completely inhibited in the presence of dithioesters as RAFT agents, excellent control over molar masses and PDIs is achieved with a MADIX agent such as X_1 in the polymerization of vinyl acetate, vinyl neodecanoate and vinyl stearate (Table 3). In such cases, X_1 is entirely consumed in the early stage of the polymerization resulting in a linear increase of the molar masses with conversion of the monomer. The MALDI TOF mass spectrum of a poly(vinyl acetate) sample prepared with X_1 shows the perfect agreement of the chain structure with the expected one (Figure 4). As already reported (*1,2,7,8*), the fact that xanthates (or particular dithiocarbamates as RAFT agents) are agents of choice for controlling the polymerization of vinyl acetate is explained by the stability of the intermediate radical (scheme 1). The latter species may undergo a β-scission much faster than when a dithioester is employed as a RAFT agent. Indeed, a stable xanthate-capped polymer chain is produced after fragmentation due to the conjugation effect mentioned above.

Table 3. Polymerization of Vinyl Esters in the Presence of X_1.

Entry	M	$[AIBN]_0/[X_1]_0/[M]_0$	t (h)	$M_{n\,th}$	$M_n{}^a$	M_w/M_n	Conv (%)
1	VOAc[b]	0.03/1/80	2.66	1750	1400	1.32	22.3
2	VOAc[b]	0.03/1/17	8	1100	1300	1.21	60.3
3	VneD[c]	0.15/1/10	3	2000	1600	1.12	90
4	VneD[c]	0.15/1/30	3	5300	4550	1.22	86
5	Vste[d]	0.15/1/10	21	3250	3750	1.18	93
6	Vste[d]	0.15/1/40	21	11250	10500	1.31	85

[a] measured by SEC in THF. Molar mass measurements were based on PS standards.
[b] Polymerizations of vinyl acetate performed in bulk, at 60 °C.
[c] Polymerization of vinyl decanoate performed in bulk, at 70 °C.
[d] Polymerization of vinyl stearate performed in 50% solution (ethyl acetate/cyclohexane=30/70), at 70 °C.
[e] Polymerization of vinyl stearate performed in a 70% cyclohexane solution, at 70°C.

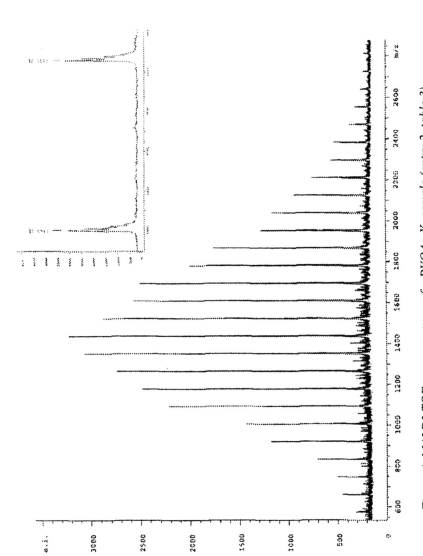

Figure 4. MALDI-TOF mass spectrum of a PVOAc-X₁ sample (entry 2, table 3).

Case of hydrophilic monomers: Acrylic Acid and Acrylamide

The controlled nature of MADIX is preserved when applied to functional monomers such as acrylic acid and acrylamide without resorting to protective groups (*16,18*). Radicals deriving from these two monomers have little streric bulk and are highly reactive. Therefore, xanthates are well-suited to control the polymerization of these monomers. Mono- and difunctional xanthates have been successfully used in the sequential polymerization of acrylic acid and acrylamide in aqueous medium. Double hydrophilic statistical, diblock and triblock copolymers made of polyacrylamide and poly(acrylic acid) have been synthesized in this way (*18*).

Conclusion

The MADIX technology relies on the interchange of dithiocarbonate groups at the polymer chain ends providing control for free radical polymerizations of miscellaneous monomers, including styrenics, acrylics and vinyl esters. The design of the xanthate CTA is crucial for an optimal control of molar masses evolution and polydispersity. In O-ethyl xanthates, the R leaving group only affects the first transfer event, that is the chain transfer constant to the xanthate, $C_{tr}(X)$. Xanthate reactivity is optimal with tertiary leaving groups like the cyanoisopropyl group. The polydispersity is dictated by the chain transfer constant to the $P_nS(C=S)OEt$ chain end, like in any "living" process involving slow equilibria. The moderate reactivity of O-ethyl xanthates is related to the conjugation between the oxygen atom of the O-ethyl group and the C=S double bond. However, dramatic increase of the reactivity of xanthates is achieved through the incorporation of electron-withdrawing groups in the β-position to the oxygen atom, minimizing the conjugation effect and resulting in much narrower distributions of molar masses and structurally well-defined materials.

References

1. (a) Moad, G.; Chiefari, J.; Chong, B. Y. K.; Krstina, J.; Mayadunne, R. T. A.; Postma, A.; Rizzardo, E.; Thang, S. H. *Polym. Int.* **2000**, *49*, 993 ; (b) Rizzardo, E. ; Chiefari, J. ; Mayadunne, R. T. A. ; Moad, G. ; Thang, S. H. in *"Controlled/Living Radical Polymerization"*, *ACS Symposium Series 768* Ed. K. Matyjaszewski **2000**, ch. 20, p. 278.
2. Charmot, D.; Corpart, P.; Adam, H.; Zard, S. Z.; Biadatti, T.; Bouhadir, G. *Macromol. Symp.* **2000**, *150*, 23.
3. (a) Delduc, P.; Tailhan, C.; Zard, S. Z. *J. Chem. Soc., Chem. Comm.* **1988**, 308; (b) Zard, S. Z. *Angew. Chem. Int. Ed. Engl.* **1997**, 37, 672.
4. Boivin, J.; Ramos, L.; Zard, S. Z. *Tetrahedron Lett.* **1998**, *39*, 6877.

5. WO. 9858974 (1998), Rhodia Chimie, invs.: Corpart, P.; Charmot, D.; Biadatti, T; Zard, S. Z.; Michelet; D. [*Chem. Abstr.* **1999**, *130*, 82018].

6. Chiefari, J.; Chong, B. Y. K.; Ercole, F.; Krstina, J.; Jeffery, J.; Le, T. P. T.; Mayadunne, R. T. A.; Meijs, G. F.; Moad, C. L.; Moad, G.; Rizzardo, E.; Thang, S. H. *Macromolecules* **1998**, *31*, 5559-5562.

7. Mayadunne, R. T. A.; Rizzardo, E.; Chiefari, J.; Chong, Y. K.; Moad, G.; Thang, S. H. *Macromolecules* **1999**, *32*, 6977.

8. Destarac, M.; Charmot, D.; Franck, X.; Zard, S. Z. *Macromol. Rapid Commun.* **2000**, *21*, 1035.

9. Mayadunne, R. T. A.; Rizzardo, E.; Chiefari; .; Krstina, J.; Moad, G.; Postma, A.; Thang, S. H. *Macromolecules* **2000**, *33*, 243.

10. Barner-Kowollik, C.; Quinn, J. F.; Morsley, D. R.; Davis, T. P. *J. Polym. Sci. Part A : Polym. Chem.* **2001**, *39*, 1353.

11. Barner-Kowollik, C.; Quinn, J. F.; Nguyen, T. L. U.; Heuts, J. P. A.; Davis, T. P. *Macromelecules* **2001**, *34*, 7849.

12. de Brouwer, H.; Schellekens, M. A. J.; Klumperman, B.; Monteiro, M. J.; German, A. L. *J. Polym. Sci. Part A : Polym. Chem.* **2000**, *38*, 3596.

13. Monteiro, M. J.; de Brouwer, H. *Macromolecules* **2001**, *34*, 349.

14. Kwak, Y.; Goto, A.; Tsujii, Y.; Murata, Y.; Komatsu, K.; Fukuda, T. *Macromolecules* **2002**, *35*, 3026.

15. Destarac, M.; Brochon, C.; Catala, J. M.; Wilczewska, A. ; Zard, S. Z. *Macromol. Chem. Phys.*, **2002**, *203*, 2281.

16. Ladavière, C.; Dörr, N.; Claverie, J. P. *Macromolecules* **2001**, 34, 5370.

17. Sumerlin, B. S.; Donovan, M. S.; Mitsukami, Y.; Lowe, A. B.; McCormick, C. L. *Macromolecules*, **2001**, *34*, 6561.

18. Taton, D.; Wilczewska, A.-Z.; Destarac, M. *Macromol. Rapid. Comm.* **2001**, *22*, 1497.

19. Monteiro, M. J.; Sjöberg, M.; Van der Vlist, J.; Göttgens, C. M. *J. Polym. Sci.* **2000**, *38*, 4206.

20. Prescott, S. W.; Ballard, M. J.; Rizzardo, E.; R. G. Gilbert, *Macromolecules*, **2002**, *35*, 5417.

21. Müller, A. H. E.; Zhuang, R.; Yan, D.; Litvinenko, G. *Macromolecules* 1995, *28*, 4326.

22. Gaynor, S.; Wang, J.-S.; Matyjaszewski, K. *Macromolecules*, **1995**, *28*, 8051.

23. Brochon, C. ; Catala, J. M. *ACS Boston, Polym. Prepr.* **2002**, 43(2), 303.

24. Destarac, M. ; Bzducha, W. ; Taton, D. ; Gauthier-Gillaizeau, I. ; Zard, S. Z. *Macromol. Rapid Commun.* **2002**, *23*, 1049.

Chapter 38

Influences of the Structural Design of RAFT Agents on Living Radical Polymerization Kinetics

Thomas P. Davis, Christopher Barner-Kowollik, T. L. Uyen Nguyen, Martina H. Stenzel, John F. Quinn, and Philipp Vana

Centre for Advanced Macromolecular Design, School of Chemical Engineering and Industrial Chemistry, The University of New South Wales, Sydney, New South Wales 2052, Australia

The influence of a *p*-substituent on the Z-group of cumyl dithiobenzoate (CDB) was investigated with respect to the rate retardation of CDB mediated RAFT polymerizations. The rate retardation is significantly decreased indicating the possible involvement of the *p*-position in a reversible termination reaction. CDB mediated copolymerizations of styrene and 2-hydroxyethyl methacrylate (HEMA) may indicate a change of the reactivity ratios when compared to conventional copolymerizations. The usage of macromolecular RAFT agents showed unexpected kinetic behavior, such as acceleration.

The reversible addition fragmentation chain transfer (RAFT) process is among the most versatile of the novel living free radical polymerization techniques allowing access to complex macromolecular architectures ranging from stars to block and comb polymers of narrow polydispersity.(*1*) While the synthetic potential of the RAFT process is well developed, a complete mechanistic and kinetic picture is only starting to emerge. A detailed understanding of the RAFT process is needed to develop strategies for rational RAFT agent design that allows for the formation of very uniform polymeric material without additional side products. Highly pure polymers of defined macromolecular architecture are a necessity in applications ranging from pharmaceutical to bio-material applications.

Various research groups are currently investigating the mechanism and kinetics of the RAFT process in detail.(2-13) Our group has been on the forefront of the mechanistic debate and we comprehensively modeled the RAFT process for cumyl dithiobenzoate mediated styrene polymerizations for the first time.(6) In this initial study, the RAFT model proposed by the *CSIRO* group (14) was employed to estimate the rate coefficients governing the core equilibrium of the RAFT process (see Scheme 1). Inevitably, the rate coefficients obtained from such a kinetic analysis are model dependent quantities. Several authors – including our own research group – have since suggested additions and modifications to the initial RAFT mechanism to accommodate their experimental findings.(5,9-12) However, direct evidence i.e. isolation of intermediates from an actual RAFT polymerization has yet to be identified.

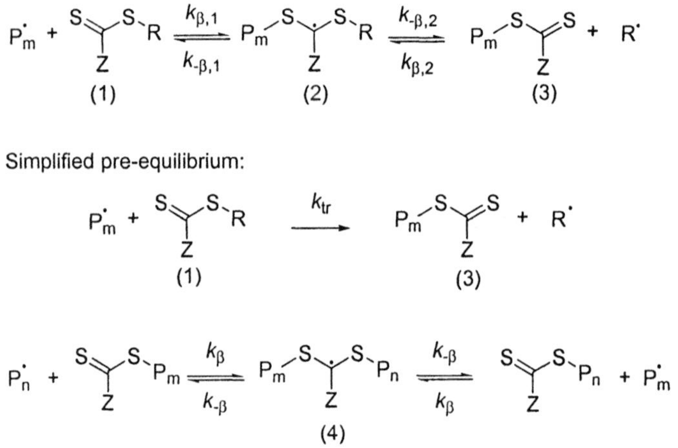

$$P_m^{\bullet} + \overset{S}{\underset{Z}{\diagdown}}\overset{S}{\diagup}R \underset{k_{-\beta,1}}{\overset{k_{\beta,1}}{\rightleftharpoons}} P_m\overset{S}{\diagdown}\overset{S}{\underset{Z}{\diagup}}R \underset{k_{\beta,2}}{\overset{k_{-\beta,2}}{\rightleftharpoons}} P_m\overset{S}{\diagdown}\overset{S}{\underset{Z}{\diagup}} + R^{\bullet}$$

$$\quad(1)\qquad\qquad\qquad(2)\qquad\qquad\qquad(3)$$

Simplified pre-equilibrium:

$$P_m^{\bullet} + \overset{S}{\underset{Z}{\diagdown}}\overset{S}{\diagup}R \overset{k_{tr}}{\longrightarrow} P_m\overset{S}{\diagdown}\overset{S}{\underset{Z}{\diagup}} + R^{\bullet}$$

$$\quad(1)\qquad\qquad\qquad(3)$$

$$P_n^{\bullet} + \overset{S}{\underset{Z}{\diagdown}}\overset{S}{\diagup}P_m \underset{k_{-\beta}}{\overset{k_{\beta}}{\rightleftharpoons}} P_m\overset{S}{\diagdown}\overset{S}{\underset{Z}{\diagup}}P_n \underset{k_{\beta}}{\overset{k_{-\beta}}{\rightleftharpoons}} \overset{S}{\underset{Z}{\diagdown}}\overset{S}{\diagup}P_n + P_m^{\bullet}$$

$$(4)$$

Scheme 1. The basic RAFT Mechanism.

Typical mediating agents, i.e. the RAFT agents, are thiocarbonylthio compounds (see Scheme 2). The ability of a RAFT agent to control the polymerization activity is governed by two characteristic groups, the Z- and R-group, also denoted as the 'stabilizing' and the 'leaving-group' (see Scheme 2). The core of the RAFT process is the main equilibrium, in which the two rate coefficients k_{β} and $k_{-\beta}$ control the position of the equilibrium: k_{β} (corresponding to the addition step) controls the bimolecular reaction between free polymeric radicals and polymeric RAFT-agent (species (3) in Scheme 1), which leads to the formation

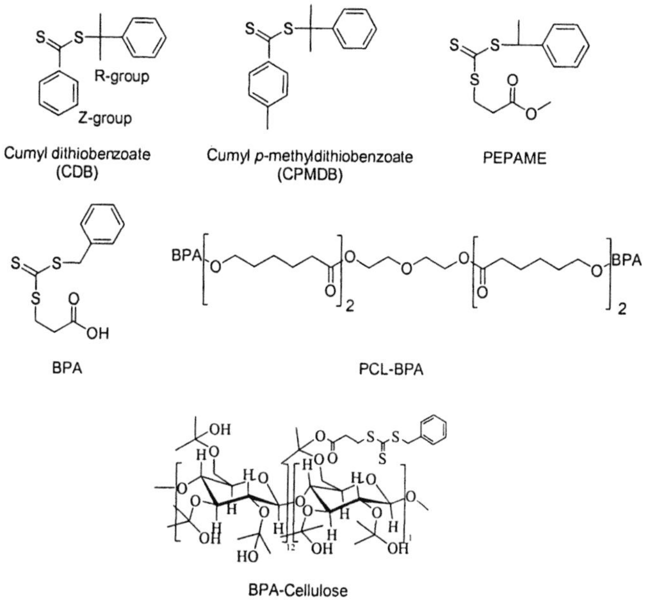

Scheme 2. *RAFT agents used in this study.*

of the macroRAFT radical (4). Finally, bimolecular termination reactions of free polymeric radicals occur and lead to the formation of 'dead' polymer.

The majority of research groups investigating the RAFT process observed both rate retardation (i.e. a decrease in the overall rate of polymerization up to high conversions with increasing initial RAFT agent concentration) and inhibition (i.e. a time period without any polymerization activity at early reaction times, dependent in extent on the initial RAFT agent concentration) phenomena in some RAFT systems. Such effects – while impeding synthetic efforts – contain valuable mechanistic information about the process. Two schools of thought have originated to explain such rate altering effects: (i) Rate retardation may be explained by slow fragmentation of the macroRAFT radical intermediates ((2) and (4) in Scheme 1)(*5,6,15*) or – alternatively – (ii) by the irreversible termination of species (2) and (4) with other radicals present in the system.(*11,12*) Evidence has been collected to underpin both theories, with experimental outcomes apparently supporting either mechanism. In an attempt to quantify the values of the kinetic coefficients describing the main equilibrium of the RAFT process (i.e. k_β and $k_{-\beta}$ in Scheme 1) our group modeled kinetic rate data and full molecular weight distributions in a combined experimental/simulation approach.(*5,6*) For the studied systems (i.e. cumyl dithiobenzoate (CDB) and cumyl phenyldithioacetate (CPDA) mediated styrene polymerizations) considerable lifetimes of the macroRAFT radical (4) in the order of seconds were identified on the basis of the original kinetic scheme proposed by the CSIRO group. Surprisingly, the predicted rather high radical concentrations for the intermediate (4) were not confirmed experimentally via ESR spectroscopy and led to the modification of Scheme 1 with additional irreversible termination reactions of free radicals and species (4).(11,12) However, to this date the hypothetical termination products have not been identified in polymerizing systems via state of the art analytical methods, i.e. electrospray ionization mass spectrometry (16) and size exclusion chromatography (17). In order to clarify whether we were observing long-lived intermediates, we designed a novel experiment to detect the presence of transient intermediates in the RAFT process via γ-radiation.(10) These experiments demonstrated that it is possible to store γ-generated radicals – either as the stable free radicals (2) and (4) themselves or in a reversibly terminated form – in a RAFT system containing cumyl dithiobenzoate (CDB) and styrene or methyl acrylate (MA). The general concept of stable intermediate radicals of the form (2) and (4) has recently been underpinned by quantum mechanical calculations.(15,18)

The kinetic and mechanistic data currently available lead to a modification of the original RAFT scheme with respect to the main equilibrium. The modified scheme (see Scheme 3) seeks to accommodate the different experimental and theoretical findings, by including possible reversible termination reactions of the macroRAFT intermediate. Whereas the formation of the three-arm star is believed to be irreversible (19) (upper part of Scheme 3) the reversible coupling of two intermediates may proceed via resonance structures of the phenyl-stabilized radical (lower part of Scheme 3).(9,20) It can not be excluded that irreversible termination reactions do occur to some extent. However, such reactions are more likely to be a consequence of the slow fragmentation rather than being the primary cause of the rate retardation.

Experimental

Synthesis

Cumyl dithiobenzoate (CDB)
   Synthesis has been described in ref.(21)

3-(1-Phenyl-ethylsulfanylthiocarbonylsulfanyl)-propionic acid methyl ester (PEPAME)
   Mercaptopropanoic acid (10.6 g, 100 mmol) was added dropwise to a solution of KOH (11.2 g, 200 mmol) in water (200 ml). The mixture was allowed to stir for 30 minutes and subsequently cooled in an ice bath, after which carbon disulfide (7.6 g, 100 mmol) was added, and the mixture stirred overnight. 1-(Bromoethyl)benzene (18.3 g, 100 mmol) was added, and the resulting solution stirred for 24h at 100 °C. The reaction mixture was then transferred to a 1 L

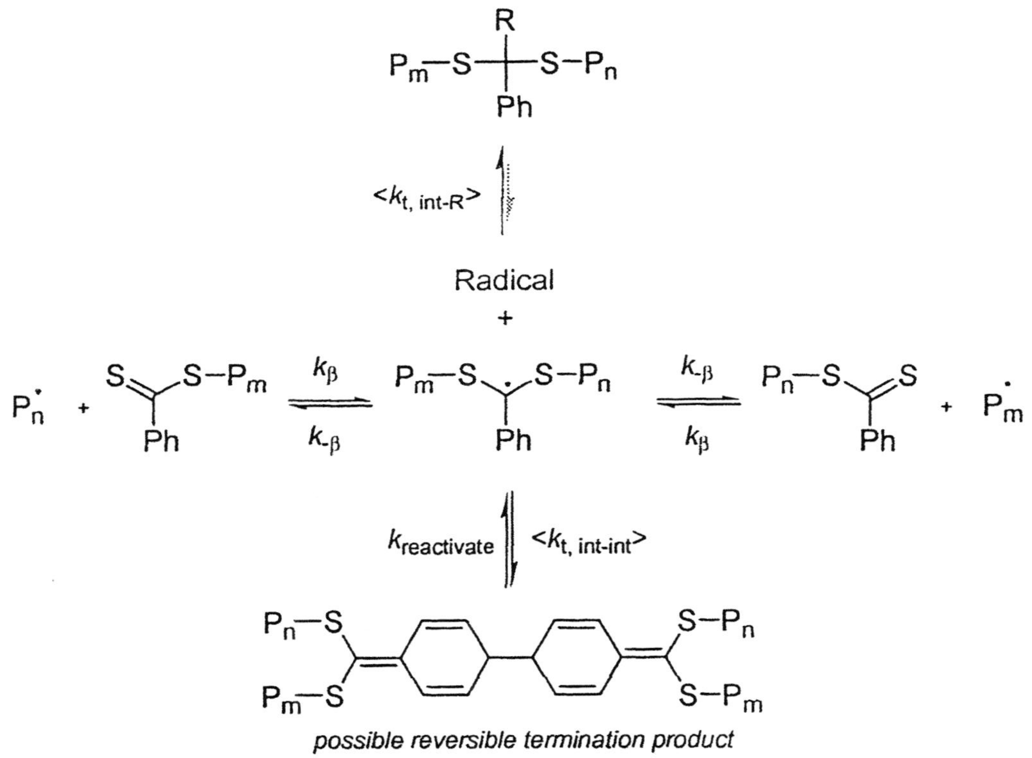

Scheme 3. Modified reaction scheme including reversible and irreversible termination of the macroRAFT intermediate.

556

separating funnel, containing 300 ml of chloroform (CHCl₃). The mixture was
acidified until the pH of the aqueous phase reached 4 (measured using universal
indicator paper), and the aqueous portion was washed a further two times using
100 ml chloroform. The organic layer was dried, (Na₂SO₄), filtered and the solvent removed using reduced pressure. The intermediate was identified as 3-(1-
phenylethylsulfanyl-thiocarbonylsulfanyl)-propanoic acid (PEPA) via ¹H-NMR
(CDCl₃, 25 C, δ = 1.76 d (3H), δ = 2.81 t (2H), δ = 3.58 t (2H), δ = 5.33 q (1H),
δ = 7.32 m (5H)). PEPA (2g, 7 mmol) was dissolved in carbon tetrachloride (30
ml), and distilled thionyl chloride was added dropwise (6.5 g, 55 mmol). After
all the thionyl chloride was added, the reaction mixture was heated to 55 °C for
1-2 hrs until gas formation ceased. The solvent was then removed under vacuum,
and the product ((phenylethylsulfanylthiocarbonylsulfanyl)-propanoic acid chloride (PEPAC)) dried under vacuum. The structure of the acid chloride was verified using ¹H NMR (CDCl₃, 25 °C, δ = 1.78 d (3H), δ = 3.35 t (2H), δ = 3.56 t
(2H), δ = 5.33 q (1H), δ = 7.32 m (5H)). PEPAC (2 g, 6.6 mmol) was refluxed
for 12 hours in methanol (30 ml, 818 mmol). The solvent was then evaporated
and the product (3-(1-phenylethylsulfanylthiocarbonylsulfanyl)-propanoic acid
methyl ester) isolated using column chromatography on silica gel (200-400
mesh, 60A, BET surface area = 500 m²g⁻¹) with toluene as the eluent. The final
overall product yield was close to 25 %. The structure was confirmed via ¹H-
NMR (CDCl₃, 25 C, δ = 1.75 d (3H), δ = 2.75 t (2H), δ = 3.57 t (2H), δ = 3.69 s
(3H), δ = 5.33 q (1H), δ = 7.32 m (5H)).

Cumyl p-methyldithiobenzoate (CPMDB)
    4-Methylbenzyl chloride (10.0 g, 71 mmol) was added dropwise over one
hour to a round bottomed flask containing elemental sulfur (4.63 g, 144 mmol) ,
25% sodium methoxide solution in methanol (32.0 g, 148 mmol) and methanol
(40 ml). The resulting brown solution was heated and allowed to reflux at 80°C
overnight. After cooling to room temperature, the mixture was filtered to remove
the white solid (sodium chloride). Subsequently, the methanol was removed via
rotary evaporation. The resulting brown solid was redissolved in distilled water
(100 ml), and washed three times with diethyl ether (150 ml total). A final layer
of ether was added to the solution and the two-phase mixture was acidified with
32% aqueous HCl until the aqueous layer lost its characteristic brown color and
the top layer was deep purple. The etherous layer was dried over calcium chloride and the residual ether removed via rotary evaporation to leave the deep purple solid of p-methyldithiobenzoic acid (ca. 5 g, 30 mmol). This acid was then
dissolved in CCl₄ and refluxed with α-methylstyrene (4.2 g, 36 mmol) overnight
in the presence of a small amount of acid catalyst (paratoluenesulfonic acid)
(0.05 g, 0.3 mmol). The product was isolated from excess α-methylstyrene and
4-methyldithiobenzoic acid residue by column chromatography, using silica gel
(200-400 mesh, 60A, BET surface area = 500 m²g⁻¹) with 2% diethyl ether in
hexane as the eluent. Final product yields were close to 42 %. The product iden-
tity was verified using ¹H-NMR (CDCl₃, 25 °C, δ = 2.0 s (6H), δ = 2.3 s (3H), δ
= 7.3 m (9H)).

3-Benzylsulfanylthiocarbonylsufanyl-propionic acid (BPA)
Synthesis has been described in ref.(*22*)

Hydroxyisopropyl cellulose-co-(3-Benzylsulfanylthiocarbonyl-sufanyl-propionyl)-hydroxyisopropyl cellulose (BPA-cellulose)
Synthesis has been described in ref.(*23*)

Polycaprolactone di-(3-Benzylsulfanylthiocarbonylsufanyl-propionate) (PCL-BPA)
PCL-BPA has been synthesized via the acid chloride of BPA and the subsequent reaction with polycaprolactone diol (M_n= 530) in chloroform (with pyridine in stochiometric amount to acid) and purified with chromatography on silica gel with chloroform as eluent.

Polymerizations

Stock solutions, with varying initial RAFT agent concentrations (variation by approximately a factor of 8) were prepared. The initiator (AIBN) concentration was the same and close to $4 \cdot 10^{-3}$ mol L^{-1} in each polymerization series. Four samples of each stock solution were transferred to individual ampoules, which were thoroughly deoxygenated by purging with nitrogen for approximately 10 min. The sealed ampoules were then placed in a constant temperature water bath at 60 °C, and an ampoule was removed after pre-set time intervals. The time intervals were selected such that the overall monomer conversion did not exceed 30 %. The reactions were stopped by cooling the solutions in an ice bath followed by the addition of hydroquinone. The polymer was isolated by evaporating off the residual monomer; initially in a fume cupboard to remove the bulk of the liquid, and then in a vacuum oven at 30 °C. Final conversions were measured by gravimetry. Each experiment was performed in duplicate using reaction mixture from the same stock solution.

Molecular weight analysis

Molecular weight distributions were measured by size exclusion chromatography (SEC) on a Shimadzu modular system, comprising an auto injector, a Polymer Laboratories 5.0 µm bead-size guard column (50 x 7.5 mm), followed by three linear PL columns (10^5, 10^4 and 10^3 Å) and a differential refractive index detector. The eluent was tetrahydrofuran (THF) at 40 °C with a flow rate of 1 ml min^{-1}. The system was calibrated using narrow polystyrene standards ranging from 500 to 10^6 g mol^{-1}.

Experimental Uncertainties

The experimental error for the weight average molecular weight, M_w, is believed to be no more than ±12 %. The uncertainty in the measured polydispersity and the gravimetrically determined monomer conversion are assumed to be also approximately 12 %.

Results and Discussion

Reversible macroRAFT intermediate termination – a structural approach

The current kinetic experimental data available for the RAFT process led to our postulation of a possible reversibly accessible radical sink in the from of two para-self-terminated macroRAFT radicals as depicted in the lower part of Scheme 3. To date, attempts to isolate such reversibly terminated species proved very difficult. In an effort to probe RAFT agent mediated polymerizations for evidence of such a mechanism, we decided to take a structural approach: The para-position of the phenyl ring – the site of the proposed termination reaction – is made less available for radical attack by a substituent more bulky than hydrogen, with the effect that the resulting RAFT agent should show less retardation effects. The novel RAFT agent was realized in the form of CPMDB, which displays a methyl group in the para-position of the Z-group phenyl ring, but is otherwise identical to CDB (see Scheme 2). CPMDB is excellently suited to mediate styrene bulk free radical polymerizations and effect polystyrene with narrow polydispersities as indicated by the obtained full molecular weight distributions given in Figure 1.

In order to assess whether CPMDB causes indeed less retardation than CDB, the rate of polymerization was determined for several initial RAFT agent concentrations of CPMDB and CDB, respectively. To complement the study, a RAFT agent with no stabilizing ability for the macroRAFT radical – PEPAME – was also tested.

In order to quantify the retardation effects generated by the three RAFT agents, a retardation parameter, ρ, is defined as

$$\rho = -\frac{d \ln R_p}{d[\text{RAFT}]_0} \qquad (1)$$

where R_p is the (mean) rate of polymerization and $[\text{RAFT}]_0$ is the initial RAFT agent concentration (R_p is obtained via linearly fitting the conversion vs time plots). The retardation parameter was found to be non-dependent on the initial RAFT agent concentration – at least in the examined concentration regimes – , that is, the $\ln R_p$ vs $[\text{RAFT}]_0$ plots are linear. The retardation parameter thus quantifies the relative decrease in the rate of polymerization. The data given in Figure 3 (for each individual RAFT agent) has been analyzed via Equation (1) and the resulting ρ have been depicted for the three RAFT agents in Figure 3.

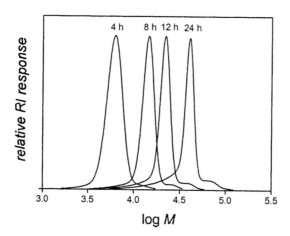

Figure 1. Evolution of molecular weight distribution for the CPMDB mediated polymerization of styrene at 60 °C with an initial RAFT agent concentration of 8.1·10⁻³ mol L⁻¹.

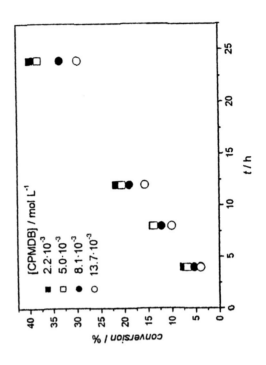

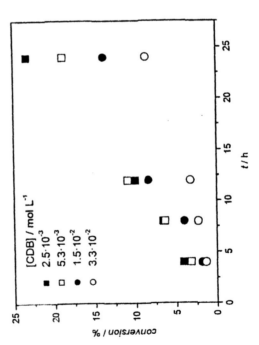

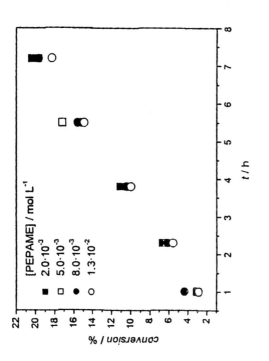

Figure 2. Time vs. conversion plots for CDB, CPMDB and PEPAME mediated styrene bulk free radical polymerization at 60 °C at varying initial RAFT agent concentrations. The AIBN concentration was close to $4 \cdot 10^{-3}$ mol L^{-1} in all cases.

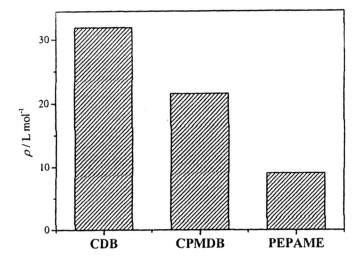

Figure 3. Retardation parameter, ρ, for three RAFT agents at 60 °C.

Inspection of Figure 3 indicates that the introduction of a methyl group at the para-position of the phenyl Z-group leads to a significant decrease in the retardation parameter. This observation is in agreement with the proposed mechanism of the RAFT process given in Scheme 3: If the para substituent is sterically more demanding, e.g. hydrogen is replaced by methyl, the para position is less available for radical attack. As a consequence, the reversible termination pathway given in Scheme 3 is (at least partially) suppressed and the RAFT agent mediated polymerization is less susceptible for retardation phenomena. Not surprisingly, the retardation parameter for PEPAME is substantially lower than for both CDB and CPMDB, due to the non-stabilized intermediate macroRAFT radical and the unavailability of the reversible termination pathway.

RAFT agent mediated statistical copolymerization

The development of the RAFT process has provided the ability to synthesize polymers with good control over the molecular weight and narrow molecular weight distributions. This is not only valid for linear homopolymers, but also for the preparation of more complex polymer structures. The preparation of statistical copolymers via the RAFT process is of interest because of the wide application of such polymers. We focused on the copolymerization of 2-hydroxyethyl methacrylate (HEMA) and styrene, due to the interesting applications of HEMA containing copolymers for the biomedical area such as contact lenses and drug delivery systems.

The copolymerisation of styrene and HEMA was carried out in N,N-dimethylformamide (DMF) in the presence of cumyl dithiobenzoate (CDB). DMF was chosen as a solvent for comparison of the data with free radical copolymerization reactivity ratios from literature.(*24*)

In contrast to conventional free radical copolymerization, the macroradical species in RAFT mediated copolymerizations may not only react with the monomer units, but also with the RAFT agent (Scheme 4).

Cumyl dithiobenzoate is capable of controlling the polymerization of both monomers separately. However, using a RAFT agent which is only suitable for the homopolymerization of one comonomer may still control the copolymerization.

The kinetics of both RAFT homopolymerizations as well as the RAFT mediated copolymerization with different feed ratios show living behavior, leading to polymers with molecular weights increasing linearly with conversion – in agreement with theoretical predicted molecular weights – and narrow molecular weight distributions (PDI<1.25). The overall rate of polymerization is dependent on the feed ratio as seen in Figure 4.

As mentioned above, the transfer of the macroradical to the RAFT agent may have an influence on the copolymer composition. To test this hypothesis, the polymerization was carried out with different feed ratios of styrene and HEMA to a conversion well below 5 %. The copolymer was isolated and the composition was determined via ^{1}H-NMR measurements in d-DMSO. Assuming the terminal model, the reactivity ratios r_{HEMA} and $r_{STYRENE}$ were obtained via a linear least square fit, resulting in values of 1.13 and 0.93, respectively. The values reported for the conventional free radical copolymerization of HEMA and styrene in DMF are r_{HEMA}= 0.527 and $r_{STYRENE}$= 0.411, indicating an enrichment of HEMA in the copolymer at lower feed concentrations of HEMA. The copolymerization, when mediated with cumyl dithiobenzoate, may proceed nearly idealy. Figure 5 shows both a *Lewis-Mayo* plot for the RAFT agent mediated copolymerization and the conventional free radical polymerization. The differences are rather minor and may be within experimental errors. We are aware that the terminal model is an oversimplification of the copolymerization mechanism and therefore one might expect disparate reactivity ratios under different experimental conditions.(*25,26*) However, a change in the copolymerization composition when using RAFT agent as the controlling agent may be envisaged. It should be noted that deviations from conventional reactivity ratios may also be observed in copolymerization processes with intermittent activation.(*27*) Further studies are underway in our laboratories.

Kinetics of macromolecular RAFT agent mediated polymerizations

The RAFT polymerization raises interest due to easy access of different polymer architectures such as block copolymer, comb polymers and star polymers. The experimental design is a crucial step, which determines the success of the synthesis of such architectures.(*28*) It is not only the setup of the RAFT polymerization strategy and the structural design of the RAFT agent that makes the preparation of these architectures more complex. A prerequisite for the macromolecular RAFT agent – which can be the RAFT agent for a comb or the first block prepared of the macro RAFT agent – is that it exhibits the same transfer

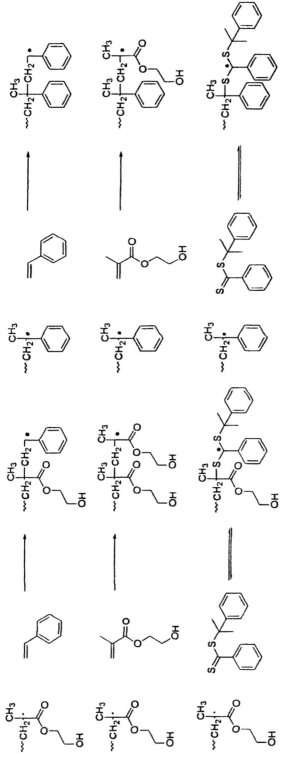

Scheme 4. The four possible propagation steps and the two different transfer steps appearing during RAFT polymerization of 2-hydroxyethyl methacrylate and styrene.

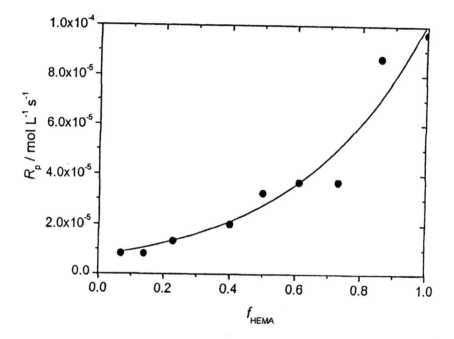

Figure 4. The rate of polymerization of a STY/HEMA copolymerization vs the 2-hydroxyethyl methacrylate feed ratio in N,N-dimethylformamide at 60 °C ([HEMA]= 2.05 mol/L, [AIBN]= 6.09·10⁻³ mol/L, [CDB]= 7.34·10⁻³ mol·L⁻¹).

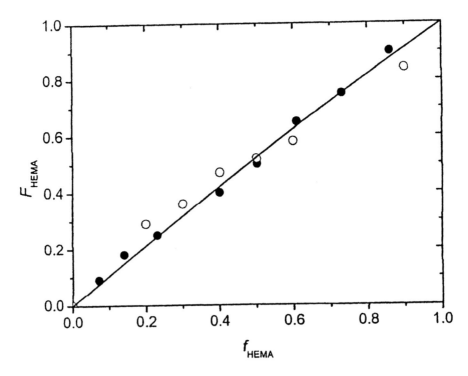

Figure 5. Dependency of the copolymer composition (F_{HEMA}) on the feed composition (f_{HEMA}) of the RAFT mediated copolymerization determined by ^1H-NMR measurements (closed dots) and the fit with the terminal model ($r_{HEMA}= 1.13$, $r_{STYRENE}= 0.93$). The values for the free radical polymerization (open dots) are obtained from ref(24).

ability toward macroradicals as low molecular weight RAFT agents to effectively control the polymerization. Nevertheless, dissolving a macromolecular RAFT agent in a monomer solution can cause different effects, such as increased viscosity or change of solvent polarity (especially, when a hydrophobic macromolecular RAFT agent is been added to a hydrophilic monomer). Depending on the system, this influence can often be neglected. However, we observed polymerization systems displaying acceleration, that is, the rate of polymerization increases with increasing RAFT agent concentration. We prepared a macromolecular RAFT agent based on polycaprolactone diol (M_n= 530 g·mol^{-1}), which was esterified with BPA. The controlling ability of the resulting macromolecular RAFT agent was compared with that of BPA in styrene bulk polymerization at 100°C. While the initial concentration of BPA has no significant influence on the conversion-time profiles (Figure 6), we observed a slight increase in conversion of styrene at higher concentrations of PCL-BPA.

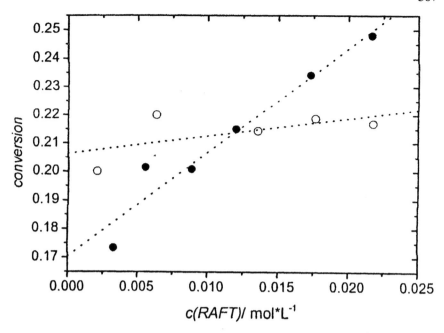

Figure 6. Conversion of the bulk polymerization of styrene vs the RAFT agent concentration of BPA (open dots) and PCL-PBA (closed dots) at 100 ℃ after 7 hrs.

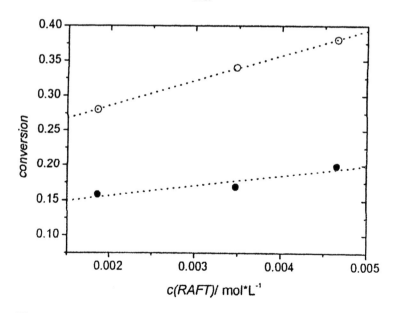

Figure 7. Conversion of the styrene polymerization (2.9 M in N-Methyl pyrrolidone, AIBN) vs the RAFT agent concentration of BPA-cellulose at 60 ℃ after 24 hrs (closed dots) and 48 hrs (open dots).

A stronger influence is found during the polymerization of styrene in the presence of a cellulose based RAFT agent.(*23*) Here, we can clearly observe the increase in the rate of polymerization with increasing initial RAFT agent concentration. This is in sharp contrast to the observed retardation of controlled polymerizations using low molecular weight RAFT agents, e.g. CDB (Figure 7).

Acknowledgments. The authors would like to thank the *Australian Research Council*, the *Austrian Science Fund*, and the *CRC for Polymers* for financial support.

References

1. Mayadunne, R. T. A.; Rizzardo, E.; Chiefari, J.; Chong, Y. K.; Moad, G.; Thang, S. H. *Macromolecules* **1999**, *32*, 6977; Moad, G.; Chiefari, J.; Chong, Y. K.; Krstina, J.; Mayadunne, R. T. A.; Postma, A.; Rizzardo, E.; Thang, S. H. *Polymer Int.* **2000**, *49*, 993; Stenzel-Rosenbaum, M.; Davis T. P.; Fane A. G.; Chen V. *J. Polym. Sci., Part A: Polym. Chem.* **2001**, *39*, 2777; Stenzel-Rosenbaum, M.; Davis T. P.; Fane A. G.; Chen V. *Angew. Chem. Int. Ed.* **2001**, *40*, 3428.

2. Quinn, J. F.; Barner, L.; Barner-Kowollik, C.; Rizzardo, E.; Davis, T. P. *Macromolecules* **2002**, *35*, 7620.

3. Quinn, J. F.; Rizzardo, E.; Davis, T. P. *Chem. Comm.* **2001**, 1044.

4. Quinn, J. F.; Barner, L.; Rizzardo, E.; Davis, T. P. *J. Polym. Sci., Part A: Polym. Chem.* **2002**, *40*, 19.

5. Barner-Kowollik, C.; Quinn, J. F.; Nguyen, T. L. U.; Heuts, J. P. A.; Davis, T. P. *Macromolecules* **2001**, *34*, 7849.

6. Barner-Kowollik, C.; Quinn, J. F.; Morsley, D. R.; Davis, T. P. *J. Polym. Sci., Part A: Polym. Chem.* **2001**, *39*, 1353.

7. Barner-Kowollik, C. *Aust. J. Chem.* **2001**, *54*, 343.

8. Perrier, S.; Barner-Kowollik, C.; Quinn, J. F.; Vana, P.; Davis, T. P. *Macromolecules* **2002**, *35*, 8300.

9. Barner-Kowollik, C.; Vana, P.; Davis, T. P. *Polym. Prepr. (ACS)* **2002**, *43*, 321.

10. Barner-Kowollik, C.; Vana, P.; Quinn, J. F.; Davis, T. P. *J. Polym. Sci., Part A: Polym. Chem.* **2002**, *40*, 1058.

11. Monteiro, M. J.; de Brouwer, H. *Macromolecules* **2001**, *34*, 349; Monteiro, M. J.; Hodgson, M.; de Brouwer, H. *J. Polym. Sci., Part A: Polym. Chem.* **2000**, *38*, 3864.

12. Kwak, Y.; Goto, A.; Tsujii, Y.; Murata, Y.; Komatsu, K.; Fukuda, T. *Macromolecules* **2002**, *35*, 3026.

13. Zhang, M.; Ray, W. H. *Ind. Eng. Chem. Res.* **2001**, *40*, 4336.

14. Moad, G.; Chiefari, J.; Chong, Y. K.; Krstina, J.; Mayadunne, R. T. A.; Postma, A.; Rizzardo, E.; Thang, S. H. *Polymer Int.* **2000**, *49*, 993.

15. Farmer, S. C.; Patten, T. E. *J. Polym. Sci., Part A: Polym. Chem.* **2002**, *40*, 555.

16. Vana, P.; Davis, T. P.; Albertin, L.; Barner, L.; Barner-Kowollik, C. *J. Polym. Sci., Part A: Polym. Chem.* **2002**, *40*, 4032.

17. Vana, P.; Davis, T. P.; Quinn, J. F.; Barner-Kowollik, C. *Aust. J. Chem.* **2002**, *55*, 425.

18. High level quantum mechanical calculations confirm the stability of the macroRAFT radicals (species (2) and (4) in Scheme 1). Coote, M. L.; Radom, L., private communication).

19. Goto, A.; Kwak, Y.; Tsujii, Y.; Fukuda, T. *Polym. Prepr. (ACS)* **2002**, *43*, 311.

20. Monteiro, M. J.; Bussels, R.; Beuermann, S.; Buback, M. *Aust. J. Chem.* **2002**, *55*, 433.

21. Oae, S.; Yagihara, T.; Okabe, T. *Tetrahedron* **1972**, *28*, 3203.

22. Stenzel, M. H.; Davis, T. P *J. Polym. Sci., Part A: Polym. Chem.* **2002**, *40*, 4498.

23. Stenzel, M. H.; Davis, T. P.; Fane A. G. *Ind Chem Eng Res* **2002**, *submitted*.

24. M. Sanchez-Chaves, G. Martinez, E. L. Madruga, *J. Polym. Sci., Part A: Polym. Chem.* **1999**, *37*, 2941.

25. Coote, M. L.; Davis, T. P. "Copolymerization Kinetics" In "Handbook of Radical Polymerization"; Matyjaszewski, K.; Davis, T. P., Eds.; Wiley-Interscience: New York, 2002.

26. Coote, M. L.; Davis, T. P. *Progress in Polymer Science* **1999**, *24*, 1217.

27. Matyjaszewski, K. *Macromolecules* **2002**, *35*, 6773.

28. C. Barner-Kowollik, T. P. Davis, J. P.A. Heuts, M. H. Stenzel, P. Vana, M. Whittaker *J. Polym. Sci., Part A: Polym. Chem.* **2003**, *41*, 365.

Chapter 39

Vinylidene Chloride Copolymerization with Methyl Acrylate by Degenerative Chain Transfer

P. Lacroix-Desmazes, R. Severac, and B. Boutevin

Laboratoire de Chimie Macromoléculaire, UMR-CNRS 5076,
Ecole Nationale Supérieure de Chimie de Montpellier,
8 rue de l'Ecole Normale, 34296 Montpellier Cedex 5, France

Degenerative chain transfer copolymerization of vinylidene chloride (VC_2) with methyl acrylate (MA) was investigated at 70°C in benzene. Different dithiocompounds ZC(S)SR were tested as chain transfer agents in the RAFT process (Reversible Addition-Fragmentation Chain Transfer) while 1-phenylethyl iodide was tested as chain transfer agent in the ITP process (Iodine Transfer Polymerization). Dithioesters (Z= Ph) proved to be much more efficient to control VC_2/MA copolymerization than both the xanthate (Z= OC_2H_5) and 1-phenylethyl iodide. The higher apparent chain transfer constant was found for the dithioester with R= $CH(CH_3)C(O)OC_2H_5$. Dithioesters had a pronounced effect on the kinetics, R= $C(CH_3)_3$ leading to the most important retardation effect. As illustrated by using the Predici® simulation package, the transfer to VC_2 was thought to be responsible for the limitation of the attainable molecular weight in a living fashion. In spite of this side reaction, chain extension as well as a block copolymerization with styrene were successfully performed.

Introduction

The development of several living free radical polymerization processes (LFRP) in the two last decades opens the route to a wide range of well-defined polymers (predetermined molecular weight, narrow distribution and tailored architecture) (1). In this field, the polymerization of styrenics, acrylics, methacrylics, and dienes in a living fashion has been extensively described in the literature. In contrast, monomers bearing halogen atoms on the reactive double bond have been only scarcely studied in living radical polymerization. In this work, we were interested in controlling the polymerization of vinylidene chloride (VC₂).

Vinylidene chloride bears two halogen atoms at the alpha position. Few studies were reported in the literature on this class of halogenated monomers in living free radical polymerization. For instance, vinylidene fluoride was successfully copolymerized with hexafluoropropene by ITP (Iodine Transfer Polymerization), leading to commercial fluoroelastomers (2-4). Vinylidene chloride was used in ATRP (Atom Transfer Radical Polymerization) by Matyjaszewski *et al.*(5) as a comonomer in the polymerization of acrylonitrile. It was shown that the polymerization was limited to low conversion, but the amount of vinylidene chloride was only 5 mol%, so it is difficult to deduce some information on the ATRP of vinylidene chloride alone. In the nineties, some success was claimed by the Geon Company in the polymerization of vinyl chloride by ITP (6). More recently, vinyl chloride was also studied by metal catalyzed radical polymerization by Percec *et al.*(7), best results being obtained with iodo compounds, making possible a combination with the ITP process. Finally, in a recent paper, we have reported the successful polymerization of butyl *alpha*-fluoroacrylate by ATRP (8). From this analysis of the literature and from our own preliminary screening, we decided to focus on degenerative transfer processes, namely the RAFT process (Reversible Addition-Fragmentation Chain Transfer) (9) and the ITP process for the polymerization of vinylidene chloride (Scheme 1).

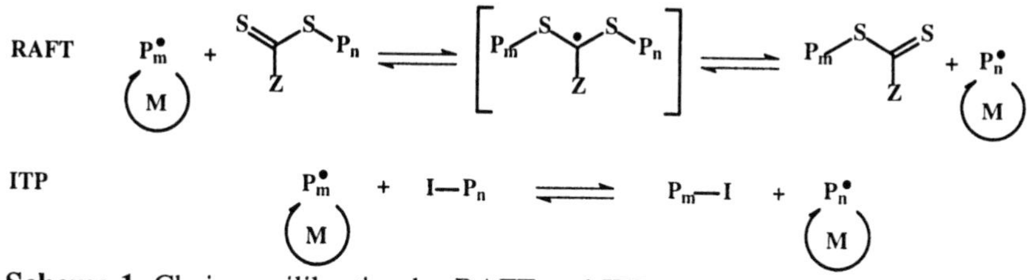

Scheme 1. Chain equilibration by RAFT and ITP processes (degenerative chain transfer)

Because poly(vinylidene chloride) homopolymer has a low solubility in conventional solvents, we used methyl acrylate (MA) as a comonomer. The good solubility of the copolymer facilitates the characterizations by proton NMR and size exclusion chromatography. Reactivity ratios are close to one (r_1=0.9, r_2=0.95, with monomer 1=VC$_2$) (10), indicating a statistical copolymer with almost no deviation in composition during the polymerization. It also means that the conversion is almost the same for both monomers. Therefore, in this work, we will refer to the monomer conversion without specifying the monomer. Furthermore, methyl acrylate is known to be compatible with both the RAFT (9) and the ITP (11) processes, so it should not have detrimental effect on the living copolymerization. In summary, this work aims at investigating the efficiency of RAFT and ITP processes for the copolymerization of vinylidene chloride with methyl acrylate. Special emphasis will be on the RAFT process because it is known to be a versatile and efficient process.

The efficiency of chain transfer agents (CTA) depends on their structure (9). Especially, for RAFT agents Z-C(S)S-R, the nature of the activating group Z and the leaving group R strongly influences the reactivity of the transfer agent. Accordingly, we have tested three dithioesters (Z= Ph) with different leaving groups: benzyl derivative 1, *tert*-butyl derivative 2, and 1-(ethoxycarbonyl)-ethyl derivative 3 (Figure 1). We have also used a xanthate (Z= OC$_2$H$_5$) 4 with a similar leaving group so that it can be compared to dithioester 3, giving an indication on the effect of the activating group Z which is either a phenyl or an ethoxy group. Lastly, we have used 1-phenylethyl iodide 5 as a chain transfer agent in ITP.

Herein, the kinetics of polymerization, the evolution of molecular weight and polydispersity with conversion, as well as the ability to prepare block copolymers will be discussed. Predici® will be used for numerical simulation in order to illustrate the possibilities and limitations of the living process for this system.

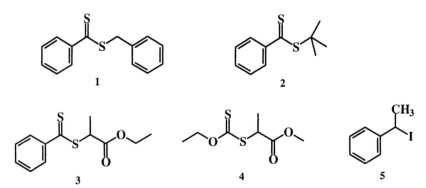

Figure 1. Structure of the reversible chain transfer agents 1-5 used in this work.

Experimental

Materials

Vinylidene chloride (VC_2, Aldrich, 99%), methyl acrylate (MA, Aldrich, 99%), and styrene (STY, Aldrich, 99%) were purified by vacuum distillation over anhydrous CaH_2. 2,2' – azobisisobutyronitrile (AIBN, Fluka, 98%) was recrystallized from 95% ethanol. S-(thiobenzoyl)thioglycolic acid (Aldrich, 99%), benzyl mercaptan (Aldrich, 99%), 2-methyl-2-propanethiol (Aldrich, 99%), ethyl 2-mercaptopropionate (Lancaster, 98%), octamethylcyclotetrasiloxane (D4, Aldrich, 98%), and benzene (SDS, 99.9%) were used as received. Xanthate α-(O-ethylxanthyl) methyl propionate 4 and 1-phenylethyl iodide 5 were synthesized in our laboratory according to the procedure of Charmot et al.(12) and Matyjaszewski et al.(11), respectively.

General procedure for the synthesis of dithioesters 1-3

S-(thiobenzoyl)thioglycolic acid (5.31g, 25 mmol) was dissolved in 30 mL of NaOH 1N, in a 100 mL, three-necked, round-bottom flask equipped with a magnetic stirrer, under argon. Thiol (25 mmol) was added dropwise to the reaction mixture at room temperature. The reaction mixture was stirred for 4-8 hours and was extracted with benzene. The organic layer was washed once with aqueous NaOH 1N solution and then three times with water, and dried over anhydrous sodium sulfate. Evaporation of the solvent under vacuum afforded the desired dithioester as a red liquid. ^{1}H NMR δ (CDCl$_3$) : 1 (yield 71%) : 7.9(2H, d), 7.4(8H, m), 4.7(2H, s); 2 (yield 75%) : 7.9(2H, d), 7.4(3H, m), 1.7(9H, s); 3 (yield 65%) : 8.0(2H, d), 7.4(3H, m), 4.7(1H, q, J=7.46), 4.3(2H, q, J=7.24), 1.7(3H, d, J=7.46), 1.3(3H, t, J=7.24).

Polymerizations

RAFT solution polymerizations were carried out in a 300 cm^3 inox autoclave (Parr instrument). The reaction mixture of vinylidene chloride (64.80 g, 6.68×10^{-1} mol), methyl acrylate (14.39 g, 1.67×10^{-1} mol), transfer agent (3.96×10^{-3} mol), and benzene (77.60 g, 9.95×10^{-1} mol) was introduced in the autoclave under nitrogen atmosphere. To launch the polymerization, a solution of AIBN (0.1445 g, 8.81×10^{-4} mol) in benzene (19.36 g, 2.48×10^{-1} mol) was added. Then, the reaction mixture was heated up to 70°C and the mechanical stirring speed was maintained at 200 rpm. The overall conversion was determined on aliquots, either by gravimetry (on samples quenched with hydroquinone and dried under vacuum at 40°C) or by ^{1}H NMR using octamethylcyclotetrasiloxane (D4) as internal standard. All RAFT copolymerizations of VC_2/MA reported in this work were performed at 70°C with $[VC_2]$= 3.74 mol.L^{-1}, [MA]= 9.35×10^{-1} mol.L^{-1}, [AIBN]= 5.00×10^{-3} mol.L^{-1}, [benzene]= 6.95 mol.L^{-1}. Block copolymerization with styrene monomer was performed in a glass schlenk reactor under argon.

ITP polymerizations were carried out in benzene at 70°C with AIBN as initiator, in 10 mL Carius tubes sealed under vacuum after purging with argon. After appropriate time, tubes were removed from the oven (shaking frame), frozen, and opened. Conversion was determined by gravimetry.

Analysis

Size Exclusion Chromatography (SEC) was performed on crude samples with a Waters Associates pump equipped with a Shodex RIse-61 refractometer detector and two 300 mm columns mixed-D PL-gel 5 μm from Polymer Laboratories (30°C). Tetrahydrofuran was used as eluent at a flow rate of 1.0 mL.min^{-1}. Calibration was performed with polystyrene standards from Polymer Laboratories. ^{1}H NMR spectra were recorded on a Bruker 200MHz instrument, chemical shifts are given in ppm using tetramethylsilane as reference and coupling constants are in Hz.

Numerical simulation

Numerical simulations of the copolymerization of VC_2/MA were performed with the Predici® software package (13), version 5.35.1, used in moments mode. We used the copolymerization module of the software with the following rate constants (T=70°C): dissociation rate constant of AIBN $k_{d(AIBN)}$=3.166×10^{-5} s^{-1} (14), efficiency f_{AIBN}=0.8 (rough estimation); propagation rate constants $k_{p,VC2}$=1785 L.mol^{-1}.s^{-1} (15) and $k_{p,MA}$=27700 L.mol^{-1}.s^{-1} (16); cross-propagation rate constants (10) $k_{p,VC2/MA}$=1983 L.mol^{-1}.s^{-1} and $k_{p,MA/VC2}$=29160 L.mol^{-1}.s^{-1}; termination rate constants $k_{t,VC2}$=4.16×10^8 L.mol^{-1}.s^{-1} (15) (dismutation mode) and $k_{t,MA}$=6.44×10^8 L.mol^{-1}.s^{-1} (17, 18)(combination mode); cross-termination rate constant was estimated by the mean value of the individual rate constants $k_{t,VC2/MA}$=($k_{t,VC2}$+$k_{t,MA}$)/2=5.3×10^8 L.mol^{-1}.s^{-1} (dismutation mode); transfer rate constant to VC_2 was calculated by an Arrhenius extrapolation from the work of Stockmayer (T=50-60°C) $k_{tr,VC2}$=11.35 L.mol^{-1}.s^{-1} (15).

Results and Discussion

Synthesis of dithioesters

The synthesis of dithioesters is usually tricky (*19*). The usual way involves the addition of phenyl magnesium bromide on carbon disulfide, and a nucleophilic substitution reaction on an alkyl halide. We decided to by-pass this tedious step by using a very straightforward transesterification method, in biphasic conditions, adapted from the work of Leon et al. (*20*). The selected thiol reacts almost instantaneously with the sodium salt of the commercially available dithioester. The course of the reaction is visible thanks to the red color of the dithioester: the water phase quickly changes from red to uncolored and the red product separates from water and is recovered in high yields for all dithioesters **1-3**. This is a very easy and quantitative synthetic route for primary, secondary and tertiary dithioesters in comparison with conventional methods. Of course, it is especially attractive when the appropriate thiol is readily available.

Effect of the nature of the activating group Z on RAFT copolymerization

RAFT agents **3** and **4** were tested to study the effect of the nature of Z (Figure 2). As expected, without transfer agent (blank experiment), the molecular weight is almost constant and the polydispersity is close to 2. With dithioester **3** as transfer agent, we observed an increase of the molecular weight with conversion while the polydispersity decreased down to about 1.5. This accounts for a control of the copolymerization by this dithioester. In contrast, with xanthate **4** as chain transfer agent, the molecular weight is rather high from the beginning of the polymerization (although lower than for the blank experiment, indicating a limited ability for transfer) and increases only slightly with conversion, with a polydispersity index still higher than 1.8. So, dithioester **3** is much more efficient than xanthate **4** as a RAFT agent. This result agrees well with the general knowledge on RAFT (*9, 21*) : the ethoxy group is not a very good activator for the radical addition to the thiocarbonyl (lower apparent chain transfer constant for xanthates in comparison with dithioesters).

Effect of the nature of the leaving group R on RAFT copolymerization

Dithioesters **1-3** were tested to study the effect of the nature of the leaving group R (Figure 3). With the benzyl derivative **1**, the molecular weight increases with conversion and the polydispersity index remains close to 1.6. With the *tert*-butyl derivative **2**, the overall behavior is essentially the same except that the polydispersity decreases slightly at the beginning of the polymerization. Dithioester **3** leads to a better control of the molecular weight and a smaller polydispersity index. The smaller molecular weight is obtained with dithioester **3**, indicating that CTA **3** has the highest apparent transfer constant in this series. The higher slopes of M_n versus conversion for derivatives **1** and **2** may arise from the lower ability of the expelled radicals to reinitiate the polymerization. Indeed, the benzyl radical slowly adds to monomers such as VC_2 and MA (k_{add}=430 $M^{-1}.s^{-1}$ at 23°C), and the *tert*-butyl radical quickly adds to monomers such as MA (k_{add}=1.1×10^6 $M^{-1}.s^{-1}$ at 27°C) but it suffers from possible side-product formation by disproportionation (isobutylene formation) (*22*).

Kinetics of RAFT copolymerization

The kinetics of RAFT copolymerization in the presence of CTA's **1-4** is shown in Figure 4 for a targeted molecular weight of 20 000 g.mol^{-1}. A blank experiment without transfer agent is also given as a reference. Xanthate **4** has almost no effect on the kinetics, but we have also previously shown that it is a rather poor reversible transfer agent. Concerning dithioesters, the benzyl derivative **1** shows a retardation effect while the *tert*-butyl derivative **2** is even slower. Finally, among dithioesters **1-3**, CTA **3** gives the fastest polymerization. Moreover, for higher targeted molecular weight (dithioester **2**, targeted molecular weight 50 000 g.mol^{-1}), the retardation effect is no longer visible. Thus, dithioesters cause an important retardation effect depending on their structure and concentration. This behavior has also been reported by others for RAFT polymerizations (*9, 23-25*). It is difficult to rationalize this retardation effect because the RAFT process involves many equilibria, especially when copolymerization is concerned.

ITP copolymerization

1-phenylethyl iodide **5** was tested in ITP copolymerization (Table I). A good correlation was found between experimental and theoretical molecular weight at high conversion, but the polydispersity index was higher than for RAFT copolymerization with dithioesters. This accounts for a lower apparent transfer constant for **5** in comparison with dithioesters **1-3**, as encountered in the case of styrene polymerization (*26*).

Limitations of the living copolymerization

In this part, we will illustrate the limitations of the living process related to vinylidene chloride. A numerical simulation of the blank experiment, using the

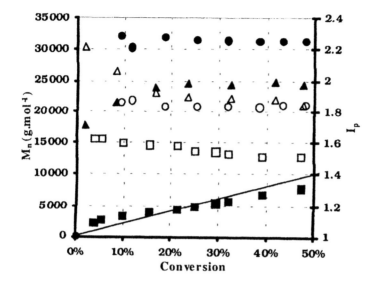

Figure 2. Evolution of M_n (black) and I_p (white symbols) versus conversion for the RAFT copolymerization of VC_2/MA : without CTA ($\bullet$,o), in the presence of CTA 3 ($\blacksquare$,□) and CTA 4 ($\blacktriangle$,$\triangle$) for theoretical $M_n = 20\ 000$ g.mol^{-1} (———).

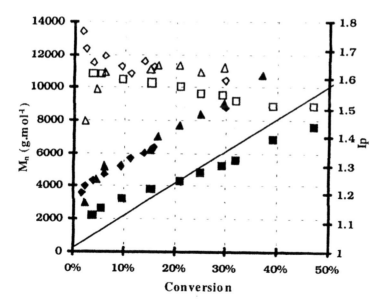

Figure 3. Evolution of M_n (black symbols) and I_p (white symbols) versus conversion for the RAFT copolymerization of VC_2/MA : in the presence of CTA 1 ($\blacktriangle$,$\triangle$), CTA 2 ($\blacklozenge$, $\lozenge$), CTA 3 ($\blacksquare$,□) for theoretical $M_n = 20\ 000$ g.mol^{-1} (———).

578

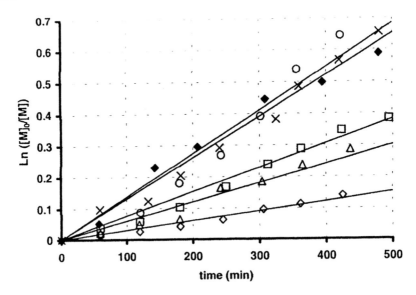

Figure 4. Evolution of Ln([M]₀/[M]) versus time for the RAFT copolymerization of VC₂/MA : without CTA (×), in the presence of CTA 1 (Δ), CTA 2 (◊), CTA 3 (□), CTA 4 (○) for theoretical M_n = 20 000 g.mol⁻¹, and CTA 2 (♦) for theoretical M_n= 50 000 g. mol⁻¹.

Table I. Copolymerization of VC₂/MA at 70°C by ITP

Run	RI	[AIBN]/[RI]	Time	Conversion	$M_{n, th}$	$M_{n, exp}$	I_p
A	5 [a]	0.27	15h	88%	7 102	8 679	2.06
B	A [b]	0.35	15h	82%	26 494	17 300	2.01

[a] [Benzene]= 4.87 M, [VC₂]= 5.31 M, [MA]= 1.59 M, [AIBN]= 2.25×10^{-2} M, [5]= 8.36×10^{-2} M;

[b] [Benzene]= 5.73 M, [VC₂]= 4.34 M, [MA]= 1.60 M, [AIBN]= 9.10×10^{-3} M, [A]= 2.58×10^{-2} M.

rate constants from the literature, is given in Figure 5. When transfer to monomer is neglected ($k_{tr,M}$=0), the obtained molecular weight is much higher than the experimental values. In contrast, when transfer to monomer is taken into account ($k_{tr,M}$=11.35 $L.mol^{-1}.s^{-1}$), the molecular weight approaches the experimental results ($k_{tr,M}$ seems overestimated). Although the numerical simulation should be considered with caution due to conflicting propagation rate constant values reported in the literature for VC₂ (27), Figure 5 suggests that there is an important effect of transfer to vinylidene chloride on the molecular weight. This should have a detrimental impact on the living copolymerization.

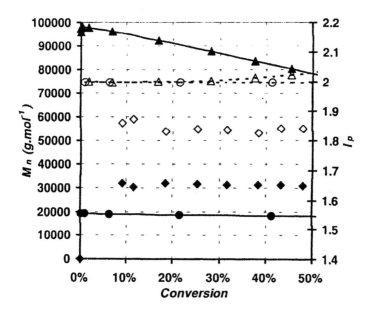

Figure 5. Evolution of M_n (black) and I_p (white symbols) versus conversion for the copolymerization of VC_2/MA at $70°C$ in the absence of CTA : experimental data ($\blacklozenge$, $\lozenge$), numerical simulation with $k_{tr,M}=0$ (M_n : $—\blacktriangle—$, I_p : $- - \vartriangle - -$), and numerical simulation with $k_{tr,M}=11.35$ $L.mol^{-1}.s^{-1}$ (M_n : $—\bullet—$, I_p : $- - \circ - -$).

A simple RAFT copolymerization model was constructed: a minimum of six transfer reactions was needed to complete the copolymerization module of Predici®. For simplicity, in order to be consistent with the copolymerization scheme, the apparent transfer rate constants were calculated from an overall apparent transfer constant $C_{tr,CTA}$. Each transfer rate constant is related to the corresponding homo- or cross-propagation rate constant: $k_{tr,i,j}=C_{tr,CTA} \times k_{p,i,j}$ (Scheme 2). Of course, such a simple kinetic scheme is not able to describe the retardation effect, but it is expected to give useful indications on the general trends of RAFT copolymerization for our system.

The effect of transfer to monomer on RAFT copolymerization was investigated for a targeted molecular weight of 20 000 g.mol^{-1} with dithioester 3. The simulation was performed with an arbitrary value $C_{tr,CTA}=10$ chosen to approach the experimental results at low conversion for an easier comparison (Figure 6, left). The difference between the two simulations (with or without transfer to monomer) is relatively small, indicating that there is a rather low effect of transfer to vinylidene chloride on the molecular weight up to about 50% monomer conversion for a targeted molecular weight of 20 000 g.mol^{-1}. Figure 6 (right) depicts the situation for various targeted molecular weights, taking into

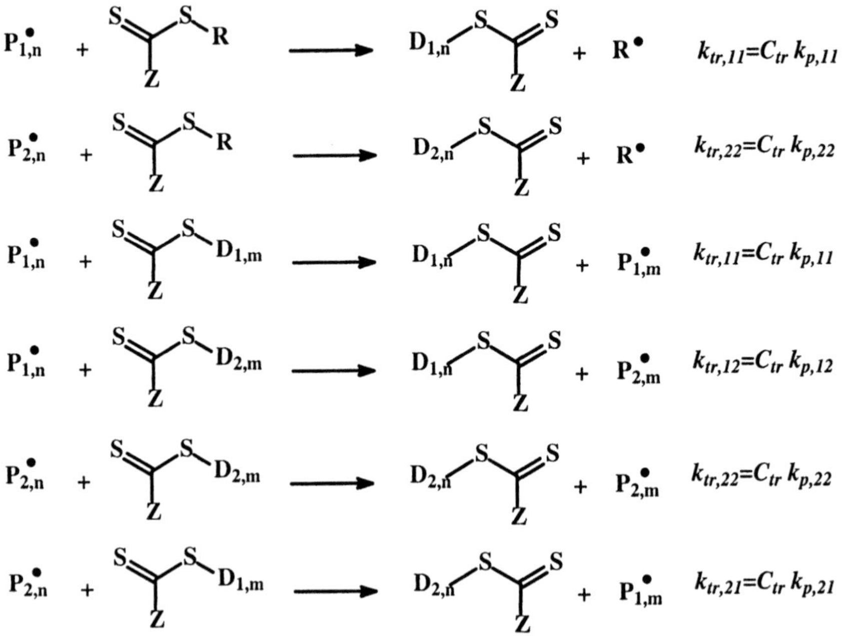

$P_{1,n}^{\bullet}$ + [structure: S=C(Z)-S-R] $\longrightarrow$ $D_{1,n}$-S-C(Z)=S + R$^{\bullet}$ $\qquad k_{tr,11}=C_{tr}\,k_{p,11}$

$P_{2,n}^{\bullet}$ + [structure: S=C(Z)-S-R] $\longrightarrow$ $D_{2,n}$-S-C(Z)=S + R$^{\bullet}$ $\qquad k_{tr,22}=C_{tr}\,k_{p,22}$

$P_{1,n}^{\bullet}$ + [structure: S=C(Z)-S-$D_{1,m}$] $\longrightarrow$ $D_{1,n}$-S-C(Z)=S + $P_{1,m}^{\bullet}$ $\qquad k_{tr,11}=C_{tr}\,k_{p,11}$

$P_{1,n}^{\bullet}$ + [structure: S=C(Z)-S-$D_{2,m}$] $\longrightarrow$ $D_{1,n}$-S-C(Z)=S + $P_{2,m}^{\bullet}$ $\qquad k_{tr,12}=C_{tr}\,k_{p,12}$

$P_{2,n}^{\bullet}$ + [structure: S=C(Z)-S-$D_{2,m}$] $\longrightarrow$ $D_{2,n}$-S-C(Z)=S + $P_{2,m}^{\bullet}$ $\qquad k_{tr,22}=C_{tr}\,k_{p,22}$

$P_{2,n}^{\bullet}$ + [structure: S=C(Z)-S-$D_{1,m}$] $\longrightarrow$ $D_{2,n}$-S-C(Z)=S + $P_{1,m}^{\bullet}$ $\qquad k_{tr,21}=C_{tr}\,k_{p,21}$

Scheme 2. Transfer reactions introduced in the RAFT copolymerization model for numerical simulation ($P_{i,n}$ and $D_{i,n}$ stand for propagating chains and dormant chains respectively, with monomer i as the last unit).

account transfer to monomer. For low targeted molecular weight (M_n= 5000 g.mol^{-1}), the control is good all along the polymerization. For intermediate targeted molecular weight (M_n=20 000 g.mol^{-1}), the increase of molecular weight is no longer linear above about 50% monomer conversion. Finally, for a targeted molecular weight of 50 000 g.mol^{-1}, the molecular weight clearly reaches a plateau at high conversion. The final molecular weight is about four times lower than the targeted molecular weight and the polydispersity is high. In the same time, percentage of dead chains increases up to high values. So, Figure 6 (right) shows that there is a strong limitation of the attainable molecular weight in the living radical copolymerization of VC$_2$ with MA. This was confirmed experimentally with dithioester **2**: the experimental molecular weight at about 50% conversion was only 15 000 g.mol^{-1} (instead of the theoretical value of 25 000 g.mol^{-1} for a truly living polymerization in the absence of transfer to monomer) and the polydispersity remained high (I_p=1.7-1.8).

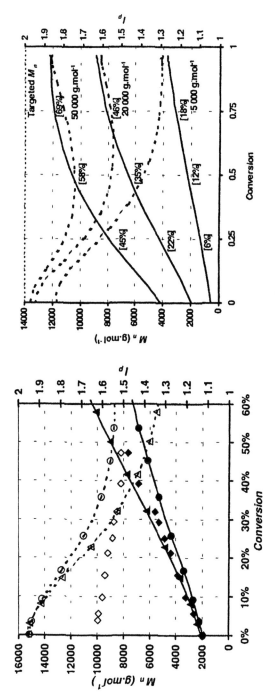

Figure 6. Left : Evolution of M_n (black symbols) and I_p (white symbols) versus conversion for the RAFT copolymerization of VC_2/MA in the presence of CTA 3 ($\bullet$, $\diamond$): numerical simulation with $k_{tr,M}=0$ (M_n : —▲—, I_p : - -△- -), numerical simulation with $k_{tr,M}=11.35$ L.mol^{-1}.s^{-1} (M_n : —●—, I_p : - - o - -). Right : Numerical simulation for the evolution of M_n (bold lines) and I_p (dashed lines) versus conversion for the RAFT copolymerization of VC_2/MA with $k_{tr,M}=11.35$ L.mol^{-1}.s^{-1} for three targeted M_n : 5 000 g.mol^{-1}, 20 000 g.mol^{-1} and 50 000 g.mol^{-1}. Percentage of dead chains is indicated in square brackets.

Possibilities of the living copolymerization

In spite of the limitations discussed above, we were interested in the synthesis of block copolymers. In a first approach, we checked the possibility of chain extension. The copolymer prepared by ITP was introduced as a macro transfer agent in copolymerization of VC_2 with MA (Table I). The SEC analysis clearly indicates an increase of the molecular weight, confirming the living nature of the ITP process for this system.

In a second approach, we first synthesized a short block by RAFT copolymerization of VC_2/MA with dithioester **3** (targeted molecular weight 15 000 g.mol^{-1}). The amount of dead chains could be estimated by numerical simulation of RAFT copolymerization. The polymerization was stopped at 55% monomer conversion, corresponding to 10% or 30% of dead chains without and with transfer to monomer, respectively. This first block **6** (M_n=6800 g.mol^{-1}, I_p=1.34) was then used as a macro transfer agent in RAFT polymerization of styrene at 70°C in bulk. According to the molecular weight of the first block, the targeted molecular weight of the second block was 23 000 g.mol^{-1}. Figure 7 shows that the polymerization follows pseudo-stationary conditions (left) whith a linear increase of the molecular weight and a small increase of the polydispersity index (right). Furthermore, as shown on Figure 8, the peak of the final diblock copolymer clearly shifts toward higher molecular weight and remains monomodal, whatever the detector (UV or RI). The tail in the low molecular weight region possibly accounts for the dead chains which were expected from the simulation. So, the successful reinitiation of the second block confirms the living nature of RAFT copolymerization of VC_2/MA in the first step.

Conclusions

We have established that degenerative chain transfer processes such as RAFT and ITP were able to control copolymerization of VC_2 with MA. Among iodo compounds, xanthates, and dithioesters, the last ones proved to be the more efficient reversible chain transfer agents (higher apparent chain transfer constant) although they may cause a decrease of the polymerization rate (retardation effect). Dithioester **3** ($PhC(S)SCH(CH_3)C(O)OC_2H_5$) gave the best results whereas dithioester **2** ($PhC(S)SC(CH_3)_3$) lead to the most important retardation effect. Transfer to vinylidene chloride leads to a limitation of the attainable molecular weight, as illustrated by numerical simulations and corresponding experiments at high targeted molecular weight. In spite of this limitation, the living nature of the copolymerization by ITP and RAFT is confirmed by the possible chain extension and formation of block copolymers.

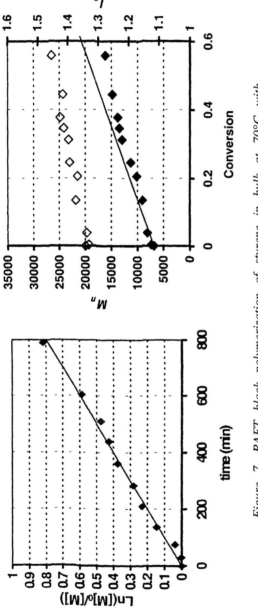

Figure 7. RAFT block polymerization of styrene in bulk at 70°C with 6/AIBN/styrene (1/0.6/222 in mol). Left : evolution of Ln([M]₀/[M]) versus time : experimental data (♦), best fit through data (——). Right : evolution of M_n (black symbols) and I_p (white symbols) versus conversion : experimental data (♦, ◊), theoretical line (——).

584

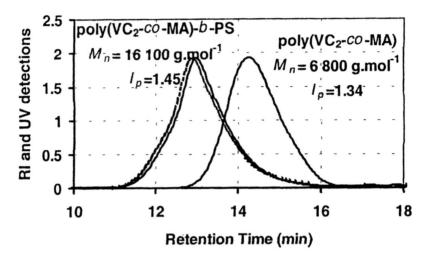

Retention Time (min)

Figure 8. SEC chromatograms of the polymer precursor poly(VC$_2$-co-MA) 6 and the diblock copolymer poly(VC$_2$-co-MA)-b-PS : refractive index detector (bold line), UV detector at 254nm (dashed line).

Acknowledgments. We thank Vincent Bodart and Jean-Raphaël Caille for valuable discussions and SOLVAY (Brussels) for financial support of this work. We thank B. Colomer and Y. Bastaraud for their contribution to this work.

References
1. Matyjaszewski, K. In *Controlled/Living Radical Polymerization;* Matyjaszewski, K., Ed.; ACS Symp. Series 768, American Chemical Society: Washington, DC, 2000, pp 2-26.
2. Tatemoto, M. In *Polymeric Materials Encyclopedia;* Salsmone, J. C., Ed.; CRC, Boca Raton, Florida, USA, 1996, Vol. 5, pp 3847-3860.
3. Hung, M.-H. U.S. Patent 5,231,154, 1993.
4. Apostolo, M.; Arcella, V.; Storti, G.; Morbidelli, M. *Macromolecules* **2002,** *35,* 6154-6166.
5. Jo, S. M.; Paik, H.-J.; Matyjaszewski, K. *Polym. Prepr. (Am. Chem. Soc., Div. Polym. Chem.)* **1997,** *38(1),* 699-700.
6. Bak, P. I.; Bidinger, G. P.; Cozens, R. J.; Klich, P. R.; Mayer, L. A. U.S. Patent 5,455,319, 1995.
7. Asandei, A.D.; Percec, V. *J. Polym. Sci. : Part A: Polym. Chem.* **2001,** *39,* 3392-3418.
8. Otazaghine, B.; Boutevin, B.; Lacroix-Desmazes, P. *Macromolecules* **2002,** *35,* 7634-7641.
9. Moad, G.; Chiefari, J.; Chong, Y. K.; Krstina, J.; Mayadunne R. T. A.; Postma, A.; Rizzardo, E.; Thang, S. H. *Polym. Int.* **2000,** *49,* 993-1001.

10. Collins, S.; Yoda, K.; Anazawa, N.; Birkinshaw, C. *Polymer Degradation and Stability* **1999**, *66*, 87-94.
11. Matyjaszewski, K.; Gaynor, S.; Wang, J.-S. *Macromolecules* **1995**, *28*, 2093-2095.
12. Charmot, D.; Corpart, P.; Adam, H.; Zard, S. Z.; Biadatti, T.; Bouhadir, G. *Macromol. Symp.* **2000**, *150*, 23-32.
13. Wulkow, M. *Macromol. Theory Simul.* **1996**, *5*, 393-416.
14. Bawn, C. E. H.; Verdin, D. *Trans. Faraday Soc.* **1960**, *56*, 815-822.
15. Matsuo, K.; Nelb, G. W.; Nelb, R. G.; Stockmayer, W. H. *Macromolecules* **1977**, *10*, 654-658.
16. Buback, M.; Kurz, H.; Schmaltz, C. *Macromol. Chem. Phys.* **1998**, *199*, 1721-1727.
17. Beuermann, S.; Buback, M.; Schmaltz, C. *Ind. Eng. Chem. Res.* **1999**, *38*, 3338-3344.
18. Buback, M.; Kowollik, C. *Macromolecules* **1999**, *32*, 1445-1452.
19. Ramadas, S. R.; Srinivasan, P. S.; Ramachandran, J.; Sastry, V. V. S. K. *Synthesis* **1983**, *August*, 605-622.
20. Leon, N. H.; Asquith, R. S. *Tetrahedron* **1970**, *26*, 1719-1725.
21. Rizzardo, E. *et al.* In *Controlled/Living Radical Polymerization;* Matyjaszewski, K., Ed.; ACS Symp. Series 768, American Chemical Society: Washington, DC, 2000, pp 278-296.
22. Walbiner, M.; Wu, J. Q.; Fischer, H. *Helvetica Chimica Acta* **1995**, *78*, 910-924.
23. Monteiro, M. J.; de Brouwer, H. *Macromolecules* **2001**, *34*, 349-352.
24. Kwak, Y.; Goto, A.; Tsujii, Y.; Murata, Y.; Komatsu, K.; Fukuda, T. *Macromolecules* **2002**, *35*, 3026-3029.
25. Barner-Kowollik, C.; Vana, P.; Quinn, J. F.; Davis, T. P. *J. Polym. Sci.: Part A: Polym. Chem.* **2002**, *40*, 1058-1063.
26. Fukuda, T.; Goto, A. In *Controlled/Living Radical Polymerization;* Matyjaszewski, K., Ed.; ACS Symp. Series 768, American Chemical Society: Washington, DC, 2000, pp 27-38.
27. Sakai, H.; Kihara, Y.; Fujita, K.; Kodani, T.; Nomura, M. *J. Polym. Sci.: Part A: Polym. Chem.* **2001**, *39*, 1005-1015.

Chapter 40

RAFT Polymerization in Homogeneous Aqueous Media

Andrew B. Lowe[1,*], Brent S. Sumerlin[2], Michael S. Donovan[2],
David B. Thomas[2], Pierre Hennaux[2], and Charles L. McCormick[1,2,*]

Departments of [1]Chemistry and Biochemistry and [2]Polymer Science,
University of Southern Mississippi, Hattiesburg, MS 39406

Controlled radical polymerization has been the focus of intense research during the last decade. However, to date, research has focused primarily on polymerizations conducted in homogeneous organic media or bulk with common monomers such as styrene, methyl methacrylate or butyl acrylate. The ability to conduct polymerizations in *homogeneous aqueous* solution is advantageous from both environmental and commercial viewpoints. In this chapter we will present a summary of our research efforts in the area of aqueous Reversible Addition-Fragmentation chain Transfer (RAFT) polymerization. We will demonstrate the broad versatility of RAFT, showing its applicability to neutral, anionic, cationic, and zwitterionic monomers from a range of monomer classes, including styrenics, acrylamides and (meth)acrylates.

Reversible Addition-Fragmentation chain Transfer (RAFT) polymerization is a controlled free radical polymerization (CRP) technique based on the principle of degenerative chain transfer as a means of conferring 'living' characteristics (*1*). The key additive in RAFT polymerizations is a suitable thiocarbonylthio compound (chain transfer agent or CTA), which is typically a dithioester (*1*), or other suitable species including dithiocarbamates (*2,3*), xanthates (*4,5*), trithiocarbonates (*6*), and phosphoryl-/(thiophosphoryl)dithioformates (*7*), The generally accepted mechanism for RAFT is shown in Scheme 1. Step **I** is initiation, with the generation of radicals from a suitable source such as an azo compound and subsequent addition of the radicals to monomer. Steps **II** and **III** constitute the so-called RAFT pre-equilibrium, during which all the RAFT CTA is consumed/activated with some propagation. Step **IV** is the main RAFT equilibrium involving chain equilibration and propagation. Step **V**, like any of the CRP methods, represents possible termination pathways.

Ideally, the degenerative transfer of the thiocarbonylthio species is fast relative to the rate of propagation and thus living behavior is observed. As with the traditional living polymerization techniques, as well as other CRP methods, RAFT facilitates the synthesis of (co)polymers with complex structures such as block (AB, ABA, ABC), graft, statistical, comb, gradient and star architectures. One advantage of RAFT is its versatility with respect to monomer choice. In principle, any monomer that is susceptible to polymerization by traditional free radical methods may be polymerized by RAFT, although judicious choice of the RAFT CTA is often required. Many examples already detail the application of RAFT polymerizations in organic media (*8-11*), heterogeneous aqueous media (*12-17*), and bulk (*18-20*).

We have a long-standing interest in water-soluble polymers and recently have been examining RAFT as a technique for the synthesis of novel amphiphilic polymers directly in aqueous media. At present there are very few reports detailing homogeneous aqueous polymerizations by RAFT, or utilizing other CRP methods for that matter (*21*).

Neutral Monomers/Polymers

N,N-Dimethylacrylamide (DMA) is a commercially important hydrophilic monomer used in applications such as pharmaceutics and personal care products. The CRP of DMA has proven difficult by both stable free radical and atom transfer radical polymerization techniques. Recently we disclosed the CRP, via RAFT, of DMA in *organic* media utilizing novel CTAs that had been designed such that the R-fragments were structurally and electronically similar to DMA and thus were expected to be highly efficient mediating species (these were synthesized by a novel thioacylation reaction) (*22*). Indeed, the application of N,N-dimethyl-s-thiobenzoylthiopropionamide (TBP - **B**) as the RAFT CTA

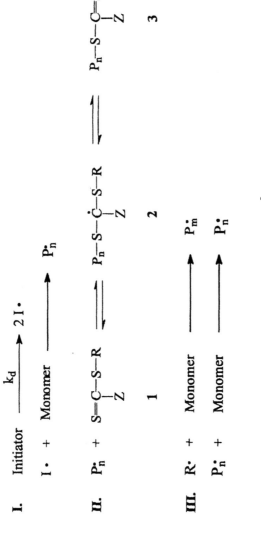

I. Initiator $\xrightarrow{k_d}$ 2 I·

    I· + Monomer $\longrightarrow$ P$_n^{\bullet}$

II. P$_n^{\bullet}$ + S=C—S—R $\rightleftharpoons$ P$_n$—S—$\overset{\bullet}{C}$—S—R $\rightleftharpoons$ P$_n$—S—C=S + R·
                $\underset{Z}{|}$ $\underset{Z}{|}$

                1   **2**   **3**   **4**

III. R· + Monomer $\longrightarrow$ P$_m^{\bullet}$

     P$_n^{\bullet}$ + Monomer $\longrightarrow$ P$_n^{\bullet}$

IV. P$_m^{\bullet}$ + S=C—S—P$_n$ $\rightleftharpoons$ P$_m$—S—$\overset{\bullet}{C}$—S—P$_n$ $\rightleftharpoons$ P$_m$—S—C=S + P$_n^{\bullet}$
               $\underset{Z}{|}$ $\underset{Z}{|}$ $\underset{Z}{|}$

               3   **5**   **6**

     Monomer   Monomer

V.   I•, R•, $P_n^•$, $P_m^•$, 2, 5 $\longrightarrow$ Dead polymer

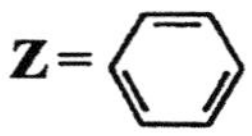

Z=

Scheme 1. The RAFT Mechanism

yielded polyDMA with good molecular weight control and narrow molecular weight distributions.

Figure 1. Chemical structures of N,N-dimethyl-s-thiobenzoylthioacetamide (TBA - A) and N,N-diethyl-s-thiobenzoylthiopropionamide (TBP – B)

Given the encouraging results obtained for the polymerization of DMA in benzene using TBP, we decided to explore the possibility of polymerizing DMA in aqueous media employing TBP and 4-cyanopentanoic acid dithiobenzoate (CTP) as the RAFT CTAs (*23*). While CTP proved effective at 60, 70 and 80 °C, TBP yielded little-to-no polymer at 60 and 70 °C and exhibited an induction period of ~ 80 min at 80 °C prior to displaying kinetics similar to those observed for the CTP-mediated polymerization. The addition of DMF as a cosolvent to the TBP-mediated polymerizations reduced the induction period to ~ 35 min, but it was not eliminated. The difference in behavior between the organic-based and aqueous-based polymerizations of DMA with TBP is thought to be related to the relative hydrophobicity of the CTA.

Anionic Monomers/Polymers

Our initial interest in anionic monomers focused on two styrenic species, namely sodium 4-styrenesulfonate (NaSSO$_3$) and 4-vinylbenzoic acid (VBZ) (*24*). VBZ is particularly interesting because of its tunable hydrophilicity/hydrophobicity as a function of solution pH. NaSSO$_3$ and VBZ homopolymers were synthesized directly in water employing 4,4'-azobis(4-cyanopentanoic acid) (V-501) as the radical source and CTP as the RAFT CTA. The combination of V-501 and CTP ensures that all the initiating fragments (whether initiator- or CTA-derived) are identical, see Scheme 2.

These conditions led to rapid rates of polymerization with near-quantitative conversion observed for NaSSO$_3$ after only ~ 100 min at 70 °C. The resulting experimentally determined molecular weights (aqueous size exclusion chromatography in 80% 0.05 M NaNO$_3$/0.01 M Na$_2$HPO$_4$ / 20% CH$_3$CN, calibrated with PNaSSO$_3$ standards) were in excellent agreement with theory: $M_{ntheory}$ = 21,000. M_{nsec} = 19,800. M_{wsec} = 22,200. M_w/M_n = 1.12 for example. We also demonstrated the ability to use the NaSSO$_3$ homopolymers as macro-CTAs for the block copolymerization with VBZ. The resulting NaSSO$_3$-VBZ

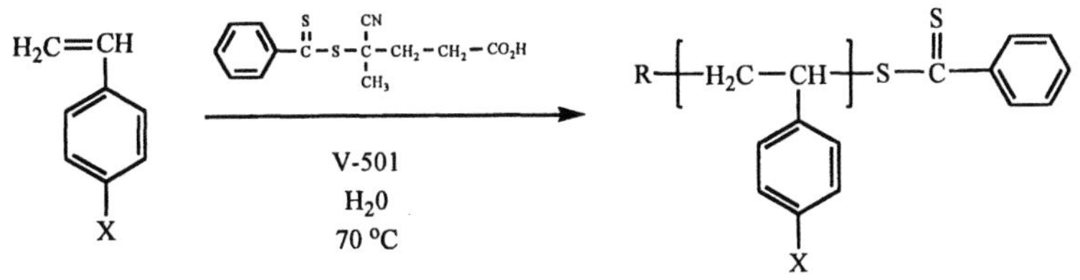

591

Scheme 2. *RAFT homopolymerization of NaSSO₃ and VBZ*

AB diblock copolymer was shown to undergo reversible pH-induced micellization due to the 'smart' properties of the polyVBZ block. Subsequently we turned our attention to the acrylamido family of monomers and investigated the RAFT polymerization of sodium 2-acrylamido-2-methylpropanesulfonate (AMPS) and sodium 3-acrylamido-3-methylbutanoate (AMBA) (*25*). As with the NaSSO₃ and VBZ homopolymers, AMPS and AMBA were homopolymerized in water employing V-501 as the radical source and CTP as the RAFT CTA.

Figure 2 shows the linear increase in molecular weight with conversion for an AMPS homopolymer, indicative of a controlled polymerization. Polydispersity increases with conversion but remains relatively low at < 1.30. Figure 3 shows the kinetic plot for the same polymerization. After an initial induction period of ~ 65 min, the polymerization exhibits pseudo-first order kinetics, implying a constant concentration of radicals. The observed induction period is not an uncommon feature of RAFT polymerizations and has been observed previously. The ability to form block copolymers employing either AMPS or AMBA macro-CTAs for the subsequent block copolymerization of the second monomer, was also demonstrated, see Figure 4. We are currently in the process of examining the aqueous solution properties of a series of AMPS-AMBA AB diblock copolymers, but we expect them to show similar pH-induced aggregation as the analogous styrenic-based diblocks since the AMBA blocks are tunably hydrophilic/hydrophobic while the AMPS blocks are permanently hydrophilic.

Cationic Monomers/Polymers

During our early research involving the anionic styrenic monomers, we also examined two cationic styrenic derivatives: N,N-dimethylvinylbenzylamine (DMVBAC) and (*ar*-vinylbenzyl)trimethylammonium chloride (VBTAC) (*24*), see Figure 5.

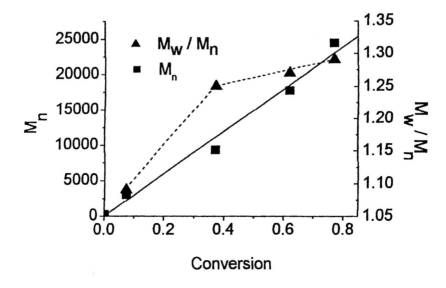

Figure 2. Molecular weight versus conversion plot for an AMPS homopolymerization. Reproduced with permission from *Macromolecules* **2001**, *34*, 6561-6564. Copyright 2001 Am. Chem. Soc.

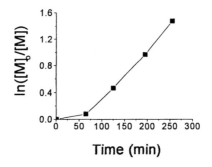

Figure 3. Pseudo-first order rate plot for an AMPS homopolymerization. Reproduced with permission from *Macromolecules* **2001**, *34*, 6561-6564. Copyright 2001 Am. Chem. Soc.

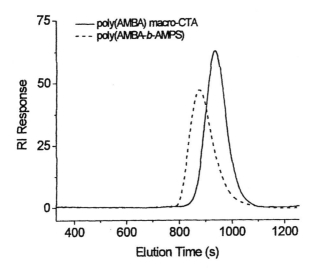

Figure 4. ASEC chromatograms demonstrating diblock copolymer formation. Reproduced with permission from Macromolecules **2001**, *34*, 6561-6564. Copyright 2001 Am. Chem. Soc.

As with NaSSO$_3$ and VBZ, the DMVBAC (**A**) is tunably hydrophilic/hydrophobic via reversible protonation of the tertiary amine residues, while VBTAC is permanently hydrophilic. Homopolymers of VBTAC were prepared in aqueous media employing the V-501/CTP initiator/CTA pair at 70 °C. The PVBTAC was then used as a macro-CTA for the block polymerization of *N,N*-dimethylvinylbenzylammonium chloride. The resulting AB diblock copolymer had a unimodal molecular weight distribution (M$_{ntheory}$: 11,700. M$_{nSEC}$ = 11,400. M$_{wSEC}$ = 12,500. M$_w$/M$_n$ = 1.10 – based on poly(2-vinylpyridine) standards). Also the observed block copolymer composition, as determined by ^{1}H NMR spectroscopy (49:51 mol basis), was in excellent agreement with the theoretical composition (50:50 mol basis). It was also shown that these diamine AB diblock copolymers were capable of reversible pH-induced micellization, with aggregates of ~ 38 nm being observed by dynamic light scattering under high pH conditions (here the DMVBAC block is deprotonated and hydrophobic and thus the block copolymer self assembles with the DMVBAC block in the core, surrounded by the hydrophilic VBTAC coronal chains).

More recently we have been examining the synthesis of hydrophilic-hydrophilic AB diblock copolymers comprised of DMA with either DMVBAC or VBTAC (*26*). It is known that when synthesizing AB diblock copolymers by sequential monomer addition, especially for blocks comprised of monomers from two different families, that the order of polymerization can be extremely important (*1*).

594

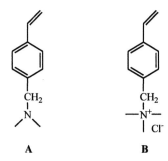

A   B

Figure 5. Chemical structures of N,N-dimethylvinylbenzylamine (A) and (ar-vinylbenzyl)trimethylammonium chloride (B)

Figure 6 shows the observed ASEC chromatograms for the synthesis of a P(DMVBAC-block-DMA) copolymer employing PDMVBAC as a macro-CTA. While block copolymer formation is observed, there is also clearly residual homopolymer and significant high molecular weight impurity. Indeed, even less well-defined block copolymers were obtained when the permanently charged PVBTAC homopolymer was employed as the macro-CTA. On the other hand, Figure 7 shows the ASEC chromatograms obtained for the block polymerization of DMVBAC employing PDMA as the macro-CTA. Under these conditions quantitative blocking efficiency is observed with the resulting AB diblock copolymers having unimodal and narrow molecular weight distributions, see Table I. These results further emphasize that the order of polymerization for the synthesis of AB diblock copolymers can be extremely important and indicates for the preparation of acrylamido/styrenic block copolymers that the acrylamido monomer should be polymerized first and used as the macro-CTA.

Zwitterionic Monomers/Polymers

Polymeric betaines are interesting for numerous reasons that include their anti-polyelectrolyte behavior (*27*) and their bio/hemocompatibility (*28*). At present very few examples exist of controlled structure polybetaines (*29-36*) and most were prepared via group transfer polymerization followed by post-polymerization modification (*29-32*). Recently, we reported the ability to polymerize styrenic, methacrylate and acrylamido-sulfopropylbetaines directly in aqueous salt employing CTP as the RAFT CTA (*37*). The order for the rate of polymerization decreases in the order methacrylate > styrenic > acrylamido

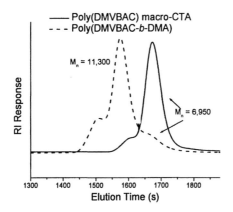

Figure 6. SEC Chromatograms for the preparation of AB diblock copolymers employing PDMVBAC as a macro-CTA

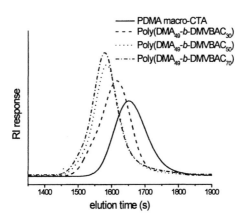

Figure 7. Observed ASEC chromatograms for the synthesis of a P(DMA-block-DVBAC) copolymer employing PDMA as a macro-CTA

Table I. Molecular Weight and Polydispersity Data for the P(DMA-*block*-DMVBAC) Copolymers

Theoretical DP DMVBAC Block	Theoretical M_n DMVBAC Block	Expt. M_n DMVBAC Block*	Copolymer M_n*	PDI*
30	5,900	7,100	12,000	1.20
50	9,900	9,400	14,300	1.17
70	13,800	10,000	14,900	1.17

*As determined by ASEC (0.1 M Na_2SO_4 / 1% acetic acid as eluent, calibration with P2VP standards)

The slower rate of polymerization for MAEDAPS is interesting considering that the rate constant for propagation is typically higher for acrylamido-based monomers than either styrenics or methacrylates under classical free radical polymerization conditions. The slower kinetics exhibited by MAEDAPS may be indicative of a higher rate constant of addition for the propagating MAEDAPS macro-radical towards a macro-CTA or, possibly, a lower rate constant for fragmentation of the macro-RAFT intermediate radical (*38,39*). Considering the higher reactivity and lower bulk of the MAEDAPS acrylamido radical compared to the DMAPS methacrylate radical, contributions from both possibilities are likely. The results are listed in Table II.

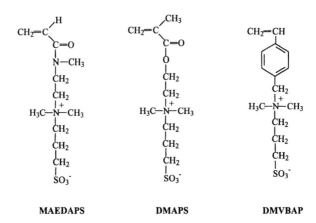

Figure 8. Chemical structures of 3-[2-(N-methylacrylamido)-ethyldimethyl ammonio]propane sulfonate (MAEDAPS), 3-[N-(2-methacroyloyethyl)-N,N-dimethylammonio]propane sulfonate (DMAPS), and 3-(N,N-dimethylvinylbenzylammonio)propane sulfonate (DMVBAPS).

Table II. Summary of the Molecular Weights and Polydispersities for the Homopolymers of MAEDAPS, DMAPS and DMVBAPS. Reproduced with permission from *Macromolecules* **2002**, *35*, 8663-8666. Copyright 2002 Am. Chem. Soc.

Sample	Conversion (%)[b]	dn/dc[c]	Theoretical M_n	Observed M_n[d]	M_w/M_n[d]
PMAEDAPS[a]	91	0. 1533	44,100	58,250	1.08
PDMAPS[a]	93	0.1293	45,700	47,500	1.04
PDMVBAPS[a]	90	0.1578	44,900	47,200	1.06

a. Prepared using 4-cyanopentanoic acid dithiobenzoate as the RAFT CTA
b. As determined by the residual monomer concentration employing the RI detector
c. Measured using Wyatt's Optilab Interferometric refractometer in 80% 0.5 M NaBr / 20% CH_3CN.
d. As determined by aqueous size exclusion chromatography in 80% 0.5 M NaBr / 20% CH_3CN using Wyatt's DAWN EOS multi-angle laser light scattering detector.

Modification of Gold Surfaces

By virtue of the RAFT mechanism, (co)polymers prepared via this technique bear thiocarbonylthio end-groups. We recently demonstrated the ability to utilize RAFT-synthesized (co)polymers for the effective stabilization of gold nanoparticles (*40,41*). The method takes advantage of the facile *in-situ* reduction of the dithioester end-groups with preformed gold colloids (Scheme 3).
We successfully stabilized gold nanoparticles with neutral, anionic, cationic and zwitterionic (co)polymers. The simultaneous reduction was performed in aqueous media using $NaBH_4$. Interestingly, when a MAEDAPS-DMA block

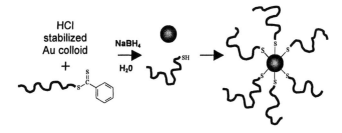

Scheme 3. Preparation of (co)polymer-stabilized Au nanoparticles. Reproduced with permission from *J. Am. Chem. Soc* **2002**, *124*, 11562-11563. Copyright 2002 Am. Chem. Soc.

copolymer was used, the colloidal solutions were only stable in aqueous salt solution (here the DMA block was covalently attached to the Au surface and thus was stabilized by outer betaine chains). This stability is consistent with the solubility characteristics of polymeric betaines which are generally insoluble in pure water but become soluble upon the addition of salt. Transmission electron microscopy (TEM) was used to verify the presence of the attached polymers as evidenced by their stabilization of the Au nanoparticles, see Figure 9.

This surface modification chemistry has been extended to flat Au surfaces as well. The dithioester end-groups of RAFT-synthesized (co)polymers were reduced with aqueous $NaBH_4$ in the presence of gold-coated silica slides. Successful surface modification was confirmed by atomic force microscopy (AFM) (Figure 10) and contact angle measurements.

Conclusions

Herein we have summarized some of our research concerning homogeneous aqueous RAFT polymerizations. We have demonstrated the versatility of RAFT by detailing its applicability to a wide range of functional monomers using a variety of RAFT CTAs. We have also shown that it is possible to covalently

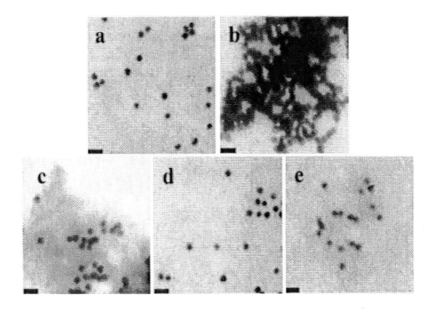

Figure 9. TEM micrographs of Au sol (a), reduced Au sol (b), PAMPS-stabilized Au-NPs (c), PVBTAC-stabilized Au-NPs (d), and PDMA-stabilized Au-NPs (e). Scale bars correspond to 40 nm. Reproduced with permission from *J. Am. Chem. Soc* **2002**, *124*, 11562-11563.

599

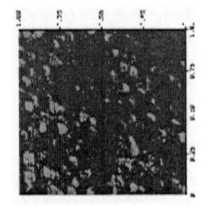

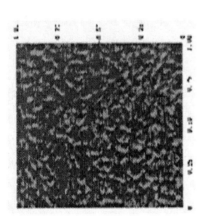

B

A

*Figure 10. AFM phase images of gold surfaces modified with PVBTAC (**A**) and PNaSSO$_3$ (**B**). (z range = 120°)*

attach RAFT-synthesized polymers in a facile manner to Au surfaces by a simple *in situ* reduction process.

Acknowledgements

We would like to thank Energizer Battery Company, GelTex Pharmaceuticals, and the U. S. Department of Energy for financial support.

References

(1) Chiefari, J.; Chong, Y. K.; Ercole, F.; Krstina, J.; Jeffery, J.; Le, T., P. T.; Mayadunne, R. T. A.; Meijs, G. F.; Moad, C. L.; Moad, G.; Rizzardo, E.; Thang, S. H. *Macromolecules* **1998**, *31*, 5559-5562.

(2) Destarac, M.; Charmot, D.; Franck, X.; Zard, S. Z. *Macromol. Rapid Commun.* **2000**, *21*, 1035-1039.

(3) Mayadunne, R., T. A.; Rizzardo, E.; Chiefari, J.; Chong, Y. K.; Moad, G.; Thang, S., H. *Macromolecules* **1999**, *32*, 6977-6980.

(4) Taton, D.; Wilczewska, A.-Z.; Destarac, M. *Macromol. Rapid Commun.* **2001**, *22*, 1497-1503.

(5) Ladavière, C.; Dörr, N.; Claverie, J. P. *Macromolecules* **2001**, *34*, 5370-5372.

(6) Mayadunne, R., T. A.; Rizzardo, E.; Chiefari, J.; Krstina, J.; Moad, G.; Postma, A.; Thang, S., H. *Macromolecules* **2000**, *33*, 243-245.

(7) Laus, M.; Papa, R.; Sparnacci, K.; Alberti, A.; Benaglia, M.; Macciantelli, D. *Macromolecules* **2001**, *34*, 7269-7255.

(8) de Brouwer, H.; Schellekens, M. A. J.; Klumperman, B.; Monteiro, M. J.; German, A. L. *J. Polym. Sci., Polym. Chem.* **2000**, *38*, 3596-3603.

(9) Ganachaud, F.; Monteiro, M. J.; Gilbert, R. G.; Dourges, M.-A.; Thang, S. H.; Rizzardo, E. *Macromolecules* **2000**, *33*, 6738-6745.

(10) He, T.; Zou, Y.-F.; Pan, C.-Y. *Polymer Journal* **2002**, *34*, 138-143.

(11) Schilli, C.; Lanzendörfer, M. G.; Muller, A. H. E. *Macromolecules* **2002**, *35*, 6819-6827.

(12) de Brouwer, H.; Tsavalas, J. G.; Schork, F. J.; Monteiro, M. J. *Macromolecules* **2000**, *33*, 9239-9246.

(13) Charmot, D.; Corpart, P.; Adam, H.; Zard, S. Z.; Biadatti, T.; Bouhadir, G. *Macromol. Symp.* **2000**, *150*, 23-32.

(14) Monteiro, M. J.; Hodgson, M.; de Brouwer, H. *J. Polym. Sci., Polym. Chem.* **2000**, *38*, 3864-3874.

(15) Monteiro, M. J.; Sjoberg, M.; van der Vlist, J.; Göttgens, C., M. *J. Polym. Sci., Polym. Chem.* **2000**, *38*, 4206-4217.

(16) Kanagasabapathy, S.; Sudalai, A.; Benicewicz, B. C. *Macromol. Rapid Commun.* **2001**, *22*, 1076-1080.

601

(17) Lansalot, M.; Davis, T. P.; Heuts, J. P. A. *Macromolecules* **2002**, *35*, 7582-7591.
(18) Quinn, J. F.; Rizzardo, E.; Davis, T. P. *Chem. Commun.* **2001**, 1044-1045.
(19) Quinn, J. F.; Barner, L.; Barner-Kowollik, C.; Rizzardo, E.; Davis, T. P. *Macromolecules* **2002**, *35*, 7620-7627.
(20) Quinn, J. F.; Barner, L.; Davis, T. P.; Thang, S. H.; Rizzardo, E. *Macromol. Rapid Commun.* **2002**, *23*, 717-721.
(21) Lowe, A. B.; McCormick, C. L. *Aust. J. Chem.* **2002**, 367-380.
(22) Donovan, M. S.; Lowe, A. B.; Sumerlin, B. S.; McCormick, C. L. *Macromolecules* **2002**, *35*, 4123-4132.
(23) Donovan, M. S.; Sanford, T. A.; Lowe, A. B.; Sumerlin, B. S.; Mitsukami, Y.; McCormick, C. L. *Macromolecules* **2002**, *35*, 4570-4572.
(24) Mitsukami, Y.; Donovan, M. S.; Lowe, A. B.; McCormick, C. L. *Macromolecules* **2001**, *34*, 2248-2256.
(25) Sumerlin, B. S.; Donovan, M. S.; Mitsukami, Y.; Lowe, A. B.; McCormick, C. L. *Macromolecules* **2001**, *34*, 6561-6564.
(26) Sumerlin, B. S.; Lowe, A. B.; McCormick, C. L. *unpublished results*.
(27) Lowe, A. B.; McCormick, C. L. In *Stimuli-Responsive Water Soluble and Amphiphilic Polymers*; McCormick, C. L., Ed.; ACS: Washington DC, 2001; Vol. 780, pp 1-13.
(28) Lowe, A. B.; Vamvakaki, M.; Wassall, M. A.; Wong, L.; Billingham, N. C.; Armes, S. P.; Lloyd, A. W. *J. Biomed. Mater. Res.* **2000**, *52*, 88-94.
(29) Lowe, A. B.; Billingham, N. C.; Armes, S. P. *Chem. Commun.* **1996**, 1555-1556.
(30) Bütün, V.; Bennett, C. E.; Vamvakaki, M.; Lowe, A. B.; Billingham, N. C.; Armes, S. P. *J. Mater. Chem.* **1997**, *7*, 1693-1695.
(31) Tuzar, Z.; Pospisil, H.; Plestil, J.; Lowe, A. B.; Baines, F. L.; Billingham, N. C.; Armes, S. P. *Macromolecules* **1997**, *30*, 2509-2512.
(32) Lowe, A. B.; Billingham, N. C.; Armes, S. P. *Macromolecules* **1999**, *32*, 2141-2148.
(33) Weaver, J. V. M.; Armes, S. P.; Bütün, V. *Chem. Commun.* **2002**.
(34) Arotçaréna, M.; Heise, B.; Ishaya, S.; Laschewsky, A. *J. Am. Chem. Soc.* **2002**, *124*, 3787-3793.
(35) Ma, Y.; Armes, S. P.; Billingham, N. C. *Polym. Mater. Sci. Eng.* **2001**, *84*, 143-144.
(36) Lobb, E. J.; Ma, I.; Billingham, N. C.; Armes, S. P.; Lewis, A. L. *J. Am. Chem. Soc.* **2001**, *123*, 7913-7914.
(37) Donovan, M. S.; Sumerlin, B. S.; Lowe, A. B.; McCormick, C. L. *Macromolecules* **2002**, *35*, 8663-8666.
(38) Moad, G.; Chiefari, J.; Chong, Y. K.; Krstina, J.; Mayadunne, R. T. A.; Postma, A.; Rizzardo, E.; Thang, S. H. *Polymer International* **2000**, *49*, 993-1001.

(39) Barner-Kowollik, C.; Quinn, J. F.; Nguyen, T. L. U.; Heuts, J. P. A.; Davis, T. P. *Macromolecules* **2001**, *34*, 7849-7857.
(40) Lowe, A. B.; Sumerlin, B. S.; Donovan, M. S.; McCormick, C. L. *J. Am. Chem. Soc* **2002**, *124*, 11562-11563.
(41) Sumerlin, B. S.; Donovan, M. S.; Mitsukami, Y.; Lowe, A. B.; McCormick, C. L. *Poly. Prep.* **2002**, *43*, 313-315.

Chapter 41

RAFT Polymers: Novel Precursors for Polymer– Protein Conjugates

Christine M. Schilli[1], Axel H. E. Müller[1,*], Ezio Rizzardo[2], San H. Thang[2], and (Bill) Y. K. Chong[2]

[1]Makromolekulare Chemie II, Universität Bayreuth, D–95440 Bayreuth, Germany
[2]CSIRO Molecular Science, Bag 10, Clayton South, Victoria 3169, Australia

Temperature- and pH-sensitive polymers and block copoly-mers have been synthesized via reversible addition-fragmenta-tion chain transfer (RAFT) polymerization. Thiocarbonylthio compounds used as chain transfer agents in the polymerization lead to end-functionalized polymers that give the correspond-ing thiols upon hydrolysis. Both block copolymers with short blocks of oligo(active ester)s and thiol-terminated polymers can be used for polymer-protein conjugation. 2-Vinyl-4,4-dimethyl-5-oxazolone (VO), N-hydroxysuccinimide methacrylate (NHSM), diacetone acrylamide (DAA), N-iso-propylacrylamide (NIPAAm), and acrylic acid (AA) have been polymerized to homo- and block copolymers with narrow MWD in most cases, and their conjugation to model peptides has been investigated. Block copolymers of NIPAAm and AA were investigated for their response to combined external stimuli. Thus, thermo- and pH-responsive systems can be cre-ated for control of enzyme activity or molecular recognition processes.

Recently, the synthesis of polymers via reversible addition-fragmentation chain transfer (RAFT) polymerization has gained importance due to its great versatility, its compatibility with a wide range of monomers, the control of the molecular weights, and the low polydispersities of the resultant polymers (*1,2*). Besides, it offers all advantages of conventional free radical polymerization. RAFT polymerizations are carried out in the presence of thiocarbonylthio compounds of general structure Z-C(=S)S-R and result in the formation of end-functionalized polymers (*3*).

Poly(*N*-isopropylacrylamide) (polyNIPAAm) exhibits a lower critical solution temperature (LCST) of 32 °C in aqueous solutions. With the conjugation of the LCST polymer to a peptide or protein, thermoresponsive systems are created. Poly(acrylic acid) can be used to enable response to pH and ionic strength. Proteins conjugated to block copolymers of NIPAAm and acrylic acid are especially interesting due to their response to combined external stimuli. Potential applications of these "smart" polymer systems are manifold and range from the controlled release of enzymes/drugs to bioseparations and sensor or actuator systems.

Two different routes for polymer-protein conjugation have been envisaged: (a) RAFT polymerization in order to obtain dithiocarbonyl endgroups which are hydrolyzed to the corresponding thiols and can be conjugated to thiol functions of proteins; (b) synthesis of block copolymers with short blocks of oligo(active ester)s for conjugation to the amino functions of proteins. Poly(2-vinyl-4,4-dimethyl-5-oxazolone) (polyVO) and poly(*N*-hydroxysuccinimide methacrylate) (polyNHSM) consist of active ester units that are frequently used for conjugation to amino moieties of proteins (*4,5*).

Experimental Section

Materials

N-Isopropylacrylamide (Aldrich, 97 %) was recrystallized twice from benzene/hexane 3:2 (v:v) and dried under vacuum prior to use. Acrylic acid was purified according to the standard procedure. The monomers 2-vinyl-4,4-dimethyl-5-oxazolone (TCI Tokyo), diacetone acrylamide (Lubrizol Co.) and *N*-hydroxysuccinimide methacrylate (donation from Dr. Steve Brocchini, School of Pharmacy at the University of London) were used as received. Azobisisobutyronitrile (AIBN, Fluka, purum) was recrystallized from methanol. 1,1'-Azobis(cyclohexanecarbonitrile) (VAZO88, DuPont) was recrystallized from ethanol and 4,4'-azobis(4-cyanopentanoic acid) (ACP, Fluka, ≥ 98 %) was recrystallized from methanol prior to use. The initiator V-70 (2,2'-azobis(4-methoxy-2,4-dimethyl valeronitrile), donation from Wako Chemicals GmbH Germany) was used as received. The syntheses of the chain transfer agents are described elsewhere (*6-11*).

General Polymerization Procedure

The reagents were mixed in a vial and aliquots were transferred to ampoules, which were degassed by three freeze-thaw-evacuate cycles and then flame sealed under vacuum. The ampoules were immersed completely in an oil bath at the specified temperature.

For GPC measurements in THF, poly(acrylic acid) was methylated using methyl iodide and 20 wt.-% Me_4NOH solution in methanol for 1 h at 80 °C.

Instrumentation

Gel permeation chromatography (GPC) of the THF-soluble polymers was performed on PSS SDVgel columns (30 x 8 mm, 5 μm particle size) with 10^2, 10^3, 10^4, and 10^5 Å pore sizes using RI and UV detection (λ=254 nm). THF was used as an eluent (flow rate 0.5 mL/min) at a temperature of 25 °C. For the polyNIPAAm samples, GPC was performed using THF + 0.25 wt.-% of tetrabutylammonium bromide as an eluent (flow rate 0.5 mL/min). The injection volume was 100 μL. As an internal standard, o-dichlorobenzene was used. Polystyrene standards were used for calibration.

GPC on the DMF-soluble polymers poly(NHSM) and PNIPAAm-b-PAA was performed using a series of four Styragel columns HT2, HT3, HT4, and HT5 and an oven temperature of 80 °C. The solvent was DMF + 0.05 M LiBr at a flow rate of 1.0 mL/min. A Dawn EOS light scattering detector with Optilab DSP interferometer (both set at 690 nm) was used.

GPC using water + 0.05 M sodium azide was conducted on PSS Suprema columns (300 x 8 mm, 10 μm particle size) with 10^2, 10^3, and 10^4 Å pore sizes. Poly(methacrylic acid) standards were used for calibration. The measurements were carried out at a flow rate of 1 mL/min at 25 °C or 60 °C, respectively, using RI and UV detection (λ=254 nm).

Cloud-point measurements were performed in 0.2 wt.-% (buffered) aqueous polymer solutions on a Hewlett Packard HP 8453 UV-visible chem station at a wavelength λ = 500 nm using a thermostatted cell with a heating rate of 0.5 K/min. Spectra were recorded in transmission and cloud points were determined as the inflection points of the transmission versus temperature plots.

MALDI-TOF mass spectrometry of the poly(NIPAAm) samples was performed on a Bruker Reflex III spectrometer equipped with a 337 nm N_2 laser in the reflector mode and 20 kV acceleration voltage. Dithranol (Aldrich, 97 %) was used as the matrix material. Sodium or potassium trifluoroacetate was added for ion formation. Samples were prepared from THF solution by mixing matrix (20 mg/mL), sample (10 mg/mL) and salt (10 mg/mL) in a ratio of 10:1:1. The number-average molecular weights, M_n, of the polymer samples were determined in the linear mode.

^{1}H-NMR spectra were recorded on a Bruker 250 MHz instrument using TMS as internal standard. UV measurements were performed on a Lambda15 UV-vis spectrophotometer (Perkin-Elmer) in the wavelength range from 190 to 550 nm.

Results and Discussion

Figures 1 and 2 give an overview of the chain transfer agents and monomers, respectively, used in the different RAFT polymerizations reported herein.

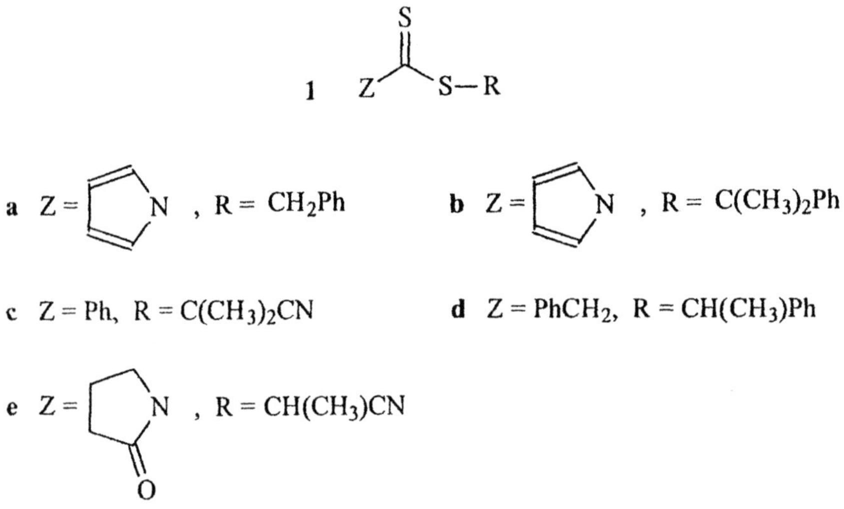

Figure 1. Chain transfer agents used in the RAFT polymerizations.

RAFT Polymerization of NIPAAm

NIPAAm was polymerized using two different chain transfer agents, i.e. benzyl 1-pyrrolecarbodithioate **1a** and cumyl 1-pyrrolecarbodithioate **1b**. In both cases, low polydispersities were obtained with good agreement between calculated and experimental molecular weights (7). The results are shown in Table I.

The GPC characterization of poly(NIPAAm) in THF involves various problems (12,13) due to irreversible chain aggregation after complete drying of the polymer samples (14). Nevertheless, we have obtained good results by the addition of 0.25 wt.-% tetrabutylammonium bromide (TBAB) to the THF solu-

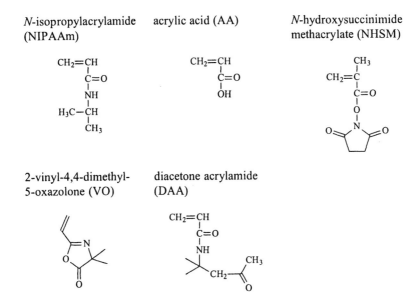

N-isopropylacrylamide (NIPAAm)

acrylic acid (AA)

N-hydroxysuccinimide methacrylate (NHSM)

2-vinyl-4,4-dimethyl-5-oxazolone (VO)

diacetone acrylamide (DAA)

Figure 2. Monomers used in the RAFT polymerizations.

tion and using PSS SDVgel columns, whereas with pure THF no analysable results could be obtained (*7*).

The GPC traces of poly(NIPAAm) obtained with chain transfer agent **1b** show a high-molecular weight shoulder only at conversions higher than 90 %. This is usually seen for RAFT polymers at high monomer conversions (*14*), which is most likely due to combination of the growing macroradicals. GPC evaluation of the molecular weights using polystyrene standards for calibration gives significantly higher molecular weights than those obtained from MALDI-TOF analysis. Therefore, M_n values obtained by MALDI-TOF were used for evaluation of the effectiveness of the RAFT process. In this case, good to excellent agreement between calculated and experimental molecular weights was found (entries 1+2 in Table I) (*7*). Considering the two chain transfer agents used, **1a** seems slightly more effective in the RAFT polymerization of NIPAAm with respect to polydispersities and control of molecular weight. This can be explained in terms of the lower stability of the benzyl radical as compared to the cumyl one, making the benzyl radical more active towards monomer addition. In

Table I. Experimental Data for Synthesized Homopolymers

en-try	mono-mer (M)	CTA (mM)	initiator (mM)	time (min)	conv (%)	$M_{n,GP}$ c	$M_{n,the}$ o	PDI
1	NIPAAm[a] (1.742)	1a (39.2)	AIBN (6.9)	415	73	2900	3900	1.13
				2000	99	3900	5300	1.28
		1a (19.6)		315	87	5600	9000	1.19
				2000	99	6400	10300	1.37
2	NIPAAm[a] (1.742)	1b (19.6)	AIBN (6.9)	462	71	10100	7300	1.21
				1085	99	15200	10200	1.07
3	NHSM (1.373)	1a (78.0)	AIBN (10.0)	600	89	43500	3100	2.11
				960	74	41300	2600	2.34
4	NHSM (1.376)	1b (74.8)	AIBN (10.0)	600	83	24100	3100	1.71
				960	60	22400	2300	1.78
5	NHSM (1.360)	1c (70.3)	AIBN (5.08)	600	81	24500	3100	1.52
				960	70	24200	2700	1.47
6	NHSM (1.367)	1d (77.0)	V-70 (12.16)	600	63	36200	2300	2.06
				960	58	31300	2200	1.95
7	AA (5.84)	1e (41.41)	VAZO88 (0.66)	60	18	1800	2000	1.48
8	AA (5.84)	1e (42.64)	VAZO88 (0.66)	180	80	7900	8100	1.19
9	VO (3.52)	1c (159.0)	AIBN (14.63)	360	35	1200	1300	1.09
				960	64	2200	2200	1.09
10	VO (3.52)	1e (200.0)	AIBN (13.84)	4140	99	2300	2600	1.46
11	DAA (1.18)	1e (21.27)	AIBN (3.57)	120	49	4000	4800	1.29
				240	71	5900	6900	1.23
12	DAA (1.20)	1e (21.69)	ACP (4.15)	120	70	5600	6800	1.23
				240	89	6600	8500	1.21

Experimental conditions: entry 1+2, dioxane at 60 °C; entry 3-5, DMF at 60 °C; entry 6, DMF at 30 °C; entry 7+8, MeOH/H$_2$O (4:1) at 90 °C; entry 9+10, benzene at 65 °C; entry 11+12, MeOH at 65 °C. For initiator abbreviations, see experimental part.

[a] M_n values determined by MALDI-TOF mass spectrometry.

contrast to that, fragmentation of the intermediate radical bearing the cumyl moiety is expected to be faster than the one bearing the benzyl moiety. Nevertheless, the radical stability factor seems to govern the process.

Poly(NIPAAm) is a thermoresponsive polymer, i.e. it shows LCST behaviour in aqueous solutions (*15*). It was of interest to find out to what extent the polymer molecular weight influences the exact temperature of this transition from hydrophilic to hydrophobic state. As can be seen from Figure 3, the cloud point, T_c, decreases virtually linearly with increasing reciprocal molecular weight and approaches T ≈ 32 °C for M_n ≥ 20,000 g/mol. This result seems reasonable as the hydrophobic endgroups should lower the cloud point due to an increase of overall hydrophobicity and a collapse of the intramolecular hydrogen bonds at lower temperature. This effect becomes less and less important the larger the molecular weight of the polymer, i.e. the smaller the fraction of the hydrophobic endgroups in the polymer. Nakahama et al. reported a dependence of LCST on endgroup structure for poly(N,N-diethylacrylamide) (16).

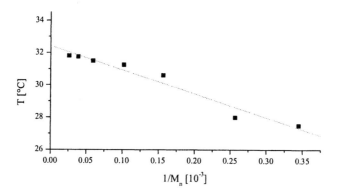

Figure 3. Cloud points of poly(NIPAAm) versus reciprocal molecular weight.

RAFT Polymerization of Acrylic Acid

RAFT is the only known process for the controlled polymerization of acrylic acid (AA) without the need for a protecting group (2). As AA is a quite active monomer in terms of RAFT polymerization, chain transfer agents with moderate activities can be used. Rizzardo et al. used 1c as chain transfer agent and methanol as solvent (17). Here, the polymerization was conducted in a mixture of methanol/water (4:1) in order to increase solvent polarity and thereby accelerate the polymerization. 1-Cyanoethyl 2-pyrrolidone-1-carbodithioate 1e was used as chain transfer agent. The pyrrolidone moiety shows a moderate reactivity as for activating the C=S double bond towards radical addition. Under these conditions, best results are obtained at medium to high conversions (cf. entry 8 in Table I). Samples taken at low conversions show high polydispersities, indicating a rather slow equilibration between dormant and active polymer chains.

Synthesis of Block Copolymers of NIPAAm and Acrylic Acid

For the synthesis of the block copolymers, poly(acrylic acid) (PAA) was used as a macromolecular chain transfer agent (see Table II). PAA was purified by reprecipitating its methanol solution into ethyl acetate in order to remove residual monomer and avoid formation of gradient copolymers in the subsequent block copolymerization with NIPAAm. The polydispersities obtained were quite low.

Table II. Experimental Data for Synthesized Block Copolymers

monomer (M)	CTA (mM)	initiator (mM)	time (min)	conv (%)	$M_{n,GPC}$	$M_{n,theo}$	PDI
NIPAAm/AA[a]	PAA-1e[c]	AIBN	600	64	$2.37 \cdot 10^6$	15100	1.11
(1.49)	(14.87)	(7.0)	960	82	$2.75 \cdot 10^6$	17200	1.09
NIPAAm/VO[b]	PVO-1c[c]	AIBN	960	59	5800	10400	1.22
(2.24)	(18.20)	(1.94)	1320	79	6200	13200	1.14

[a] Polymerization in MeOH at 60 °C; [b] polymerization in benzene at 65 °C; [c] macromolecular chain transfer agents, M_n(PAA) = 7900 and M_n(PVO) = 2200; cf. Table I

In order to investigate the influence of the PAA block on the LCST behaviour of poly(NIPAAm), cloud-point measurements were conducted at different pH values in buffered polymer solutions. Figure 4 shows that transmission decreases only slightly at pH ≥ 5 when the temperature is increased above the LCST of poly(NIPAAm). This may suggest the presence of micelles with poly(NIPAAm) forming the micellar core at $T > T_c$ and PAA forming the corona. Micelle formation has been confirmed by preliminary dynamic light scattering experiments. Transmission decreases substantially at pH 4.5, indicating the formation of a gel due to increasing insolubility of the protonated PAA corona. Thus, the formation of this type of micelles is dependent on both pH and temperature. A doubly-responsive behaviour in aqueous media has been described by Armes et al. (*18*) and by Laschewsky et al. (*19*).

Both DMF and aqueous GPC were used to determine the molecular weights and polydispersities of the polymer samples. DMF GPC gave M_n values that were about two orders of magnitude higher than expected (see Table II). Therefore, it was assumed that some sort of micelle formation takes place. This is not quite expected as DMF should be a good solvent for both blocks. In order to clarify this assumption, aqueous GPC of the polymer samples was performed at 25 °C and at 60 °C. At 25 °C, only one peak was observed. At 60 °C, two peaks were observed, one in the high-molecular weight region and the second one at the same elution volume as the peak observed at 25 °C (Figure 5).

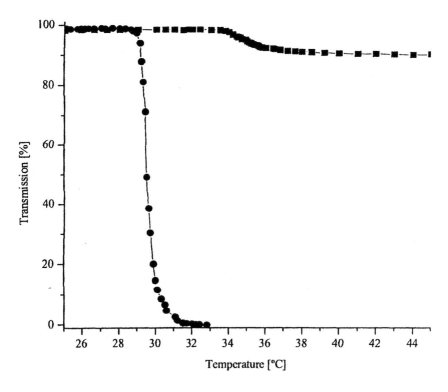

Figure 4. Turbidimetry of buffered aqueous solutions of PNIPAAm-b-PAA; (,) pH 4.5, (!) pH 5-7.

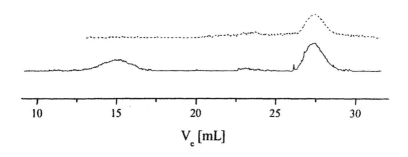

Figure 5. Aqueous GPC of PNIPAAm-b-PAA at 25 °C (top) and 60 °C (bottom).

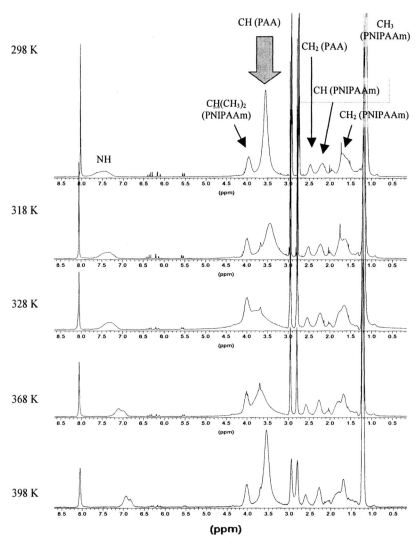

Figure 6. Temperature-sweep ^{1}H-NMR spectrum of PNIPAAm-b-PAA in DMF/LiBr.

Temperature-sweep ^{1}H-NMR spectroscopy in d_7-DMF + 0.05 M LiBr shows an interesting behaviour of these block copolymers (see Figure 6): The methine proton ascribed to acrylic acid disappears at temperatures around 328 K and then reappears above 358 K. This phenomenon has not been explained yet.

RAFT Polymerization of Active Esters

To the best of our knowledge, the active ester monomers 2-vinyl-4,4-dimethyl-5-oxazolone (VO) and N-hydroxysuccinimide methacrylate (NHSM) have been polymerized by RAFT for the first time.

Figure 7 shows the plots of M_n (PS calibration) and PDI versus conversion for the RAFT polymerization of VO using **1c** as a chain transfer agent. The agreement between experimental and calculated molecular weights (shown as solid line in the plot with M_{CTA} as intercept) are excellent as well as the obtained polydispersities which are equal or smaller than 1.10. With chain transfer agent **1e**, a broad molecular weight distribution is obtained. This result might be attributed to the pyrrolidone moiety in **1e** which slightly deactivates the monomer in the RAFT process as compared to the phenyl substituent in **1c** (*2,6*).

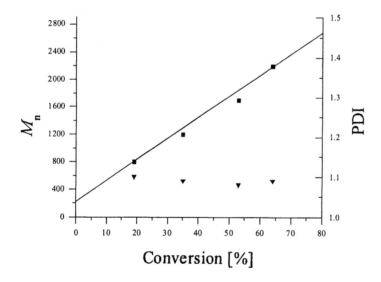

Figure 7. Plot of M_n (▪) and PDI (▼) vs. conversion for RAFT polymerization of vinyl oxazolone using CTA 1c (entry 9 in Table I); the straight line represents the calculated molecular weights.

The RAFT polymerization of NHSM is much more difficult and so far, no satisfactory results have been obtained regarding polydispersities and control of molecular weight. NHSM is a rather active monomer in controlled radical polymerization and reactive radicals are produced upon initiation. Therefore, a chain transfer agent containing an equally active leaving group has to be used to ensure effective polymerization. For this purpose, RAFT agent **1c** seemed to be a good choice as it consists of a reactive cyanopropyl leaving group. Nevertheless, the obtained polymers showed polydispersities around 1.5 and their molecular weights were much higher than predicted (see entry 5 in Table I). The mother liquors of the precipitated polymers showed residual unreacted chain transfer agent in the ^{1}H-NMR spectra. The same findings were made for the polymerizations using RAFT agents **1a**, **1b**, and **1d**. Chain transfer agent **1d** decomposes at ambient temperature (8), and the RAFT polymerization was conducted at 30 °C using the low-temperature initiator V-70. It was assumed that a lower polymerization temperature might slow down the polymerization process and a controlled polymerization would take place. This was not the case, though.

Synthesis of Block Copolymers of VO and NIPAAm

For the synthesis of the block copolymers, poly(VO) was used as a macro-chain transfer agent for the polymerization of NIPAAm. The observed polydispersities were 1.22 and 1.14, respectively, the molecular weights are somewhat lower than expected (see Table II), which might be ascribed to GPC calibration with polystyrene standards.

Containing a thermoresponsive NIPAAm block on one side and an active ester block on the other side, these block copolymers are of interest for the synthesis of polymer-protein conjugates. These conjugates also show thermoresponsive behaviour due to the poly(NIPAAm) block. Cloud-point measurements on the block copolymers were carried out in order to determine the LCST of this system. It is expected that the active ester block influences the LCST behaviour of NIPAAm through its polar nature. Generally speaking, hydrophobic substituents lower the LCST of poly(NIPAAm) whereas hydrophilic substituents raise it (20,21). Turbidity measurements show that the LCST is shifted to a higher temperature (37 °C) relative to the LCST reported for the homopolymer poly(NIPAAm) (32 °C)(15), accounting for a more hydrophilic nature of the poly(VO) block.

RAFT Polymerization of Diacetone Acrylamide

Diacetone acrylamide was polymerized by RAFT for the first time using **1e** as chain transfer agent. Two different initiators, namely AIBN and 4,4'-azobis(4-cyanopentanoic acid) ACP, were employed but no significant influence was found on molecular weights and polydispersities even though ACP gives rise to faster polymerizations (entry 11 and 12 in Table I, 70 % conversion after 2 h compared to same conversion after 4 h with AIBN). The monomer is especially interesting for polymer-analogous reactions as it possesses the reactivity of an activated double bond, an *N*-substituted amide and a methyl ketone (*22*).

Synthesis of Polymer-Protein Conjugates Using RAFT Polymers

RAFT polymers can be utilized for protein conjugation in two ways: (i) hydrolysis of the dithiocarbonyl-terminated polymers to the corresponding thiol-terminated polymers to provide functional groups for disulfide formation or for bismaleimide crosslinkers between polymer and protein (Figure 8), (ii) short active ester blocks synthesized via RAFT containing functional groups that allow for amide formation between polymer and protein amino groups (Figure 9)(*4,5*).

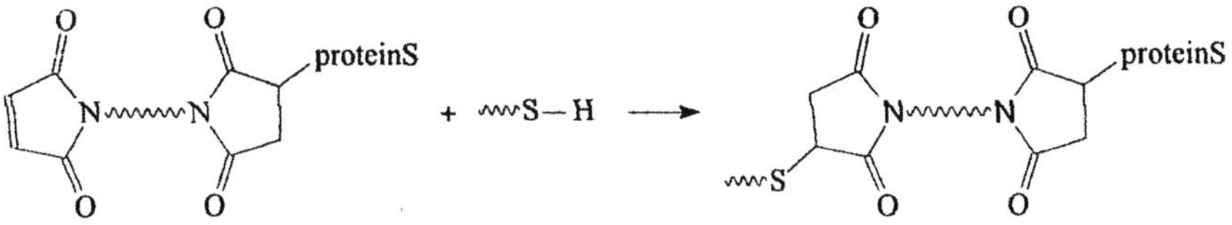

Figure 8. Bismaleimide crosslinker for protein conjugation.

The hydrolytic cleavage of the dithiocarbonyl endgroup with aqueous NaOH was confirmed by UV spectroscopy: The dithiocarbamate moiety absorbs in the UV-vis region with $\lambda_{max} = 296$ nm. After hydrolysis, only the thiol moiety is left, which does not absorb in the range measured. Additional proof of the thiol endgroup was provided by MALDI-TOF mass spectrometry and by the formation of a thioester with 2-naphthoyl chloride giving rise to an absorption shift in the UV spectrum.

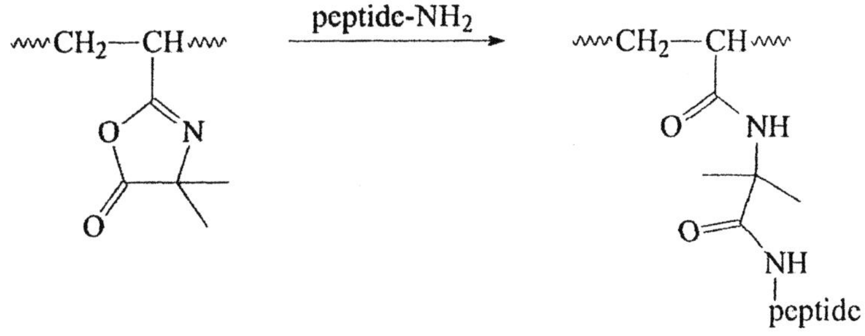

Figure 9. Conjugation of primary amino groups to poly(VO).

The active ester block poly(VO) undergoes ring opening on addition of compounds containing active hydrogen atoms. Azlactone ring opening is especially facile with primary amines. The conjugation of a peptide with its primary amino group to poly(VO) is illustrated in Figure 9. As a model peptide, glycine-leucine (Gly-Leu) was used, and the reaction was performed in dry DMF under reflux. Figure 10 shows the ^{1}H-NMR spectra of poly(VO) and its conjugate with Gly-Leu. Conjugation to the active ester NHSM has been described elsewhere (5).

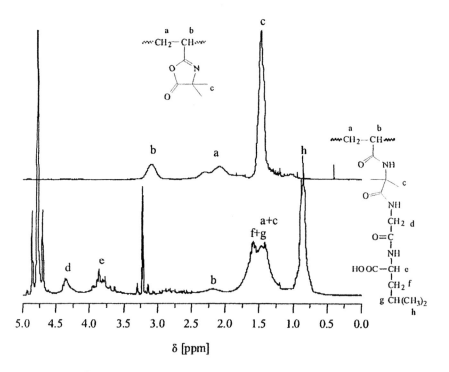

Figure 10. ^{1}H-NMR spectra of poly(VO) (top) and poly(VO) conjugate to Gly-Leu (bottom) in C$_6$D$_6$ and CD$_3$OD, respectively.

Conclusions

The synthesized RAFT homopolymers and block copolymers are promising candidates for protein or drug conjugation as they show narrow molecular weight distributions. This is especially important in the field of polymer therapeutics where narrow MWDs are required for defined retention times in the body.

Active esters copolymerized with PNIPAAm represent thermoresponsive systems that can also be used for protein conjugation. The system PNIPAAm-*b*-PAA is both pH- and thermoresponsive and forms micelles in aqueous solutions. These micelles can be used for drug encapsulation and controlled release thereof by utilizing both pH and temperature stimuli.

Acknowledgement. This work was supported by the *Deutsche Forschungsgemeinschaft* (Mu 896/13-1). The authors would also like to thank Dr. Michael Lanzendörfer for the MALDI-TOF measurements.

References

1. Chong, Y. K.; Le, T. P. T.; Moad, G.; Rizzardo, E.; Thang, S. H. *Macromolecules* **1999**, *32*, 2071-2074.
2. Rizzardo, E.; Chiefari, J.; Mayadunne, R. T. A. In *Controlled/Living Radical Polymerization - Progress in ATRP, NMP, and RAFT, ACS Symposium Series 768*; Matyjaszewski, K., Ed.; American Chemical Society: Washington, DC, 2000; Vol. 768, pp 278-295.
3. Chiefari, J.; Chong, Y. K.; Ercole, F.; Krstina, J.; Jeffery, J.; Le, T. P. T.; Mayadunne, R. T. A.; Meijs, G. F.; Moad, C. L.; Moad, G.; Rizzardo, E.; Thang, S. H. *Macromolecules* **1998**, *31*, 5559-5562.
4. Zimmermann, J.; Mülhaupt, R. *Polym. Prepr. (Am. Chem. Soc., Div. Polym. Chem.)* **2000**, *41(1)*, 764-765.
5. Godwin, A.; Hartenstein, M.; Müller, A. H. E.; Brocchini, S. *Angew. Chem. Int. Ed.* **2001**, *40*, 594.
6. Chiefari, J.; Mayadunne, R. T. A.; Moad, G.; Rizzardo, E.; Thang, S. H. invs.: DuPont de Nemours Company USA, 1999, WO 99/31144.
7. Schilli, C.; Lanzendörfer, M. G.; Müller, A. H. E. *Macromolecules* **2002**, *35*, 6819-6827.
8. Quinn, J. F.; Rizzardo, E.; Davis, T. P. *Chem. Commun.* **2001**, 1044-1045.
9. Takeshima, T.; Ikeda, M.; Yokoyama, M.; Fukada, N.; Muraoka, M. *J. Chem. Soc., Perkin Trans. 1* **1979**, 692-695.
10. Mayadunne, R. T. A.; Rizzardo, E.; Chiefari, J.; Chong, Y. K.; Moad, G.; Thang, S. H. *Macromolecules* **1999**, *32*, 6977-6980.
11. Moad, G.; Chiefari, J.; Chong, Y. K.; Krstina, J.; Mayadunne, R. T. A.; Postma, A.; Rizzardo, E.; Thang, S. H. *Polym. Int.* **2000**, *49*, 993-1001.
12. Cole, C.-A.; Schreiner, S. M.; Monji, N. *Polym. Prepr. (Am. Chem. Soc., Div. Polym. Chem.)* **1986**, *27(1)*, 237-238.
13. Yang, H. J.; Cole, C.-A.; Monji, N.; Hoffman, A. S. *J. Polym. Sci., Part A: Polym. Chem.* **1990**, *28*, 219-226.
14. Ganachaud, F.; Monteiro, M. J.; Gilbert, R. G.; Dourges, M.-A.; Thang, S. H.; Rizzardo, E. *Macromolecules* **2000**, *33*, 6738-6745.
15. Schild, H. G. *Prog. Polym. Sci.* **1992**, *17*, 163-249.

16. Kobayashi, M.; Okuyama, S.; Ishizone, T.; Nakahama, S. *Macromolecules* **1999**, *32*, 6466-6477.
17. Le, T. P.; Moad, G.; Rizzardo, E.; Thang, S. H. invs.: Du Pont De Nemours and Co., USA; Le, Tam Phuong; Moad, Graeme; Rizzardo, Ezio; Thang, San Hoa, 1998, WO 98/01478.
18. Liu, S.; Billingham, N. C.; Armes, S. P. *Angew. Chem., Int. Ed.* **2001**, *40*, 2328-2331.
19. Arotcarena, M.; Heise, B.; Ishaya, S.; Laschewsky, A. *J. Am. Chem. Soc.* **2002**, *124*, 3787-3793.
20. Licea-Claverie, A.; Salgado-Rodriguez, R.; Arndt, K.-F. *Polym. Prepr. (Am. Chem. Soc., Div. Polym. Chem.)* **2001**, *42(1)*, 564-565.
21. Stile, R. A.; Healy, K. E. *Biomacromolecules* **2001**, *2*, 185-194.
22. Coleman, L. E.; Bork, J. F.; Wyman, D. P.; Hoke, D. I. *J. Polymer Sci., Part A* **1965**, *3*, 1601-1607.

Chapter 42

New Concepts for Controlled Radical Polymerization: The DPE System

P. C. Wieland[1], O. Nuyken[1,*], Y. Heischkel[2], B. Raether[2], and C. Strissel[1]

[1]Lehrstuhl für Makromolekulare Stoffe, Technical University of München, D–85747 Garching, Germany
[2]BASF AG, Ludwigshafen, Germany

The DPE-System is a new, versatile method for controlled radical polymerization which consists of a conventional radical polymerization system where a small amount of 1,1-diphenylethylene (DPE) is added. It allows the facile synthesis of block copolymers with numerous industrial important monomers. The mechanism is believed to involve the combination of two stable diphenylmethylradicals to a thermolabile quinoid unit as a key step. The exact mechanism is not clear in all details yet, but some results shown in this article give good evidence on the suggested mechanism.

Introduction

The synthesis of new block copolymer structures is of great interest in modern polymer chemistry. Next to ionic polymerization techniques, which require highly pure monomers and solvents, living radical polymerization has been established since several years for the synthesis of all kind of polymer architectures. Atom transfer radical polymerization (ATRP) (*1-3*), nitroxide mediated polymerization (NMP) (*4-6*) or the reversible addition fragmentation transfer (RAFT) (*7-9*) process are well known techniques in this field. Additionally, degenerative transfer with alkyl iodides is a potential method for controlled radical polymerization (*10*). Nevertheless, no broad industrial application of these techniques took place due to some important disadvantages.

These are the use of toxic metal catalysts, an unsatisfactory universality or the formation of thiols during the polymerization.

Due to this problem, there are extensive investigations for new systems which are free of these limitations. Recently, we reported a new additive for controlled radical polymerization , which is especially suitable for an easy synthesis of block copolymers and free of the above mentioned disadvantages (11,12). Addition of 1,1-diphenylethylene (DPE) to a conventional radical polymerization system changes the polymerization characteristics into a controlled like system. Furthermore, the polymers which were synthesized in the presence of DPE are able to initiate the polymerization of a second monomer at elevated temperatures which leads to block copolymers.

The DPE-System

In general, the DPE-system consists of a conventional system for free radical polymerization where a small amount of 1,1-diphenylethylene (DPE) is added. The amount of DPE is typically in the range of 0.3 mol-% in respect to the monomer and in a 1:1 mole-ratio to the initiator. Suitable initiators are azo isobutyronitrile (AIBN) or benzoyl peroxide (BPO).

The polymers which were synthesized in the presence of DPE can be re-initiated at elevated temperatures from 70 to 110 °C. If the re-initiation is carried out in the presence of a second monomer block copolymers are obtained in good yields. There is almost no limitation in the choice of monomers. Styrenes, acrylates, methacrylates N-vinylmonomers etc. can be polymerized and combined to block copolymers. The polymerization can be carried out in water-based systems like emulsion or suspension polymerization systems (13), in solution and in bulk.

Homopolymerization

The DPE system shows characteristics which are different from a free radical polymerization. Figure 1 shows the time-conversion plot for the bulk polymerization of styrene at 80 °C with and without DPE. The DPE-free polymerization was performed with AIBN as initiator in the same amount as in the polymerization in presence of DPE. A retarding effect on the conversion is observed by adding a small amount of DPE in comparison to the free radical polymerization. Nevertheless, the conversions are rather good in the given time compared to other systems e.g. like the 2,2,6,6-tetramethyl-piperidin-1-oxyl (TEMPO) controlled polymerization (4).

Figure 2 shows the molecular weight and the polydispersity index (PDI) versus conversion of the bulk polymerization of styrene with and without DPE. The DPE controlled polymerization shows a slightly increasing molecular weight with conversion after a short period of uncontrolled polymerization. During this period polymers with a relatively high molecular weight are formed. The DPE free polymerization shows the expected behavior. The increase of the molecular weight at higher conversion is due to the increasing viscosity of the system and the higher probability for radical combination.

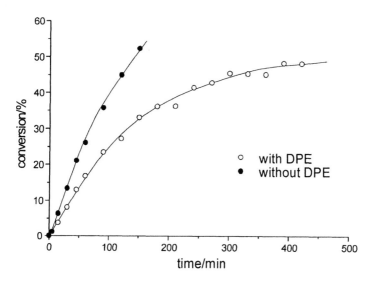

Figure 1: Time-conversion plot of the radical bulk polymerization of styrene in the presence of DPE and without DPE at 80 °C; conc.(AIBN) = 0.3 mol-%, conc.(DPE) = 0.3 mol-%.

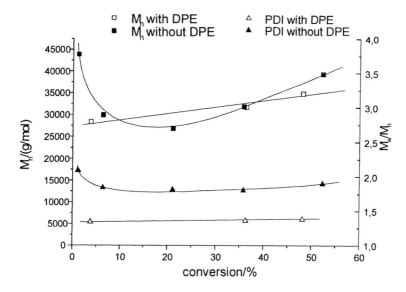

Figure 2: Plot of molecular weight and PDI versus conversion for the radical bulk polymerization of styrene in the presence of DPE and without DPE at 80 °C; conc.(AIBN) = 0.3 mol-%, conc.(DPE) = 0.3 mol-%.

The PDI of the polymers is clearly lower if the polymerization is carried out in the presence of DPE than for the DPE free system. That means that the number of termination reactions is decreased by the added DPE which is important evidence for control during the polymerization process.

Additionally, we investigated the temperature dependence of the polymerization of methyl methacrylate in the presence of 0.3 mol-% DPE with 0.3 mol-% AIBN as initiator. The polymerization was performed in toluene to avoid viscosity effects. Figure 3 shows the time conversion plot in a temperature range of 70 to 90 °C. The conversion increases clearly with increasing temperature but is always in a good range from 40 to 75 %.

Figure 4 shows the plot of the molecular weight and the PDI with the conversion for the different temperatures. The molecular weight shows no increase at 70 °C and 80 °C. In contrast, there is an increase of the molecular weight versus the conversion at 90 °C. This effect can be ascribed to a more frequent exchange between a dormant and an active species due to higher temperature. Apparently, lower temperatures are not high enough for an effective exchange between the two species.

The PDI in the range between 1.5 and 2.0 is much smaller than in common free radical polymerization at all temperatures for a monomer with such a strong transfer tendency as methyl methacrylate. This indicates again the decrease of termination and transfer reactions due to the addition of DPE.

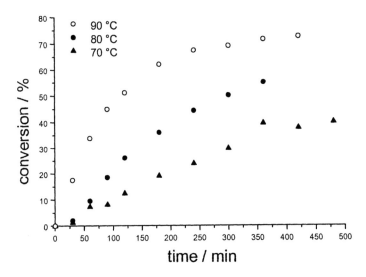

Figure 3: Time-conversion plot of the radical polymerization of methyl methacrylate in the presence of DPE at different temperatures; conc.(AIBN) = 0.3 mol-%, conc.(DPE) = 0.3 mol-%.

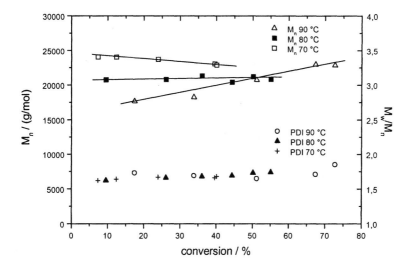

Figure 4: Plot of molecular weight and PDI versus conversion for the radical polymerization of methyl methacrylate in the presence of DPE at different temperatures; conc.(AIBN) = 0.3 mol-%, conc.(DPE) = 0.3 mol-%.

Block Copolymerization

A very important feature of the DPE-system is the facile synthesis of block copolymers. Polymers which have been synthesized in the presence of DPE are able to initiate the polymerization of a second monomer at temperatures from 70 – 110 °C. The block copolymerization can be carried out with the second monomer as solvent or in common solvents like e.g. toluene.

The system seems to be quite universal with regard of the block formation. Numerous technically important monomers can be converted to block copolymers as shown for some examples in Table 1. All block structures have been proven by several methods like GPC, ¹H-NMR spectroscopy and specific extraction and precipitating.

Table 1. Block copolymers P1-*b*-P2 synthesized by the DPE method. PMMA = poly(methyl methacrylate); PBA = poly(n-butyl acrylate); PS = poly(styrene); PVAc = poly(vinyl acetate); PVP = poly(N-vinyl pyrrolidone); Y = (g block copolymer after extraction of homopolymer / g block copolymer before extraction of homopolymer) · 100 %

P1	P2	T/°C	t/h	Mn/(g/mol)	PDI	Y/%
PMMA	PBA	85	3	41670	1,63	>95
PS	PVAc	110	4	129800	1,80	53
PVAc	PS	110	4	17950	1,52	58
PVP	PS	110	5	289400	1,42	65

Obviously, the block copolymer yield Y changes with the monomer type of P_1 from 50 % for styrene to >95 % for methyl methacrylate. The reason for this result is not clear yet but the investigation of the mechanism of the polymerization will open ways for improvements especially of the block yield.

Next to common monomers like styrene or (meth)acrylates it is also possible to synthesize block copolymers with functional monomers like 4-chloromethyl styrene. This result is interesting for the synthesis of new macroinitiators for numerous polymerization techniques like cationic polymerization or ATRP (*14-17*).

Mechanism

The emphasis of our investigations is to clear the mechanism of this polymerization type which is important for further improvements of this system.

Figure 5 shows how we suggest that the polymerization and the reinitiation occurs.

The first step is the formation of an active chain end $P_1^{\cdot}$ by common radical initiation. $P_1^{\cdot}$ reacts with a DPE molecule under formation of a stabilized diphenylmethyl radical $P_2^{\cdot}$. This radical can be seen as a dormant species which can avoid transfer reactions. If the formation of $P_2^{\cdot}$ is reversible all needs for an equilibrium between active and dormant species are given. This equilibrium is required for all controlled polymerizations.

According to our view, the final dormant species is built by combination of two $P_2^{\cdot}$ radicals to a quinone like structure as have been shown for other hindered tetraphenyl ethanes (*18,19*). Such units are thermolabile, as has been shown by different authors (*20-22*). Furthermore, it is also possible to get combination between $P_2^{\cdot}$ and $P_1^{\cdot}$ under formation of a triphenylmethane unit which should also be thermolabile. If the polymers formed by combination of $P_2^{\cdot}$ and $P_2^{\cdot}$ are heated in the presence of a second monomer, they split into two $P_2^{\cdot}$ radicals. The $P_2^{\cdot}$ radicals split into $P_1^{\cdot}$ and free DPE. $P_1^{\cdot}$ can initiate the polymerization of the second monomer whereby block copolymers are formed.

Proof of the mechanism

The proof of this mechanism is under intensive investigation by the synthesis of model compounds and the investigation of their initiator properties. Furthermore, spectroscopic investigations can help to clear the exact mechanism.

Figure 6 shows a part of the ^{1}H-NMR spectrum of low molecular weight poly(methyl methacrylate) with M_n = 4450 g/mol which has been synthesized in the presence of DPE. Next to the signals for the aromatic protons are four broad signals at 6.70 ppm, 6.25 ppm, 5.90 ppm and 5.65 ppm detectable. These signals are characteristic for quinoid protons as we suggest them for the combination of two $P_2^{\cdot}$ radicals in Figure 5.

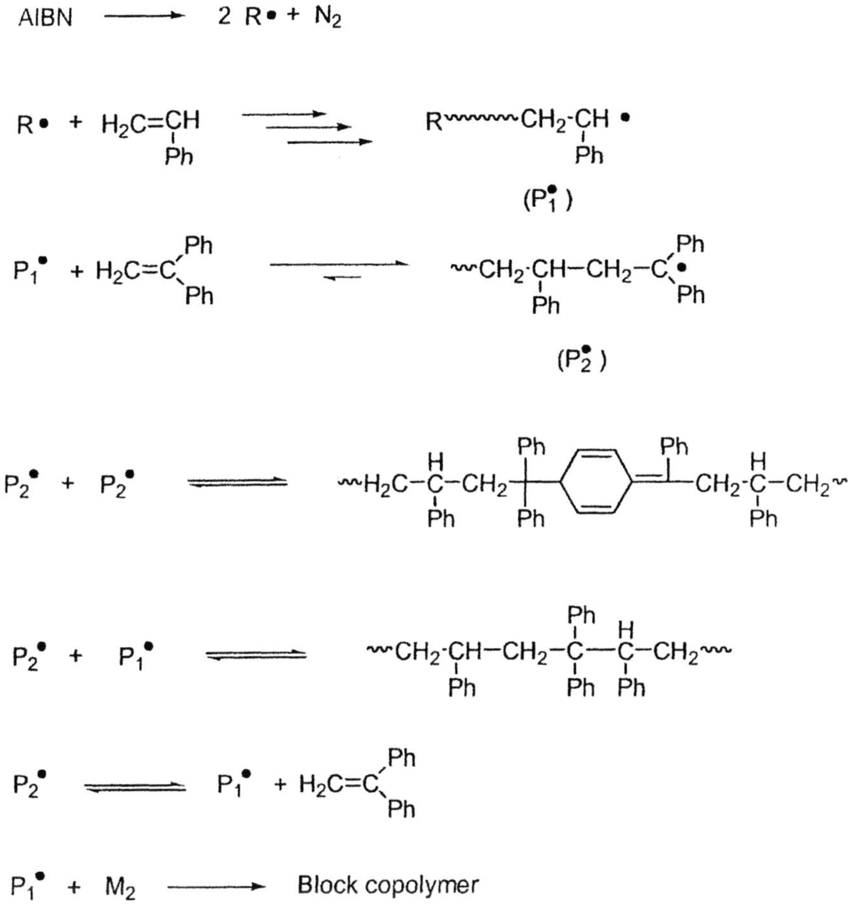

$$AIBN \longrightarrow 2\ R\bullet + N_2$$

$$R\bullet + H_2C=\underset{\underset{Ph}{|}}{CH} \underset{\longleftarrow}{\longrightarrow} R\text{\textasciitilde\textasciitilde\textasciitilde}CH_2\text{-}\underset{\underset{Ph}{|}}{CH}\ \bullet$$

$$(P_1^\bullet)$$

$$P_1^\bullet + H_2C=C\underset{Ph}{\overset{Ph}{\diagdown}} \longrightarrow \text{\textasciitilde}CH_2\text{-}\underset{\underset{Ph}{|}}{CH}\text{-}CH_2\text{-}C\underset{Ph}{\overset{Ph}{\diagup}}\bullet$$

$$(P_2^\bullet)$$

$$P_2^\bullet + P_2^\bullet \rightleftharpoons \text{\textasciitilde}H_2C\text{-}\underset{\underset{Ph}{|}}{\overset{\overset{H}{|}}{C}}\text{-}CH_2\text{-}\left[\underset{\underset{Ph}{|}}{\overset{\overset{Ph}{|}}{}}\right]=\underset{\underset{Ph}{|}}{\overset{\overset{Ph}{|}}{}}CH_2\text{-}\underset{\underset{Ph}{|}}{\overset{\overset{H}{|}}{C}}\text{-}CH_2\text{\textasciitilde}$$

$$P_2^\bullet + P_1^\bullet \rightleftharpoons \text{\textasciitilde}CH_2\text{-}\underset{\underset{Ph}{|}}{CH}\text{-}CH_2\text{-}\underset{\underset{Ph}{|}}{\overset{\overset{Ph}{|}}{C}}\text{-}\underset{\underset{Ph}{|}}{\overset{\overset{H}{|}}{C}}\text{-}CH_2\text{\textasciitilde}$$

$$P_2^\bullet \rightleftharpoons P_1^\bullet + H_2C=C\underset{Ph}{\overset{Ph}{\diagdown}}$$

$$P_1^\bullet + M_2 \longrightarrow \text{Block copolymer}$$

Figure 5: *Suggested mechanism of the radical polymerization with small amounts of DPE.*

The signals at 5.45 ppm and 6.20 ppm can be attributed to olefinic endgroups in poly(methyl methacrylate) as the result of disproportionation reactions.

Signals in the same range of the spectrum can be detected if the combination product of AIBN and DPE is investigated by ^{1}H-NMR spectroscopy. The reaction product between these two compounds after 14 h reaction time at 80 °C is clearly the quinoid structure as shown in Figure 7. This is a clear indication for the presence of such groups in the re-initiatable polymers.

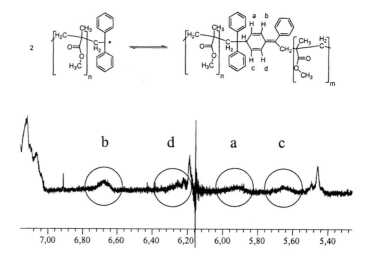

Figure 6. *Part of the ^{1}H-NMR spectrum of poly(methyl methacrylate) (M_n = 4450 g/mol) which has been polymerized in the presence of DPE.*

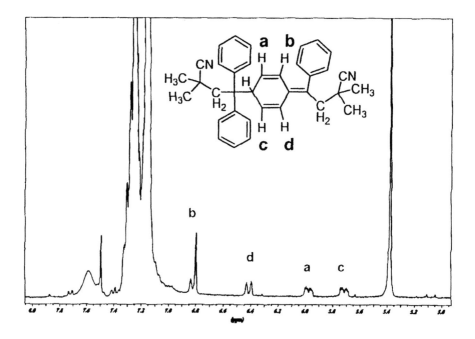

Figure 7. *Combination product of the reaction between AIBN and DPE.*

Next to spectroscopic investigations we studied the thermal behavior of the DPE-containing polymers. We heated DPE-containing poly(methyl methacrylate) in the presence of radical scavengers to 85 °C for 3 h and observed if a decrease of the molecular weight occurs. We have chosen the two radical scavengers 2,2,6,6-tetramethyl-piperidin-1-oxyl (TEMPO) and galvinoxyl. Both are excellent radical scavengers for alkyl radicals which are widely used in the investigation of radical reactions. The results are shown in Figure 8.

It is clearly indicated that the polymer chains break at the chosen temperatures and the resulting radicals are capped by the radical scavengers resulting in a decrease of the molecular weight of ca. 20 %. That supports our view of the incorporation of DPE in the polymer cain under formation of a thermolabile unit. We repeated this experiment under the same conditions with poly(methyl methacrylate) which has been synthesized by conventional free radical polymerization with AIBN as initiator. The results are shown in Figure 9.

For this polymer no decrease of the molecular weight occurs under the chosen conditions. That means that DPE is clearly responsible for the cleavage of the polymer chains at elevated temperature.

Conclusions

It has been shown that the DPE-system is a new, versatile method for controlled radical polymerization which is especially useful for a facile synthesis of numerous block copolymers under mild conditions. The polymerization control is not as good as in the established systems like ATRP, NMP or RAFT, but the versatility of the DPE system makes it interesting especially for the synthesis of block copolymers even on a industrial scale. The exact mechanism is not clear in all details yet, but some results give a good indication on what occurs during the polymerization. We suggest that the mechanism is probably based on the reversible addition of DPE on the growing chain ends under formation of a stabilized diphenylmethyl radical. This radical has the function of a dormant species. Two diphenylmethyl radicals can combine under formation of a thermolabile quinoid unit. Evidences of such quinoid units have been found in the ^{1}H-NMR spectra of low molecular weight DPE-containing poly(methyl methacrylate). Furthermore, we observed a decrease of the molecular weight of poly(methyl methacrylate) which has been synthesized in the presence of DPE if these polymers are heated in the presence of radical scavengers. This indicates clearly that the groups which are responsible for the reinitiation are incorporated in the polymer chain.

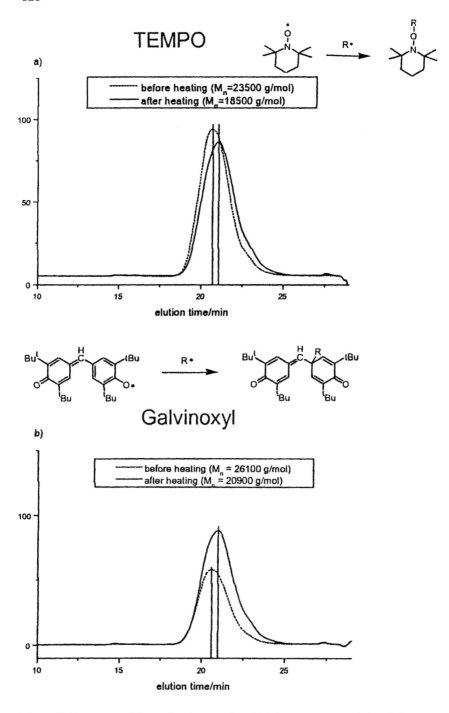

Figure 8. *Decrease of the molecular weight of DPE-containing poly(methyl methacrylate) in presence of radical scavengers at 85 °C.*

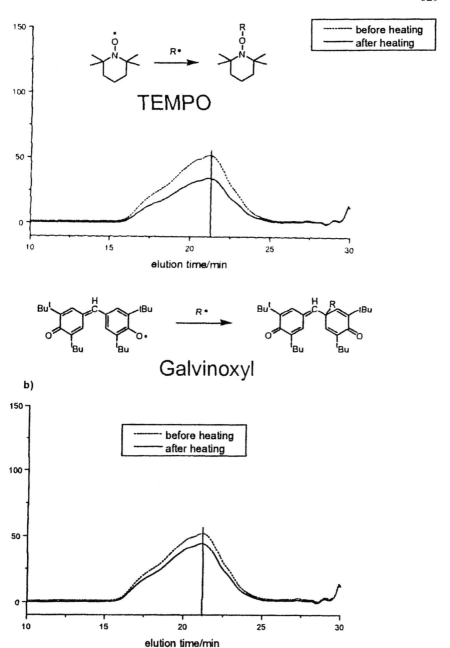

Figure 9. *Behavior of the molecular weight of poly(methyl methacrylate) after heating the polymer to 85 °C for 3 h.*

References

1. Kato, M.; Kamigaito, M; Sawamoto, M; Higashimura, T. *Macromolecules* **1995**, *28*, 1721.
2. Wang, J.S.; Matyjaszewski, K *J. Am. Chem. Soc.* **1995**, *117*, 5614.
3. Wang, J.S.; Matyjaszewski, K. *Macromolecules* **1995**, *28*, 7901.
4. Georges, M.K.; Veregin, R.P.N.; Kazmeier, P.; Hamer, G.K. *Macromolecules* **1993**, *26*, 2987.
5. Moad, G.; Solomon, D.H.; Johns, S.R.; Willing, R.I. *Macromolecules* **1982**, *15*, 1188.
6. EP135280 (1985), Solomon, D.H.; Rizzardo, E.; Cacioli, P.; Chem. Abstr. **1985**, *28*, 221335q.
7. Mayadunne, R.T.A.; Rizzardo, E.; Chiefari, J.; Chong, Y.K.; Moad, G.; Thang, S.H. *Macromolecules* **1999**, *32*, 6977.
8. Chong, Y.K.; Moad, G.; Rizzardo, E.; Thang, S.H. *Macromolecules* **1999**, *32*, 2071.
9. Hawthorne, D.G.; Moad, G.; Rizzardo, E.; Thang, S.H. *Macromolecules* **1999**, *32*, 5457.
10. Matyjaszewski, K.; Gaynor, S.; Wang J.-S. *Macromolecules* **1995**, *28*, 2093.
11. Wieland, P.C.; Raether, B.; Nuyken, O. *Macromol. Rapid Commun.* **2001**, *22*, 700.
12. Raether, B.; Nuyken, O.; Wieland, P.C.; Bremser, W. *Macromol. Symp.* **2002**, *177*, 25.
13. Tauer, K. et al., unpublished results
14. Schaefer, M.; Wieland, P.C.; Nuyken, O. *J. Polym. Sci.: Part A: Polym. Chem.* **2002**, *40(21)*, 3725.
15. Wieland, P.C.; Stoeckel, N.; Nuyken, O. *J. Macromol. Sci. – Pure Appl. Chem.* **2003**, *A40(1)*, in press.
16. Wieland, P.C.; Schaefer, M.; Nuyken, O. *Macromol. Rapid. Comm.* **2002**, *23*, 809.
17. Stoeckel, N.; Wieland, P.C.; Nuyken, O. *Polym. Bull.*, in print.
18. Beckhaus, H.D.; Schaetzer, J.; Rüchardt, C. *Tetrahedron Lett.* **1983**, *24*, 3307.
19. Rüchardt, C.; Beckhaus, H.D. *Angew. Chem.* **1980**, *92*, 417.
20. Borsig, E.; Lazar, M.; Capla, M; Florian, S. *Angew. Makromol. Chem.* **1969**, *105*, 89.
21. Tharanikkarasu, K.; Radhakrishnan, G. *J. Appl. Polym. Sci.* **1997**, *66*, 1551.
22. Braun, D. *Angew. Makromol. Chem.* **1994**, *223*, 69.

Chapter 43

Organotellurium-Mediated Living Radical Polymerization

Shigeru Yamago, Kazunori Iida, and Jun-ichi Yoshida

Department of Synthetic Chemistry and Biological Chemistry,
Graduate School of Engineering, Kyoto University, Kyoto 606–8501, Japan

Polymer-end mimetic organotellurium compounds initiate
controlled/living radical polymerization that allows accurate
molecular weight control with defined end-groups. A variety
of monomers including styrene, acrylate, and methacrylate
derivatives are polymerized under mild thermal conditions
using same initiators. AB-Diblock, ABA- and ABC-triblock
copolymers composed of different families of monomers are
also synthesized with highly controlled manner.
Transformations of the end-groups via radical and ionic
reactions provide a variety of end-group modified polymers
with defined structure with various functional groups.

The synthesis of new nanostructural organic materials by controlled
polymerization has attracted a great deal of attention, because these materials
would lay essential foundations for nanoscience and nanotechnologies (*1,2*).
Living radical polymerization (LRP) is becoming increasingly important for this
goal because of its potential applicability to different types of monomers with
various polar functional groups, which do not lend themselves to ionic and
metal-catalyzed polymerization conditions. While impressive developments in
LRP systems have emerged such as nitroxide-mediated polymerization (NMP)
(*3*), atom transfer radical polymerization (ATRP) (*4*), and reversible addition-
fragmentation chain transfer (RAFT) (*5*), the invention of a more versatile
system is clearly needed to achieve polymerization of different families of
monomers with control of molecular structure and with defined polymer end-
groups.

We have already reported that organotellurium compounds undergo
reversible carbon-tellurium bond cleavage upon thermolysis and photolysis (*6*),

631

and that the resulting carbon-centered radicals can react with a variety of radical acceptors (7,8). Since NMP also relies on reversible generation of carbon-centered radicals and persistent nitroxyl radicals at the polymer end (9), we decided to investigate the use of organotellurium compounds as unimolecular radical initiators for living radical polymerization. We report here a highly versatile method for the synthesis of structurally defined polymers based on organotellurium mediated living radical polymerization (TERP). We have found that TERP is extremely general and can be applicable for the polymerization of different families of monomers, such as styrene, acrylate, and methacrylate derivatives, using the same initiators in a highly controlled manner. Furthermore, the versatility of TERP allows the synthesis of various AB-, ABA- and ABC-block copolymers starting from a single monofunctional initiator (Scheme 1) (10,11).

Scheme 1.

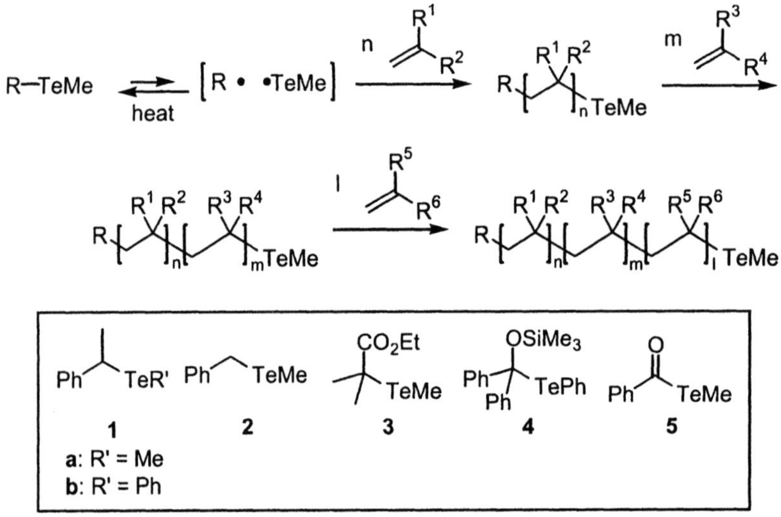

Bond dissociation energies of initiators.

Previous reports on the NMP indicate that the efficiency of the initiators is closely related to their bond dissociation energies (BDE) (12). Therefore, we first calculated the BDEs of the organotellurium compounds 1-5 (Table 1). Density functional theory (DFT) calculations indicated that the BDEs of 1 and 3 are 112 – 123 kJ/mol, the values of which are very similar to those of the corresponding TEMPO analogue (119 kJ/mol).

Table 1. Calculated bond dissociation energies of initiators.[a]

Compound	BDE (kJ/mol)	Compound	BDE (kJ/mol)
1a	123	**3**	114
1b	112	**4**	25
2	142	**5**	182

[a]Obtained by B3LYP DFT calculations with LANL2DZ basis set for tellurium atom and 6-31G(d) basis set for the rest.

Polymerization of vinyl monomers.

Bulk polymerization of styrene (X = H) was initially examined at 105 °C for 16-18 h, and the results are shown in Table 2 (entries 1-8). The polymer-end mimetic initiator **1a** (R' = Me) initiated the polymerization efficiently, and afforded polystyrene with the predicted molecular weight and low polydispersity (M_n = 9200, PD = 1.17) in 96% yield (Table 1, entry 1). The initiator **1b** (R' = Ph) also promoted polymerization, but the control of the molecular weight was less efficient. Benzyl telluride **2** also initiated polymerization with acceptable polydispersity. The result is in sharp contrast to the NMP polymerization, in which benzyl derivatives are far less efficient than the 1-phenylethyl derivatives (cf. **1a** vs. **2**).The ester **3** also initiated polymerization efficiently with low polydispersity. The ability to initiate polymerization of **4** and **5** was found to be unsatisfactory. These results may suggest that both the BDEs and the reactivity of the initiating radicals toward styrene are important factors in controlling the polymerization process. It is also worth mentioning that, while the first-generation initiators for NMP required high temperature and long reaction times, e.g., 130 °C for 72 h, the initiators **1a** and **3** promoted polymerization under much milder conditions. Molecular weight increased linearly with the increase of the amount of styrene, and the products were obtained with low polydispersity (entries 7 and 8). The observed linearlity as well as low polydispersity (PD < 1.3) strongly suggested the living character of the current polymerization (see below).

Because TERP proceeds under neutral conditions, styrenes possessing a variety of functional groups, such as chloro- and methoxy groups, could also be polymerized using **1a** as the initiator (entries 9 and 10). It is worth noting that *p*-methoxy-substituted styrene, which is a poor monomer for ATRP (*13*), was successfully polymerized by this method.

To understand the generality of TERP, we next examined the polymerization of acrylates by heating at 100 °C for 24 h with the initiators **1a** or **3**, which are excellent initiators for the polymerization of styrenes. We found that both **1a** and **3** worked efficiently and afforded polyMA with predictable molecular weight and low polydispersity (entries 11 and 12). The initiators also polymerized a variety of acrylate derivatives efficiently (entries 13 – 17). It is worth noting that all the monomers gave the desired polymers with low polydispersity (PD < 1.23) and in high yield. The successful polymerization of

Table 2. TERP of styrene, acrylate and methacrylate derivatives.

Entry	Initiator	Monomer[a]	Condns (°C/h)	Yield (%)	M_n^b	PD^b
1	**1a**	St	105/17	96	9200	1.17
2	**1b**	St	105/17	91	15900	1.45
3	**2**	St	105/16	89	9000	1.46
4	**3**	St	105/20	79	9000	1.15
5	**4**	St	105/16	76	50700	1.80
6	**5**	St	105/18	83	25400	1.58
7[c]	**1a**	St	105/27	78	35700	1.21
8[d]	**1a**	St	105/29	84	62600	1.30
9	**1a**	ClSt	100/17	88	8800	1.41
10	**1a**	MeOSt	100/36	94	10900	1.17
11	**1a**	MA	100/24	86	8800	1.12
12	**3**	MA	100/24	70	6400	1.11
13	**1a**	nBA	100/24	89	10300	1.13
14	**3**	tBA	100/24	85	9800	1.18
15[e]	**1a**	DMAEA	100/96	81	12000	1.23
16	**1a**	DMA	105/23	100	10100	1.22
17[e]	**1a**	AN	100/24	53	20800	1.07
18	**3**	MMA	80/15	67	11800	1.77
19[f]	**3**	MMA	80/13	84	8200	1.16
20[f]	**1a**	MMA	80/13	92	9700	1.18
21[f,g]	**1a**	MMA	80/19	83	16200	1.14
22[c,h]	**1a**	MMA	80/18	79	36300	1.18
23[d,h]	**1a**	MMA	80/24	83	79400	1.14
24[f]	**1a**	EMA	105/2	97	10600	1.12
25[e,f]	**3**	HEMA	80/17	97	22300	1.18

[a]St: styrene, ClSt: *p*-chlorostyrene, MeOSt: *p*-methoxystyrene, MA: methyl acrylate, nBA: *n*-butyl acrylate, tBA: *t*-butyl acrylate, DMAEA: 2-dimethylaminoethyl acrylate, DMA: *N*, *N*-dimethylacrylamide, AN: acrylonitrile, MMA: methyl methacrylate, EMA: ethyl methacrylate, HEMA: 2-hydroxyethyl methacrylate. [b]Molecular weight (M_n) and polydispersity (PD) were determined by size exclusion chromatography calibrated by polySt or polyMMA standards. [c]500 Equiv of monomer was used. [d]1000 Equiv of monomer was used. [e]The reaction was carried out in DMF. [f]One equiv of dimethyl ditelluride was added. [g]200 Equiv of monomer was used. [h]Two equiv of dimethyl ditelluride were added.

2-dimethylaminoethyl acrylate (DMAEA), N,N-dimethyl acrylamide (DMA), and acrylonitrile (AN) is particularly noteworthy, since the polar functional groups of these monomers often disturb the precise control of the polymerization using other methods.

We next examined the polymerization of methyl methacrylate (MMA, 100 equiv) with initiator **3**, but initial attempts revealed that the control of the reaction was insufficient (PD = 1.77, entry 18). The result could be attributed to the high reactivity of MMA toward the polymer-end radicals, and we anticipated that the addition of an agent to cap the radical species would enhance controllability. Because ditellurides are excellent radical capturing reagents (*11,14*), we thought that they would serve as capping reagents for polymer-end radicals. Indeed, polyMMA with low polydispersity was obtained by the addition of one equiv of dimethyl ditelluride (PD = 1.16, entry 19). Initiator **1a** also polymerized MMA with highly controlled manner in the presence of one equiv of dimethyl ditelluride (PD = 1.18, entry 20). The molecular weight increased predictably with the amount of MMA used, and high molecular-weight polyMMA formed with precise molecular weight control by the addition of one or two equiv of dimethyl ditelluride (entries 21 - 23). The polymerization completed within 2 h by carring out the reaction at 105 °C with ethyl methacrylate (EMA), and the desired polymer formed with low polydispersity (entry 24). 2-Hydroxyethyl methacrylate (HEMA) could also be polymerized in the presence of dimethyl ditelluride in a highly controlled manner (entry 25).

Confirmation of living character.

The "living" character of the current polymerization was ascertained by several control experiments. First, the molecular weight (M_n) increased linearly with an increase in the amount of styrene or MMA used as shown above (entries 7, 8 and 21 - 23). Secondly, the molecular weight also increased linearly with an increase of the conversion of styrene (Figure 1). The same experiments with MA and MMA also showed excellent linear relation between the molecular weight and the conversion.

Finally, the existence of active carbon-tellurium bond in the polymer end was confirmed by labeling experiments. Thus, treatment of the polymer block **6** prepared from **1a** and 30 equiv of styrene with either tributyltin hydride or tributyltin deuteride afforded **8** or **8**-d$_1$ quantitatively through the radical

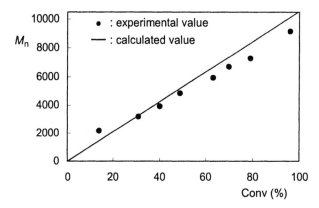

Figure 1. Molecular weight (M_n) of polySt obtained by the bulk polymerization of styrene (100 equiv) with **1a** as a function of the conversion of styrene.

636

intermediate **7** (Scheme 2) (*8a*). Selective incorporation of the deuterium atom in the polymer was ascertained by the MALDI-TOF mass spectroscopy by observing an increase of one mass numbers in **8-d₁** compared to those in **8** (Figure 2). The ^{2}H NMR of **8-d₁** further supported the selective incorporation of deuterium at the benzylic position (δ = 2.36 ppm, broad singlet). The results clearly revealed the existence of organotellurium group at the polymer end,

Scheme 2.

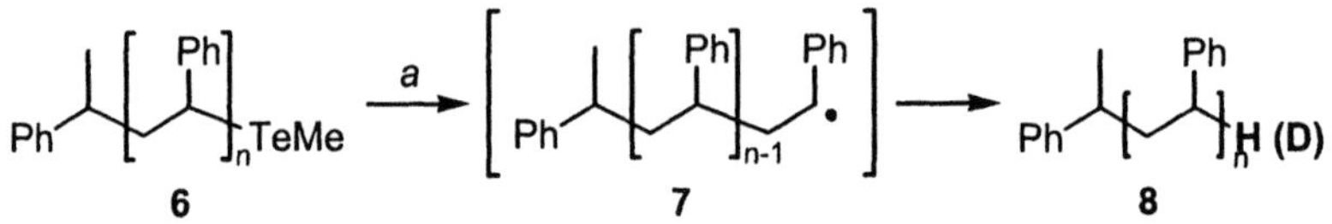

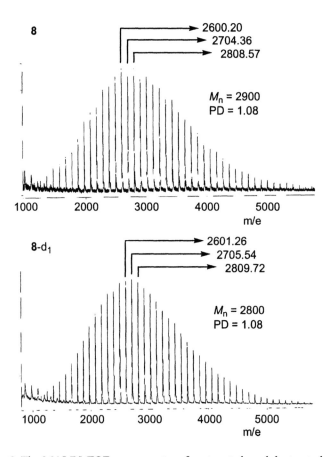

^{a}AIBN (0.1 equiv), Bu₃SnH(D) (3 equiv), $C_6H_5CF_3$, 80 °C, 1 h.

Figure 2. The MALDI-TOF mass spectra of protonated- and deuterated polySt **8** and **8-d₁**.

which was reduced by tributyltin hydride (or deuteride). It is also worth mentioning that the differences of each mass peaks in the major series of peaks are 104, which corresponds to the molecular mass of styrene, and that there are essentially no peaks derived from impurities in between the major series of peaks. The results must be attributed to the highly controlled character of TERP, in which the polymerization is initiated by the definite initiating radical generated from **1a** and proceeds with least unfavorable side reactions. The same deuterium labeling experiments with polyMMA, which was prepared from **1a** with 30 equiv of MMA, resulted in the virtually same results.

Block copolymer synthesis.

The results presented above suggest that the TERP would be suitable for a tailored synthesis of block copolymers composed of different families of monomers using macroinitiators, because the same initiators can control the polymerization of different types of monomers under similar thermal conditions (*15*). To clarify this point, we next examined the synthesis of AB-block copolymers with all possible combinations of styrene, MMA, and tBA starting from **1a** as an initiator. We were pleased to find that TERP could synthesize the desired AB-diblock copolymers regardless of the order of first and second monomers (Table 2, entries 1-6). Thus, the AB diblock copolymer of styrene and MMA could be efficiently prepared starting from either the polystyrene block (prepared by **1a** and styrene) or the polyMMA block (prepared by **1a** and MMA) with MMA or styrene, respectively (entries 1 and 3). The AB-diblock copolymers of styrene and tBA, and of MMA and tBA were also synthesized regardless of the order of the added monomers (entries 2 and 4-6) (*16*). The desired diblock copolymers were obtained in all cases with predictable molecular weight with low polydispersity. Due to the stronger carbon-tellurium bond in poly(tBA) compared with the one in polystyrene and polyMMA, the controllability of the diblock copolymers initiated by the poly(tBA) macroinitiator was slightly less efficient (entries 5 and 6), but is still at an acceptable level (PD < 1.35).

Because the order of monomer addition is less important in TERP compared to that of other LRP systems, ABA and ABC triblock copolymers could also be prepared starting from diblock macroinitiators (entries 7-11). Thus, successive treatment of **1a** with MMA and styrene afforded poly-MMA-*b*-polySt macroinitiator, which was further treated with MMA to give the desired poly-MMA-*b*-polySt-*b*-polyMMA triblock copolymer with narrow molecular weight distribution (entry 7).The GPC trace of the each block polymers clearly indicates the increase in the molecular weight as the progress of the reaction (Figure 3). ABA triblock copolymer of MMA and tBA could be also prepared by the successive polymerization of MMA, tBA followed by MMA starting from **1a** (entry 8). The ABC-triblock copolymer of styrene, MMA, and tBA, of MMA, styrene, and tBA, and of MMA, tBA, and styrene could be also synthesized by successive addition of each monomers starting from **1a** (entries 9

Table 3. Synthesis of AB di- and ABA tri- and ABC triblock copolymers using macroinitiators.

Entry	Macroinitiator $(M_n/PD)^a$	Monomer	Yield (%)	M_n^b	PD^b
AB Diblock copolymer					
1	PolySt (9000/1.15)	MMA^c	85	13900	1.25
2	PolySt (9000/1.15)	tBA	50	11300	1.18
3	PolyMMA (8500/1.12)^d	St	85	18800	1.13
4	PolyMMA (8500/1.12)^d	tBA	57	17100	1.11
5	Poly(tBA) (9600/1.10)^d	St	77	19200	1.32
6	Poly(tBA) (8200/1.19)^d	MMA^c	88	19500	1.35
ABA Triblock copolymer					
7	PolyMMA-b-polySt (18700/1.18)	MMA^c	65	28100	1.22
8	PolyMMA-b-poly(tBA) (11000/1.11)	MMA^c	83	18600	1.30
ABC Triblock copolymer					
9	PolyST-b-polyMMA (12600/1.30)	tBA	32	16100	1.27
10	PolyMMA-b-polySt (19000/1.13)	tBA	45	21800	1.18
11	PolyMMA-b-poly(tBA) (11500/1.09)	St	69	21600	1.27

^{a}The macroinitiator was prepared from 1a and the corresponding monomer according to the conditions shown in Table 1. 100 equiv and 200 equiv of monomers were used for the diblock and triblock copolymer synthesis, respectively. ^{b}Molecular weight (M_n) and polydispersity (PD) were determined by size exclusion chromatography calibrated by polySt standards for crude samples. ^{c}One equiv of dimethyl ditelluride was added. ^{d}Calibrated using polyMMA standards.

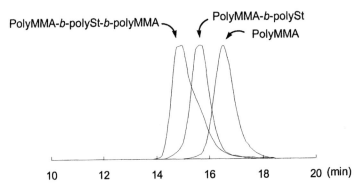

PolyMMA-b-polySt-b-polyMMA

PolyMMA-b-polySt

PolyMMA

10   12   14   16   18   20 (min)

Figure 3. Comparison of GPC traces of the polyMMA, polyMMA-b-polySt, and polyMMA-b-polySt-b-polyMMA block copolymers.

– 11). These triblock copolymers were obtained in all cases in a highly controlled manner.

Possible mechanisms.

Detailed mechanisms of the current polymerization are not clear at the present time. One possible explanation is that tellurium radicals have sufficient lifetime to act as persistent radicals (*17*). However, tellurium-centered radicals are less persistent than nitroxyl radicals and usually form ditellurides. Indeed, The DFT calculated bond dissociation energy of dimethyl ditelluride is 147 kJ/mol (B3LYP density functionals with LANL2DZ basis set for tellurium atom and 6-31G(d) basis set for the rest), suggesting that the equilibrium between the methyltellanyl radical and dimethyl ditelluride completely shifts toward dimethyl ditelluride. Therefore, it is unlikely that the methyltellanyl radical acts as a persistent radical. An alternative mechanism is the ditelluride capping mechanism, in which ditellurides serve as the capping reagent for the reactive polymer ends to give the dormant species (Scheme 3). The effect of the dimethyl ditelluride in the polymerization of MMA strongly suggests that the current polymerization proceeds via the ditelluride-capping mechanism. In this case, the high reactivity of ditellurides toward carbon-centered radicals must be responsible for the high control of the polymerization process (*14*). However, other possibilities involving the degenerative chain transfer could not be rigorously excluded (*18*). Further experimental as well as theoretical investigations are required to clarify the mechanism.

Scheme 3.

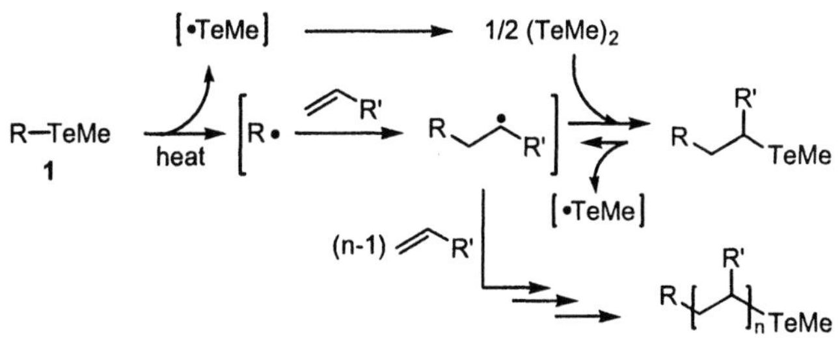

End-group transformations.

Another characteristic advantage of TERP is the versatility of the end-group transformations (Scheme 4), because organotellurium compounds are excellent precursors for carbon-centered radicals (*6-8*), carbanions (*19*), and carbocations (*20,21*). Thus, the radical-mediated transformation through **7** with ethyl tributylstannylmethylacrylate afforded the enoate-functionalized polymer **9** with 61% end-group functionalization (See also Scheme 2). Furthermore, the tellurium-lithium transmetallation by treatment of **6** with buthyllithium followed by trapping of the resulting lithium species **10** with carbon dioxide gave lithium carboxylate **11**, which was treated with aqueous HCl to give carboxylic acid **12**. The esterification of **11** with pyrenebutanol under standard Yamaguchi conditions (*22*) afforded **13**, UV spectra of which revealed 86% incorporation of the carboxylate residue to the polymer end. The functional groups in **9**, **11** and **12** would afford good foundations for the further end-group modifications.

Scheme 4.

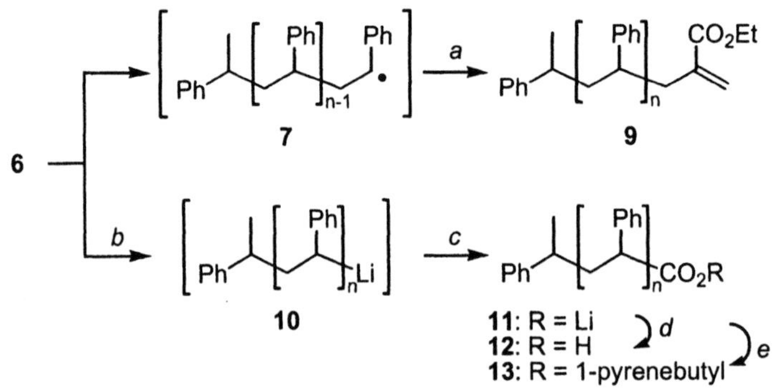

^{a}AIBN (0.1 equiv), Ethyl tributylstannylmethylacrylate (4 equiv), $C_6H_5CF_3$, 80 °C, 6 h, ^{b}BuLi (1.5 equiv), THF, -72 °C, 3 min, ^{c}CO$_2$ (excess), ^{d}aq. HCl (excess), ^{e}2,4,6-Cl$_3$C$_6$H$_2$COCl (2 equiv), Et$_3$N (2 equiv), THF, rt, 1.5 h, then 1-pyrenebutanol (4 equiv), DMAP (4 equiv), CH$_2$Cl$_2$, rt, 3h.

Summary

Organotellurium compounds initiate living radical polymerization with a variety of vinyl monomers to give structurally defined macromolecules. A variety of AB-, ABA-, and ABC-block copolymers with defined structures are also prepared using macroinitiators. As an analogy to the successful synthesis of the block copolymers in Table 2, combined with the data in Table 1, various combinations of multiblock copolymers could possibly be synthesized. Furthermore, random copolymerization using different families of monomers could be also possible. In addition, the versatility of the end-group transformations of the polymer would be also beneficial to add new functions to the polymers. These features clearly indicate that TERP provides a powerful method for the synthesis of functionalized macromolecules with defined structures.

Acknowledgement
This work is partly supported by a Grant-in-Aid for Scientific Research from the Japan Society for the Promotion of Science. Experimental assistance of Mitsuru Nakajima is gratefully acknowledged.

References

1. Matyjaszewski, K. *Controlled/Living Radical Polymerization. Progress in ATRP, NMP, and RAFT*; American Chemical Society: Washington, DC, 2000. *Chem. Rev.* **2001**, *101*, issue 12. Frechet, J. M. J. *Science* **1994**, *263*, 1710.
2. Lee, M.; Cho, B.; Zin, W.-C. *Chem. Rev.* **2001**, *101*, 3869. Klox, H.-A.; Lecommandoux, S. *Adv. Mater.* **2001**, *13*, 1217. Förster, S.; Antonietti, M. *Adv. Mater.* **1998**,*10* , 195.
3. Hawker, C. J.; Bosman, A. W.; Harth, E. *Chem. Rev.* **2001**, *101*, 3661.
4. (a) Matyjaszewski, K.; Xia, J. *Chem. Rev.* **2001**, *101*, 2921. (b) Kamigaito, M.; Ando, T. Sawamoto, M. *Chem. Rev.* **2001**, *101*, 3689. (c) Percec, V.; Popov, A. V.; Ramirez-Castillo, E.; Monteiro, M.; Barboiu, B.; Weichold, O.; Asandei, A. D.; Mitchell, C. M. *J. Am. Chem. Soc.* **2002**,*12 4*, 4940, and references therein. (d) Ma, Q.; Wooley, K. L. *J. Polym. Sci.: Part A: Polym. Chem.* **2000**, *38*, 4805.
5. Destarac, M.; Charmot, D.; Franck, X.; Zard, S. Z. *Macromol. Rapid. Commun.* **2000**,*21* , 1035. Chiefari, J.; Chong, Y. K.; Ercole, F.; Krstina, J.; Jeffery, J.; Le, T. P. T.; Mayadunne, R. T. A.; Meijs, G. F.; Moad, C. L.; Moad, G.; Rizzardo, E.; Thang, S. H. *Macromolecules* **1998**,*31* , 5559.
6. Yamago, S.; Miyazoe, H.; Yoshida, J. *Tetrahedron Lett.* **1999**, 40, 2339. Miyazoe, H.; Yamago, S.; Yoshida, J. *Angew. Chem. Int. Ed.* **2000**, *39*, 3669.
7. Yamago, S.; Miyazoe, H.; Nakayama, T.; Miyoshi, M.; Yoshida, J. *Angew. Chem. Int. Ed.* **2003**, *42*, 117. Yamago, S.; Hashidume, H.; Yoshida, J. *Tetrahedron* **2002**, *58*, 6805. Yamago, S.; Miyoshi, M.; Miyazoe, H.; Yoshida, J. *Angew. Chem. Int. Ed.* **2002**, *41*, 1407. Yamago, S.; Miyazoe, H.; Goto, R.; Hashidume, M.; Sawazaki, T.; Yoshida, J. *J. Am. Chem. Soc.* **2001**, *123*, 3697. Yamago, S.; Hashidume, M.; Yoshida, J. *Chem. Lett.* **2000**, 1234. Yamago, S.; Miyazoe, H.; Sawazaki, T.; Goto, R.; Yoshida, J. *Tetrahedron. Lett.* **2000**, *41*, 7517. Yamago, S.; Miyazoe, H.; Yoshida, J. *Tetrahedron Lett.* **1999**, *40*, 2343. Yamago, S.; Miyazoe, H.; Goto, R.; Yoshida, J. *Tetrahedron Lett.* **1999**,*40* , 2347.
8. Organotellurium-mediated radical reactions. See, (a) Clive, D. L. J.; Chittattu, G. J.; Farina, V.; Kiel, W.; Menchen, S. M.; Russell, C. G.; Singh, A.; Wong, C. K.; Curtis, N. *J. Am. Chem. Soc.* **1980**, *102*, 4438. (b) Barton, D. H. R.; Ramesh, M. *J. Am. Chem. Soc.* **1990**, *112*, 891. (c) Han, L.-B.; Ishihara, K.; Kambe, N.; Ogawa, A.; Ryu, I.; Sonoda, N. *J. Am. Chem. Soc.*

1992, *114*, 7591. (d) Crich, D.; Chen, C.; Hwang, J.-T.; Yuan, H.; Papadatos, A.; Walter, R. I. *J. Am. Chem. Soc.* **1994**, *116*, 8937. (e) Engman, L.; Gupta, V. *J. Org. Chem.* **1997**, *62*, 157. (f) Lucas, M. A.; Schiesser, C. H. *J. Org. Chem.* **1996**,*61* , 5754.

9. Fischer, H. *Chem. Rev.* **2001**,*10 1*, 3581.

10. Yamago, S.; Iida, K.; Yoshida, J. *J. Am. Chem. Soc.* **2002**, *124*, 2874. Yamago, S.; Iida, K.; Yoshida, *J. Am. Chem. Soc.* **2002**, *124*, 13666.

11. The use of diphenyl ditelluride for the radical capturing reagent in the AIBN-initiated radical polymerization of styrene has been reported. See, Takagi, K.; Soyano, A.; Kwon, T. S.; Kunisada, H.; Yuki, Y. *Polym. Bull.* **1999**, *43*, 143.

12. Skene, W. G.; Belt, S. T.; Connolly, T. J.; Hahn, P.; Scaiano, J. C. *Macromolecules* **1998**, *31*, 9103. Moad, G.; Rizzardo, E. *Macromolecules* **1995**, *28*, 8722. Hu, Y.; Wang, S. Q.; Jamieson, A. M. *Macromolecules* **1995**, *28*, 1847.

13. Qiu, J.; Matyjaszewski, K. *Macromolecules* **1997**, *30*, 5643.

14. Russell, G. A.; Tashtoush, H. *J. Am. Chem. Soc.* **1983**,*10 5*, 1398.

15. (a) Hawth, E.; Bosman, A.; Benoit, D.; Helms, B.; Fréchet, J. M. J.; Hawker, C. J. *Macromol. Symp.* **2001**, *174*, 85. Benoit, D.; Chaplinski, V.; Braslau, R.; Hawker, C. J. *J. Am. Chem. Soc.* **1999**, *121*, 3904. (b) Davis, K. A.; Matyjaszewski, K. *Macromolecules* **2000**,*33* , 4039. Davis, K. A.; Charleux, B.; Matyjaszewski, K. *J. Polym. Sci., Part A: Polym. Chem.* **2000**, *38*, 2274. Cassebras, M.; Pascual, S.; Polton, A.; Tardi, M.; Vairon, J.-P. *Macromol. Rapid. Commun.* **1999**, *20*, 261. (c) Rizzardo, E.; Chiefari, J.; Mayadunne, R.; Moad, G.; Thang, S. *Macromol. Symp.* **2001**, *174*, 209. Chong, Y. K.; Le, T. P. T.; Moad, G.; Rizzardo, E.; Thang, S. H. *Macromolecules* **1999**, *32*, 2071.

16. As initiators **1-3** and the macroinitiators are slightly sensitive toward oxygen, the handling of them should be carried out under inert atmosphere using standard Schlenk/Syringe techniques or in a glove box.

17. Yoshikawa, C.; Goto, A.; Fukuda, T. *Macromolecules* **2002**, *35*, 5801. See also ref. 9.

18. Matyjaszewski, K.; Gaynor, S.; Wang, J.-S. *Macromolecules* **1995**, *28*, 2093. Goto, A.; Ohno, K.; Fukuda, T. *Macromolecules* **1998**, *31*, 2809. See also, Oka, M.; Tatemoto, M. In *Contemporary Topics in Polymer Science*; Bailey, W. J.; Tsuruta, T., Eds.; Plenum: New York, 1984; pp 763-777.

19. Kanda, T.; Kato, S.; Sugino, T.; Kambe, N.; Sonoda, N. *J. Organomet. Chem.* **1994**, *473*, 71. Hirao, T.; Kambe, N.; Ogawa, A.; Miyoshi, N.; Murai, S.; Sonoda, N. *Angew. Chem. Int. Ed. Engl.* **1987**,*26* , 1187.

20. (a) Uemura, S.; Fukuzawa, S. *J. Chem. Soc. Perkin Trans. 1* **1985**, 471. (b) Yamago, S.; Kokubo, K.; Hara, O.; Masuda, S.; Yoshida, J. *J. Org. Chem.* **2002**, *67*, 8584. Yamago, S.; Kokubo, K.; Murakami, H.; Mino, Y.; Hara, O.; Yoshida, J. *Tetrahedron Lett.* **1998**, *39*, 7905. Yamago, S.; Kokubo, K.; Yoshida, J. *Chem. Lett.* **1997**, 111.

21. Petragnani, N. *Tellurium in Organic Synthesis;* Academic Press: London, 1994. Comasseto, J. V.; Barrientos-Astigarraga, R. E. *Aldrichimica Acta;* **2000**, *33*, 66.

22. Inanaga, J.; Hirata, K.; Saeki, H.; Katsuki, T.; Yamaguchi, M. *Bull. Chem. Soc. Jpn.* **1979**, *52*, 1989.

Indexes

Author Index

Subject Index